Lecture Notes in Earth Sciences

Edited by Somdev Bhattacharji, Gerald M. Friedman, Horst J. Neugebauer and Adolf Seilacher

25

F. Sansò R. Rummel (Eds.)

Theory of Satellite Geodesy and Gravity Field Determination

Springer-Verlag
Berlin Heidelberg GmbH

Editors

Prof. Dr. Fernando Sansò
Polytecnico di Milano
Istituto di Topografia, Fotogrammetria e Geofisica
Piazza Leonardo da Vinci 32, I-Milano 20133, Italy

Prof. Dr. Reiner Rummel
Delft University of Technology, Faculty of Geodesy
Thijsseweg 11, NL-2629 JA Delft, The Netherlands

ISBN 978-3-540-51528-9 ISBN 978-3-540-48223-9 (eBook)
DOI 10.1007/978-3-540-48223-9

Originally published by Springer-Verlag Berlin Heidelberg New York in 1989

2132/3140-543210 – Printed on acid-free paper

FOREWORD

This book is the collection of the Lecture Notes of an International Summer School of Theoretical Geodesy held in Assisi (Italy) from May 23 to June 3 -1988. The School was sponsored by the International Association of Geodesy and organized by R. Rummel and by the writer.

I still remember when I first conceived the idea of organizing such a School and I realized that the goal was really very ambitious. Talking for the first time to some of the persons who afterwards became lecturers of the School, I got very different reactions: some very positive, from Reiner Rummel for instance, who was enthusiastic and agreed to supply his substantial support to organize the school, some very cautious or even sceptical.

I am proud to say that at the end of the courses all the lecturers claimed they were very happy about the work they did.

Therefore, I must first of all give my sincere thanks to all the teachers for the fine job they did which, I believe, will remain for long in the form of these Lecture Notes, as basic material for scientists studying satellite geodesy.

Thanks are also due to the students who really participated, concentrating their interest and intelligence on the subject, cooperating with the teachers in a very pleasant atmosphere.

Special thanks are due to Jozsef Adam who prepared the Introduction to the school that immediately follows this Foreward.

We must also recognize the foundamental role of the secretary of the school, Miss. Sandra Marescalchi, who helped so much in finding the right organization and equilibrium between scientific labour and social events, solving all the practical problems, some of them very hard like an unexpected transport strike, right on the last day of the school.

In this work of organization, we have been greatly supported by all the staff of our host organization, "La Cittadella", and primarily by Mr. Marco Marchini. Their active presence and their cordiality have facilitated establishing the right mood among people during the School.

We must also recognize that many organizations have substantially contributed to the realization of the School by supporting it in various forms, from fellowships to grants. To **IUGG**, **IAG**, **ESA**, **PSN** (Italian Space Plan), **CISET** Co., **Salmoiraghi** Co., **Wild** Co., **Zeiss** Co., go our gratitude and our thanks.

Moreover, sincere thanks are due to the Municipality of Assisi, to the "Azienda di Promozione Turistica" of Assisi, Todi, and Orvieto, who provided on different occasions, support in the organization of excursions and specially the beautiful concert of Medioeval music and songs.
Last but not least I want to add personal thanks to my friend Reiner Rummel, who did so much for the scientific program of the School; it was a pleasure for me to work with him and I hope it will last for long.

Fernando Sansò

CONTENTS

Four Lectures on Special and General Relativity
E.W. Grafarend

Reference Coordinate Systems: An Update
I.I. Mueller

Gravity Field Recovery from Satellite Tracking Data
C. Reigber

Fundamentals of Orbit Determination
B.D. Tapley

Combination of Satellite, Altimetric and Terrestrial Gravity Data
R.H. Rapp

Summer School Lectures on Satellite Altimetry
C.A. Wagner

Advanced Techniques for High-Resolution Mapping of the Gravitational Field
O. Colombo

SEMINARS

The Integrated Approach to Satellite Geodesy
B. Betti, F. Sansò

INTRODUCTION

From May 23rd to June 3rd, 1988 the International Summer School of Theoretical Geodesy on "Theory of Satellite Geodesy and Gravity Field Determination" took place in Assisi (Italy) under the sponsorship of the International Association of Geodesy. Professor Fernando Sansò ("Istituto di Topografia, Fotogrammetria e Geofisica" - "Politecnico di Milano") and Professor Reiner Rummel ("Afdeling der Geodesie" - "Technische Hogeschool Delft") were the initiators and organizers of this meeting. The aim of this School was to simulate the interest among young geodesists in satellite geodesy, and to intensify the communication between the world of satellite geodesists and other scientists working on different branches of geodesy.
The International Summer School was attended by 63 participants from 14 Nations, mainly from Europe. Students came from Austria (4), Belgium (1), Czechoslovakia (2), England (2), Federal Republic of Germany (12), Hungary (1), Italy (23), The Netherlands (7), Norway (1), Poland (1), Portugal (2) and Spain (1).
The scientific program of the School was organized by Professors Reiner Rummel and Fernando Sansò. It comprised a series of nine main lectures supplemented by seminars on specific topics. The main lectures were designed to systematically cover all aspects of dynamic satellite geodesy and Earth's gravity field determination by special satellite methods.
The International Summer School of Theoretical Geodesy, 4th Course, was opened by Prof. F. Sansò who reminded us of the previous courses of the International School of Advanced Geodesy founded and directed by Antonio Marussi. The previous three Courses were organized at Erice (Sicily, Italy): 1st Course on Advanced Geodetic Problems (October 1-26, 1974); 2nd Course on Space-Time Geodesy, Differential Geodesy and Geodesy in the Large (May 18-June 2, 1978, Directors of the Course: Professors E.W. Grafarend and A. Marussi); and 3rd Course on Optimization and Design of Geodetic Networks (April 25-May 10, 1984, Directors of the Cours: Professors E.W. Grafarend and F. Sansò).
The main lectures were subdivided into three groups: Foundations, Dynamic Satellite Geodesy and Special Methods. Topics of main lectures are given below.

Lectures on Foundations

1) **H. Moritz**: Introduction to Classical Mechanics.
   The aim of the Lectures was to introduce the students to classical mechanics from its beginning with Newton and Euler through the classical theories of Lagrange

and Hamilton to the modern theory of general dynamical systems, which goes back to Poincaré but has became highly fashionable only during the last twenty years.

2) **E.W. Grafarend**: Four Lectures on Special and General Relativity.
Flat spacetime, pseudo-Euclidean space, the Lorentz transformation. Curved spacetime, pseudo-Riemann space, the affine transformation. PPN formalism. The Kerr metric. Schwarzschild metric in isotropic coordinates.

3) **J. Kovalevsky**: Lectures in Celestial Mechanics.
The two body problem. Equations of perturbed motion. General perturbation techniques. Motion of an artificial satellite. Resonnances. Numerical methods.

Lectures on Dynamic Satellite Geodesy

4) **B.D. Tapley**: Orbit Determination.
Initial and boundary value formulation. Numerical problems of orbit integration. Orbit adjustment. Non gravitational effects. Analysis of a twelve-year long arc.

5) **I.I. Mueller**: Reference Coordinate Systems: An Update.
Conventional Inertial Systems (CIS) of reference. Conventional Terrestrial Systems (CTS) of reference, transformations between frames. Modeling the deformable Earth. The International Earth Rotation Service.

6) **C. Reigber**: Lecture Notes on Gravity Field Recovery from Satellite Tracking Data.
Principles of gravity parameter determination. Gravity induced linear orbit perturbations. Adjustment procedures. Tracking data. Processing steps. Special topics. Global gravity field models.

Lectures on Special Methods

7) **C.A. Wagner**: Summer School Lectures on Satellite Altimetry.
Purposes and motivation: the altimetric equation, radial perturbations. Frequency classification and observability of radial variations. Determination of permanent sea topography from altimetry 1) removal of orbit error. Determination of PST from altimetry 2) simulation of a subtraction method. Determination of PST from altimetry 3) simulation of a simultaneous solution for the geoid. A footnote on new results from the subtraction method.

8) **O.L. Colombo**: Advanced Techniques for High-Resolution Mapping of the Gravitational Field.
Basic techniques for gathering data on a global basis: satellite-to-satellite tracking and satellite gravity gradiometry. Gravity gradiometers: global data analysis. Mission error analysis for a 10^{-2} E.U., full-tensor instrument. Implications for the study of the Earth of the results of a global error analysis of a full-tensor gradiometer mission.

9) **R.H. Rapp**: Combination of Satellite, Altimeter and Terrestrial Gravity Data.
Representation of the gravitational potential. Data definition. Data combination Observation equation formation. The development of high degree potential coeffi-

cients models. The role of satellite altimeter data. Comparisons of satellite and terrestrial gravity anomaly fields.

Besides these main lectures, which took several hours each, special seminars were given concentrating particularly on areas of recent geodetic research interest. The lecturers were: **M. Mariani** from Italy (Experiences with GPS observations. Theory of GPS data processing), **V. Schwarze** from F.R.Germany (Influence of Special Relativiity on Geodetic Measurements), **G. Bianco** (The activity of research of the PSN in the field of Satellite Geodesy) and **M. Crespi, B. Crippa, G. Forlani** and **L. Mussio** from Italy (Selections from the I.T.M. Computing Facilities), **E. Schrama** from the Netherlands (The treatment of Altimetry Data and related datum problems), **P. Holota** from Czechoslovakia (Boundary Value Problems and Invariants of Gravitational Tensor in Satellite Gradiometry), **R. Rummel** from the Netherlands (Basic Concepts of Gradiometry), **B. Betti** and **F. Sansò** (The Integrated Approach to Satellite Geodesy) and **A. Milani** and **E. Melchioni** from Italy (Determination of a local geodetic network by multi-arc processing of satellite Laser ranges), **S. Hieber** from France, ESA Headquarters (Earth Observation Program of ESA), **G. Avanzi** from Italy (Simulation of signals in a Spaceborne gradiometry mission), **A. Caporali** from Italy (Studies of Polar Motion and Long Baselines by Analysis of SLR Data) and **J. Adam** from Hungary (A Possible Application of the Space VLBI observations for Establishment of a new Connection of Reference Frames).

All the presentations were well prepared, and the lecture notes and materials of some seminars were distributed during the School. However, deeper studies will be possible after publication of the Proceedings, which will be available in time. The final versions of the lecture notes and the materials of seminars will be published by the Springer-Verlag Company.

An excellent social and touristic program, organized by Mrs. Sandra Marescalchi and Prof. Fernando Sansò and further members of his staff, provided welcome diversions from long days of theoretical lectures in geodetic science. The Major of Assisi welcomed the lecturers and participants in the old palace of the town. After the Major's party an official dinner was a culinary highlight for all participants. A guided tour and walks in the wonderful Assisi showed an extraordinarily rich town in tourist attractions. An "Ensamble Micrologus" concert at St. Maria Maggiore in honour of the participants was an excellent event serving both the cultural interest and the need of recovering from hard work. A guided visit to the towns of Etruscan origin, Todi and Orvieto, was a touristic highlight of the School's program.

Looking back after some months, and summarizing one may state that the participants, both lecturers and students enjoyed a well-organized Summer School. There is no doubt that the rich tradition of these summer schools initiated by Antonio Marussi will be continued. Sincere thanks are to Professors R. Rummel and F. Sansò organizing the outstanding scientific part of the school, and last but not least to Mrs. Sandra Marescalchi and Prof. F. Sansò and his staff, who organized the social and touristic part of the school. Thanks are also due to the organizers of "La Cittadella" that provided the necessary material support to all the activities of the School with friendship and modesty. The participants had the chance to get acquainted with a part of the country, Umbria, the green heart of Italy, with its beautiful landscape and rich culture. Our stay in Assisi was so interesting and stimulating.

Jòzsef Adàm

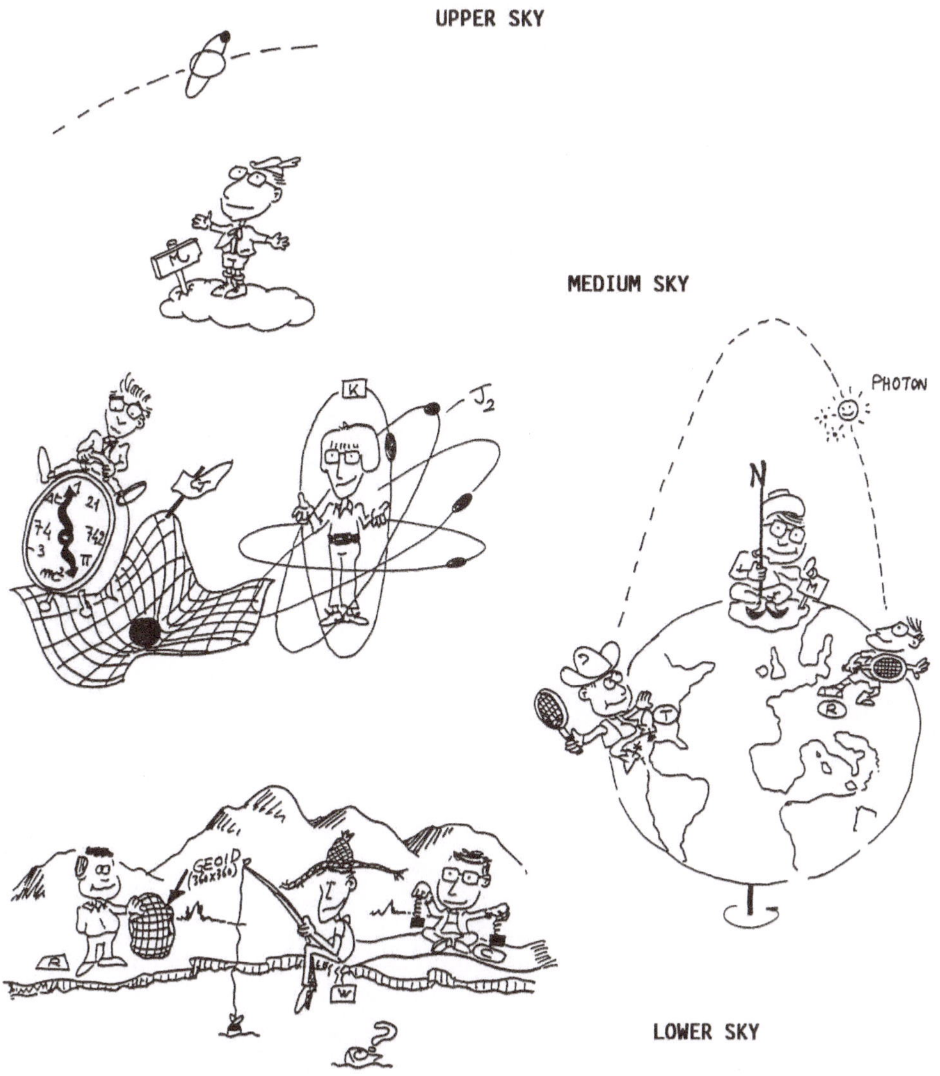

Lecturers
Colombo, Grafarend, Kovalevsky, Moritz, Mueller, Rapp, Reigber, Tapley, Wagner

Participants
Adam, Agrotis, André, Avanzi, Barzaghi, Benciolini, Betti, Bianco, Breuer Brovelli, Caporali, Carpino, Crespi, Dare, Delikaraoglou, Dominici, Ekholm Faustino, Feltens, Fermi, Forlani, Gerstl, Hehl, Hieber, Holota, Klees Knickmeyer, Koop, Kuehtreiber, Łyszkowicz, Mager, Manzino, Massmann, Melchioni Milani, Mussio, Noomen, Pierozzi, Radicioni, Rautz, Reinking, Ricardo Romay-Merino, Rossi, Rueger, Rummel, Sacerdote, Sansò, Schrama, Schuyer, Schwarz Schwarze, Siegerstetter, Sìma, Smeets, Solheim, Stolfa, Stornelli, Strykowski, Suenkel, Teunissen, Van der Marel, Van Gelderen, Vespe, Weber

LECTURES

INTRODUCTION TO CLASSICAL MECHANICS

Helmut Moritz
Institute of Theoretical Geodesy
Technical University, Steyrergasse 17, A-8010 Graz

INTRODUCTION

The aim of these lectures is to introduce the reader to classical mechanics from its beginning with Newton and Euler through the classical theories of Lagrange and Hamilton to the modern theory of general dynamical systems, which goes back to Poincaré but has become highly fashionable only during the last twenty years.

It seems a ridiculous enterprise indeed to treat within the scope of four lectures such a broad material, which is hardly covered completely by a single textbook.

In fact, there is a huge standard textbook literature on the classical topics considered in Lectures 1 - 3, and an equally voluminous literature on the topic of Lecture 4, but all of the latter written within the last decade or two, sometimes under impressive headings such as deterministic chaos, ergodic theory, or synergetics.

Our aim is to give an <u>introduction</u>, which means concentration on the fundamentals and on logical coherence, which sometimes relentlessly excludes standard topics which, to the author, do not fit into the main stream.

This mean stream is the <u>elementary</u> treatment of <u>conservative</u> dynamical systems by means of <u>ordinary</u> differential equations with emphasis on <u>geometry</u>.

The restriction to ordinary differential equation excludes the <u>partial differential equation of Hamilton-Jacobi</u> which we feel does not fit well into our system and is not required for our purposes. It is found in all classical textbooks. An adequate treatment such as given in the beautiful works by Lanczos (1970) or Synge (1960) would have required at least another lecture.

The restriction to conservative systems unfortunately exludes the fascinating modern topic of <u>strange attractors</u> which is treated, e.g., in (Thompson and Stewart, 1986) or in (Abraham and Shaw, 1982-1985); the latter work is extraordinary because it almost entirely consists of geometric pictures and succeeds in explaining the most difficult dynamical structures (and the subject is difficult!) without formulas.

In order to be elementary, we particularly regret that this excludes a treatment by means of Cartan's calculus of exterior differential forms, which provides the most elegant access to the geometrical structure of Hamiltonian systems, symplectic geometry. Still, our emphasis being on geometry, we have attempted to give an elementary treatment of the symplectic geometry of phase space, as well as of the Riemannian geometry of configuration space, also to provide a contact with Erik Grafarend's lectures on General Relativity.

The same restriction of elementary also has forbidden a systematic treatment in terms of the calculus of variations, which the reader may enjoy in Lanczos' book.

We have tried to look at the subject from the point of view of a geodesist, exhibiting cross-connections to geodesy wherever feasible, from earth rotation to hard inverse function theorems (employed by Hörmander in his treatment of Molodensky's problem), to ergodic stationary stochastic processes in collocation, and to the convergence of the spherical-harmonic expansion of the geopotential at the earth's surface. Somewhat surprisingly, this is possible without deviating essentially from our main course.

Illustrative examples have been taken from celestial mechanics, in spite of, or rather with a view to, the lectures of Jean Kovalevsky on Celestial Mechanics. I always find it quite instructive if the same problem is regarded from two different perspectives.

The list of references is entirely restricted to textbooks. The purpose of these lectures would be achieved if the reader is encouraged to look into appropriate books to some detail.

It would be impossible to quote original papers or even to touch on questions of priority. Names are to be regarded only as conventional (and convenient) labels attached to formulas or theorems, beginning with the titles of the lectures: this is not a course on the history of the subject.

LECTURE 1

NEWTON

This lecture will cover "simple" classical mechanics of point masses and rigid bodies in our familiar threedimensional Euclidean space. Its title is typically a "convenient label" since, although contemporary mechanics was founded by Newton, rigid-body dynamics was fully developed only later by Euler and others.

1.1. Motion of a Mass Point

As a matter of fact, a mass point, or point mass, is a fiction, but a useful one if the size of a body, in some sense, is small. For instance, planets may be regarded as mass points since their size is so much smaller than their mutual distances.

Let the vector $\underline{x} = (x,y,z) = (x_1,x_2,x_3)$ denote the Cartesian coordinates, in some coordinate system, of our mass point in three-dimensional Euclidean space. The time derivative of $\underline{x}$ is the velocity

$$\underline{v} = \frac{d\underline{x}}{dt} = \dot{\underline{x}} \; ; \tag{1}$$

the dot will always denote d/dt , t being the time. The second time derivative is acceleration

$$\underline{b} = \frac{d\underline{v}}{dt} = \frac{d^2\underline{x}}{dt^2} = \ddot{\underline{x}} \; . \tag{2}$$

The path of our mass point will be a curve in space, and the description of the motion of such a point is the subject of kinematics (briefly, geometry in time).

Dynamics arises by the introduction of the notion of force. This is done by Newton's law of motion

$$m\underline{b} = \underline{K} \; . \tag{3}$$

Here $\underline{K}$ denotes the force vector and m the mass of our point. This law now is so familiar to us that one hesitates even to mention it, but to find it was one of the greatest breakthroughs in human thinking, and its exact interpretation is sometimes still a matter of philosophical discussion, into which, of course we shall not enter.

We may also introduce the momentum vector $\underline{G}$ by defining

$$\underline{G} = m\underline{v} \; . \tag{4}$$

Then (3) takes the form

$$\frac{d\underline{G}}{dt} = \underline{K} \; , \tag{5}$$

which in spite (or because) of its simplicity contains a tremendous power for generalization, as we shall see in the sequel.

Knowing the force $\underline{K}$, we may use Newton's law in the form

$$m\ddot{\underline{x}} = \underline{K} \tag{6}$$

to get the position vector $\underline{x}$ by integrating twice.

Strictly speaking, (3) is more general than (6): the latter holds only in an inertial system. What is an inertial system? The circular answer, an inertial system is a coordinate system in which (6) holds, already shows that simplicity is only apparent and that there are enormous conceptual problems behind our elementary mathematical apparatus.

It may happen that the force K can be expressed as the gradient of a function $U(x,y,z)$:

$$\underline{K} = -\operatorname{grad} U = -\left(\frac{\partial U}{\partial x}, \frac{\partial U}{\partial y}, \frac{\partial U}{\partial z}\right) \; . \tag{7}$$

Physicists use the minus sign, geodesists the plus sign; we shall here follow the physical sign convention. The function U is called potential energy, and forces satisfying (7), conservative forces (no political implication!). Since $\underline{K}$, together with U, is a function of position, we speak of a force field.

Then we have the equations of motion, from (6),

$$m\ddot{\underline{x}} = -\operatorname{grad} U \; . \tag{8}$$

As (5) shows, momentum may be considered a time integral of force. What is the path integral of $\underline{K}$? We call it work A:

$$A = \int_{\underline{x}_1}^{\underline{x}_2} \underline{K} \cdot d\underline{x} \; , \tag{9}$$

the dot (in the present position) denoting the inner, or scalar, product of two vectors. Using (6) we have

$$A = \int_{t_1}^{t_2} \underline{K}\cdot\frac{d\underline{x}}{dt}\,dt = m\int_{t_1}^{t_2} \ddot{\underline{x}}\cdot\dot{\underline{x}}\,dt = \frac{1}{2}m\int_{t_1}^{t_2}\frac{d}{dt}(\dot{\underline{x}}^2)\,dt =$$

$$= \frac{1}{2}m(\dot{\underline{x}}_2^2 - \dot{\underline{x}}_1^2) = \frac{1}{2}mv_2^2 - \frac{1}{2}mv_1^2 , \tag{10}$$

by (1), since $v = |\underline{v}| = \sqrt{\underline{v}^2}$ is the scalar velocity.

The quantity

$$T = \frac{1}{2}mv^2 = \frac{1}{2}m\dot{\underline{x}}^2 \tag{11}$$

is called kinetic energy.

In the case of a conservative force field, with (7), we have by (9)

$$dA = \underline{K}\cdot d\underline{x} = -\,\text{grad}\,U\cdot d\underline{x} = -\,dU , \tag{12}$$

so that

$$A = -\,(U_2 - U_1) . \tag{13}$$

By means of (11) we may write (10) as

$$A = T_2 - T_1 , \tag{14}$$

and the comparison with (13) shows that

$$T_2 - T_1 = -U_2 + U_1$$

or

$$T_2 + U_2 = T_1 + U_1 ; \tag{15}$$

the sum of kinetic energy T and potential energy is constant:

$$T + U = E = \text{const.} \tag{16}$$

Calling the sum the total energy E, this expresses the conservation of the total energy throughout the motion (this accounts for the term, conservative forces).

This is the simplest case of a conservation law, or an integral of the motion (why?). Remember these expressions: we shall meet them quite often.

1.2. Central Forces

Assume that the force vector has the form

$$\underline{K} = f(x,y,z)\,\underline{x}\ , \tag{17}$$

with an arbitrary smooth function $f(x,y,z)$. Its direction obviously is that of the position vector $\underline{x}$; $\underline{K}$ always points exactly towards the origin O $(f<0)$ or exactly away from it $(f>0)$. Such a force (17) is thus called a central force.

Then Newton's law (6) becomes

$$m\ddot{\underline{x}} = f(x,y,z)\,\underline{x}\ . \tag{18}$$

Take the vector product (denoted by ×) of this equation with $\underline{x}$:

$$m\ddot{\underline{x}} \times \underline{x} = f(x,y,z)\,\underline{x} \times \underline{x}\ .$$

Now it is a well-known fact that the vector product of a vector with itself is zero. This gives

$$m\ddot{\underline{x}} \times \underline{x} = 0 \tag{19}$$

or on integration (verify by differentiation!)

$$m\underline{x} \times \dot{\underline{x}} = \text{const.}\ , \tag{20}$$

another conservation law or integral of the motion.

The vector

$$\underline{H} = m\underline{x} \times \dot{\underline{x}} \tag{21}$$

is called the angular momentum; thus, in the case of a central force we have

$$\underline{H} = \text{const.} \tag{22}$$

This is the conservation of angular momentum which furnishes three integrals of the motion (all three components of $\underline{H}$ are constant!).

1.3. Planetary Motion

Newton's law of gravitational attraction is well known to be

$$\underline{K} = -G\frac{Mm}{r^3}\underline{x} \ , \tag{23}$$

where

$$r = |\underline{x}| = \sqrt{x^2+y^2+z^2} \ . \tag{24}$$

The "large mass" M is at the origin O and denotes, e.g., the sun; the "small mass" m may be a planet.

$\underline{K}$ obviously is a central force and derives from a potential energy

$$U = -G\frac{Mm}{r} \tag{25}$$

by (7) (verify!).

Energy E and angular momentum $\underline{H}$ are conserved. The integration of (8) in this case gives the three laws of Kepler for planetary (or satellite) motion which are well known and which will be derived in the lectures of Jean Kovalevsky in this Volume : the orbits are ellipses, hyperbolas or parabolas with one focus at the origin O .

Let me only mention that the constancy of the angular momentum $\underline{H}$ immediately implies that the orbit is a plane curve, its plane being normal to the constant vector $\underline{H}$; cf. (21) and (22).

By the way, Newton started from Kepler's laws to rigorously derive his law of gravitation (23) by means of the just invented differential calculus; he did not only observe a falling apple!

1.4. Free Motion

In the absence of any force, (6) reduces to

$$\ddot{\underline{x}} = 0 \ , \tag{26}$$

which can be directly integrated twice:

$$\dot{\underline{x}} = \underline{v} = \text{const.} \ , \tag{27}$$

$$\underline{x} = \underline{x}_0 + \underline{v}t \ , \tag{28}$$

$\underline{x}_0$ denoting another constant vector. This expresses uniform motion along a straight line.

By (4) and (27) we have

$$\underline{G} = \text{const.} \,, \tag{29}$$

implying conservation of momentum $\underline{G}$, and (28) shows that the coordinates are linear functions of time.

The energy is

$$E = T + U = T = \frac{1}{2} m v^2 = \frac{1}{2m} \underline{G}^2 = \text{const.} \,, \tag{30}$$

so that the energy integral here is a consequence of the angular momentum integral (29); the various integrals need not be independent!

This case is extremely simple, but it is by no means trivial. Because of its basic importance we shall have occasion later on to refer back to it.

1.5. The Many-Body Problem

Newton's law of motion (6) also works for several (n) mass points which attract each other by Newton's law of gravitation (23). We use the form (8) with U given by (25):

$$m\ddot{\underline{x}}_i = -\sum_{j=1}^{n}{}' \operatorname{grad}_i U_{ij} \tag{31}$$

with

$$U_{ij} = -G \frac{m_i m_j}{|\underline{x}_j - \underline{x}_i|} \tag{32}$$

and grad_i denotes that the gradient consists of derivatives with respect to the coordinates of the point P_i of position vector $\underline{x}_i$ and mass m_i. The sum $\sum'$ omits the term $j=i$ and thus represents the combined attractional force, on m_i, of all the other point masses m_j.

For $n = 2$ we have the two-body problem. Its solution again leads to Kepler's equations, but the focus of the ellipses, etc., is now at the common center of mass of m_1 and m_2, rather than at the origin O.

For $n = 3$ we have a three-body problem which already is so difficult that its general solution is not known. There are, however, important partial results, and Poincaré's (1892-1899) fundamental investigations, on which the modern theory of dynamical systems (Lecture 4) is based, issue from the "restricted three body problem".

Above all, we know the existence of the "classical integrals": conservation of total energy, total momentum, and total angular momentum.

1.6. Motion of a Rigid Body

There are two basic types of rigid-body motion: translation and rotation. Translation is parallel displacement (in general, along a space curve). Rotation leaves a point of the body unchanged, for which we may take its center of mass. In fact, an arbitrary rigid-body motion can always and uniquely be regarded as the superposition of a translation and a rotation around the center of mass.

With respect to translation, a rigid body behaves very similarly as a mass point. In particular, the translation of the center of mass S is described by the law

$$M\ddot{\underline{x}}_S = \underline{K} \; , \tag{33}$$

which is formally identical to Newton's law of motion (6) for a mass point. Now, M denotes the total mass of the body, $\underline{x}_S$ is the position vector of the center of mass, and $\underline{K}$ denotes the total force: the resultant of all external forces acting on the body.

Introducing the velocity of S ,

$$\underline{V} = \dot{\underline{x}}_S \; , \tag{34}$$

and defining the momentum $\underline{G}$ by

$$\underline{G} = M\underline{V} \; , \tag{35}$$

eq. (33) may be written in the alternative form

$$\frac{d\underline{G}}{dt} = \underline{K} \; , \tag{36}$$

formally identical to (5).

For rotation we have a similar formula:

$$\frac{d\underline{H}}{dt} = \underline{L} \; , \tag{37}$$

where $\underline{H}$ denotes the angular momentum defined by a generalization of (21):

$$\underline{H} = \iiint \underline{x} \times \dot{\underline{x}} \, dM , \tag{38}$$

the integral being extended over the whole body.

The vector $\underline{L}$ in (37) represents the <u>torque</u> acting on the body. It is given by

$$\underline{L} = \sum \underline{x} \times \underline{k} \tag{39}$$

where $\underline{k}$ is the external force acting on a molecule of the body, and $\sum$ denotes the sum over all molecules of which the body is made up.

Note that we are not very consistent: in (38) the body is considered continuous, whereas in (39) it is regarded as discrete -- both are mathematical idealizations anyway.

Almost all textbook authors try to derive the equations of translation (33) or (36), and of rotation, (37), from Newton's law for a mass point (3). Such "derivations" are based on additional, entirely unnatural, assumptions, as Truesdall and Toupin (1960, p.534) and others have pointed out: (3) alone is not sufficient.

The only rigorous derivation of the equations of motion of a rigid body is by means of d'Alembert's principle (to be treated in Lecture 2), with which Helmert opened his classical treatise on physical geodesy; see also (Lanczos, 1970,pp.103-106). Otherwise it is best to consider (36) and (37) as independent axioms ("Euler's laws").

We now proceed heuristically to find further analogies between translation and rotation. We have seen that momentum $\underline{G}$ and angular momentum $\underline{H}$ correspond, as well as force $\underline{K}$ and torque $\underline{L}$. The rotational analog to the velocity vector $\underline{V}$ is the vector of rotational velocity $\underline{\omega}$ whose direction is the rotation axis and whose magnitude is the angular velocity ω of the rotation.

Eq. (35) shows that there is a linear relation between velocity $\underline{V}$ and momentum $\underline{G}$. There is also a linear relation between the corresponding rotational quantities $\underline{\omega}$ and $\underline{H}$:

$$\underline{H} = \underline{C}\,\underline{\omega} . \tag{40}$$

The difference is that the mass M is a scalar, whereas the "rotational mass" depends on direction so that $\underline{C}$ is a matrix, the <u>inertia tensor</u>. (This difference is less fundamental than it looks: for a spherically symmetric body, the inertia tensor is a multiple of the unit matrix,

$\underline{C} = C\underline{I}$, so that then (40) reduces to $\underline{H} = C\underline{\omega}$ in complete analogy to (35).)

The expressions for the kinetic energy of translation,

$$E_{trans} = \frac{1}{2} M \underline{V}^T \underline{V} \tag{41}$$

($\underline{V}^T$ is the transpose of $\underline{V}$ as usual), and of rotation,

$$E_{rot} = \frac{1}{2} \underline{\omega}^T \underline{C}\, \underline{\omega} \tag{42}$$

complete the analogies between translation and rotation. More about this can be found in (Moritz and Mueller, 1987, sec.2.1.).

Let us finally mention that the inertia tensor $\underline{C}$ is symmetric and can, therefore, be brought into diagonal form by a suitable choice of the coordinate system:

$$\underline{C} = \begin{bmatrix} A & 0 & 0 \\ 0 & B & 0 \\ 0 & 0 & C \end{bmatrix} . \tag{43}$$

The (positive) quantities are called principal moments of inertia, and the corresponding coordinate axes, principal axes.

1.7. Inertial Navigation and Surveying

The measuring apparatus consists of three accelerometers measuring acceleration along three mutual perpendicular axes, i.e., the vector $\underline{b}$. If the motion is purely translational (this can be achieved directly or indirectly by an inertial platform), then the equations of motion (33) take the Newtonion form (3) or (6):

$$\ddot{\underline{x}} = \underline{b}(t) . \tag{44}$$

As $\underline{b}$ has been measured as a function of time t , we can integrate once:

$$\underline{v}(t) = \int_{t_0}^{t} \underline{b}(\tau)\, d\tau \tag{45}$$

(we assume $\underline{v}(t_0) = 0$, and a second time:

$$\underline{x}_1 - \underline{x}_0 = \int_{t=t_0}^{t_1} \int_{t_0}^{t} \underline{b}(\tau)\, d\tau\, dt \ . \tag{46}$$

The result thus is relative position $\underline{x}$. This method is used routinely for inertial navigation of ships and airplanes, and, if the apparatus is installed in an automobile or a helicopter and if adequate measures are taken to ensure high precision, it also has considerable geodetic application (inertial surveying).

1.8. Error propagation

Considering one component b of the vector $\underline{b}$ only, let $b(t)$ be affected by a random error $\varepsilon(t)$. The corresponding error in relative position, by (46), is

$$\eta(t) = \int_{t_0}^{t} \int_{t_0}^{u} \varepsilon(\tau)\, d\tau\, du \ . \tag{47}$$

Because of the double integration, $\eta(t)$ has a much smoother and more regular behavior than $\varepsilon(t)$, so as almost to give the impression of a systematic error. The explanation is that not only neighboring points of $\eta(t)$, but also neighboring tangential directions $\dot{\eta}(t)$ are strongly correlated, so that the curve $\eta(t)$ has a tendency not to change its course.

The discrete analog of double integration is double summation, which occurs with long and rather straight polygonal traverses and, above all, in aerial triangulation. This is the "systematic behavior of random errors" in aerial triangulation, which was much discussed some 30 years ago.

1.9. Unstable Convergence in Integration

Integration will play a basic role throughout these lectures. So far we have considered iterated (double) integration, but also single integration sometimes exhibits astonishing features.

In celestial mechanics (treated by methods described in the following lectures), we often meet with integration of trigonometric series, a typical form of which is

$$\dot{f}(t) = \sum_{n=1}^{\infty} \sum_{m=1}^{\infty} \mu^{n+m} \cos(n - \alpha m)t \tag{48}$$

where $|\mu| < 1$ and α is certain positive number. If we assign $f(0) = 0$, then the integration gives

$$f(t) = \sum_{n=1}^{\infty} \sum_{m=1}^{\infty} \frac{\mu^{n+m}}{n-\alpha m} \sin(n-\alpha m)t \ . \tag{49}$$

Since $|\mu| < 1$, $|\cos x| < 1$, the double series (48) is absolutely and uniformly convergent. The integrated series (49) behaves quite differently, however.

If α is a rational number, then, for certain m and n, the denominator $n - \alpha m$ is zero because any rational α can be represented in the form

$$\alpha = \frac{p}{q} , \tag{50}$$

with integer p and q (take $m = q$ and $n = p$), and the representation (49) breaks down ("case of resonance").

Now, any real α can be arbitrarily closely approximated by a real number (cut off the 10th, or 1000th decimal!). Then a term in (49), destroying convergence, will occur.

If this happens "far away" in the series (e.g., by cutting off after the 1000th decimal only), then the difficulty is more theoretical than practical since the first terms of the series (49) may still give a good approximation. Poincaré (1893, Chapter VIII,secs.118 and 119) studies this problem and points out the analogy with what nowadays is known as asymptotic series which, though theoretically divergent, may well be "practically convergent".

It has been suggested that the series expansion of the geopotential at the earth's surface and Molodensky's series for the geodetic boundary problem have a similar convergence behavior, cf. (Moritz, 1980, secs. 6 to 8 and 47).

From a theoretical point of view, even for irrational values of α the question of convergence or divergence is not easy; it leads to a deep problem of Diophantine approximations in number theory. The result is that (49) converges for "almost all" values of α while still diverging on an uncountable set of α's!

This was first proved by Heinrich Bruns in 1884 (the same who wrote "The Figure of the Earth" in 1878!); cf. (Wintner, 1941, pp.407-410).

In astronomy, this is known as the "problem of small denominators"; it shows that the convergence behavior in our case is unstable: an arbitrarily small change of α may cause a change from convergence to divergence and vice versa; note again the geodetic analog, the "grain-of-sand argument" (Moritz, 1980, pp.64-65).

Remember the "small-denominator problem": it will come up again in Lecture 4.

LECTURE 2

LAGRANGE

Contrary to an often-heard opinion, Newton's law of motion of a point mass, (6), or its many-point equivalents such as (31), are not sufficient as a general basic of classical mechanics, as we have already seen with rigid-body motion.

The reason is that, even considering an extended body to consist of mass points (such as molecules), we cannot in general presuppose the knowledge of the nature of the "internal forces" that keep them together. They are certainly not simple central forces, but are based on complicated quantum-mechanical interactions.

The engineer who calculates a bridge, surely is not interested in the quantum structure of the material of the bridge, but only in the real "external forces" acting on it.

So we wish to construct a theory of analytical mechanics which contains "real" or "external" forces only.

2.1. Statics; Principle of Virtual Displacements

Take the example of a rigid lever rotating around its fulcrum O (Fig. 1). Let two forces $\underline{K}_1$, and $\underline{K}_2$ act on its ends, regarded as points P_1 and P_2. We wish to know the condition of equilibrium of

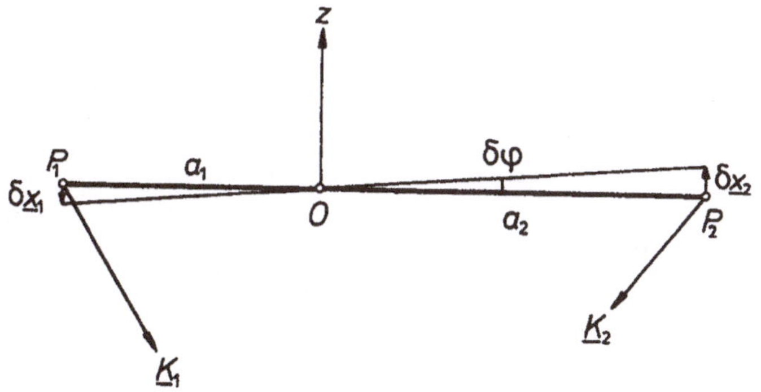

Figure 1. Equilibrium of a rigid lever.

the lever, without caring about the infinity of particles of which it is physically composed, and about the inner forces that act between them.

The basis for such problem is the principle of virtual work, or principle of virtual displacements. A virtual displacement is not a real movement performed by the system, but a very small fictitious displacement which, however, is compatible with the given constraints. In the present case, the lever can only rotate as a rigid body around its fulcrum O, so that its the "virtual displacements" $\delta\underline{x}_1$ of P_1 and $\delta\underline{x}_2$ of P_2 can only have the form shown in Fig. 1.

The "virtual work" associated with the infinitesimal displacement $\delta\underline{x}_1$, by (12), is

$$\underline{K}_1 \cdot \delta\underline{x}_1 ,$$

so we have the principle

$$\delta A = \sum_{i=1}^{2} \underline{K}_i \cdot \delta\underline{x}_i = 0 . \tag{51}$$

Since

$$\underline{K}_i \cdot \delta\underline{x}_i = \pm K_i' \delta z_i$$

and

$$\frac{\delta z_1}{a_1} = -\frac{\delta z_2}{a_2} \; (= -\delta\phi) , \tag{52}$$

K_i' denoting the "vertical" component of $\underline{K}_i$ and δz_i the vertical component of $\delta\underline{x}_i$, we get from

$$K_1' \delta z_1 - K_2' \delta z_2 = 0$$

and

$$a_2 \delta z_1 + a_1 \delta z_2 = 0$$

immediately (vanishing of the determinant!)

$$a_1 K_1' = a_2 K_2' ,$$

the well-known law for the equilibrium of a lever.

This example is intended to motivate the general principle of virtual work

$$\sum_{i=1}^{n} \underline{K}_i \cdot \delta \underline{x}_i = 0 \tag{53}$$

for the equilibrium of n mass points which are linked by certain geometrical constraints, and $\delta \underline{x}_i$ are small "virtual displacements", arbitrary but compatible with the constraints. Note that both $\delta \underline{x}_i$ and the "external" or "impressed" forces $\underline{K}_i$ are vectors, and that (53) is a sum of inner products of these vectors.

Note that in statics, for equilibrium, there is no motion and hence no kinetic energy T . Then the total energy E is potential energy, and virtual work δA corresponds to a change in U . Hence (53) is equivalent to

$$\delta U = 0 \ , \tag{54}$$

which means that for equilibrium, the potential energy is stationary (maximum or minimum).

2.2. Mass Point in Equilibrium on a Surface

Consider a mass point which is free to move on a smooth surface

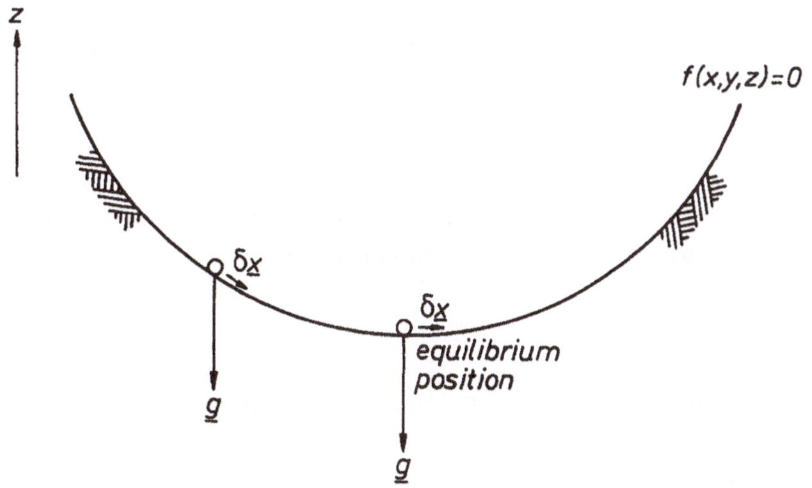

Figure 2. Mass point sliding on a surface.

under the effect of gravity represented by the vector $\underline{g}$ (vertical, along the z-axis).

There is $n=1$, and the principle of virtual displacements (53) gives

$$\underline{g}\cdot\delta\underline{x} = 0 \ . \tag{55}$$

Since $\delta\underline{x}$, by geometry, is tangential to the surface, equilibrium is reached in the deepest position, for which $\delta\underline{x}$ is horizontal so that (55) is satisfied.

Writing the surface in the form

$$f(x,y,z) = 0 \ , \tag{56}$$

then the surface normal is well known to be proportional to the vector

$$\text{grad } f \ .$$

Thus equilibrium is reached if

$$\underline{g} = \lambda \, \text{grad } f \ , \tag{57}$$

with a certain scalar multiplyer λ .

Regarding $\underline{g}$ as the gradient of potential energy

$$\underline{g} = -\text{grad } U \tag{58}$$

we can write (58) in the form

$$\text{grad}(U + \lambda f) = 0. \tag{59}$$

On forming the inner product with any infinitesimal space vector $\delta\underline{x}$ (now <u>unrestricted</u> by the condition (56)!) we get

$$\delta(U + \lambda f) = 0 \ , \tag{60}$$

corresponding to an extremum of U with side condition (56) and Lagrange multiplyer λ.

The physical interpretation is that λ grad f represents the <u>reaction force</u> exerted by the surface, which is always normal to the surface, and (57) expresses the equality between the external force $\underline{g}$ and this reaction force.

Even more: λf may be interpreted as the potential energy of the reaction force, and (60) states that the total potential energy (U plus energy of reaction force) is stationary for equilibrium. I took this beautiful physical interpretation from (Lanczos, 1970, p.84).

2.3. D'Alembert's Principle

By a stroke of genius, d'Alembert was able to generalize the principle of virtual displacements from statics to dynamics.

Rewrite the fundamental Newton equation (3) in the form

$$\underline{K} - m\underline{b} = 0 \tag{61}$$

and define

$$\underline{I} = -m\underline{b} \tag{62}$$

to be the "force of inertia". Then (61), written as

$$\underline{K} + \underline{I} = 0 ,$$

has the form of a "static" equation of equilibrium; the sum of the external force and the force of inertia is zero. The "total force" is now $\underline{\bar{K}} = \underline{K} + \underline{I}$.

The corresponding analog to the principle of virtual displacements (53) is then

$$\sum_{i=1}^{n} \underline{\bar{K}}_i \cdot \delta\underline{x}_i = \sum_{i=1}^{n} (\underline{K}_i + \underline{I}_i) \cdot \delta x_i = \sum_{n=1}^{n} (\underline{K}_i - m_i \underline{b}_i) \cdot \delta\underline{x}_i = 0 \tag{63}$$

or finally

$$\sum_{i=1}^{n} (\underline{K}_i - m_i \ddot{\underline{x}}_i) \cdot \delta\underline{x}_i = 0 . \tag{64}$$

This is d'Alembert's principle. Note that the "derivation" by which we have found it, is purely heuristic and is by no means a logical deduction.

Thus d'Alembert's principle is a new and very general principle. It is much more general than Newton's law of motion (6) and can serve as a basis of most of classical analytical mechanics. (Note again that the "virtual displacements" $\delta\underline{x}_i$ must be compatible with the geometrical constraints and play an essential role in (64)!) In particular, the equations of motion of a rigid body (sec. 1.6.) can be derived by it (Lanczos, 1970, pp.103-106).

Besides its fundamental conceptual importance, it serves as a starting point for deriving Lagrange's equations.

2.4. Lagrange's Equations of First Kind

Let the n mass points be restricted by ℓ condition equations

$$F_k(x_1, y_1, z_1, x_2, y_2, z_2, \ldots, x_n, y_n, z_n) = 0$$

$$(k = 1, 2, \ldots \ell < 3n) \quad . \tag{65}$$

We differentiate them to get

$$\mathrm{grad}_i F_k \cdot \delta\underline{x}_i = 0 \qquad (i = 1, 2, \ldots, n) \; ; \tag{66}$$

multiply these equations by a factor λ_k (as yet undetermined) and add to (64):

$$\sum_{i=1}^{n} (\underline{K}_i - m_i \ddot{\underline{x}}_i + \sum_{k=1}^{\ell} \lambda_k \, \mathrm{grad}_i F_k) \cdot \delta\underline{x}_i = 0 . \tag{67}$$

Now, however, we can take the $\delta\underline{x}_k$ arbitrary, no longer subjected to the geometrical constraints. Then all quantities between parentheses ( ) must be zero:

$$m_i \ddot{\underline{x}}_i = \underline{K}_i - \sum_{k=1}^{\ell} \lambda_k \, \mathrm{grad}_i F_k \quad . \tag{68}$$

These are Lagrange's equations of the first kind.

Let us first give the reason for the transition from (67) to (68). In the sum (67), which contains $3n$ (scalar) terms, we can make the last ℓ terms vanish by a suitable choice of the ℓ parameters λ_k. The remaining first $3n - \ell$ terms, however, contain only $3n - \ell$ displacement components, i.e., precisely as many as are necessary to define the system of $f = 3n - \ell$ "degrees of freedom" uniquely. Since these f independent displacements are arbitrary, the remaining first

$3n-\ell$ terms within parentheses must also vanish, which leads to (68). Note the subtlety of this argument, familar from the Lagrange multiplyer method for finding maxima or minima with side conditions.

Thus we have formally restored the equations of motion to the Newtonian form, by introducing, mathematically, Lagrangian multiplyers λ_k and, physically, "reaction forces" $-\lambda_k \operatorname{grad}_i F_k$; cf. also sec.2.2.

2.5. Lagrange's Equations of Second Kind

As we have just seen, the number of independent variables that describe our dynamical system is $f = 3n - \ell$. Taking this fact seriously, it is only natural to introduce f variables $q_1, q_2, \ldots, q_f$ in terms of which the Cartesian coordinate vectors $\underline{x}_i$ $(i = 1,2,\ldots,n)$ are expressed:

$$\underline{x}_i = \underline{x}_i(q_1, q_2, \ldots, q_f) \; . \tag{69}$$

Then

$$\delta\underline{x}_i = \sum_{r=1}^{f} \frac{\partial \underline{x}_i}{\partial q_r}\,\delta q_r = \frac{\partial \underline{x}_i}{\partial q_r}\,\delta q_r \; , \tag{70}$$

using the summation convention: a subscript occurring twice (in our case, r) automatically implies summation with respect to that subscript.

Similarly we have

$$\dot{\underline{x}}_i = \frac{\partial \underline{x}_i}{\partial q_r}\,\dot{q}_r \; , \tag{71}$$

and the following expression that occurs in (64) may be transformed as shown:

$$\ddot{\underline{x}}_i \cdot \delta\underline{x}_i = \left[\frac{d}{dt}\left(\dot{\underline{x}}_i \cdot \frac{\partial \underline{x}_i}{\partial q_r}\right) - \dot{\underline{x}}_i \cdot \frac{d}{dt}\left(\frac{\partial \underline{x}_i}{\partial q_r}\right)\right]\delta q_r \; . \tag{72}$$

Now we formally introduce $\dot{q}_r$ as additional variables, and (71) gives immediately

$$\frac{\partial \dot{\underline{x}}_i}{\partial \dot{q}_r} = \frac{\partial \underline{x}_i}{\partial q_r} \; , \tag{73}$$

so that the first term on the right-hand side becomes

$$\frac{d}{dt}\left(\dot{\underline{x}}_i \cdot \frac{\partial \underline{x}_i}{\partial q_r}\right) = \frac{d}{dt}\left(\dot{\underline{x}}_i \cdot \frac{\partial \dot{\underline{x}}_i}{\partial \dot{q}_r}\right) = \frac{1}{2}\frac{d}{dt}\left(\frac{\partial \dot{\underline{x}}_i^2}{\partial \dot{q}_r}\right) \tag{74}$$

since $\underline{x}_i^2 = \underline{x}_i \cdot \underline{x}_i$.

In the second term we have

$$\frac{d}{dt}\left(\frac{\partial \underline{x}_i}{\partial q_r}\right) = \frac{\partial^2 \underline{x}_i}{\partial q_r \partial q_s}\frac{dq_s}{dt} = \frac{\partial^2 \underline{x}_i}{\partial q_r \partial q_s}\dot{q}_s = \frac{\partial}{\partial q_r}\left(\frac{\partial \underline{x}_i}{\partial q_s}\dot{q}_s\right)$$

$$= \frac{\partial}{\partial q_r}\left(\frac{d\underline{x}_i}{dt}\right) = \frac{\partial \dot{\underline{x}}_i}{\partial q_r} , \tag{75}$$

which shows that the order of differentiation d/dt and $\partial/\partial q_r$ can be interchanged. Hence

$$\dot{\underline{x}}_i \cdot \frac{d}{dt}\left(\frac{\partial \underline{x}_i}{\partial q_r}\right) = \dot{\underline{x}}_i \cdot \frac{\partial \dot{\underline{x}}_i}{\partial q_r} = \frac{1}{2}\frac{\partial}{\partial q_r}(\dot{\underline{x}}_i^2) . \tag{76}$$

Thus (72) takes the form

$$\ddot{\underline{x}}_i \cdot \delta \underline{x}_i = \left[\frac{d}{dt}\left(\frac{1}{2}\frac{\partial \dot{\underline{x}}_i^2}{\partial \dot{q}_r}\right) - \frac{1}{2}\frac{\partial \dot{\underline{x}}_i^2}{\partial q_r}\right]\delta q_r \tag{77}$$

and

$$\sum_{i=1}^{n} m_i \ddot{\underline{x}}_i \cdot \delta \underline{x}_i = \left[\frac{d}{dt}\left(\frac{\partial T}{\partial \dot{q}_r}\right) - \frac{\partial T}{\partial q_r}\right]\delta q_r , \tag{78}$$

where

$$T = \sum_{i=1}^{n} \frac{1}{2} m_i \dot{\underline{x}}_i^2 \tag{79}$$

represents the total kinetic energy of the system; cf. (11).

Assuming that the forces $\underline{K}_i$ can be derived from a potential energy U which is a function of position only:

$$U = U(q_1, q_2, \ldots, q_f) , \tag{80}$$

we have

$$\sum_{i=1}^{n} \underline{K}_i \cdot \delta \underline{x}_i = -\sum_{r=1}^{f} \frac{\partial U}{\partial q_r} \delta q_r \ , \tag{81}$$

so that (64) gives (we introduce the summation sign over r explicitly although it would not be necessary because of the summation convention):

$$\sum_{r=1}^{f} \left[\frac{d}{dt}\left(\frac{\partial T}{\partial \dot{q}_r}\right) - \frac{\partial T}{\partial q_r} + \frac{\partial U}{\partial q_r} \right] \delta q_r = 0 \ . \tag{82}$$

Since the δq_r are arbitrary and independent, all terms within brackets must vanish. This gives

$$\frac{d}{dt}\left(\frac{\partial T}{\partial \dot{q}_r}\right) - \frac{\partial T}{\partial q_r} = -\frac{\partial U}{\partial q_r} \tag{83}$$

or, noting that U does not depend on $\dot{q}_i$ and introducing

$$L = T - U = L(q_r, \dot{q}_r) \ , \tag{84}$$

the brief form

$$\frac{d}{dt}\left(\frac{\partial L}{\partial \dot{q}_r}\right) - \frac{\partial L}{\partial q_r} = 0 \ . \tag{85}$$

These are Lagrange's equations of the second kind, or since they are much more important than (68), briefly Lagrange's equations. The function $L = T - U$, being the difference between potential and kinetic energy, is called the Lagrangian.

After performing the differentiation, (85) becomes a system of f ordinary differential equations of the second order for the f variables $q_r = q_r(t)$.

2.6. Geometrical Interpretation

Introduce (71) into (79):

$$T = \frac{1}{2}\sum_{i=1}^{n} m_i \dot{\underline{x}}_i \cdot \dot{\underline{x}}_i = \frac{1}{2}\left[\sum_{i=1}^{n} m_i \frac{\partial \underline{x}_i}{\partial q_r} \cdot \frac{\partial \underline{x}_i}{\partial q_s}\right] \dot{q}_r \dot{q}_s \ . \tag{86}$$

Denoting the quantities between brackets by

$$a_{rs} = a_{rs}(q_1, q_2, \dots, q_f) \, , \tag{87}$$

we get

$$T = \frac{1}{2} a_{rs}\dot{q}_r\dot{q}_s \tag{88}$$

as a quadratic form of $\dot{q}_r$ (summation convention!).

The q_r may be considered curvilinear (generalized) coordinates of an f-dimensional curved manifold, or Riemannian space, the <u>configuration space</u>.

The line element of this configuration space has the usual form

$$ds^2 = a_{rs}dq_r dq_s \tag{89}$$

(there is no danger to confuse the line element ds with the subscript s !), so that (88) becomes

$$T = \frac{1}{2}\frac{ds^2}{dt^2} = \frac{1}{2}v^2 \, , \tag{90}$$

representing a fictitious "particle" of unit mass, moving with velocity v in this configuration space.

<u>Examples</u>. In the case of the <u>lever</u>, it can only rotate in the plane of Fig. 1 around an angle ϕ which can be taken as generalized coordinate: there is only one degree of freedom and the configuration space is onedimensional.

Imagine a particle that can move freely along the <u>surface of a sphere</u>, but is constrained never to leave the surface (this is analogous to the case of Fig. 2). Its "generalized coordinates" are latitude ϕ and longitude λ :

$$q_1 = \phi \, , \qquad q_2 = \lambda \, , \tag{91}$$

the line element is

$$ds^2 = d\phi^2 + \cos^2\phi \, d\lambda^2 = dq_1^2 + \cos^2 q_1 \, dq_2^2 \, , \tag{92}$$

and the kinetic energy by (89) and (90)

$$T = \frac{1}{2}(\dot{q}_1^2 + \cos^2 q_1 \dot{q}_2^2) \, . \tag{93}$$

The rotation of a rigid body can be described by three Euler angles ϕ , θ , ψ such that

$$0 \leq \phi < 2\pi , \quad 0 \leq \theta \leq \pi , \quad 0 \leq \psi < 2\pi . \tag{94}$$

There is an important classical theory of earth rotation (from Poisson to Woolard) that is based on the Lagrangian equations (85) applied to the present problem (Moritz and Mueller, 1987, sec.2.5).

The line element (ibid.,p.85, eq. (2-253)) can be shown to be

$$ds^2 = A(d\theta^2 + \sin^2\theta \, d\psi^2) + C(d\phi + d\psi \cos\theta)^2 , \tag{95}$$

$A = B$ and C being the earth's principal moments of inertia, cf.(43).

Clearly, the configuration space is three-dimensional, but it is not our familiar Euclidean space: it is curved. It is identical to rotation group space discussed in (Moritz, 1980, sec.36). What is it topologically? See (94): ϕ and θ describe a sphere, and ψ describes a circle. So it is the "topological product of a sphere with a circle".

This is easy to understand abstractly, but hardly possible to visualize directly in the same way as we can visualize our Euclidean space; cf. (Synge and Schild, 1978, pp. 181-183).

Geodesic motion. If there are no forces acting in configuration space, then $U = 0$ and (83) reduces to

$$\frac{d}{dt}\left(\frac{\partial T}{\partial \dot{q}_r}\right) - \frac{\partial T}{\partial q_r} = 0 . \tag{96}$$

This may be shown to be the equation of a geodesic in configuration space. A geodesic is the "straightest possible" line in configuration space; among all possible curves connecting two points, a geodesic gives the shortest connection.

Let us verify this for ordinary Euclidean space and a free particle (we may introduce $m \neq 1$). Now

$$q_1 = x , \quad q_2 = y , \quad q_3 = z ,$$

$$ds^2 = dx^2 + dy^2 + dz^2 = dq_1^2 + dq_2^2 + dq_3^2 , \tag{97}$$

$$T = \frac{1}{2} m(\dot{x}^2 + \dot{y}^2 + \dot{z}^2) = \frac{1}{2} m(\dot{q}_1^2 + \dot{q}_2^2 + \dot{q}_3^2) .$$

Then (96) reduces to

$$\ddot{x} = \ddot{y} = \ddot{z} = 0 \tag{98}$$

as it should: we get a uniform motion along a straight line, cf. sec.1.4.

Less trivially, free rotation (without external torques) corresponds to geodesic motion in rotation group space.

Most important of all: motion under the effect only of gravitation corresponds to a geodesic in the curved fourdimensional space time of General Relativity; see the lectures of E. Grafarend.

Interpretation of Lagrange's equations. Eq. (98) means that the acceleration of our particle in Euclidean space is zero. This suggests to regard, generally, the left-hand side of (96) as the acceleration of our fictitious "particle" of mass 1 in configuration space. Then (83) expresses "Newton's law for configuration space": acceleration times mass 1 equals the force $-\partial U/\partial q_r$, which is a nice generalization of (8). See also (McConnell,1957 pp.248-249) or (Synge and Schild (1978,pp.173-174))

2.7. The Principle of Least Action

Taking the Lagrangian (84), Hamilton's principle of least action states that

$$A_H = \int_{t_1}^{t_2} L\,dt = \int_{t_1}^{t_2} (T - U)\,dt = \text{extremum} \,. \tag{99}$$

The action A_H (not to be confused with (9)!) thus is defined as the difference: kinetic minus potential energy) integrated along the trajectory from an initial point (time t_1) to an end point (time t_2). The actual motion will be such that the action takes a minimum (or generally, a stationary) value.

Finding the extremum of (99) is a problem of the so-called calculus of variations, a branch of mathematics dealing with extrema of functionals (A_H is a nonlinear functional depending on the functions $q_r(t)$ and involving the functional operations of differentiation and integration; cf.(88)).

Using the calculus of variations, Lagrange's equations (85) are a direct consequence of (99). Having followed a more elementary road to get (85), we refer the reader for this to books such as (Lanczos, 1970).

Using the variational notation δ (employed on such occasions rather than the ordinary differential d), we may write (99) in the form

$$\delta A_H = \delta \int_{t_1}^{t_2} L\,dt = 0 \;, \tag{100}$$

whose formal similarity to the principle of virtual work (51) is purely accidental: there is no direct relation between the two principles (they are only indirectly related by the entire subject matter of the present Lecture).

Finally I should like to point out, however, that free motion with $U = 0$ leads to

$$A_H = \int_{t_1}^{t_2} T\,dt = \frac{1}{2} \int_{t_1}^{t_2} \left(\frac{ds}{dt}\right)^2 dt = \frac{1}{2} \int_{P_1}^{P_2} ds$$

if we assume $dt = ds$, which is justified. Thus we immediately obtain geodesic motion (shortest distance).

LECTURE 3

HAMILTON

It was Hamilton's merit to transform Lagrange's equations (85), which are a system of f differential equations of second order, to a particularly elegant and simple system of $2f$ differential equations of the first order. Hamilton's equation constitute by far the most important and useful formulation of analytical dynamics. For instance, the theory of dynamical systems discussed in Lecture 4 generally presupposes first-order differential equations and makes large use of Hamiltonian systems. The Hamiltonian formulation is also fundamental for quantum mechanics.

The transition from Lagrange's to Hamiltonian's equations is done by a Legendre transformation.

3.1. The Legendre Transformation

Consider a function of two variables:

$$f = f(x,y) \quad . \tag{101}$$

Its differential may be written

$$df = u\,dx + v\,dy \tag{102}$$

where

$$u = \frac{\partial f}{\partial x} \,, \qquad v = \frac{\partial f}{\partial y} \; . \tag{103}$$

The right-hand side of the second equation (103) is a function of x and y, symbolically

$$v = v(x,y) \quad . \tag{104}$$

By solving this equation for y, we can represent y as a function of x and v:

$$y = y(x,v) \tag{105}$$

and substituting this expression into (101) we obtain f as a function of x and v :

$$f = f^*(x,v) \ . \tag{106}$$

Thus we may use x,v as new independent variables, in place of x,y .

This transition to x,v becomes particularly elegant if we also introduce a new function g by

$$g = g(x,v) = yv - f \tag{107}$$

(thus transforming the dependent variable f as well as the independent variables x,y), f and y being expressed in terms of x,v by (106) and (105). The differential of (107) is

$$dg = v\,dy + y\,dv - df \ .$$

In view of (102) this reduces to

$$dg = -u\,dx + y\,dv \ . \tag{108}$$

The comparison with

$$dg = \frac{\partial g}{\partial x}dx + \frac{\partial g}{\partial v}dv$$

shows that

$$u = -\frac{\partial g}{\partial x} \ , \qquad y = \frac{\partial g}{\partial v} \ . \tag{109}$$

This expresses x and v in terms of our new independent variables u, y .

Besides their mathematical importance, such Legendre transformations play a considerable role in thermodynamics (Goldstein, 1980, p.341). Geodesists may know them from Sansò's formulation of Molodensky's boundary-value problem in gravity space (Moritz, 1980, p.452).

3.2. Hamilton's Canonical Equations

The Lagrangian equations (85) are formulated in terms of $2f$ independent variables q_i, $\dot{q}_i$ (they are formally considered independent although, of course, $\dot{q}_i = dq_i/dt$). In place of $\dot{q}_i$, let us introduce new variables p_i by

$$p_i = \frac{\partial L}{\partial \dot{q}_i} . \qquad (110)$$

(Since our former index i is no more needed for numbering particles, it can now be used instead of r). This is indeed a Legendre transformation; there correspond:

q_i, $\dot{q}_i$ to x, y ,

$L(q_i,\dot{q}_i)$ to $f(x,y)$,

p_i to v ,

and (107) takes the form (writing H in place of g):

$$H = H(q_i,p_i) = p_i\dot{q}_i - L . \qquad (111)$$

H is called Hamiltonian function, or Hamiltonian. What corresponds to u ? Lagrange's equations (85) may be written, using (110),

$$\dot{p}_i = \frac{\partial L}{\partial q_i} . \qquad (112)$$

This is the equivalent of the first equation of (103); hence $\dot{p}_i$ corresponds to u . Now (109) gives directly

$$\dot{q}_i = \frac{\partial H}{\partial p_i} ,$$

$$\dot{p}_i = - \frac{\partial H}{\partial q_i} . \qquad (113)$$

These are Hamilton's equations, also called canonical because of their basic importance and their elegance ("canon" denotes a basic standard, a measure of perfection). They are readily seen to be a system of $2f$ first order differential equations for q_i,p_i , as opposed to the Lagrangian equations, a system of f second-order differential equations for q_i .

The fact that the single variable x in the above Legendre transformation corresponds to f variables q_i , and similarly for y , u , and v , is taken into consideration simply by the summation convention in (111). If he wishes, the reader may also directly derive (113) from (110) by repeating the development leading from (101) to (109).

The quantities q_i , p_i are called canonical coordinates, the p_i also being denoted as canonical impulses.

The Hamiltonian. Since U does not depend on $\dot{q}_i$, we get from (110) with (88) and (84)

$$p_i = \frac{\partial L}{\partial \dot{q}_i} = \frac{\partial T}{\partial \dot{q}_i} = a_{ij}\dot{q}_j \ . \tag{114}$$

Hence

$$p_i\dot{q}_i = a_{ij}\dot{q}_i\dot{q}_j = 2T \ , \tag{115}$$

so that (111), also using (84), gives

$$H = T + U \ . \tag{116}$$

Thus the Hamiltonian is nothing else than the total energy: kinetic plus potential energy.

Conservation of energy. Form the time derivative of H :

$$\frac{dH}{dt} = \frac{\partial H}{\partial q_i}\dot{q}_i + \frac{\partial H}{\partial p_i}\dot{p}_i \tag{117}$$

and substitute (113). The immediate result is

$$\frac{dH}{dt} = 0 \tag{118}$$

or, on integration,

$$H = \text{const.} = E \ . \tag{119}$$

More explicitly, we have the energy integral

$$H(q_1,q_2,\ldots,q_f;p_1,p_2,\ldots,p_f) = E = \text{const.} \tag{120}$$

<u>Example</u>. Again we shall use the trivial example of a free mass point (sec.1.4) to illustrate the Hamiltonian equations. From (97) and (114) we get

$$p_i = \frac{\partial T}{\partial \dot{q}_i} = m\dot{q}_i \qquad (i = 1,2,3) \ , \tag{121}$$

(which explains the name, canonical <u>impulses</u>). Expressed in terms of these p_i , (97) becomes

$$L = \frac{1}{2m} (p_1^2 + p_2^2 + p_3^2) = \frac{1}{2m} p_i p_i \ ,$$

so that the Hamiltonian (111) takes the form

$$H = \frac{1}{m} p_i p_i - \frac{1}{2m} p_i p_i = \frac{1}{2m} p_i p_i \ .$$

The Hamiltonian equations (113) yield

$$\dot{q}_i = \frac{1}{m} p_i \ , \tag{122}$$

$$\dot{p}_i = 0$$

since H does not depend on q_i now. The first equation is identical to (121) as it should; and the combination of both equations of (122) gives

$$\ddot{q}_i = 0 \ ,$$

again equivalent to straight-line motion (98).

This simple example is only to illustrate the <u>mechanism</u> of Hamilton's equations. They give no idea of the tremendous <u>power</u> of the Hamiltonian approach. They are standard tools in celestial mechanics, as we shall see later, and the most modern and precise treatment of the rotation of a rigid earth by Kinoshita is based on it, cf. (Moritz and Mueller, 1987, sec.2.6).

Note finally that the second equation of (113) is a "grown-up" version of the elementary momentum theorem (5); this is particularly evident if we also look at (7).

3.3. Canonical Transformations

It is often required to find coordinate transformations

$$q_i = q_i(Q_1,Q_2,\dots,Q_f;P_1,P_2,\dots,P_f) \ ,$$
$$p_i = p_i(Q_1,Q_2,\dots,Q_f;P_1,P_2,\dots,P_f) \ , \tag{123}$$

introducing new generalized coordinates Q_i and generalized impulses P_i in such a way that the Hamiltonion structure, expressed by (113), is preserved. Thus for such a "canonical transformation" there should hold

$$\dot{Q}_i = \frac{\partial \bar{H}}{\partial P_i} \ ,$$
$$\dot{P}_i = -\frac{\partial \bar{H}}{\partial Q_i} \ . \tag{124}$$

This will be the case if there is a function $S(q_i,Q_i)$ of the old and the new position coordinates such that

$$p_i = \frac{\partial S}{\partial q_i} \ ,$$
$$P_i = -\frac{\partial S}{\partial Q_i} \ . \tag{125}$$

This function $S(q_i,Q_i)$ is called the generating function of our canonical transformation. Taking

$$\bar{H}(P_i,Q_i) = H(p_i,q_i) \tag{126}$$

we then find that (123) are satisfied together with (113).

This can be seen in the following way. Transform the coordinates q_i only (a slight but very useful restriction):

$$q_i = f_i(Q_1,Q_2,\dots,Q_f) \ , \qquad i = 1,2,\dots,f \ , \tag{127}$$

and express the Lagrangian in terms of Q_i, such that

$$\bar{L}(Q_i,\dot{Q}_i) = L(q_i,\dot{q}_i) \ , \tag{128}$$

i.e., $\overline{L}$ is obtained by substituting (127) into (84). Now Lagrange's equations are invariant with respect to a change of coordinates: express $\underline{x}_i$ in terms of Q_i by substituting (127) into (69); this will change nothing in the process leading to (85). (This invariance also is a direct consequence of the variational principle (99).) Therefore, if we define new generalized impulses following (110):

$$P_i = \frac{\partial \overline{L}}{\partial \dot{Q}_i} , \tag{129}$$

Lagrange's equations give

$$\dot{P}_i = \frac{\partial \overline{L}}{\partial Q_i} , \tag{130}$$

in full analogy to (112).

Now also the Hamiltonians (111) will be equal in the sense of (126) if and only if

$$P_i \dot{Q}_i = p_i \dot{q}_i . \tag{131}$$

This, however, is where equations (125) enter. Since by definition S is a function of q_i and Q_i only and does not explicitly contain the time t, there must be

$$0 = \frac{dS}{dt} = \frac{\partial S}{\partial q_i} \dot{q}_i + \frac{\partial S}{\partial Q_i} \dot{Q}_i = p_i \dot{q}_i - P_i \dot{Q}_i$$

by (125). This immediately gives (131), completing the proof.

3.4. Symplectic Geometry

In sec. 2.6 we have introduced the natural space for Lagrangian mechanics, the f-dimensional configuration space parametrized by the coordinates q_i $(i = 1,2,...,f)$.

The natural space for Hamiltonian mechanics is 2f-dimensional phase space parametrized by the 2f canonical coordinates $q_1, q_2, .., q_f$; $p_1, p_2, ..., p_f$, which we may regard as position coordinates and impulse coordinates, respectively.

The trajectory described by Hamilton's equations (113) thus is a curve in phase space. The energy integral (120) shows the basic fact that all trajectories corresponding to a given energy E must lie on a surface in phase space, whose equation is just (120) and which is called the energy surface.

Combining all q_i and all p_i into a vector z (which we do not bother any more to underline):

$$z = (q_1, q_2, \ldots, q_f; p_1, p_2, \ldots, p_f)^T \tag{132}$$

we may write (113) in the concise form

$$\dot{z} = J \text{ grad } H \tag{133}$$

where the $2f \times 2f$ matrix J has the form

$$J = \begin{bmatrix} 0 & I \\ -I & 0 \end{bmatrix} , \tag{134}$$

I denoting the $f \times f$ unit matrix, and 0 the $f \times f$ zero matrix. Thus J is a very simple skew-symmetric square matrix.

By direct computation it follows that (similar to $i^2 = -1$!)

$$J^2 = -I \tag{135}$$

(I is now the $2f \times 2f$ unit matrix), and J is also orthogonal:

$$J^T J = I , \tag{136}$$

so that

$$J^{-1} = J^T = -J . \tag{137}$$

The matrix J allows us to introduce a skew-symmetric "scalar product"

$$(x,y) = x^T J y = -(y,x) . \tag{138}$$

It is easy to verify that (x,y) expresses the sum of the areas of the projections of the parallelogram spanned by the $2f$-vectors x and y onto the coordinate planes $p_i q_i$ $(i = 1,2,\ldots,f)$.

Linear transformations which preserve the skew-scalar product (138),

$$(Sx,Sy) = (x,y) \quad , \tag{139}$$

are called symplectic transformations.

They satisfy the condition

$$S^T J S = J \ . \tag{140}$$

In fact, by (138) and (140)

$$(Sx,Sy) = (Sx)^T JSy = x^T S^T JSy = x^T Jy = (x,y) \ .$$

The symplectic condition (140) is a beautiful analog of ordinary orthogonal transformations A :

$$A^T A = A^T I A = I \ .$$

Symplectic structure. Let us now look more closely into the relation between phase space (p.s.) and configuration space (c.s.). Consider a point $P = (q_i)$ in configuration space and consider its tangent space to P, which obviously is spanned by the vector $(\dot{q}_1, \dot{q}_2, \ldots, \dot{q}_f)$ at P. In the language of tensor calculus, $\dot{q}_i$ is a contravariant vector, and the vector p_i (114) is the corresponding covariant vector, considering a_{ij} as the metric tensor of c.s. in agreement with (89). Thus the p_i can be said to span the cotangent space, or cotangent manifold, to c.s. at P. Both the tangent and the cotangent spaces abviously are linear spaces (think of the tangent plane to a curved surface!). Thus phase space (q_i, p_i) is nothing else than the set of all cotangent spaces to all the points P making up the underlying configuration space, which we may express in the simple form: phase space is the cotangent space of configuration space.

(For those trained in modern mathematics we mention that p.s. is a fibre bundle, and the most natural description is in terms of exterior differential forms. In this elementary introduction, we cannot treat exterior forms, but not that the skew-scalar product (138) is an exterior form in disguise.)

The structure induced on p.s. from the Riemannian structure of c.s. by the method described in sec. 3.3. is called symplectic structure. Coordinate transformations (123) in p.s., briefly

$$z = z(Z) \tag{141}$$

that leave the symplectic structure, in particular the canonical equations (133), invariant, are nothing else than the canonical transformations considered in sec. 3.3.

Symplectic transformations may be regarded as infinitesimal, or linearized, canonical transformations.

Poisson brackets. Consider two functions $u(q_i, p_i)$ and $v(q_i, p_i)$ of the canonical variables. Their Poisson bracket is defined as (summation convention!)

$$[u,v] = \frac{\partial u}{\partial q_i}\frac{\partial v}{\partial p_i} - \frac{\partial u}{\partial p_i}\frac{\partial v}{\partial q_i} = -[v,u] \tag{142}$$

which shows a typical skew-symmetric symplectic structure. In fact, it can readily be reduced to matrix form:

$$[u,v] = \left(\frac{\partial u}{\partial z}\right)^T J \left(\frac{\partial v}{\partial z}\right) \tag{143}$$

with the obvious notation

$$\frac{\partial u}{\partial z} = \text{grad}\, u = \left(\frac{\partial u}{\partial q_1}, \frac{\partial u}{\partial q_2}, \ldots, \frac{\partial u}{\partial p_f}\right) . \tag{144}$$

In terms of the skew-scalar product (138) this becomes simply

$$[u,v] = \left[\frac{\partial u}{\partial z}, \frac{\partial v}{\partial z}\right] . \tag{145}$$

It is trivial but important that

$$\begin{aligned} [q_j, q_k] &= 0 = [p_j, p_k] , \\ [q_j, p_k] &= \delta_{jk} = -[p_j, q_k] , \end{aligned} \tag{146}$$

as one verifies immediately by substituting p_j and q_j for u and v in (142); the Kronecker delta δ_{ij} denotes the elements of the unit matrix as usual. Equations (146) may be summarized in the simple form

$$[z,z] = J . \tag{147}$$

Time derivatives. Consider any function $F(q_i, p_i)$ not depending explicitly on time t (only through p_i and q_i). Its time derivative is, by (113),

$$\frac{dF}{dt} = \frac{\partial F}{\partial q_i}\dot{q}_i + \frac{\partial F}{\partial p_i}\dot{p}_i = \frac{\partial F}{\partial q_i}\frac{\partial H}{\partial p_i} - \frac{\partial F}{\partial p_i}\frac{\partial H}{\partial q_i} .$$

In Poisson bracket notation (142) this is simply

$$\frac{dF}{dt} = [F,H] . \tag{148}$$

In particular,

$$\frac{dH}{dt} = [H,H] = 0$$

in agreement with (118).

The Poisson bracket formalism permits a direct transition to quantum mechanics; for us it will be important for perturbation theory (sec. 3.6.).

Lagrange brackets. Besides the Poisson bracket, there is also the Lagrange bracket, defined as

$$\{u,v\} = \frac{\partial q_i}{\partial u}\frac{\partial p_i}{\partial v} - \frac{\partial q_i}{\partial v}\frac{\partial p_i}{\partial u} = -\{v,u\}. \tag{149}$$

It is quite similar to (142), is closely related to the Poisson bracket, and admits a similar interpretation in terms of symplectic geometry. Both the Poisson and the Lagrange brackets are invariant with respect to canonical transformations.

3.5. Cyclical Variables

If the Hamiltonian H does not explicitly contain a coordinate q_i, then the second of Hamilton's equations (113) becomes

$$\dot{p}_i = \frac{\partial H}{\partial q_i} = 0 , \tag{150}$$

which means that the corresponding canonical impulse p_i is constant. Such a coordinate q_i is called cyclic, for reasons to be seen later.

If all position coordinates q_i are cyclic, then

$$H = H(p_1,p_2,...,p_f) \tag{151}$$

is a function of the impulse coordinates only. Then (150) shows that all p_k are constant, and the first of Hamilton's equations (113) gives

$$\dot{q}_i = \frac{\partial H}{\partial p_i} = F_i(p_k) = \text{const.} = \omega_i \ , \tag{152}$$

since all p_k are constant. Thus all q_i will be linear functions of time t :

$$q_i = \omega_i t + \alpha_i \tag{153}$$

with constant ω_i and α_i .

Two cases may arise. First, q_i may increase indefinitely with time; this corresponds to uniform motion along a straight line in phase space. Second, q_i has the character of an angle; then it will be a periodic function of time; this corresponds to uniform motion along a circle (always considering one particular q only). (Also the "mixed" case exists: a motion may be rectilinear in one coordinate and circular in another.)

An example for the first case is, of course, free motion of a particle in Euclidean space, our favorite "trivial example".

Much more important, especially in celestial mechanics where all interesting motions are more or less periodic, including earth rotation, is the second case where all coordinates q_i are cyclic, or angle variables. The conjugated (i.e., corresponding) impulse p_i is called an action variable and is frequently denoted by J_i (it is easy to see that it has the dimension of action as defined by (99)).

One advantage of the angle-action variables is simplicity: J_i is constant and q_i is a linear function of time. Another advantage is that the frequencies ω_i can be obtained immediately by differentiation of the Hamiltonian: (152) gives

$$\omega_i = \frac{\partial H(J_1, J_2, \ldots, J_f)}{\partial J_i} \quad . \tag{154}$$

The geometrical interpretation of the orbit in the onedimensional case $(f = 1)$ is a circle; in two dimensions the orbit lies on a torus, which is the "topological product of two circles" (Fig. 3); and for f degrees of freedom we have an "f-dimensional torus". Mathematically this corresponds to the expansion of q_i into a f-dimensional Fourier series.

Angle-action variables play an important role in classical mechanics: the Delaunay variables in Keplerian motion (sec.3.7.), the Andoyer

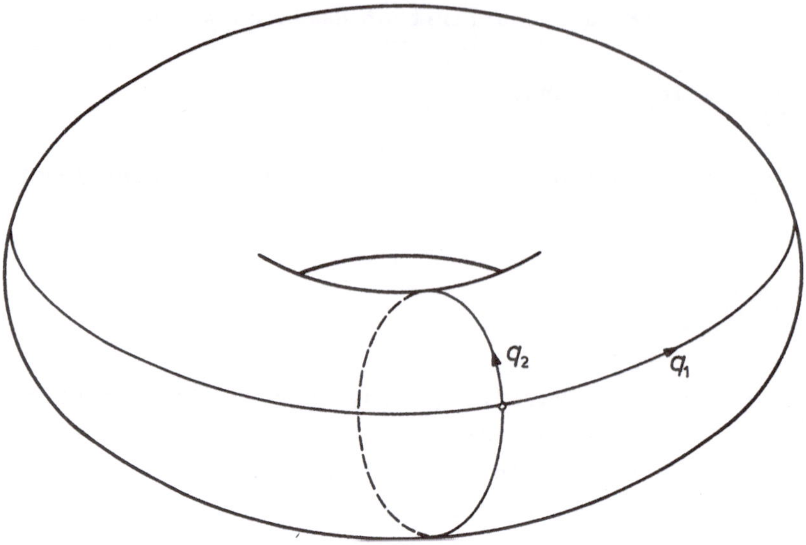

Figure 3. The torus as product of two circles .

variables in earth rotation (Moritz and Mueller, 1987, p.96), etc.

Their use is by no means restricted to force-free motion, as (151) seems to indicate. In fact, canonical transformations obliterate the neat distinction between "coordinate" and "impulse" variables. Indeed, if the original Hamiltonian has the form $H(p_i, q_i)$, it is often possible to find a canonical transformation that reduces it to the form $\bar{H}(P_i)$ depending on the new "impulses" only. This can be done, e.g., by solving the so-called partial differential equation of Hamilton-Jacobi which we cannot treat here but which is found in all relevant textbooks, or by the more specific and esoteric method of Delaunay (Brouwer and Clemence, 1961, pp.541-559).

We also mention that the motion may degenerate by one or several of the ω_i in (153) becoming zero; then the corresponding coordinate $q_i = \alpha_i$ will reduce to a constant. This is the case with the angular variable h in earth rotation (Moritz and Mueller, 1987, p.102) and even more so in Keplerian motion, as we shall see in sec. 3.7.

3.6. Perturbation Theory

Let the coordinates q_i and p_i be cyclic for an "unperturbed" motion with H given by (151), and let H be perturbed by an additional term R, which has the character of an additional small potential energy and is called <u>disturbing function</u>. Thus H is replaced by

$$\bar{H}(p_i,q_i) = H(p_i) + R(p_i,q_i) \ . \tag{155}$$

Then Hamilton's equations (113) give with (152)

$$\begin{aligned} \dot{q}_i &= \frac{\partial \bar{H}}{\partial p_i} = \omega_i + \frac{\partial R}{\partial p_i} \, , \\ \dot{p}_i &= -\frac{\partial \bar{H}}{\partial q_i} = -\frac{\partial R}{\partial q_i} \, . \end{aligned} \tag{156}$$

This is a case of "variation of constants" or "variation of elements": the canonical elements p_i and the frequences ω_i are no longer constant, but undergo small variations expressed by (156).

For elements A which are not canonical but functions of p_i and q_i, the variation is governed by (148):

$$\dot{A} = [A,\bar{H}] = \frac{\partial A}{\partial q_i}\dot{q}_i + \frac{\partial A}{\partial p_i}\dot{p}_i \, , \tag{157}$$

$\dot{q}_i$ and $\dot{p}_i$ being given by (156).

An application of (156) to the earth rotation problem may be found in (Moritz and Mueller, 1987, pp.103-105), and (157) will be applied to Keplerian motion in the next section.

3.7. Keplerian Motion

In order to illustrate these abstract developments by means of a concrete and useful example, let us apply some of the foregoing considerations to Keplerian motion. Excluding hyperbolas and parabolas as possible orbits, we concentrate on elliptic motion. Since this case is considered in practically all textbooks on classical mechanics and in all books on celestial mechanics, and particularly since there is a Course on Celestial Mechanics by J. Kovalevsky in this School, we can limit ourselves to stating some important relations. We understand that this situation comprises the motion of a planet around the sun and the motion of a satellite around the earth, both subject to various perturbations. Since the mass of the central body is so much larger than that of the other, we may assume that one focus of the ellipse (in the unperturbed case) lies at the center of mass of the sun or of the earth, respectively. For simplicity, let us concentrate on the earth-satellite case.

The orbital ellipse is shown in Fig. 4. Kepler's first law has the

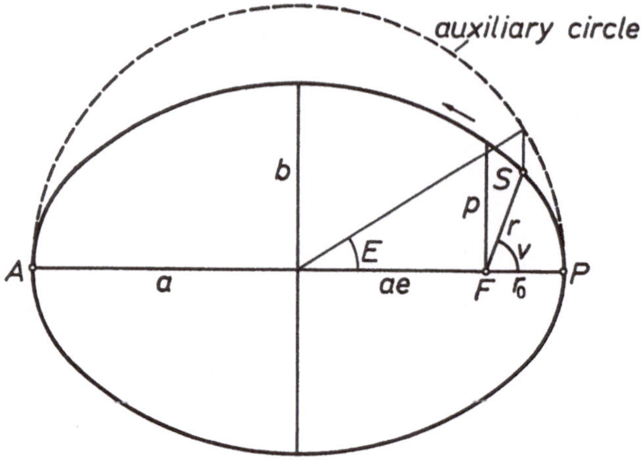

Figure 4. The orbital ellipse. P = perigee, A = apogee, F = earth's center of mass, S = instantaneous position of satellite.

mathematical expression

$$r = \frac{p}{1 + e \cos v} \tag{158}$$

which is the equation of an ellipse in polar coordinates r, v with semiaxes a, b and excentricity

$$e = \frac{\sqrt{a^2 - b^2}}{a} \tag{159}$$

and with

$$p = \frac{b^2}{a} = a(1 - e^2) \ . \tag{160}$$

Besides the radius vector r, we have the other polar coordinate v, called true anomaly for historical reasons. We may also write

$$r = a(1 - e \cos E) \ , \tag{161}$$

expressing r in terms of another angle E called excentric anomaly (also very historical!). Comparing the right-hand sides of (158) and (161) we find a relation between v and E; the geometrical situation is shown in Fig. 4.

The excentric anomaly is related to the mean anomaly M by Kepler's equation

$$E - e \sin E = M \ . \tag{162}$$

This equation has no direct geometric interpretation, but M is a linear function of time:

$$M = n(t - T) \tag{163}$$

and thus has the character of an angular canonical variable q_i in the sense of sec. 3.5. Then, E and v are indirectly functions of t through (162), (158) and (161).

We only briefly mention Kepler's second law

$$r^2 \frac{dv}{dt} = \sqrt{GM a(1 - e^2)} \ , \tag{164}$$

the area law related to the constancy of the magnitude of the angular momentum vector (22), and the third law

$$n^2 a^3 = GM \ , \tag{165}$$

fundamental for the highly precise determination of GM from satellite observations, n being the same quantity as in (163) and called mean motion (i.e., mean angular velocity).

Orbit in space. This is illustrated by Fig. 5. We thus have the following six orbital elements:

$$\begin{array}{ll} a & \text{semimajor axis,} \\ e & \text{excentricity,} \\ i & \text{inclination,} \\ \Omega & \text{right ascension of the ascending node} \\ \omega & \text{argument of perigee,} \\ T & \text{time of perigee passage.} \end{array} \tag{166}$$

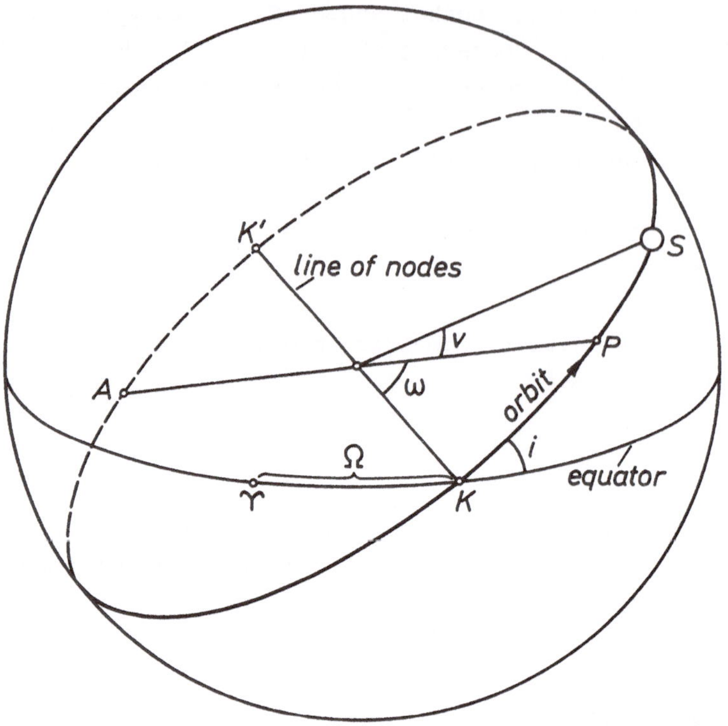

Figure 5. The satellite orbit as projected onto a unit sphere.
P = perigee, A = apogee, K = ascending node, K' = descending node,
S = instantaneous position of satellite.

For unperturbed Keplerian motion, these orbital elements, or orbital parameters, are constants.

A canonical set of elements. The six orbital elements (166) are necessary and sufficient to describe the elliptic orbit in space. They are geometrically simple and intuitive, but they are not canonical elements p_i, q_i .

At any rate, the phase space must be sixdimensional since we have six elements (166). It is easy to convert them to canonical variables, the so-called Delaunay variables, which are even angle-action variables in the sense of sec. 3.5. We have the canonical impulses (with $m = 1$)

$$p_1 = \sqrt{GMa}\ , \qquad p_2 = p_1\sqrt{1-e^2}\ , \qquad p_3 = p_2 \cos i \qquad (167)$$

and the corresponding canonical angle variables

$$q_1 = M = n(t-T) , \qquad q_2 = \omega , \qquad q_3 = \Omega , \tag{168}$$

First we note that the system is highly degenerated since q_2 and q_3 are constants for the unperturbed motion. Second, the (constant) impulses (167) are related to the angular momentum: p_2 is the total angular momentum $|\underline{H}|$, the angular momentum vector $\underline{H}$ being normal to the orbital plane, as we have already mentioned in sec. 1.2, cf.(164), and p_3 is the component of $\underline{H}$ along the earth's rotation axis.

The first impulse p_1 is related to the total energy:

$$p_1^2 = -\frac{(GM)^2}{2E} . \tag{169}$$

Thus the Hamiltonian <u>function</u> H , which <u>numerically</u> equals E , becomes

$$H = -\frac{(GM)^2}{2p_1^2} \tag{170}$$

It does not contain p_2 and p_3 , which accounts for the constancy of q_2 and q_3 in (168). For q_1 , Hamilton's equations give, using (167),

$$\dot{q}_1 = \frac{\partial H}{\partial p_1} = \frac{(GM)^2}{p_1^3} = \frac{(GM)^2}{(GM)^{3/2}a^{3/2}} = \sqrt{\frac{GM}{a^3}} = n \tag{171}$$

by (165), in agreement with (154) and (168).

Don't mind the minus sign in (169): the total energy is negative for ellipses and positive for hyperbolas.

<u>Variation of elements</u>. Finally we come to apply the perturbation theory of sec. 3.6. Since our basic orbital parameters (166) are not canonical, we should express them by $q_1, q_2, \ldots, p_3$. This is easily done by solving (167) and (168) for the parameters (166). Then we can in (157) successively put $A = a, e, i, \omega, \Omega, T$ to get equations for $\dot{a}, \dot{e}$, etc., in terms of $\partial R/\partial q_i$ and $\partial R/\partial p_i$. The latter in turn can be expressed in terms of $\partial R/\partial a$, $\partial R/\partial e$, $\partial R/\partial i$, etc. This is the principle; in practice there are simpler methods. We just quote the result from (Brouwer and Clemence, 1961, p.289): putting $-nT = \sigma$, so that the mean anomaly becomes

$$M = nt + \sigma , \tag{172}$$

and abbreviating

$$\sqrt{1-e^2} = f \tag{173}$$

(no relation to the flattening!), there is

$$\begin{aligned}
\dot{a} &= \frac{2}{na}\frac{\partial R}{\partial \sigma}\,, \\
\dot{e} &= -\frac{f}{na^2 e}\frac{\partial R}{\partial \omega} + \frac{f^2}{na^2 e}\frac{\partial R}{\partial \sigma}\,, \\
\frac{di}{dt} &= -\frac{1}{na^2 f \sin i}\frac{\partial R}{\partial \Omega} + \frac{\cot i}{na^2 f}\frac{\partial R}{\partial \omega}\,, \\
\dot{\Omega} &= \frac{1}{na^2 f \sin i}\frac{\partial R}{\partial i}\,, \\
\dot{\omega} &= \frac{f}{na^2 e}\frac{\partial R}{\partial e} - \frac{\cot i}{na^2 f}\frac{\partial R}{\partial i}\,, \\
\dot{\sigma} &= -\frac{2}{na}\frac{\partial R}{\partial a} - \frac{f^2}{na^2 e}\frac{\partial R}{\partial e}\,.
\end{aligned} \tag{174}$$

The function R depends on the nature of perturbation: irregularities of the earth's gravity field in the case of an artificial satellite, attraction of the other planets in the motion of a planet around the sun.

For planetary motion, the right-hand sides of (174) will be trigonometric series of a form similar to (48), which on integration shows the peculiar convergence behavior already discussed in sec. 1.9.

More details will be found in the lectures by J. Kovalevsky in this Volume.

LECTURE 4

POINCARE

Poincaré was the first to study systematically the global behavior of the Hamiltonian trajectories in phase space; more generally, he investigated the global ("qualitative", geometric) behavior of solutions of systems of first-order ordinary differential equations.

The results were surprising, sometimes startling: they tend to obliterate the neat distinction of determinism and randomness, of order and chaos. Birkhoff (1927), Siegel in 1956 (see (Siegel and Moser, 1971) and others continued Poincaré's investigations ("ergodic theorems" etc.), and Kolmogorov in 1954 (his fundamental article is reprinted in (Abraham and Marsden, 1978)), followed by Arnold, Abraham, Moser and others started an avalanche of research in a topic which is nowadays known to a broad scientific community as the modern theory of (nonlinear) dynamic systems and even to a more general public under spectacular headings such as deterministic chaos, catastrophe theory, etc.

It is impossible to cover this rapidly expanding topic by a single lecture. All we can do is to give some introductory examples to get the reader interested in this highly fascinating and important subject, which also has points of contact with geodetic problems.

4.1. Liouville's Theorem, Measure-Preserving Transformations and Stochastic Processes

The motion of a dynamical system corresponds to a Hamiltonian trajectory in phase space, cf. sec. 3.4. Neighboring trajectories can be regarded as stream lines of a "Hamiltonian flow", somewhat similar to the motion of a fluid in ordinary space.

Liouville's theorem which is a consequence of the symplectic structure of phase space and is proved in all classical textbooks, states that the "Hamiltonian phase fluid" is incompressible, i.e. a volume whose element is

$$dq_1 dq_2 \cdot\cdot dq_f dp_1 dp_2 \cdot\cdot dp_f \quad , \qquad (175)$$

is preserved during the motion (Fig. 6).

Let the point P_0 denote the state of the system at time t_0, and P_t the state at time t (Fig. 6). Thus the motion of our system defines a one-parameter family of transformations T_t by the relation

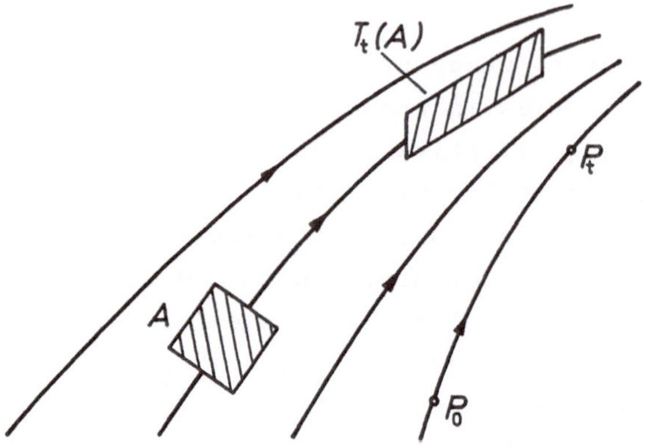

Figure 6. Hamiltonian motion preserves volume.

$$P_t = T_t(P_0) \ . \tag{176}$$

Liouville's theorem may then be expressed in the form

$$T_t(A) = T(A) \ , \tag{177}$$

the volume or "measure" of a set A is preserved: Hamilton's equations describe a measure-preserving transformation.

If a trajectory which must lie on the energy surface (120) finally covers the whole surface densely in a particularly intricate way, preserving measure, it may happen that the time average of a function $f(P)$ equals its phase average; this is the condition of ergodicity:

$$\lim_{T\to\infty} \frac{1}{T} \int_0^T f(P_t) \ dt = \frac{1}{\Omega} \int_\Omega f(P) \ d\Omega \ , \tag{178}$$

Ω denoting the total volume of the phase space and $d\Omega$ its element (175) (actually we should have averaged over the energy surface (120) only since the trajectory entirely lies on it).

A simple example will illustrate the situation (Fig. 7). Instead of phase space, consider a glass of water, which certainly is an incompressible fluid. Drop a spoonful of concentrated orange juice. By stirring (which is a measure-preserving transformation T_t!) we rapidly

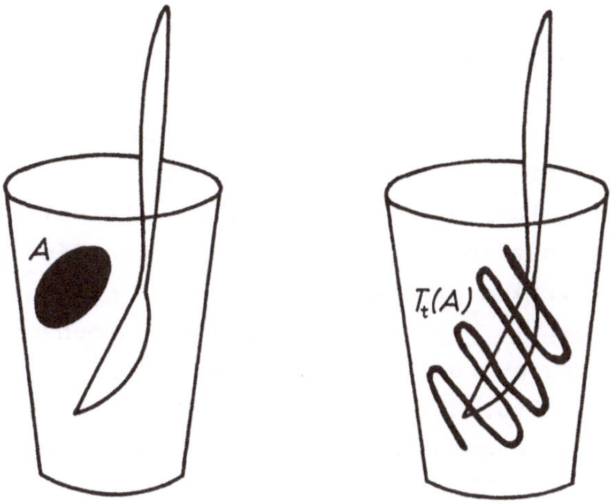

Figure 7. A spoonful of orange juice dissolves in a glass of water; after (Arnold and Avez, 1968).

make the shape of $T_t(A)$ very irregular and finally distribute it evenly throughout the whole volume.

Relation to kinetic gas theory. The laws of classical mechanics are "time reversible" (their form does not change if t is replaced by $-t$), whereas thermodynamics and kinetic gas theory are irreversible. If one end of an iron bar is heated and the bar is isolated afterwards, the temperature will become uniform throughout the bar, but the reverse situation will never happen: an iron bar of uniform temperature will never spontaneously become hotter at one end. Similarly, if a container

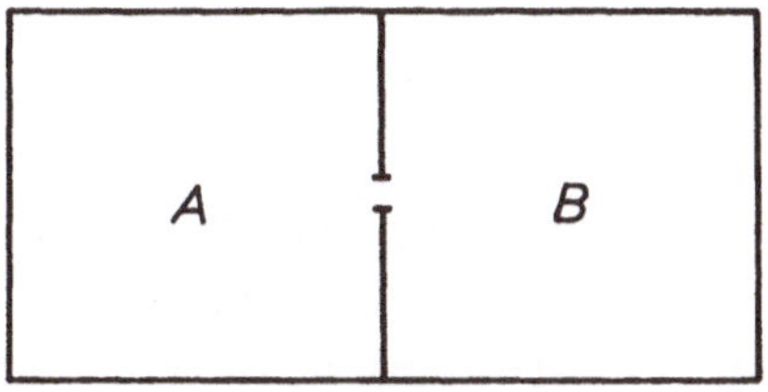

Figure 8. A container consisting of two compartments.

consists of two equal compartments, one evacuated and the other filled with air of density ρ , and if a hole is opened in the wall separating A and B , air will immediately flow from B to A and soon both compartments will be uniformly filled with air of density $\rho/2$. The reverse situation, that the air again flows back spontaneously from A to B such as to accumulate in B only, obviously will never happen.

Thus it seems that kinetic theory, which is based on classical mechanics (the gas particles are regarded as point masses moving according to the laws of mechanics) will not work: an irreversible situation cannot arise through reversible laws. There is a fundamental paradox which Boltzmann tried to resolve by the ergodic hypothesis. Think again of the stirring of the glass of water of Fig. 7.

Relation to stochastic processes. Stationary stochastic processes are also based on measure-preserving transformations of a probability space Ω, cf. (Doob, 1953, Chapter XI). Ergodicity of such a process is highly desirable. This question is also of considerable geodetic relevance, cf. (Moritz, 1980, p. 269).

There is even a relation to analysis: there is an ergodic theorem for continued fractions (Kac, 1959, p. 89). Further examples will be found, e.g., in (Arnold and Avez, 1968).

Representation in Hilbert space. I cannot refrain from mentioning an interesting representation in Hilbert space which is fairly well known for stochastic processes (Doob, 1953, p. 638), but which for classical Hamiltonian mechanics I have found only in G. Ludwig, "Die Grundlagen der Quantenmechanik", Springer, Berlin etc., 1954, p. 11). The reader who does not like Hilbert space may directly go to sec. 4.2.

Consider the Hilbert space consisting of functions

$$f(q_i, p_i) = f(P) \tag{179}$$

square-integrable in phase space. Let

$$U_t f = f(P_t) = f(T_t(P_0)) \tag{180}$$

denote a transformation of f induced by the Hamiltonian point transformation (176). Then it can be shown that (since F_t is measure preserving) U_t is a unitary transformation in our Hilbert space which can be represented in the standard spectral form

$$U_t = \int_{\omega=-\infty}^{\infty} e^{i\omega t}\, dE_\omega \ . \tag{181}$$

Finding the spectral set of operators E_ω is equivalent to the solution of Hamilton's equations (113) since we then have

$$q_i(t) = \int_{-\infty}^{\infty} e^{i\omega t} \, dE_\omega q_i(0) \;,$$
$$p_i(t) = \int_{-\infty}^{\infty} e^{i\omega t} \, dE_\omega p_i(0) \;, \tag{182}$$

and generally

$$F(q_i(t),p_i(t)) = \int_{-\infty}^{\infty} e^{i\omega t} \, dE_\omega F(q_i(0),p_i(0)) \quad , \tag{183}$$

which is simply a Fourier analysis solution. The nonlinear Hamiltonian equations give rise to a linear transformation in Hilbert space!

4.2. The Kolmogorov-Arnold-Moser Theorem

Remember a cyclic system (sec. 3.5.) with $f = 2$. We then have

$$H = H(p_1,p_2) \tag{184}$$

and the Hamiltonian equations give

$$\dot{p}_i = -\frac{\partial H}{\partial q_i} = 0 \;, \qquad \dot{q}_i = \frac{\partial H}{\partial p_i} = \omega_i \qquad (i = 1,2) \tag{185}$$

so that

$$p_i = J_i = \text{const.} \;, \qquad q_i = \omega_i t + \alpha_i \tag{186}$$

become angle-action variables. This corresponds to motion along a torus (Fig. 3).

Imagine now a perturbation of form (155). The Hamiltonian equations are then (156), different in form from (185). Thus the question arises whether there exists a canonical transformation

$$Q_i = Q_i(q_1,q_2,p_1,p_2) \;,$$
$$P_i = P_i(q_1,q_2,p_1,p_2) \tag{187}$$

$(i = 1,2)$ such that the new canonical coordinates P_i, Q_i are again angle-action variables of form (186).

Geometrically speaking, this means that a perturbation $R(p_i, q_i)$ merely "deforms" the torus of Fig. 3 without destroying it.

Formally this can be achieved by means of a series expansion, but in reality the torus character is only preserved if the series converges. This can by no means be taken for granted. In fact, Poincaré showed that many series of celestial mechanics diverge.

A positive result, that the tori are preserved for "most" frequencies ω_i if R is "sufficiently small", was first proved by Kolmogorov, and later, Arnold and Moser successively loosened the requirements. For instance, while Arnold presupposed that the Hamiltonian (155) is analytic, Moser only required the existence of 333 derivatives!

Generalizing this to f degrees of freedom one obtains a corresponding theorem of preservation of f-dimensional tori, which is now known under the name of KAM theorem (Kolmogorov-Arnold-Moser). It is widely quoted and used; among other applications, it is considered to have a bearing on the question of stability of the solar system.

Of greatest interest to geodesists, however, is the method of proof: one solves a so-called hard inverse function problem of functional analysis, which is very "hard" indeed. A similar hard inverse function theorem is used by Hörmander in his well-known proof of existence and uniqueness of Molodensky's boundary-value problem. A first idea concerning such inverse function problems is obtained in (Moritz, 1980, pp. 336 and 434-437). A very general and rigorous discussion of hard inverse function theorems is found in (Sternberg, 1969, Part II). I like best the proof in (Arnold and Avez, 1968, pp. 249-269) and that in (Siegel and Moser, 1971, pp. 183-198) on a related problem. Note that the problem is connected to the question of convergence of the series (49).

What happens in those cases where the tori are not preserved? Then the motion may become quite irregular, even chaotic, as we shall see now by means of an example.

4.3. The Example of Hénon and Heiles

This is one of the most famous examples of "deterministic chaos", therefore we give the original reference: M. Hénon and C. Heiles, "The applicability of the third integral of motion: some numerical experiments", The Astronomical Journal, 69(1), pp. 73-79, 1964. Our presentation follows (Schuster, 1988, pp. 13-15).

The Hamiltonian is assumed to be

$$H = \frac{1}{2}(p_1^2 + p_2^2) + \frac{1}{2}(q_1^2 + q_2^2) + q_1^2 q_2 - \frac{1}{3} q_2^3 \ . \tag{188}$$

which, in plane polar coordinates r , ϕ with

$$q_1 = r\cos\theta \ , \qquad q_2 = r\sin\theta \tag{189}$$

describes the motion of a particle under the influence of the potential energy

$$U(r,\phi) = \frac{1}{2} r^2 + \frac{1}{3} r^3 \sin 3\theta \ . \tag{190}$$

The question is whether, besides of the energy integral E , there exists another constant of the motion I. Then we would have an analog to the KAM case (sec. 4.2) because we could use E and I to get the two constant impulses required for two-dimensional cyclic motion (sec. 3.5).

The trajectory is a curve in fourdimensional phase space. Two integrals

$$H(q_i,p_i) = E \ , \tag{191}$$

$$F(q_i,p_i) = I \ , \tag{192}$$

would restrict the motion to a twodimensional surface, which cuts the p q - plane (say) along a smooth curve. (The p q - plane is a "surface of section" introduced by Poincaré for easier visualization.)

Figures 9a,b,c, taken from the original paper, show that this is indeed the case for low energy E (Fig. 9a). For higher energies, the smooth curves decay progressively and the motion becomes almost completely "chaotic".

These figures are so beautiful and instructive that they are shown, e.g., also in (Arnold and Avez, 1968) and (Schuster, 1988, p.10), which we recommend for further reading.

4.4. Chaos and Order

"An intelligent being which, for some given moment of time, knew all the forces by which nature is driven, and the relative position of the objects by which it is composed (provided the being's intelligence

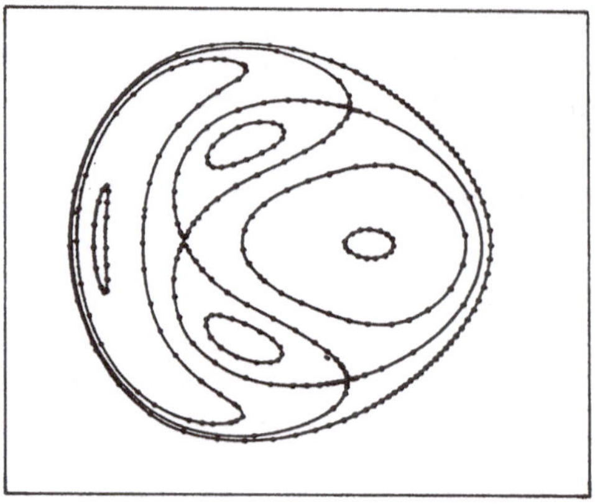

Figure 9a. Results for
E = 0.08333.

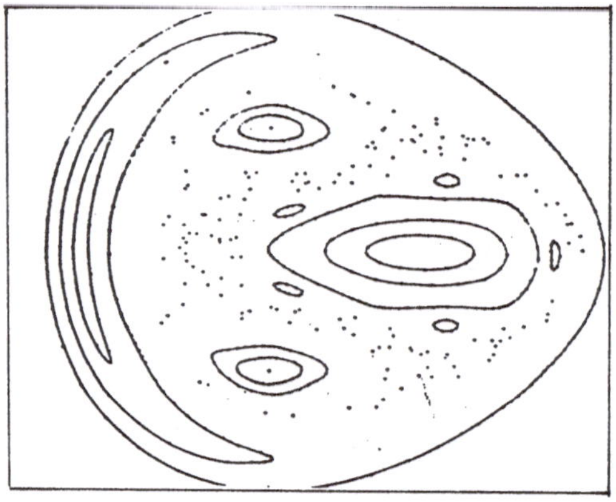

Figure 9b. Results for
E = 0.12500.

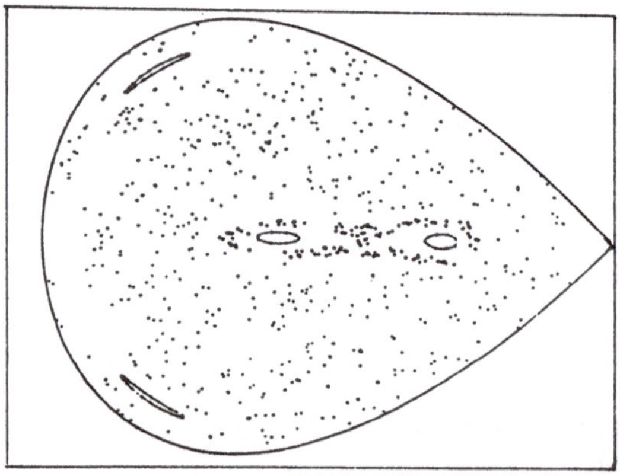

Figure 9c. Results for
E = 0.16667.

were so vast as to be able to analyze all the data), would be able to comprise, in a single formula, the movements of the largest bodies in the universe and those of the lightest atom: nothing would be uncertain to it, and both the future and the past would be present to its eyes. The human mind offers in the perfection which it has been able to give to astronomy, a feeble inkling of such an intelligence". This famous formulation of determinism has been given by Laplace in 1814 ("Laplace's demon").

In fact, this idea is based on classical mechanics: given initial conditions $\underline{x}(t_0)$ or $q_i(t_0)$ and $p_i(t_0)$, all $\underline{x}(t)$ or $q_i(t)$ and $p_i(t)$ are uniquely determined; cf. (182). This holds both for all values $t > t_0$ (future) and $t < t_0$ (past).

We cannot here offer anything like a philosophical discussion of determinism (cf. Weyl, 1949, sec. 23). We only mention that quantum theory has thoroughly shaken the foundations on which it is based, introducing a basic background of randomness.

But determinism has also been under attack from its very stronghold, classical mechanics.

Stability and instability. A stable dynamical situation can be characterized by the slogan:"Small causes - small effects". But we often meet with unstable situations. It is sufficient to consider throwing dice. Even assuming that the die moves rigorously as a rigid body in the sense of classical mechanics in the absence of atmospheric friction, a minute change of initial conditions (position of hand throwing it, initial velocity) will result in a rather different trajectory, and it will come to rest in an arbitrary way on one of its six faces. If the dice is a perfectly symmetric cubical body, the probability for each face is equal, $p = 1/6$.

So instability, so to speak, destroys dependence on the initial condition and even on the laws of motion; a new regularity, based on symmetry, takes over. This is a nice illustration of the way in which statistical "macroscopic" regularity arises out of the "microscopic" regularity of the laws of classical mechanics, the mechanism being instability.

It is believed that the "macroscopic" regularity of the laws of thermodynamics may be explained in much the same way from classical mechanics applied to the microscopic motion of the molecules; this is the subject of classical statistical mechanics, cf. (Khinchin, 1949).

Chaos in conservative systems. Chaos is called the situation when a dynamical system loses its regularity. A typical example is shown in Figures 9a,b,c. It is a result of instability (Fig. 10) combined with

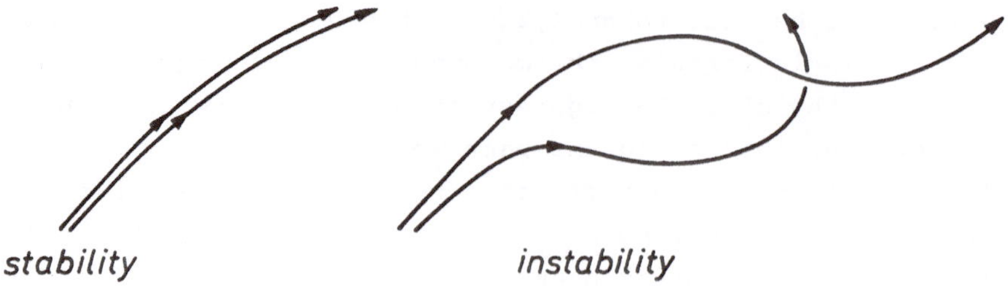

Figure 10. Stable and unstable trajectories.

a complicated topological structure of the phase space. This is geometrically illustrated by the beautiful figures of (Abraham and Shaw, 1983), which we highly recommend. E.g., almost any point in Fig. 9c may finally be reached starting from a given initial point.

Another example of "deterministic chaos" is "Arnold diffusion" of a trajectory between the invariant tori: "most" tori are preserved according to the KAM theorem (sec. 4.2), but some are broken up and then the situation may become very irregular indeed.

Such a situation was foreseen by Poincaré (1899, p.389): "Imagine the figure formed by these two curves and their infinitely many intersections...; these intersections form a kind of meshwork or infinitely dense tissue... One is struck by the complexity of this figure which I do not even attempt to draw. Nothing is better suited to give us an idea of the complexity of the three-body problem and in general of all the problems of Dynamics in which there is no uniform integral of the motion and where Bohlin's series diverge."

The series of Bohlin are typical for the series of celestial mechanics for whose peculiar convergence behavior we have given an example already in sec. 1.9.

In the passage quoted at the beginning of the present section, Laplace proudly mentions the example of astronomy, having in mind celestial mechanics. He would never have thought that his series might diverge...

Until the end of the 19th century, it was implicitly assumed that all meaningful physical processes are governed by stable laws. Classical mathematical physics was dominated by "properly posed problem", whose very definition includes stability. Only recently, the importance of improperly posed problems, from geophysical prospecting to weather forecasting, was recognized.

Strange attractors and turbulence. Dynamical systems in which energy is not conserved because of friction are called dissipative: energy is "dissipated" into heat (we are not concerned here with the question whether the microscopic random motion of the molecules associated with heat is not itself governed by the laws of dynamics...).

A typical phenomenon arising in dissipative dynamical systems is turbulence occurring in viscous fluid motion and in meteorology. The most recent attempt to explain turbulence is by means of strange, or chaotic, attractors going back to E.N. Lorenz in 1963. We cannot enter into this fascinating subject here and must refer to reader to the literature, cf. (Thompson and Steward, 1986), (Schuster, 1988), and, of course, (Abraham and Shaw, 1983).

Self-organization and synergetics. The above example of a dice shows a new structure, based on the symmetry of the die, emerging from a theoretically deterministic situation, the motion of the die, through instability. This is only a poor and trivial example to illustrate self-organization through instability, by which a higher-level structure is able to impose itself. Examples range widely: self-stabilization of a nonlinear-electronic ocscillator, laser beams, fluid patterns, chemical waves, morphogenesis in biology, the forming of public opinion and many more. This is the new subject of synergetics (Haken 1978, 1983).

Relation to quantum mechanics. Curiously enough, if we presuppes a "microscopic" background of quantum mechanics, the deterministic laws of classical mechanics themselves arise as "macroscopic" consequences from that random background. In fact, it is possible to derive Hamilton's equations of classical mechanics from quantum theory; cf. (Pauli, 1980, pp. 91-93).

Conclusion. In this way, the formerly nice boundaries between determinism and randomness themselves become fuzzy, or rather, there is a delicate interplay between determinism and randomness. To the geodesist, the inevitable presence of measuring errors comes into mind (and perhaps also the discussion on the statistical character of collocation). At the end of sec. 4.1 we have found a common mathematical superstructure for Hamiltonian mechanics and stationary stochastic processes, and according to Kac (1959), even "prime numbers play a game of chance". And "improperly posed problems" are very popular nowadays...

So Laplace's demon mentioned at the beginning of this section, will find himself out of business in our present world, and he may wish to look for another, less ambitious, job.

SUGGESTED ADDITIONAL READING

(Books in English only, with one obvious exception)

Abraham R. and Marsden J.E. (1978): Foundations of Mechanics, 2nd ed. Benjamin/Cummings, Reading, Mass.

A very comprehensive and profound treatment of dynamical systems by modern differential geometry (exterior differential forms), fascinatingly difficult.

Abraham R. and Shaw C.D. (1982, 1983, 1985): Dynamics: The Geometry of Behavior, 3 Parts. Aerial Press, P.O.Box 1360, Santa Cruz, Cal. 95061, U.S.A.

The best introduction to the geometric theory of dynamical systems, entirely by pictures and without formulas. A didactical masterpiece (for Lecture 4).

Arnold V.I. (1978): Mathematical Methods of Classical Mechanics. Springer, Berlin-Heidelberg-New York.

A modern treatise on the entire range of these lectures, including exterior forms. Eminently readable and highly recommended.

Arnold V.I. and Avez A. (1968): Ergodic Problems of Classical Mechanics. Benjamin, New York.

Colorful, well written, many examples, excellent for Lectures 4 and also 3.

Birkhoff G.D. (1927): Dynamic Systems. American Mathematical Society, Providence, Rhode Island (reprinted 1966).

An indispensable classic for advanced reading.

Brouwer D. and Clemence G.M. (1961): Methods of Celestial Mechanics. Academic Press, New York and London.

A standard textbook, thorough treatment (the other books on this subject quoted here use celestial mechanics mainly as a starting point for advanced mathematics).

Doob J.L. (1953): Stochastic Processes. Wiley, New York.

Still one of the best books on the topic, not easy, but unsurpassed in clarity (for sec. 4.1).

Flügge S., ed. (1960): Encyclopedia of Physics, Vol.III, Part 1. Springer, Berlin etc.

Contains relevant articles by Synge and Truesdall-Toupin.

Goldstein H. (1980): Classical Mechanics, 2nd ed. Addison-Wesley, Reading, Mass.

Excellent textbook for Lectures 1-3.

Haken H. (1978): Synergetics: An Introduction. Springer, Berlin etc.

Haken H. (1983): Advanced Synergetics. Springer, Berlin etc.

Basic books on self-organization in physics, chemistry and biology. Well written, emphasis on applications, many examples (for sec. 4.4).

Kac M. (1959): Statistical Independence in Probability, Analysis and Number Theory. Wiley, New York.

A masterpiece in interrelating various mathematical disciplines, delightful reading (for Lecture 4).

Khinchin A.I.(1949): Mathematical Foundations of Statistical Mechanics. Dover, New York.

A very readable introduction (for sec. 4.4).

Lanczos C. (1970): The Variational Principles of Mechanics, 4th ed. Univ. of Toronto Press.

Perhaps the best book for Lectures 1-3, to me one of the most beautiful books in science.

McConnell A.J. (1957): Application of Tensor Analysis. Dover, New York.

Advanced reading for treatment of Lagrange's equations by Riemannian Geometry (for sec. 2.6).

Moritz H. (1980): Advanced Physical Geodesy. Wichmann, Karlsruhe, and Abacus Press, Tunbridge Wells, Kent.

For cross-connections with physical geodesy.

Moritz H. and Mueller I.I. (1987): Earth Rotation: Theory and Determination. Ungar, New York.

For application of classical mechanics to earth rotation (for Lectures 1-3).

Pauli W. (1980): General Principles of Quantum Mechanics. Springer, Berlin etc.

Contains on pp. 91-93 a derivation of the classical Hamiltonian equations as a limiting case of quantum mechanics (for sec. 4.4).

Poincaré H. (1892, 1893, 1899): Les Méthodes Nouvelles de la Mécanique Céleste, 3 vols. Gauthier-Villars, Paris (reprinted 1987 by A. Blanchard, 9 rue de Médicis, F-75006 Paris).

The classic for Chapter 4. Poincaré was not only a great scientist, but also a splendid writer. It is worth learning French for the sake of reading Poincaré.

Schuster H.G. (1988): Deterministic Chaos: An Introduction, 2nd ed. VCH, D-6940 Weinheim.

Written by a physicist, excellent for Lecture 4.

Siegel C.L. and Moser J.K. (1971): Lectures on Celestial Mechanics. Springer, Berlin etc.

A mathematical classic for advanced reading (Lecture 4).

Sternberg S. (1969): Celestial Mechanics, 2 Parts. Benjamin, New York.

Typically mathematical treatment, colorful but partly very difficult. Part II contains an extensive treatment of hard implicit function theorems (for Lecture 4).

Synge J.L. (1960): Classical Dynamics. In Flügge (1960), pp. 1-225.

The most detailed geometric treatment of the subject of Lectures 1-3 known to me, very readable and highly recommended.

Synge J.L. and Schild A. (1978): Tensor Calculus. Dover, New York.

Similar in character to McConnell, but more modern. Contains on pp. 181-183 a nice topological illustration of rotation group space (for sec. 2.6), excellent textbook.

Thirring W. (1978): Classical Dynamical Systems. Springer, New York/Wien.

A typical contemporary textbook on theoretical physics requiring a good background in modern mathematics, concise and clear, more difficult than Arnold but less than Abraham/Marsden.

Thompson J.M.T. and Stewart H.B. (1986): Nonlinear Dynamics and Chaos. Wiley, New York.

Written for engineers and scientists, excellent for Lecture 4.

Truesdall C. and Toupin R. (1960): The Classical Field Theories. In Flügge (1960), pp. 226-793.

Relevant for dynamics of a rigid body in comparison with that of an elastic body (for sec. 1.6).

Weyl H. (1949): Philosophy of Mathematics and Natural Science. Princeton Univ. Press.

Contains a profound discussion on causality and determinism versus randomness and freedom, very relevant for sec. 4.4. A classic written by an eminent mathematician and physicist (the inventor of gauge field theories!).

Whittaker E.T. (1937): A Treatise on the Analytical Dynamics of Particles and Rigid Bodies, 4th ed. Cambridge Univ. Press.

First edition in 1904, still unsurpassed in wealth of material treated, highly recommended.

Wintner A. (1941): The Analytical Foundations of Celestial Mechanics Princeton Univ. Press.

A classic written by a mathematician, on the order of Birkhoff and Siegel-Moser, advanced reading.

LECTURES IN CELESTIAL MECHANICS

J. Kovalevsky
C.E.R.G.A.
Avenue Copernic, F-06130 Grasse

1 - GENERAL INTRODUCTION

The use of artificial satellites in geodesy is based upon the very general principle that their motion is governed by the forces that act on them. More precisely, it results from general theorems of Mechanics that, provided that there are no collisions, if forces acting on a body are given in a certain interval of time in all points of the space where it can move, and if the initial conditions of motion (position and velocity if the object is point-like) are known, the trajectory of the body is univocally determinable during the same period of time. Consequently the observations of the trajectory of an artificial satellite give clues to the forces acting on them and particularly to the forces originating from the Earth gravitational potential. Conversely, knowing these forces, one can improve the description of the trajectories.

To do so, it is necessary to know the relations that exist between forces and trajectories. It is the objective of Celestial Mechanics - a particular section of Mechanics devoted to the study of the motions of celestial bodies - to set up these relations and to identify the best methods to obtain a quantitative description of the trajectories from any given law of forces.

In this series of lectures, we restrict ourselves to the motion of artificial satellites around the Earth. This brings several simplifications to the general case of the motion of celestial bodies.

(i)- Satellites are very small with respect to the dimensions of the Earth and it is legimate for all geodetic applications to consider that they are point-like.

(ii)- Among the various forces that act on a satellite, the gravitational attraction of a point P assumed to have the Earth's mass and placed at its centre of mass is largely predominant. All other components of the forces can be considered as minor quantities that only perturb the motion that would occur if P was acting alone.

(iii)- Similarly, one can consider as a good first approximation that, despite the motion of the Earth around the Sun, one can refer the motion to a system of axes with fixed directions in space. The error so introduced can also be treated as a perturbation.

(iv)- The gravitation interaction can be described in terms of Newton's law of gravitation: two particles P and P' of masses m and m' attract each other along the line PP' with a force directly proportional to the product of their masses and inversely proportional to the square of their distance. The force excerted by P on P' is :

$$\mathbf{F} = - \frac{- kmm'\mathbf{PP'}}{r^3} \tag{1.1}$$

k is the constant of gravitation.

The actual law of forces is given by the theory of General Relativity. The difference introduced by this improved description of the dynamical reality is very small and can be described under the form of small corrections to the trajectories obtained using Newton's law.

These remarks justify the general approach to the study of the motion of an artificial satellite. As a first approximation, the two body problem is solved (Chapter 2). Then more general equations are written with the assumption that other forces are small (Chapter 3) and solved (Chapter 4). We shall apply these theories to artificial satellites (Chapter 5). Chapter 6 is devoted to particular cases that do not follow the general approach and Chapter 7 to numerical methods.

2 - THE TWO BODY PROBLEM

2.1. ELLIPTICAL SOLUTION OF THE TWO BODY PROBLEM

Let us consider a point P of negligible mass m representing a satellite attracted by a point situated at the origin 0 of a system of axes of coordinates and whose mass is

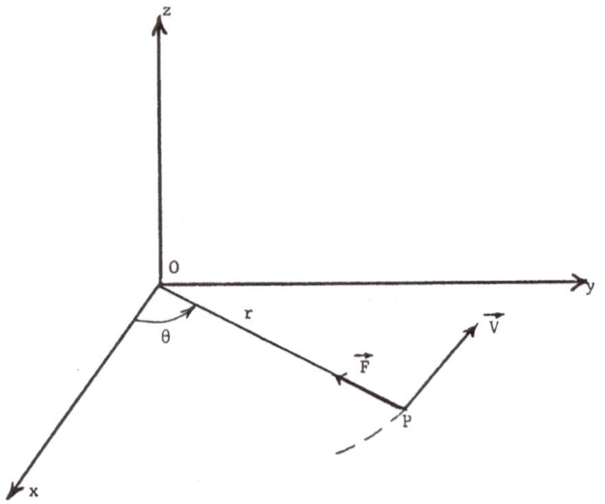

Figure 1

M. Since the mass of P is taken equal to zero, there is no force acting on O and on the system of axes Oxyz. Let us assume that P and its velocity **V** with respect to O are at a given instant (t_0) in the Oxy plane. Let us call r and θ the polar coordinates of P in this plane (see figure 1).

The acceleration γ of P, deduced from 1.1 is :

$$m\gamma = \frac{-\mu m \mathbf{OP}}{r^3} \tag{2.1}$$

where μ=kM is the geocentric constant of gravitation. The theorem of angular momentum writes as :

$$\mathbf{OP} \wedge \mathbf{V} = \mathbf{C} = \mathbf{OP}(t_0) \wedge \mathbf{V}(t_0) \tag{2.2}$$

It results that **C** is along the Oz axis. Let us call C its magnitude. Then, 2.2 reduces to :

$$r^2 \frac{d\theta}{dt} = C \tag{2.3}$$

The theorem of kinetic energy writes as :

$$\frac{d}{dt}\left[\frac{1}{2} mV^2\right] - m\gamma.\mathbf{V} = 0 \tag{2.4}$$

Integrating, one gets :

$$\frac{1}{2} m V^2 + \int \frac{\mu m}{r^2} \frac{dr}{dt} dt = C'$$

where C' is another constant. If we eliminate m by letting h=C'/m and resolve the integral, one finally gets, using also 2.3 :

$$V^2 = 2h + \frac{2\mu}{r} = \left(\frac{dr}{dt}\right)^2 + r^2\left(\frac{d\theta}{dt}\right)^2 = C^2\left[\frac{1}{r^4}\left(\frac{dr}{d\theta}\right)^2 + \frac{1}{r^2}\right] \tag{2.5}$$

Let us call u=1/r; 2.5 can then be written as :

$$\left(\frac{du}{d\theta}\right)^2 + u^2 - \frac{2\mu u}{C^2} - \frac{2h}{C^2} = \left[\frac{d(u-\mu/C^2)}{d\theta}\right]^2 + \left(u - \frac{\mu}{C^2}\right)^2 - \frac{\mu^2}{C^4} - \frac{2h}{C^2} = 0$$

This equation has the following general solution :

$$\frac{1}{r} = \frac{\mu}{C^2}\left[1 + \sqrt{1 + \frac{2C^2 h}{\mu^2}} \cos(\theta - \theta_0)\right]$$

It is a conic section whose focus is 0. It is an ellipse (the only case that will be considered here) if the coefficient of $\cos(\theta-\theta_0)$ is smaller than 1. We shall call :

$p=\frac{C^2}{\mu}$, parameter of the ellipse, hence $C=\sqrt{\mu p}$

$e=\sqrt{1+\frac{2C^2h}{\mu^2}}$, eccentricity of the ellipse, hence $2h=\frac{\mu(e^2-1)}{p}$

$v=\theta-\theta_0$, true anomaly

The equation of the elliptical orbit is, with these notations :

$$\frac{1}{r}=\frac{1+e\cos v}{p} \tag{2.6}$$

If a is the semi-major axis, one has $p=a(1-e^2)$ and $2h=-\mu/a$. With these notations, the kinetic energy integral 2.5 becomes :

$$V^2=\mu\left(\frac{2}{r}-\frac{1}{a}\right) \tag{2.7}$$

The angular momentum C can be computed by integrating 2.3 over the period P of revolution :

$$\int r^2\,d\theta=\int C\,dt$$

One gets twice the area of the ellipse :

$$2\pi a^2\sqrt{1-e^2}=CP=P\sqrt{\mu p}=P\sqrt{\mu a(1-e^2)}$$

It results :

$$\frac{4\pi^2}{P^2}a^3=\mu=n^2a^3 \tag{2.8}$$

where $n=2\pi/P$ is called mean motion. This last equation represents the Kepler's third law.

2.2 - KEPLER'S EQUATION

Let us introduce the eccentric anomaly E as shown in figure 2. If Σ is the principal circle whose diameter is the major axis of the ellipse, E is the polar angle (Cx, CP') of the intersection with Σ of the parallel to Oy drawn from P.

The rectangular coordinates of P expressed in terms of the true and eccentric anomalies are :

$$\left.\begin{aligned} x&=r\cos v=a(\cos E-e)\\ y&=r\sin v=a\sqrt{1-e^2}\sin E \end{aligned}\right\} \tag{2.9}$$

From this, one obtains :

$$r = a(1 - e\cos E) \tag{2.10}$$

Equating r as given by 2.6 and 2.10, one gets, after some computations :

$$\tan\frac{v}{2} = \sqrt{\frac{1+e}{1-e}}\tan\frac{E}{2} \tag{2.11}$$

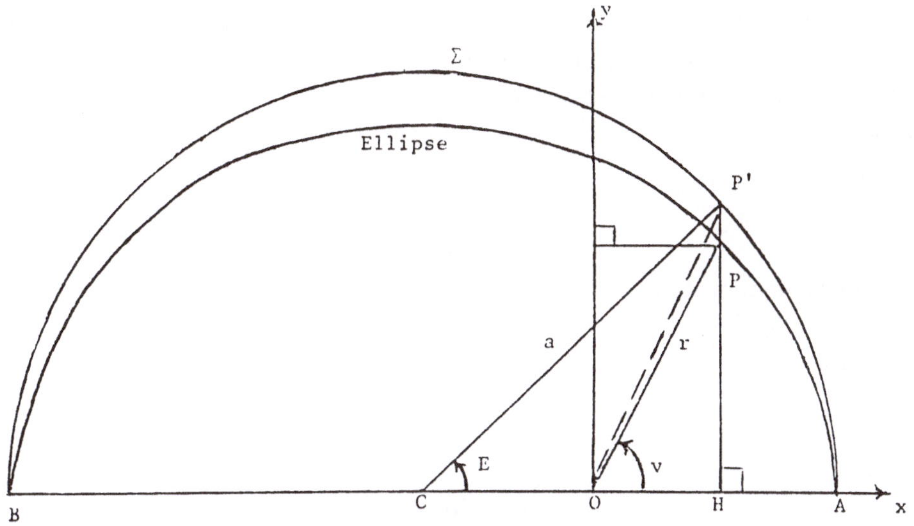

Figure 2

The angular momentum integral tells us that the area of the ellipse OAP (figure 2) is proportional to time. If t_0 is the time of passage through A - called the pericentre - this area is :

$$S = \frac{1}{2}\int_{t_0}^{t} C\,dt = \frac{1}{2} n a^2\sqrt{1-e^2}\,(t-t_0)$$

Let us introduce the mean anomaly $M = n(t-t_0)$:

$$S = \frac{1}{2} a^2\sqrt{1-e^2}\,M$$

One can also compute S as the difference between areas CAP' and COP' multiplied by $\sqrt{1-e^2}$:

$$S = \frac{1}{2}\left(a^2E - a^2 e\sin E\right)\sqrt{1-e^2}$$

Equating these two expressions, one gets after some simplifications the Kepler's equation :

$$E - e \sin E = M = n(t-t_0) \qquad (2.12)$$

Together with equations 2.9, it permits to express the coordinates of P in function of time. For this, it is necessary to solve 2.12 when t, and hence M, is given. Let us first remark, that one can reduce it to the case for which M and E are between 0 and π. Several successive approximation methods exist. A very efficient algorithm - at least for eccentricities not too close to 1 - is the following. Let E_0 be an approximate value of E and put $E = E_0 + \Delta E$.

One has $E_0 + \Delta E - e \sin(E_0+\Delta E) = M$.

Neglecting ΔE^2, one gets an approximate evaluation of ΔE :

$$\Delta E_0 = \frac{M - E_0 + e \sin E_0}{1 - e \cos E_0} \qquad (2.13)$$

The next approximation is :

$$E_1 = E_0 + \Delta E_0$$

A good starting value is :

$$E_0 = M$$

or

$$E_0 = M + e \sin M$$

The convergence is quadratic: 3 or 4 iterations insure an accuracy of 16 decimal figures.

2.3 - EXPANSIONS IN MEAN ANOMALY

Since the motion is periodic, the coordinates and any function of coordinates of P are 2π-periodic functions of M. They are therefore developable in Fourier series of M. Such developments are very useful in Celestial Mechanics and are usually quickly convergent for small eccentricities - a general case for artificial satellites used in space geodesy. Let us give an example and develop Kepler's equation 2.12. It reduces to develop e sin E in Fourier series of M. Let us actually take a more general example and expand sin kE where k is a positive integer :

$F(M) = \sin kE$

Since it is an odd function, general results on Fourier series permit us to write:

$$\sin kE = \sum_{p=1}^{\infty} a_p^k \sin kM \qquad (2.14)$$

with :

$$a_p^k = \frac{1}{\pi} \int_0^{2\pi} \sin kE \sin pM \, dM$$

Noting that sin pMdM = - 1/p d(cos pM), on can write :

$$a_p^k = \frac{-1}{p\pi} \int_0^{2\pi} \sin kE \, d(\cos pM) = \frac{-1}{p\pi} \Big[\sin kE \cos pM \Big]_0^{2\pi} + \frac{k}{p\pi} \int_0^{2\pi} \cos pM \cos kE dE$$

The first term is naught. Replacing M by E-e sin E in the second term, one gets :

$$a_p^k = \frac{k}{2p\pi} \left[\int_0^{2\pi} \cos\big((p+k)E - pe \sin E\big) \, dE + \int_0^{2\pi} \cos\big((p-k)E - pe \sin E\big) dE \right]$$

Let us remark, at this point, that one of the definitions of a Bessel function of order n is :

$$J_n(x) = \frac{1}{2\pi} \int_0^{2\pi} \cos(nt - x \sin t) dt$$

Then, 2.14 becomes :

$$\sin kE = \sum_{p=1}^{\infty} \frac{k}{p} \Big[J_{p-k}(pe) + J_{p+k}(pe) \Big] \sin pM \qquad (2.15)$$

In particular :

$$e \sin E = E - M = \sum_{p=1}^{\infty} \frac{e}{p} \Big[J_{p-1}(pe) + J_{p+1}(pe) \Big] \sin pM$$

Each of the coefficients may be developed in powers of the eccentricity using the developments of $J_n(x)$:

$$J_n(x) = \left(\frac{x}{2}\right)^n . \frac{1}{n!} \left[1 - \left(\frac{x}{2}\right)^2 \frac{1}{1!(n+1)} + \ldots + \left(\frac{x}{2}\right)^{2j} \frac{(-1)^j n!}{j!(n+j)!} + \ldots \right] \qquad (2.16)$$

For example, one has :

$$\sin E = \left(1 - \frac{e^2}{8} + \frac{e^4}{192} + \ldots\right) \sin M + \left(\frac{e}{2} - \frac{e^3}{6} + \ldots\right) \sin M +$$

$$+ \left(\frac{3}{8} e^2 - \frac{27}{128} e^4 + \ldots\right) \sin 3M + \left(\frac{e^3}{3} + \ldots\right) \sin 4M + \left(\frac{125}{384} e^4 + \ldots\right) \sin 5M + \ldots$$

Similarly, one has :

$$\cos kE = a_0^k + \sum_{p=1}^{\infty} \frac{k}{p} \Big[J_{p-k}(pe) - J_{p+k}(pe) \Big] \cos pM \qquad (2.17)$$

with $a_0^1=-e/2$ and $a_0^k=0$ for $k > 1$.

2.4 - ORBITAL ELEMENTS

In practice, the orbital plane of a satellite is not chosen as reference for coordinates. The motion has to be referred to a space-fixed system of coordinates. Actually, for artificial satellite motion, a geocentric equatorial frame of reference is used. It is defined by the Earth equator, the Ox_1 axis being directed towards the vernal equinox. The Ox_3 axis points towards the North pole. We shall assume that this reference frame is fixed and neglect the effects of its rotation due to the precession.

The plane of the orbit intersects the Ox_1x_2 plane on the line of nodes. This line cuts the orbit in two points. The particular point for which the x_3-coordinate increases is the ascending node N (see figure 3).

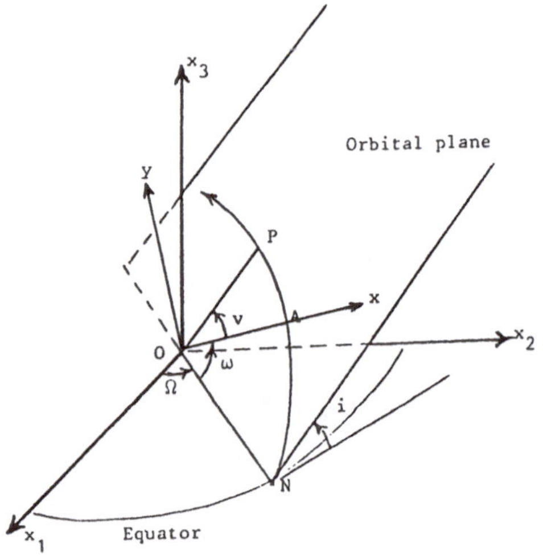

Figure 3

- The direction ON is reckoned from Ox_1 and defined by the longitude of the ascending node :

  $\Omega = (Ox_1, ON)$

- The angle between the orbital plane and the Ox_1x_2 plane is the inclination i. If $0 \leq i \leq 90°$, the motion is direct. If $90° < i < 180°$ the motion is called retrograde.
- In the plane of the orbit, the direction of the perigee A is reckoned from the ascending node and is called argument of the perigee :

  $\omega = (ON, OA)$

- The time t_0 at which the satellite crosses the perigee is the epoch or the time of passage at perigee. The mean anomaly at any time t is given by :

$M = n(t-t_0)$

- The semi-major axis a and the eccentricity e, already defined are the last two orbital elements that describe the shape and the size of the ellipse.

The six orbital parameters a, e, i, Ω, ω and t_0 define in a unique manner the orbit and the motion of the satellite in function of time.

In order to obtain the rectangular coordinates in the $Ox_1x_2x_3$ axes at a given time, one has first to compute the coordinates x, y in the orbital plane as given by 2.9. Then, the coordinates in the $Ox_1x_2x_3$ axes are :

$$\begin{vmatrix} x_1 \\ x_2 \\ x_3 \end{vmatrix} = R_3(-\Omega)\, R_1(-i)\, R_3(-\omega) \begin{vmatrix} x \\ y \\ 0 \end{vmatrix} = R \begin{vmatrix} x \\ y \\ 0 \end{vmatrix} \qquad (2.18)$$

where $R_3(\theta)$ represents a rotation matrix of the angle θ around the x_3 axis and $R_1(\theta)$ is the rotation matrix of the angle θ around the x_1 axis. The explicit form of 2.18 is :

$$x_i = P_j x + Q_i y \;\; ; \;\; j = 1,2,3$$

with

$$\begin{aligned} P_1 &= \cos\Omega \cos\omega - \cos i \sin\Omega \sin\omega \\ P_2 &= \sin\Omega \cos\omega + \cos i \cos\Omega \sin\omega \\ P_3 &= \sin i \sin\omega \\ Q_1 &= -\cos\Omega \sin\omega - \cos i \sin\Omega \cos\omega \\ Q_2 &= -\sin\Omega \sin\omega + \cos i \cos\Omega \cos\omega \\ Q_3 &= \sin i \cos\omega \end{aligned} \qquad (2.19)$$

Similarly, one can express the velocity vector in this reference frame. In the orbital axes, one has, differentiating 2.9 :

$$\frac{dx}{dt} = \dot{x} = -a \sin E \frac{dE}{dt} \;\; ; \;\; \frac{dy}{dt} = \dot{y} = a\sqrt{1-e^2} \cos E \frac{dE}{dt}$$

Differentiating 2.12, one gets :

$$(1-e \cos E)dE = ndt$$

$$\frac{dE}{dt} = \frac{n}{1-e \cos E} = \frac{an}{r} \qquad (2.20)$$

It results :

$$\dot{x} = -\frac{an \sin E}{1-e \cos E} \;\; ; \;\; \dot{y} = \frac{an\sqrt{1-e^2} \cos E}{1-e \cos E} \qquad (2.21)$$

The transformation into equatorial rectangular coordinates gives for the components of the velocity vector $y_1\ y_2\ y_3$:

$$\begin{vmatrix} y_1 \\ y_2 \\ y_3 \end{vmatrix} = R. \begin{vmatrix} \dot{x} \\ \dot{y} \\ 0 \end{vmatrix} \tag{2.22}$$

or, explicitely :

$$y_i = P_j\dot{x} + Q_j\dot{y} \tag{2.23}$$

P_i and Q_i being defined by 2.19.

3 - EQUATIONS OF PERTURBED MOTION

As already stated, the two body problem is only an approximation, and many other forces than the central force act on the satellite. We shall assume that they are significantly smaller and consider that they are of the first order in a certain small quantity. Among these forces, the most important for Space Geodesy are produced by the difference between the actual attraction by the Earth and the central force considered in the preceding chapter. But, there are other forces such as the attraction by the Moon and the Sun or the atmospheric drag.

Let $\mathbf{F}(F_1,F_2,F_3)$ be the expression of the acceleration due to all these additional forces. In the general case, they depend upon the position of the satellite $\mathbf{OP}(x_1,x_2,x_3)$, its velocity $\mathbf{V}(y_1,y_2,y_3)$ and of the time t. Using the fondamental equation of dynamics, the equation of motion is :

$$\frac{d^2\mathbf{OP}}{dt^2} = \frac{-\mu\mathbf{OP}}{OP^3} + \mathbf{F}(\mathbf{OP}.\mathbf{V}.t) \tag{3.1}$$

3.1. THE DISTURBING FUNCTION

Let us now assume that the additional accelerations are of gravitational origin like those due to external bodies (Sun, Moon, planets) or to an uneven mass distribution in the Earth. They may be expanded as a sum (or an integral) of individual accelerations :

$$\mathbf{dF} = \mu \frac{\mathbf{PM}}{PM^3} dm \tag{3.2}$$

Noting that :

$$-\frac{\mathbf{PM}}{PM^3} = \mathbf{grad}\left(\frac{1}{PM}\right)$$

and using the property that the sum of gradients is also a gradient, one has :

$$\sum \mathbf{dF} = \mathbf{grad}\, R$$

R is called the disturbing function. In vectorial notation, the equation of motion is :

$$\frac{d^2\mathbf{OP}}{dt} = -\mu \frac{\mathbf{OP}}{OP^3} + \mathbf{grad}\, R = \mathbf{grad}\, V$$

or, in rectangular coordinates with $r^2=x_1^2+x_2^2+x_3^2$:

$$\frac{d^2x_j}{dt^2} = -\frac{\mu x_j}{r^3} + \frac{\partial R}{\partial x_j} = \frac{\partial V}{\partial x_j} \quad , \quad j = 1,2,3 \tag{3.3}$$

Let us consider the reduced kinetic energy :

$$T = \frac{1}{2}\sum_{j=1}^{3} y_j^2 \quad \text{and} \quad H = T - V$$

we can write the system 3.3 as :

$$\frac{dx_j}{dt} = \frac{\partial H}{\partial y_j} \quad ; \quad \frac{dy_j}{dt} = -\frac{\partial H}{\partial x_j} \qquad j = 1,2,3 \tag{3.4}$$

This system is called canonical. The function H is the Hamiltonian or characteristic function; x_i and y_i are conjugate variables.

Let us remark if H does not depend upon t, it is an integral of the problem since, with 3.4, one has :

$$\frac{dH}{dt} = \sum_{j=1}^{3} \left(\frac{\partial H}{\partial x_j}\frac{dx_j}{dt} + \frac{\partial H}{\partial y_j}\frac{dy_j}{dt} \right) = 0$$

3.2. OSCULATING ELEMENTS

In the two-body problem, equations 2.18 and 2.21 can be viewed as biunivocal transformations between x,y,x,y in the orbital plane and the coordinates x_1, x_2, x_3 and the components of the velocity y_1, y_2, y_3 through the matrix $R(i,\Omega,\omega)$. In addition, x, y are functions of a, e and M through equations 2.8, 2.9 and 2.12 while x and y depend upon the same quantities through equations 2.21 and 2.12. This actually means that if a set of values of the six elliptic elements a, e, i, Ω, ω and M is given for a time t, there is a single possible position and velocity vectors for the body. Similarly, if the position and the velocity are known at time t, there exists a single set of elliptic elements.

Suppose now that P moves not only under the action of the central force, but is also perturbed as described by equation 3.1. Nevertheless, it is still possible to construct a two body problem solution defined by the central force $-\mu\mathbf{OP}/OP^3$ and the position and velocity vectors at time t. The elliptic elements corresponding to it are called "osculating elements". They are no more constant with time since the trajectory of P is described under the effect of the central force with additional forces. However, if at each instant osculating elements are computed, their variations define uniquely the actual variations of position and velocity. In other terms, one may consider that the relations that exist between the osculating elements and the rectangular coordinates and velocity components define a transformation of coordinates. This transformation is explicitely defined by the equations 2.8, 2.9, 2.12 and 2.21.

3.3. LAGRANGE PLANETARY EQUATIONS

It is possible - although through rather heavy algebra - to apply this transformation to equations 3.4. The result is a system of six other differential equations, called the Lagrange planetary equations. They write as follows :

$$
\begin{aligned}
\frac{da}{dt} &= \frac{2}{na}\frac{\partial R}{\partial M} \\
\frac{de}{dt} &= \frac{-\sqrt{1-e^2}}{na^2 e}\frac{\partial R}{\partial \omega} + \frac{1-e^2}{na^2 e}\frac{\partial R}{\partial M} \\
\frac{di}{dt} &= \frac{-1}{na^2\sqrt{1-e^2}\sin i}\frac{\partial R}{\partial \Omega} + \frac{\cos i}{na^2\sqrt{1-e^2}\sin i}\frac{\partial R}{\partial \omega} \\
\frac{d\Omega}{dt} &= \frac{1}{na^2\sqrt{1-e^2}\sin i}\frac{\partial R}{\partial i} \qquad (3.5) \\
\frac{d\omega}{dt} &= \frac{\sqrt{1-e^2}}{na^2 e}\frac{\partial R}{\partial e} - \frac{\cos i}{na^2\sqrt{1-e^2}\sin i}\frac{\partial R}{\partial i} \\
\frac{dM}{dt} &= n - \frac{2}{na}\frac{\partial R}{\partial a} - \frac{1-e^2}{na^2 e}\frac{\partial R}{\partial e}
\end{aligned}
$$

where n is given by $n^2a^3=\mu$.

Since it is the result of a change of variables, it is clear that R must also be expressed in terms of new variables - the osculating elements. This transformation will be described in Chapter 4.

3.4. GAUSS EQUATIONS

If the perturbing accelerations of 3.1 cannot be expressed by a gradient, in particular if **F** depends on the velocities, Lagrange equations do not exist. One has to perform the change of variables directly on equation 3.1. It is convenient to express the equations in terms of a local system of rectangular coordinates :

R : radial component along **OP**

W : along the axis perpendicular to the orbit, in direction of the angular momentum

S : perpendicular to R in the plane of the orbit in the direction of the motion

The transformed differential equations are called Gauss equations. They write as follows :

$$\frac{da}{dt} = \frac{2}{n\sqrt{1-e^2}}\left[Re \sin v + S(1+e \cos v)\right]$$

$$\frac{de}{dt} = \frac{\sqrt{1-e^2}}{na}\left[R \sin v + S(\cos E + \cos v)\right]$$

$$\frac{di}{dt} = \frac{Wr \cos(\omega+v)}{na^2\sqrt{1-e^2}\sin i} \tag{3.6}$$

$$\frac{d\Omega}{dt} = \frac{Wr \sin(\omega+v)}{na^2\sqrt{1-e^2}}$$

$$\frac{d\omega}{dt} = \frac{\sqrt{1-e^2}}{nae}\left[-R \cos v+S(1+\frac{1}{1+e \cos v})\sin v- \frac{Wr \cos i \cos(\omega+v)}{na^2\sqrt{1-e^2}\sin i}\right]$$

$$\frac{dM}{dt} = n+ \frac{1-e^2}{nae}\left[R(\frac{-2e}{1+e \cos v} + \cos v)-S(1+\frac{1}{1+e \cos v})\sin v\right]$$

As for the Lagrange equations, R, S and W must be expressed in function of a,e,i,Ω,ω and M.

3.5. CANONICAL OSCULATING ELEMENTS

Equations 3.5 can be written in matrix notation :

$$\left(\frac{da}{dt},\frac{de}{dt},\frac{di}{dt},\frac{d\Omega}{dt},\frac{d\omega}{dt},\frac{dM}{dt}\right) = \Lambda(a,e,i,\Omega,\omega)*\left(\frac{\partial R}{\partial a},\frac{\partial R}{\partial e},\frac{\partial R}{\partial i},\frac{\partial R}{\partial \Omega},\frac{\partial R}{\partial \omega}\right)$$

where Λ is an antisymmetric matrix. It is possible to simplify it even more and give it a canonical or symplectic form :

$$\Lambda = \begin{vmatrix} 0 & I_3 \\ -I_3 & 0 \end{vmatrix}$$

where I_3 is the third order unit matrix.

For this, it is sufficient to change only the first three variables. They are the Delaunay variables. With Delaunay's classical notation, they are :

$$L = \sqrt{\mu a} \quad ; \quad G = \sqrt{\mu a(1-e^2)} \quad ; \quad H = \sqrt{\mu a(1-e^2)}\cos i \tag{3.7}$$

$$l = M = n(t-t_0) \quad ; \quad g = \omega \quad ; \quad h = \Omega$$

Applying this change of variables to Lagrange equations 3.5, one easily obtains the following canonical form of equations :

$$\frac{dL}{dt} = \frac{\partial\phi}{\partial l} \quad ; \quad \frac{dG}{dt} = \frac{\partial\phi}{\partial g} \quad ; \quad \frac{dH}{dt} = \frac{\partial\phi}{\partial h}$$

$$\frac{dl}{dt} = \frac{-\partial\phi}{\partial L} \quad ; \quad \frac{dg}{dt} = \frac{-\partial\phi}{\partial G} \quad ; \quad \frac{dh}{dt} = \frac{-\partial\phi}{\partial H} \tag{3.8}$$

with

$$\phi = \frac{\mu}{2a} + R = \frac{\mu^2}{2L^2} + R$$

In this formulation, R has to be expressed in function of the six Delaunay varia bles.

4 - GENERAL PERTURBATION TECHNIQUES

In general, for most of the problems of perturbations - and this is the case of the motion of artificial satellites - one of the three systems of equations presented in the preceding chapter is used. In all the cases, it is necessary to express the right hand members in function of the osculating elements. This is not possible in closed form, and it is necessary to use series developments assuming that some quantities are small.

4.1. DEVELOPMENT OF THE DISTURBING FUNCTION

We shall restrict ourselves to the case when the disturbing forces can be expressed in terms of a disturbing function R presented in section 3.1. In this case, it is sufficient to express R in function of the osculating variables, while in other cases, one should do so with the three components of **F**.

In general, R is a function of the position of $P(x_1,x_2,x_3)$ and of the position of other masses that are function of time and/or various parameters p_j :

$$R = R(x_1,\ x_2,\ x_3,\ p_j,\ t)$$

Transformed by 2.19, one obtains :

$$R = F(A(i,\Omega,\omega)x,\ B(i,\Omega,\omega)y,\ p_j,\ t)$$

and, using 2.9 :

$$R = G(a, e, i, \Omega, \omega, E, p_j, t)$$

G is a finite function. But in order to express R in function of M, it is necessary to apply the expressions described in section 2.3 linking E to M. Substituting them into G, the result is a trigonometric series in M. In addition, since x_j are periodic functions of Ω and ω, this is also true for R. Consequently, it can be developped in multiple trigonometric series of Ω, ω and M. If some of the p_j are also periodic functions of time, the multiple trigonometric series may also include these parameters.

4.2. ARTIFICIAL SATELLITE DISTURBED BY THE SUN

As an example of the development of the disturbing function let us consider an artificial satellite P perturbed by the Sun S of mass m' and geocentric coordinates X_1, X_2, X_3. The equations of motion of the satellite P with respect to the Earth O are obtained by subtracting the accelerations undergone by the Earth from the acceleration affecting P :

$$\frac{d^2\mathbf{OP}}{dt^2} = -\frac{\mu\mathbf{OP}}{OP^3} + \frac{km'\mathbf{PS}}{PS^3} - \frac{km'\mathbf{OS}}{OS^3} \qquad (4.1)$$

The disturbing function is :

$$R = km'\left(\frac{1}{PS} - \frac{X_1x_1+X_2x_2+X_3x_3}{OS^3}\right) \qquad (4.2)$$

If r and r' are distances from the centre of the Earth to the satellite and to the Sun and if δ is the angle between the two directions :

$$\frac{km'}{PS} = \frac{km'}{\sqrt{r'^2+r^2-2rr'\cos\delta}}$$

Since r' is much larger than r, one may write :

$$\frac{km'}{PS} = \frac{km'}{r'\sqrt{1+\left(\frac{r}{r'}\right)^2-2\frac{r}{r'}\cos\delta}}$$

and develop it in power series of r/r' :

$$\frac{km'}{PS} = \frac{km'}{r'}\left[1+\frac{r}{r'}\cos\delta+\left(\frac{r}{r'}\right)^2\left(-\frac{1}{2}+\frac{3}{2}\cos^2\delta\right)+\dots\right] \qquad (4.3)$$

The coefficients of r/r' are Legendre polynomials (see Section 5.1).

The second part of 4.2 is obviously $-km'r\cos\delta/r'^2$. In addition, since km'/r' is independent of the parameters depending on δ, one may restrict R to :

$$R = \frac{km'r^2}{r'^3}\left(-\frac{1}{2}+\frac{3}{2}\cos^2\delta\right) \qquad (4.4)$$

where we have neglected higher powers of r/r'.

Let us call ε the obliquity of the ecliptic and let us assume, as a simplifying assumption, that the orbit of the Earth is circular. If a' and n' are respectively the radius of this orbit and the mean motion ($n'^2a'^3=km'$), the longitude of the Sun counted from the vernal equinox is :

$$\lambda = n'(t-t_0)$$

The coordinates of the Sun are :

$$\begin{aligned} X_1 &= a' \cos\lambda \\ X_2 &= a' \sin\lambda \cos\varepsilon \\ X_3 &= a' \sin\lambda \sin\varepsilon \end{aligned} \tag{4.5}$$

and $r' = a'$; $\dfrac{km'}{a'^3} = n'^2$

The coordinates of the satellite are given by 2.19 and they can be expressed in function of the eccentric anomaly. One has :

$$\cos\delta = \frac{1}{r}\Big[\cos\lambda(P_1x+Q_1y)+\sin\lambda\cos\varepsilon(P_2x+Q_2y)+\sin\lambda\sin\varepsilon(P_3x+Q_3y)\Big] \tag{4.6}$$

with :

$$\begin{aligned} x &= a(\cos E-e) \\ y &= a\sqrt{1-e^2}\sin E \\ r &= a(1-e\cos E) \end{aligned}$$

If e is a small quantity, applying 2.15 and 2.17, one gets x, y and $1/r$ as trigonometric series of M. Replacing systematically products of trigonometric functions by the sum of two trigonometric functions, one may transform 4.6 into a multiple trigonometric series of the form :

$$\cos\delta = \sum_{h=-1}^{+1}\sum_{j=-1}^{+1}\sum_{k=-1}^{+1}\sum_{l=0}^{\infty} A_{hjkl}(e,i)\cos(l\,M+h\Omega+j\omega+k\lambda) \tag{4.7}$$

It easy to check that $\cos\delta$ is an even function of the angles. An analogous expression will be obtained for $\cos^2\delta$ and then for $r^2\cos^2\delta$. Finally, one obtains for R an expression of the form :

$$R = n'^2a^2\sum_{h,j,k,l} B_{h,j,k,l}(e,i)\cos(l\,M+h\Omega+j\omega+k\lambda) \tag{4.8}$$

This expression is the development of the disturbing function of the problem. The limit of the summation in l is to be set up from the magnitude of the coefficients $B(e,i)$ so that only significant terms are kept. Let us also remark that R is factored by the small quantity n'^2.

The same procedure is adopted for the development of the disturbing function of the lunar motion, but it is advisable in this case to adopt an ecliptic system of coordinates.

4.3. METHOD OF SOLUTION

Once the disturbing function is developed as in the preceding sections, it is necessary to develop similarly the right hand members of the equations. Let us continue the example treated in 4.2. The partial derivatives of R as given by 4.8 are readily obtained. Let us substitute them into the Lagrange equations 3.5. One may notice that the first three equations contain only derivatives of R with respect to the angles present in the trigonometric arguments, while in the last three equations, partial derivatives are taken with respect to a, e and i present only in the coefficients. It results that da/dt, de/dt and di/dt are odd series of the form :

$$\frac{d(a,e,i)}{dt} \sim \sum C_{hjkl}(a,e,i)\sin(l M+h\Omega+j\omega+k\lambda) \tag{4.9}$$

while dΩ/dt, dω/dt and dM/dt are even series of the form :

$$\frac{d(\Omega,\omega,M)}{dt} \sim \sum D_{hjkl}(a,e,i)\cos(l M+h\Omega+j\omega+k\lambda) \tag{4.10}$$

In particular, in 4.10 there may exist - and generally do exist - terms independent of the angular arguments noted $D_{0000}(a,e,i)$.

Many different methods exist for obtaining the solution of such equations. They are generally very involved. However, in some cases, it is sufficient to obtain only the most significant terms and, in particular, neglect all the terms that are of the order of the square of the small quantity characterizing the disturbing function. Such a procedure is called a first order solution and is very simple to apply once the equations are written as in 4.9 and 4.10.

Let us designate by a_o, e_c and i_o the mean values of a, e and i, mean values that are obtained if one neglects all the terms in the equations. These values are substituted into 4.10 in which the periodic terms have also been provisionally neglected and which contain only the D_{0000} terms. Let us designate them as follows :

$$\frac{d\Omega}{dt} = n_\Omega(a_o,e_o,i_o)$$

$$\frac{d\omega}{dt} = n_\omega(a_o,e_o,i_o)$$

$$\frac{dM}{dt} = n_M(a_o,e_o,i_o)$$

Actually n_M consists of $n_o=(km/a_o{}^3)^{1/2}$ and of the term coming from the development. Integrating these equations, one obtains :

$$\begin{aligned}\bar{\Omega} &= n_{\Omega}(t-t_0) + \Omega_0 \\ \bar{\omega} &= n_{\omega}(t-t_0) + \omega_0 \\ \bar{M} &= n_M(t-t_0) + M_0\end{aligned} \tag{4.11}$$

Then, these quantities as well as a_0, e_0 and i_0 are substituted in the right hand members of the differential equations 4.9 and 4.10. The arguments of the periodic terms are linear functions of time and therefore the equations are integrable term by term. The result has the following form :

$$\begin{vmatrix} a \\ e \\ i \end{vmatrix} = \begin{vmatrix} a_0 \\ e_0 \\ i_0 \end{vmatrix} - \sum_{l,h,j,k} \frac{C(a_0,e_0,i_0)\cos(lM+h\Omega+j\omega+k\lambda)}{l\,n_M+hn_{\Omega}+jn_{\omega}+kn'} \tag{4.12}$$

for the last three equations :

$$\begin{vmatrix} \Omega \\ \omega \\ M \end{vmatrix} = \begin{vmatrix} \bar{\Omega} \\ \bar{\omega} \\ \bar{M} \end{vmatrix} + \sum_{l,h,j,k} \frac{D(a_0,e_0,i_0)\sin(l\,M+h\Omega+j\omega+k\lambda)}{l\,n_M+hn_{\Omega}+jn_{\omega}+kn'} \tag{4.13}$$

Of course, the coefficients C and D depend on l,h,j,k and are different for each of the three elements.

However, in this procedure, we have overlooked the fact that the first term of dM/dt is n and not n_0. One has :

$$n = (km/a^3)^{1/2}$$

and if, from 4.12, one writes :

$$a = a_0 - \sum_{l,h,j,k} \frac{C'(a_0,e_0,i_0)\cos(l\,M+h\Omega+j\omega+k\lambda)}{l\,n_M+hn_{\Omega}+jn_{\omega}+kn'} = a_0 - \Delta_a \quad ,$$

on gets a first order development of n as :

$$\begin{aligned} n &= n_0\,(1+\frac{3\Delta_a}{2a_0}) . \\ &= n_0 + \frac{3n_0}{2a_0}\sum_{l,h,j,k} \frac{C'(a_0,e_0,i_0)\cos(l\,M+h\Omega+j\omega+k\lambda)}{l\,n_M+hn_{\Omega}+jn_{\omega}+kn'}\end{aligned}$$

So, one must add to the solution in M given by 4.13, the integral of these additional terms, namely :

$$\frac{3n_0}{2a_0}\sum_{l,h,j,k} \frac{C'(a_0,e_0,i_0)\sin(l\,M+h\Omega+j\omega+k\lambda)}{(l\,n_M+hn_{\Omega}+jn_{\omega}+kn')^2} \tag{4.14}$$

These terms have the same form as those presented in 4.13 and can be combined with them, so that the form 4.13 holds for the complete first order solution. It is a very general feature that, in solving the equations of motion of Celestial Mechanics, one has to compute at least one double integral.

4.4. SOLUTION IN CANONICAL VARIABLES

We have introduced, Section 3.5, a canonical formulation of the equations. Let us sketch how one could solve these equations in a very elegant way - though not necessarily with less computational effort. Let us write equations 3.8 in new unified notations replacing those of 3.7 :

$$\frac{dq_i}{dt} = \frac{\partial \phi}{\partial p_i} \quad ; \quad \frac{dp_i}{dt} = -\frac{\partial \phi}{\partial q_i} \quad , \quad i = 1, 2, 3$$

with

$$\phi = \frac{\mu^2}{2q_1^2} + R(p_1, p_2, p_3, q_1, q_2, q_3, \lambda)$$

where R is given by 4.8 expressed in canonical variables. In order to have the equations totally canonical, it is necessary to give to λ also the status of a canonical variable, although we know what should be the solution :

$$p_4 = \lambda = n'(t-t_0) \quad ; \quad \frac{dp_4}{dt} = n' \tag{4.15}$$

Let us introduce a fictituous conjuguate variable q_4, which is some kind of generalized energy, so that :

$$\frac{dp_4}{dt} = -\frac{\partial \phi}{\partial q_4}$$

Condition 4.15 is satisfied if the Hamiltonian ϕ is replaced by :

$$\phi' = \frac{\mu^2}{2q_1^2} - n'q_4 + R(p_1, p_2, p_3, p_4, q_1, q_2, q_3, -) \tag{4.16}$$

In 4.16 and in further equations, the bar - means that a variable (here q_4) is not present in the function. Let us define a canonical transformation :

$$p_i, q_i \rightarrow P_i, Q_i \quad , \quad i=1,2,3,4$$

using a generating function :

$S(p_i, Q_i)$ such that

$$q_i = \frac{\partial S}{\partial p_i} = Q_i + \varepsilon_i(p_1, p_2, p_3, p_4, Q_1, Q_2, Q_3, -)$$

$$p_i = \frac{\partial S}{\partial Q_i} = p_i + \eta_i(p_1, p_2, p_3, p_4, Q_1, Q_2, Q_3, -)$$

This condition means that the transformation is an identity to the order zero. This is possible if one sets :

$$S = \sum_{i=1}^{4} Q_i p_i + \varepsilon(p_1, p_2, p_3, p_4, Q_1, Q_2, Q_3, -) \tag{4.17}$$

where ε_i η_i and ε are first order quantities. After this transformation, there is a new Hamiltonian :

$$\phi^* = \frac{\mu^2}{2Q_1^2} - n'Q_4 + R^*(P_1, P_2, P_3, P_4, Q_1, Q_2, Q_3, -)$$

The function $\varepsilon(P_i, Q_i)$ is constructed in such a way that R* is simpler than R. In Delaunay method, the largest term of R is constrained to be transformed into a term of higher order. In Poincaré-Von Zeipel method, R* should not contain one of the angular variable if one neglects higher orders of the small quantity. One gets :

$$\phi^* = \frac{\mu^2}{2Q_1^2} - n'Q_4 + R^*\left(-, P_2, P_3, P_4, Q_1, Q_2, Q_3, -\right)$$

Succesive similar transformations permit to eliminate one by one the other angular variables, so that finally, one is left with the following Hamiltonian :

$$\tilde{\phi} = \frac{\mu^2}{2\tilde{Q}_1^2} - n'\tilde{Q}_4 + \tilde{R}(-,-,-,-,\tilde{Q}_1, \tilde{Q}_2, \tilde{Q}_3, -)$$

The transformed equations are, at this stage, readily integrated :

$$\frac{d\tilde{Q}_i}{dt} = \frac{\partial\tilde{\phi}}{\partial\tilde{P}_i} = 0 \quad , \quad \text{hence } \tilde{Q}_i = Q_i^0 \text{ (constant)}$$

$$\frac{d\tilde{P}_i}{dt} = -\frac{\partial\tilde{\phi}}{\partial\tilde{Q}_i} = -\frac{\partial\tilde{\phi}}{\partial Q_i^0} = n_i^0$$

and $\tilde{P}_i = n_i^0 t + P_i^0$

After successive inverse transformations, one can return to the initial variables and obtain the final solution :

$$q_i = Q_i^0 + \sum \text{terms}(Q_i^0, n_i^0 t + P_i^0)$$
$$p_i = P_i^0 + \sum \text{terms}(Q_i^0, n_i^0 t + P_i^0) \qquad (4.18)$$

These expressions are actually of the form 4.12 and 4.13.

4.5. FORM OF THE SOLUTION

The form of the solution obtained in Section 4.3 is in fact quite general and applies to most of the perturbation problems of a gravitational nature. It is also unchanged if a second order or higher order solution is computed.

(i)- The three elements a, e and i are represented by a mean value and even multiple argument trigonometric series, the arguments being linear function of time and the coefficients depending only on a_0, e_0, i_0 (and of course on the other parameters of the problem).

(ii)- The three angular elements Ω, ω and M are odd multiple argument trigonometric series, the arguments being linear function of time and the coefficients depending only on a_0, e_0, i_0 and on other parameters of the problem. In addition they have a secular term, which is the time coefficient of the unitary angular arguments of the trigonometric series.

These results hold only if the terms retained in the development of the disturbing function are such that they do not produce a divisor $ln_M+hn_\Omega+jn_\omega+kn'$ of the order of the small quantity characterizing the disturbing function. If this is the case, the solution has another nature that will be sketched in Chapter 6.

5 - MOTION OF AN ARTIFICIAL SATELLITE

The motion of an artificial Earth satellite is essentially governed by the Earth's gravity field. Would the Earth have a spherical symmetry in the mass distribution, then the external gravitational attraction would be a central Newtonian force proportional to the mass of the Earth. However, this is not the case, and one has to take into account the actual distribution of masses within the Earth.

5.1. EARTH'S GRAVITATIONAL POTENTIAL

Let us first evaluate the acceleration undergone by a point P (x,y,z) outside the Earth. If M (ξ,η,ζ) is a point in the Earth and $\delta(\xi,\eta,\zeta)$, the density at this point (figure 4), the elementary acceleration of P due to M is given by 3.2 :

$$\mathbf{dF} = k\delta(\xi,\eta,\zeta)\,\frac{\mathbf{PM}}{PM^3}\,d\xi,d\eta,d\zeta$$

Integrating this over the Earth's volume V, one gets the total acceleration :

$$\mathbf{F} = \iiint_V k\delta(\xi,\eta,\zeta)\,\frac{\mathbf{PM}}{PM^3}d\xi,\,d\eta,d\zeta \tag{5.1}$$

As shown in Section 3, it can be written in terms the components of a gradient of a function U such that :

$$\mathbf{F} = \mathbf{grad}\,U$$

with :

$$U = \iiint_V \frac{k\delta(\xi,\eta,\zeta)}{\sqrt{(x-\xi)^2+(y-\eta)^2+(z-\zeta)^2}}d\xi d\eta d\zeta \tag{5.2}$$

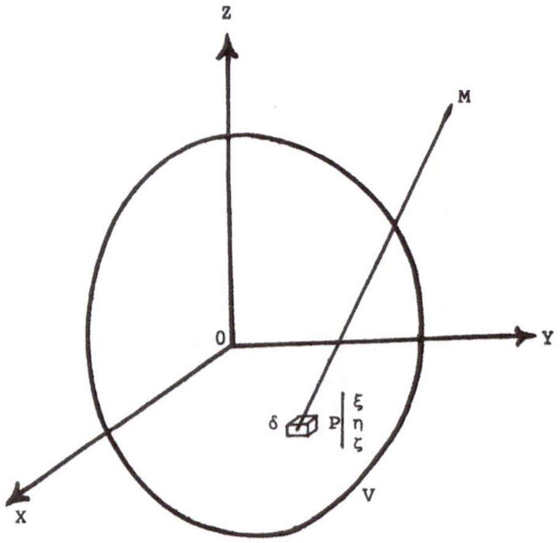

Figure 4

U is the gravitational potential of the Earth. Let us express U in spherical coordinates :

$$P : r,\lambda,\phi \quad ; \quad M : r',\lambda',\phi'$$

Let $\theta = (OM,OP)$, $\rho = r'/r$ and $dm = \delta(\xi,\eta,\zeta)d\xi,d\eta d\zeta$:

$$\cos\theta = \sin\phi\,\sin\phi' + \cos\phi\,\cos\phi'\,\cos(\lambda-\lambda')$$

$$\text{and } |MP| = r\sqrt{1-2\rho\cos\theta+\rho^2} \tag{5.3}$$

With these notations, one has :

$$U = \frac{k}{r} \iiint_V \frac{dm}{\sqrt{1-2\rho\cos\theta + \rho^2}} \tag{5.4}$$

One can apply the same procedure as in 4.2 and develop this expression in Legendre polynomials P_n (this is one of the definitions of these polynomials) :

$$U = \frac{k}{r} \iiint_V dm + \frac{k}{r} \iiint_V \rho P_1(\cos\theta)dm + \sum_{n=2}^{\infty} \iiint_V \rho^n P_n(\cos\theta)dm \tag{5.5}$$

The first integral is the mass of the Earth. The corresponding term is μ/r. If we note that $rr'\cos\theta = x\xi+y\eta+z\zeta$, the second integral becomes :

$$\frac{k}{r^3}\left[x \iiint_V \xi dm + y \iiint_V \eta dm + z \iiint_V \zeta dm\right]$$

Each of these integrals gives a coordinate of the centre of mass of the Earth. If the origin of coordinates is chosen in a way to coincide with it, the second integral of 5.5 is nought.

For the other integrals, one has to express by 5.3 $P_n(\cos\theta)$ in terms of the angles. One has to introduce the associate Legendre functions, through the formula:

$$P_n(\cos\theta) = P_n(\sin\phi')P_n(\sin\phi) + 2\sum_{k=1}^{n} \frac{(n-k)!}{(n+k)!} P_{nk}(\sin\phi)P_{nk}(\sin\phi')\cos k(\lambda'-\lambda)$$

Let us use the following definitions :

(i)- <u>Legendre polynomials</u>

$$P_n(x) = \frac{1}{2^n n!} \frac{d^n}{dx^n}(x^2-1)^n \tag{5.6}$$

(ii)- <u>Associate Legendre functions</u>

$$P_{n,k}(x) = (-1)^k (1-x^2)^{k/2} \frac{d^k P_n(x)}{dx^k} \tag{5.7}$$

Then U can be written as :

$$U = \frac{\mu}{r}\left[1 - \sum_{n=2}^{\infty} \frac{J_n a_e^n}{r^n} P_n(\sin\phi) + \sum_{n=2}^{\infty}\sum_{k=1}^{n} \frac{J_{nk} a_e^n}{r^n} P_{n,k}(\sin\phi)\cos k(\lambda-\lambda_{nk})\right] \tag{5.8}$$

In this expression :

- a_e is a normalizing factor, conventionally set equal to the Earth's equatorial radius
- J_n are called zonal harmonics and are dimensionless numbers
- $J_{n,k}$ are called tesseral harmonics and λ_{nk} are the phases of these harmonics. Sometimes :

$$C_{nk} = J_{nk}\cos\lambda_{nk}$$

$$S_{nk} = J_{nk}\sin\lambda_n$$

are used instead and C_{no} replaces $-J_n$. Then, if $P_{n,0} \equiv P_n$, one has the general form :

$$U = \frac{\mu}{r} + \sum_{n=2}^{\infty} \sum_{k=0}^{n} \frac{\mu a_e^n}{r^{n+1}} P_{n,k}(\sin\phi)(C_{nk}\cos k\lambda + S_{nk}\sin k\lambda) \tag{5.9}$$

With the exception of J_2 which is of the order of 10^{-3} ($J_2 \approx 0.00108263$), all other harmonics are of the order of 10^{-5} or smaller. This allows us to use a perturbation theory to study the motion of close Earth satellites. Refering to Section 3.1, one can see that $U-\mu/r$ is the disturbing function R. In addition, separating the first order of the small quantity represented by J_2, one can write the disturbing function as :

$$R = -\frac{\mu J_2 a_e^2}{r^3} P_2(\sin\phi) + R_2 \tag{5.10}$$

where R_2 represents the other terms of 5.8 which are at least of the second order. The first term takes into account essentially the effect of the Earth's oblateness.

5.2. DEVELOPMENT OF THE DISTURBING FUNCTION

Let us limit ourselves to the first order term of 5.10 :

$$R_1 = \frac{\mu J_2 a_e^2}{r^3} \left(\frac{1}{2} - \frac{3}{2}\sin^2\phi\right) \tag{5.11}$$

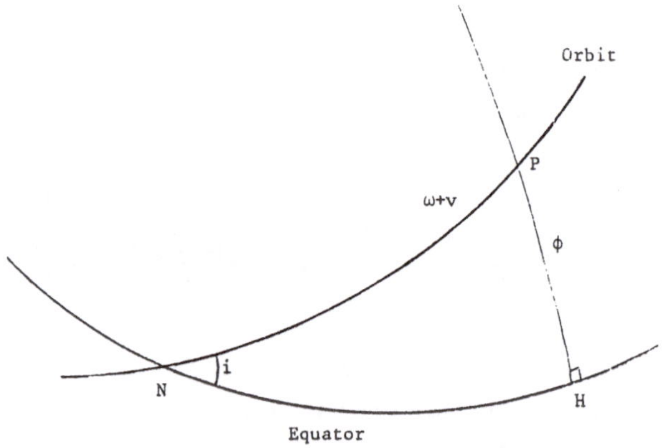

Figure 5

and let us express it in function of the elliptic elements. The following relation is evident from figure 5 :

$$\sin\phi = \sin i \sin(\omega+v) ,$$

since $\omega+v$ is the longitude of the satellite from the node N in the orbital plane. Introducing also the semi-major axis, one can write 5.10 as follows :

$$R_1 = \frac{\mu a_e^2 J_2}{a^3}\left[\left(\frac{1}{2} - \frac{3}{4}\sin^2 i\right)\frac{a^3}{r^3} + \frac{3}{4}\sin^2 i\,\frac{a^3}{r^3}\cos(2\omega + 2v)\right] \quad (5.12)$$

At this stage let us apply the theory described in Section 2.3 and develop :

$$\frac{a^3}{r^3} \quad , \quad \frac{a^3}{r^3}\cos 2v \quad \text{and} \; \frac{a^3}{r^3}\sin 2v$$

in Fourier series of the mean anomaly M. The first terms of these developments are

$$\frac{a^3}{r^3} = \left(1+\frac{3}{2}e^2\right)+\left(3e+\frac{27}{8}e^3\right)\cos M+\frac{9}{2}e^2\cos 2M+\frac{53}{8}e^3\cos 3M \quad (5.13)$$

The first term is actually the beginning of the development of $1/(1-e^2)^{3/2}$ which is its exact value.

Combining the two last expressions, one gets :

$$\begin{aligned}\frac{a^3}{r^3}\cos(2\omega-2v) = {} & \frac{e^3}{48}\cos(2\omega-M)+\left(\frac{-e}{2}+\frac{e^3}{16}\right)\cos(2\omega+M) + \\ & + \left(1-\frac{5}{2}e^2\right)\cos(2\omega+2M) + \left(\frac{7e}{2}-\frac{123}{16}e^3\right)\cos(2\omega+3M) \\ & + \frac{17e^2}{2}\cos(2\omega+4M)+\frac{845}{48}e^3\cos(2\omega+5M)\end{aligned} \quad (5.14)$$

An important feature of the last development is that there is no term in $\cos 2\omega$. In other terms :

$$\int_0^{2\pi}\frac{a^3}{r^3}\cos(2\omega+2v)dM = 0$$

The next step is to substitute 5.13 and 5.14 into 5.12 and the result is the development of R_1 which has the same form as the one given by 4.8.

5.3. FIRST ORDER SOLUTION

Let us follow the method described in Section 4.3. Using the development of R_1 that has been just obtained, we have to compute the derivatives of R_1 with respect of the elements and substitute them into the Lagrange equations 3.5. The first action is to compute the non periodic part of the solution in angular variables. The equations are obtained from the part of R_1 that is independent of ω and M :

$$R' = \frac{\mu a_e^2 J_2}{a^3}\left(\frac{1}{2} - \frac{3}{4}\sin^2 i\right)(1-e^2)^{-3/2} \tag{5.15}$$

From it and 3.5, one deduces, noting that $\mu = n^2 a^3$:

$$\frac{d\Omega}{dt} = \frac{1}{na^2\sqrt{1-e^2}\sin i}\frac{\partial R'}{\partial i} = -\frac{3}{2}na_e^2\frac{J_2\cos i}{a^2(1-e^2)^2}$$

$$\frac{d\omega}{dt} = \frac{\sqrt{1-e^2}}{na^2 e}\frac{\partial R'}{\partial e} - \frac{\cos i}{na^2\sqrt{1-e^2}\sin i}\frac{\partial R'}{\partial i} = \frac{3na_e^2 J_2}{a^2(1-e^2)^2}\left(1 - \frac{5}{4}\sin^2 i\right) \tag{5.16}$$

$$\frac{dM}{dt} = n - \frac{2}{na}\frac{\partial R}{\partial a} - \frac{1-e^2}{na^2 e}\frac{\partial R}{\partial e} = n + \frac{3n\, a_e^2 J_2}{a^2(1-e^2)^{3/2}}\left(\frac{1}{2} - \frac{3}{4}\sin^2 i\right)$$

The integration of this system gives the secular variations in the form of equation 4.10 after a substitution of some mean values a_0, e_0 and i_0 of the other three elements.

The first equation represents the rotation of the orbital plane with an angular velocity :

$$n_\Omega = -\frac{3}{2}\frac{n_0\, a_e^2 J_2 \cos i_0}{a_0^2(1-e_0^2)^2} \tag{5.17}$$

It is maximum for small inclination orbits, while polar orbits remain always in the same plane. The rotation of the nodes is retrograde for direct orbits ($i < 90°$) and direct for retrograde orbits ($i > 90°$).

The second equation describes the motion of the perigee referred to the node with an angular velocity equal to :

$$n_\omega = \frac{3n_0\, a_e^2 J_2}{a_0^2(1-e_0^2)^2}\left(1 - \frac{5}{4}\sin^2 i_0\right) \tag{5.18}$$

It is a retrograde rotation for all inclinations between i_0 and i_0' such that :

$$\sin i_0 = \frac{2}{\sqrt{5}} \quad ; \quad i_0 = 63°.435 \text{ and } i_0' = 116°.565$$

These particular values are called 'critical inclinations'. For orbits less inclined on the equator, the perigee has a direct motion. Around i_0 and i_0', the motion has a different character that will be described in Chapter 6.

Finally, the last equation indicates that another effect of J_2 is to slightly modify the period of the satellite in comparison with the period as given by the two body problem :

$$n_M = n_0 \left[1 + \frac{3}{2} \frac{a_e^2 J_2}{a_0^2 (1-e_0^2)^{3/2}} \left(1 - \frac{3}{2} \sin^2 i_0\right) \right] \tag{5.19}$$

The motion is slower for inclinations between i_1=54°.736 and i'=125°.264 and faster outside this interval.

The general solution is obtained in adding the integrated other terms of the equations after substitution of a_0, e_0, i_0 in the coefficients and $\varpi=\omega_0+n_\omega t$ and $\bar{M}=M_0+n_M t$ in the equations (Ω is absent). The general structure of such a term is :

$$T = \frac{\mu\, a_e^2 J_2}{n a^j} f(i,e) \begin{matrix}\cos\\ \sin\end{matrix} (\varepsilon\bar{\omega}+k\bar{M})$$

In this expression j=5 with the exception of the equation in da/dt for which j=4. The argument is such that ε is 0 or 2, while k is an integrer, but as it has been remarked after 5.14, one has never k=0.

The substitution transforms T into :

$$T' = \frac{\mu a_e^2 J_2}{n_0 a_0^j} f(i_0,e_0) \begin{matrix}\cos\\ \sin\end{matrix} \left[\varepsilon\omega_0+kM_0+(\varepsilon n_\omega+kn_M)t \right]$$

where ω_0 and M_0 are the constants of integration for ω and M. Integrated with respect to time, this becomes :

$$T^* = \pm \frac{\mu a_e^2 J_2 f(i_0,e_0)}{n_0 a_0^j (\varepsilon n_\omega + kn_M)} \begin{matrix}\sin\\ \cos\end{matrix} (\varepsilon\omega_0+kM_0+(\varepsilon n_\omega+kn_M)t)$$

Because of the property given above, the denominator is never of the order of J_2 and has a leading term equal to kn_M or kn_0. The coefficient of T* is therefore multiplied by :

$$\frac{n_0^2 a_0^3 a_e^2 J_2 f(i_0,e_0)}{n_0 a_0^j (kn_0+J_2 g(i_0,e_0..))}$$

Other terms, produced by the double integration of n in dM/dt (see end of Section 4.3) have $(kn_0+J_2 g(i_0,e_0,\ldots))^2$ in the denominator. This does not change the result that all the periodic terms are of the order of J_2 and have periods of the order of the period of the satellite or of its divisors by k. Finally, the solution has the following form :

$$a = a_0 + J_2 \sum a_{\varepsilon,k} \cos(\bar{\omega}+k\bar{M})$$

$$e = e_0 + J_2 \sum a_{\varepsilon,k} \cos(\varepsilon\bar{\omega}+k\bar{M})$$

$$i = i_0 + J_2 \sum i_{\varepsilon,k} \cos(\varepsilon\bar{\omega}+k\bar{M})$$
$$\Omega = \bar{\Omega} + J_2 \sum \Omega_{\varepsilon,k} \sin(\varepsilon\bar{\omega}+k\bar{M})$$
$$\omega = \bar{\omega} + J_2 \sum \omega_{\varepsilon,k} \sin(\varepsilon\bar{\omega}+k\bar{M} \qquad (5.20)$$
$$M = \bar{M} + J_2 \sum M_{\varepsilon,k} \sin(\varepsilon\bar{\omega}+k\bar{M})$$

where the summation is meant on $\varepsilon=0$ and 2, and k for all positive integers if $\varepsilon=0$ or any non zero integers if $\varepsilon=2$.

5.4. SECOND ORDER SOLUTION

A more precise solution will be obtained in substituting the first order solution in the Lagrangian equations and integrating again term by term. Since quantities in J_2 appear in the first order solution and will be multiplied by J_2 in the partial derivatives of R, the resulting expressions will have J^2_2 in factor. But the procedure described in section 4.3 remains valid and one will obtain a solution that will have the same form in 5.20 with the exception that there will be terms in $J_2{}^2$ in n_Ω, n_ω, n_M and all the periodic terms and that ε may also have the value +4. This solution is called second order solution.

However, in contrast with what happens in the first order solution, the substitution gives rise to a term T in which k=0 and $\varepsilon=2$:

$$T = \frac{\mu a_e^2 J_2^2 f(i_0,e_0)}{n_0 a_0^j} \begin{matrix}\cos\\ \sin\end{matrix} 2\bar{\omega}$$

After integration, this becomes :

$$T^* = \frac{\mu a_e^2 J_2^2 f(i_0 e_0)}{2 n_0 a_0^j n_\omega} \begin{matrix}\sin\\ \cos\end{matrix} 2\bar{\omega}$$

and, since n_ω, given by 5.18 has J_2 in factor, T* is of the order of J_2, that is of the first order. In addition, its period is longer than the period of the satellite by a factor of the order of $1/J_2$. It is called a long period term. It is general feature of the long period terms that, in any method of solving the equations of motion, terms of order 1 do not appear during the first order theory but at the next iteration.

5.5. EFFECTS OF OTHER ZONAL HARMONICS

The treatment of all the other terms of the second part of expression 5.8 is similar to the one described here. Since all the J_n with $n > 2$ are of the order of $J_2{}^2$ or smaller, they give normally rise to terms of the second order when integrating the developed equations in the way described for J_2 in 5.3. It is indeed sufficient to substitute a_0, e_0, i_0, ω and M in the equations to obtain the terms factored by J_n.

However, the property of the absence of terms in $\varepsilon\omega$ does no more exist and there are, in the equations, terms of the form :

$$\frac{\mu a_e^n J_n}{n_0 a_0^j} f(i,e) \begin{matrix}\cos\\ \sin\end{matrix} (\varepsilon\omega)$$

They will be treated in the manner described in the preceding section. The corresponding long period terms will, hence, be factored by J_n/J_2 and will be of one order smaller than J_n. Let also mention that since $P_n(\sin\theta)$ of equation 5.8 have the same parity than n, odd harmonics produce only long period terms, while even harmonics also contribute to the secular terms in Ω, ω and M.

5.6. TREATMENT OF TESSERAL HARMONICS

In dealing with the terms resulting from the third part of expression 5.8, we shall encounter more complex trigonometric arguments. While, when only ϕ was present in the expressions, the developments involved M and ω only, in this case, they will also involve $\lambda-\lambda_{nk}$ where λ is the geographic longitude in an Earth fixed reference frame. But the equations of the satellite motion are written in a space-fixed reference in which the Earth rotates with a period equal to one sideral day. It results that, as seen in figure 6, one has :

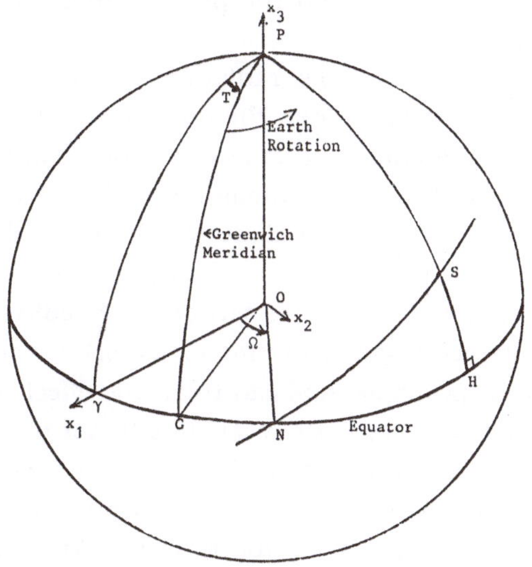

Figure 6

$$\lambda = \Omega + \overline{NH} - T = \Omega + \overline{NH} - \nu(t-t_0) \tag{5.21}$$

where T is the Greenwich sidereal time and NH can be expressed in function of i and $\omega+v$. So any quantity of 5.8 of the form :

$$P_{nk}(\sin\phi)\cos k(\lambda-\lambda_{nk})$$

will finally be developed in multiple trigonometric series whose arguments will be of the form :

$$j\omega + hM + k(\Omega-T-\lambda_{nk}) \tag{5.22}$$

with $k\neq 0$.

In general, such terms will be integrated as described in the first order theory and will have a divisor equal to :

$$D = j\, n_0 + hn_M + k(n_0 - \nu) \tag{5.23}$$

In general, it will not be small and the resulting term will remain small. However, for some particular combinations of j, h and k, this may not be the case, and some large long periodic effects may arise (see Section 6.3).

5.7. OTHER PERTURBATIONS

Several other forces act on a satellite and have often to be taken into account.

The gravitational perturbations by the Sun or the Moon are treated as already described in section 4.2. Generally, it is not necessary to take into account the cross effects, so that one can just add the perturbations computed independently for each of these two bodies.

The perturbations due to the solar radiation pressure are also often not negligible. It is actually a repulsive force from the direction of the Sun. It is proportional to the cross section of the satellite. If this cross-section can be considered as constant with time or if the rotation of the satellite is sufficiently rapid so as to average out its variations, the acceleration is constant and there is a disturbing function :

$$R = A\, r \cos \delta$$

where A is a constant and δ is the angle between the radius vector and the direction of the Sun. It can be expressed in terms of the orbital elements of the satellite and of the Sun so that, in principle, one can treat this effect in the same manner as the solar gravitational perturbations. However, this is not true if the satellite is shadowed by the Earth. Then the acceleration disappears suddenly and the above mentionned treatment is no more possible. One has to use numerical methods.

Another important perturbation is due to the residual air drag acting on the satellite. The resulting acceleration is opposite to the velocity vector of the satellite and is proportional to its square, to the cross-section of the satellite and to the density of the medium :

$$T = -B V^2$$

In this case, the force is no more a gradient and it is not possible to write Lagrange equations. Gauss equations 3.6 are to be used. Since the force is tangential, W=0. It results that the inclination and the longitude of the node are not affected by drag. This is an important result since Ω and i are often used for the determination of some geodesic parameters. Another feature is that the right hand members of the equations have simultaneously sin v and cos v terms. This means that the developments in orbital elements will present simultaneously odd and even terms.

In particular there will be a large negative constant term $-S_0$ arising from the S component of the drag. This component is of the order of BV^2 for a satellite with small eccentricity. The first Gauss equation will contain a constant term :

$$C = \frac{2\, S_0}{n\sqrt{1 - e^2}}$$

whose mean value is C_0 so that, in integrating it, one will obtain :

$$a = a_0 - C_0 t$$

The main effect of drag is a systematic decrease of the semi major axis that leads finally to the decay of the satellite.

Let us also remark that in the second equation, the development of cosE+cos v has a constant term equal to -3e/2. It results that similarly the equation in de/dt has also a constant negative term: another effect of drag is to decrease the eccentricity. The effects of these non-gravitational forces are described in Milani et al., 1987.

6 - RESONANCES

In a mechanical system, a resonance occurs when the frequency of an external action is very close to a natural frequency of the system. It results a considerable enhancing of the amplitude of oscillations. A similar phenomenon exists in Celestial Mechanics. Since, as we have seen in Chapter 4, the motion of a celestial body is characterized by a set of basic frequencies and a number of linear combinations thereof, they may be close to some external perturbing frequency and resonance will occur.

6.1. GENERAL PRESENTATION.

Let us consider a general perturbation problem and let θ be a certain linear combination of the angular variables that has a very small mean motion. If, for instance, it is a problem with three degrees of freedom with an external frequency as this is the case of the lunar theory presented in section 4.3, the solution has the general form given by the equations 4.12 and 4.13. Let us take :

$$\theta = (l n_M + h n_\Omega + j n_\omega + k n')t + \theta_0 = n_\theta (t - t_0) + \theta_0 \tag{6.1}$$

and assume that n_θ very small.

If one of the integers, for instance j, is not zero, it is always possible to change the variables of the differential equations and replace ω by :

$$\theta = lM + h\Omega + j\omega + k\lambda$$

With this change of variables, the equation in θ will take the following form :

$$\frac{d\theta}{dt} = \varepsilon x(a,e,i) + \sum_{j=1}^{\infty} Aj \cos j\theta + \sum_{h,j,k,l} a_{hjkl} \cos(hM + j\theta + l\Omega + k\lambda) \tag{6.2}$$

where ε is the small quantity of the problem and where we have separated the terms giving rise to small divisors. Similarly, one can change the other three variables in such a way that one of them is x defined by the non-periodic term of 6.2. The equation in x will have the following form :

$$\frac{dx}{dt} = \sum_{j=1}^{\infty} B_j \sin j\,\theta + \sum_{h,j,k,l} b_{hjkl} \sin(hM + j\theta + l\Omega + k\lambda) \tag{6.3}$$

The other equations, that will not be written here, will have similar forms.

In general, εx and at least one of the Aj and/or Bj are first order quantities. If one would apply the general method described in Section 4.3 in order to integrate the terms in cos $j\theta$ or sin $j\theta$, the coefficients would be :

$$\frac{Aj}{j\varepsilon x_0} \quad \text{and} \quad \frac{Bj}{j\varepsilon x_0}$$

where x_0 is the value taken by x when a, e, i have their mean values a_0, e_0 and i_0. These terms are of order zero and this is not compatible with the principles of the perturbation theory in which not only the disturbing forces, but also their effects have to be small quantities. It is necessary to treat them before entering the perturbation procedures. Instead of solving the equations 4.9 as described in Section 4.3, it is necessary to add to them the terms in θ and solve the more complex system summarized as follows :

$$\frac{dx}{dt} = \sum_{j=1}^{\infty} Bj \sin j\theta \tag{6.4}$$

$$\frac{d\theta}{dt} = \varepsilon x + \sum_{j=1}^{\infty} Aj \cos j\theta$$

and similar equations in other variables, with the special condition that the mean value of εx is of the first order, as are the coefficients Aj and Bj.

6.2. A SIMPLIFIED RESONANCE PROBLEM

In order to present the properties of the system, let us ignore the other equations and study only the two equations 6.4. Let us simplify even more and assume that there is only one significant periodic term in dx/dt corresponding to j=1. After having substituted the mean values of the other variables in 6.4, the system of equations reduces to the following form :

$$\frac{dx}{dt} = -B\sin\theta \quad \text{and} \quad \frac{d\theta}{dt} = \varepsilon x \tag{6.5}$$

or :

$$\frac{d^2\theta}{dt} = -\varepsilon B\sin\theta \tag{6.6}$$

It is the equation of a simple pendulum. Multiplying 6.6 by dθ/dt and integrating, one obtains, introducing 2c as a constant of integration :

$$\left(\frac{d\theta}{dt}\right)^2 = -2\varepsilon B\cos\theta + 2c$$

$$\left(\frac{d\theta}{dt}\right)^2 = -4\varepsilon B\sin^2\frac{\theta}{2} + 2\varepsilon B + 2c \tag{6.7}$$

Let us introduce a new variable y=sin θ/2 :

$$\frac{dy}{dt} = \frac{1}{2}\cos\frac{\theta}{2}\frac{d\theta}{dt}$$

Equation (6.7) becomes :

$$\left(\frac{dy}{dt}\right)^2 = \frac{1}{4}\left(1-y^2\right)\left(2\varepsilon B + 2c - 4\varepsilon B y^2\right)$$

that can be solved in terms of elliptic integrals. The type of motion depends upon the ratio s^2 of the coefficients of the second factor: $s^2=\varepsilon B+c/2B$.

(i)- If $s^2>1$, the motion is periodic with a period equal to :

$$P = \frac{4}{\sqrt{\varepsilon B}}\int_0^1 \frac{dy}{\sqrt{(1-y^2)(1-s^2y^2)}}$$

The argument θ is oscillating between two limiting values ±2arc sin 1/s with the period P. Such a motion is called libration. Being periodic in time, θ can be expressed in Fourier series of the argument $\psi=2\pi(t-t_0)/P$.

(ii)- If $s^2<1$, all values of sin θ/2 are possible, so that the argument θ is indefinitely increasing (or decreasing). The motion is called circulating motion.

Another approach to this simplified resonance problem is to remark that the system of equation 6.5 is canonical. If one sets :

$$\phi = B\cos\theta - \frac{\varepsilon x^2}{2} ,$$

then

$$\frac{dx}{dt} = -B\sin\theta = \frac{\partial\phi}{\partial\theta}$$

$$\frac{d\theta}{dt} = \varepsilon x = -\frac{\partial\phi}{\partial x}$$

From the theory of Hamiltonian systems, $\phi=C$ is an integral. Let us plot the curves representing it for various values of C (figure 7) in both representations θ-x and θ-x^2. One can see that there are two special value of C_0 :

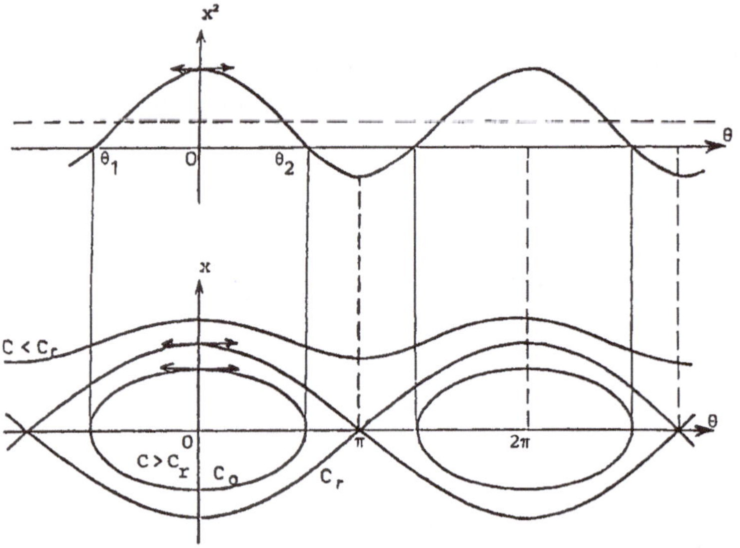

Figure 7

a) $C_0 = B$

In this case, θ can only stay equal to 0 or $2k\pi$ with x=0. It is a stable equilibrium solution. No motion is possible for C>0.

b) $C_r = -B$

The upper curve is tangent to the θ axis for $\theta=k\pi$. These are unstable equilibrium points, and θ may take all values between two such points. These solutions are called asymptotic; one can show that the equilibrium points are reached in an infinite time. It separates the two types of motion described above :

(i)- $C>C_r$; there is a libration around the equilibrium value. In this case, θ varies between two values $\pm\theta_1$ in function of time and can be expressed in terms of a Fourier series of a new time argument ψ called the libration argument. In deriving the complete solution for all the variables of the equations of motion, one will obtain expressions similar to those obtained in Section 4.3 the θ argument being replaced by the librating argument ψ.

(ii)- $C<C_r$; then the solution calls for the circulating case, θ is a circulating argument and has therefore a mean motion n_θ. The general solution is, in this case, the solution presented in section 4.3.

The general solution of equations 6.2 and 6.3 is of course more complicated and much more difficult to obtain. However, it has the same structure as presented here with both libration and circulation cases.

6.3. APPLICATIONS TO ARTIFICIAL SATELLITES

We have assumed, in the discussion of tesseral harmonics in Section 5.6 that the divisor 5.23 is not too small. However this case may happen, in particular for geosynchronous satellites for which the mean motion n_M is exactly equal to the angular velocity of the Earth rotational motion ν. Among the arguments arising in the equations one will find arguments of the type (see expression 5.22) :

$$\theta = j\omega + k(M+\Omega-T-\lambda_{nk})$$

$$n_\theta = jn_\omega + k(n_M+n_\Omega-\nu)$$

may be very close to zero, in particular for those terms in which j=0. Using 5.17 and 5.19, one has :

$$n_M + n_\Omega = n_0 + \frac{3}{2}\frac{n_0 a_e^2 J_2}{a_0^2(1-e_0^2)^2}\left[(1-\frac{3}{2}\sin^2 i_0)\sqrt{1-e_0^2} - \cos i_0\right]$$

The additional term is practically equal to zero for i_0 and e_0 small. It results that n_θ may be very small and the theory given in Section 6.2 applies. Indeed all the so-called geostationary satellites have a librational motion around some points above the Earth situated at longitudes which can be derived from the expression adopted for the geopotential.

Another important case of resonance in the motion of artificial satellites occurs for lower satellites. Among the terms produced by the tesseral harmonic J_{nk} as given by 5.22, the largest effect corresponds to h=1 since for larger h, the terms are factored by increasing powers of the eccentricity. The corresponding angular arguments are :

$$\theta = j\omega + hM + k(\Omega-T-\lambda_{nk}) \tag{6.8}$$

with a mean motion :

$$n_\theta = jn_\omega + hn_M + k(n_\Omega - \nu) \qquad (6.9)$$

which is of the order of hn_M-kν, if we neglect n_ω and n_Ω.

Noting that the motion of a real artificial satellite is necessarily smaller than 16 revolutions per day, let us consider for instance n=15. With h=1 and k=15, n_θ given by 6.9 may be small and even may desappear: a resonance occurs caused by harmonics J_{n15}. For some classes of higher satellites, J_{n14}, J_{n13}, J_{n12} ... also cause resonances. If the satellite is not exactly in a resonant condition due to J_{nk}, but close to it, the particular perturbing terms are greatly enhanced. This phenomenon has been widely used to determine tesseral harmonics of order 15, 14 and 13. Thanks to it, they are better determined than the neighbouring harmonics. Similarly, tesseral harmonics of order 30 and lower are in resonance with the mean anomaly for arguments in 6.8 with h=2. But by the fact that the coefficients are smaller, this method to determine than is less sensitive. Some enhancement of the perturbations has also been observed for third order resonances (h=3).

6.4. THE CRITICAL INCLINATION

We have seen, in Section 5.4 that a term in 2ω is created by the second order solution and (Section 5.5) by other zonal harmonics. The equation for the argument of the perigee takes the form :

$$\frac{d\omega}{dt} = \frac{3n\, a_e^2 J_2}{a^2(1-e^2)^2}\left(1-\frac{5}{4}\sin^2 i\right) + A\, J_2^2 \cos 2\omega + B\, J_3 \cos\omega + \ldots + C\, J_4 \cos 2\omega + \ldots$$

When i is very close to the critical inclination, the constant term becomes much smaller than J_2 and, although this is not physically the effect of a resonance, the equation has the structure described by equation 6.2. The analytical procedure used to find the solution in this case is consequently valid.

The result is that when the mean inclination is very close to the critical inclination, the perigee does no more circulate as described by equation 5.18, but librates around $\omega = \pm\pi/2$, with an amplitude that depends upon the initial conditions.

7 - NUMERICAL METHODS

The various methods presented in the previous chapters have in common the property that the solution is obtained in the form of analytical expressions depending on litteral parameters. Such an approach has the fundamental advantage that the origin of various features of the motion can be traced back and the actual reasons for them can be found and discussed. As an example, we could forecast, in Section 6.4, the behaviour of the motion of the perigee around the critical inclina tion. Similarly, we have shown in Section 4.3 that, in general, the variation of the elements can be represented by a sum of trigonometric functions periodic in time, and have specified the relations that exist between their periods and the secular terms of the angular elements.

However, this approach has a practical drawback: it involves, in general, a fantastic amount of computational effort even when a relatively modest accuracy objective is sought for. For this reason, more expeditive methods are usually applied in which parameters or litteral expressions are replaced by numbers. This approach indeed reduces the computational load, but at the expense of reducing drastically the set of orbits to which the expressions can be applied. If some parameters only are replaced by their numerical value and others are kept in a litteral form, the methods are called "semi-numerical". In some other approaches, no parameter is left numerically undefined: they are purely numerical methods. We shall give examples of both types.

7.1. SEMI-NUMERICAL GENERAL ANALYTICAL THEORIES

Semi-numerical theories are characterized by the fact that while the form of the solution is preserved, quantities such as the coefficients of different terms and the coefficients of time in the trigonometric arguments are numbers or analytic expressions of some of the parameters only.

The most common form is obtained replacing the coefficients of 4.12 and 4.13 by numbers, but the litteral expressions for the trigonometric arguments λj are kept. The solutions take the following form :

$$\left.\begin{aligned} L_i &= L_i^0 + \sum c_{j_1 j_2 \ldots j_n} \cos\left(j_1\lambda_1 + j_2\lambda_2 + \ldots + j_n\lambda_n\right) \\ l_i &= n_i t + l_i^0 + \sum s_{j_1 j_2 \ldots j_n} \sin\left(j_1\lambda_1 + j_2\lambda_2 + \ldots + j_n\lambda_n\right) \end{aligned}\right] \qquad (7.1)$$

where l_i $(i=1,2,3)$ represent the angular elements (Ω,ω,M) and the L_i the other three. L_i^o, l_i^o, n_i, c, s are numbers, while although the λ_i are linear functions of time with numerical parameters, they are treated litterally. Such theories are called "general semi-numerical".

To built such a semi-numerical theory, an iterative method starting in a manner analogous to the one described in Section 4.3 is used. One begins with an approximate solution that can be the two body problem solution with some preliminary values of the time expressions for the secular part of the angular variables and other trigonometric arguments. Then this solution is substituted into the right-hand members of equations 4.10. The results of this substitution are explicit trigonometric series of time and can therefore be readily integrated to obtain an improved solution in the required form.

This first step is generally not sufficient and one has to iterate the procedure, that is to substitute in the equations the solution just obtained. The difficulty is that the form 4.10 of the equations must be retained. This means that all the elements of the right hand members of the equations must be expressed in terms of L_i and l_i in an analytical form. One has to derive them from the analytical development of the disturbing function or from similar developments of the elements using the expressions presented in Section 2.3. Let us give an example.

In the theory of the motion of an artificial satellite under the J_2 force field, the disturbing function R_1 is given by 5.12. The equation in inclination as given in 3.5 is:

$$\frac{di}{dt} = \frac{-1}{na^2\sqrt{i-e^2}\sin i}\frac{\partial R_1}{\partial \Omega} + \frac{\cos i}{na^2\sqrt{1-e^2}\sin i}\frac{\partial R_1}{\partial \omega}$$

The first term desappears since R_1 does not depend on Ω and one is left with :

$$\frac{di}{dt} = \frac{-3}{2}\frac{\cos i}{na^2\sqrt{1-e^2}\sin i}\frac{\mu a_e^2 J_2}{a^3}\cdot\frac{a^3}{r^3}\sin^2 i\ \sin(2\omega+2v)$$

Using the equality $n^2a^3=\mu$, one finally obtains :

$$\frac{di}{dt} = \frac{-3}{4}\sqrt{\mu}\, a_e^2 J_2 \frac{a^{-7/2}}{\sqrt{1-e^2}}\sin 2i\left(\frac{a}{r}\right)^3\sin\left(2\omega+2v\right) \tag{7.2}$$

We can now apply the result of the expression 5.14, including the development of $1/\sqrt{1-e^2}$ and write, still in a litteral form :

$$\frac{di}{dt} = \frac{-3}{4}\sqrt{\mu a_e^2}\, J_2\, a^{-7/2}\sin 2i\sum B_j(e)\sin(2\omega+jM) = A \tag{7.3}$$

Let the result of the preceding iteration be expressed as :

$$\left.\begin{aligned} a &= a_0 + \Delta a \\ e &= e_0 + \Delta e \\ i &= i_0 + \Delta i \\ \omega &= \omega_0 + \Delta\omega \\ M &= M_0 + \Delta M \end{aligned}\right| \tag{7.4}$$

where a_0, e_0, i_0 are numerical constants, ω_0 and M_0 are linear functions of time with numerical coefficients, while Δa, Δe, Δi are cosine series of the λ_j arguments and $\Delta\omega$ and ΔM are sine series of the same arguments. These additions are, in the frame of the assumptions of perturbation theory, small quantities. It is therefore possible to develop 7.3 in power series of Δa, Δe, Δi, $\Delta\omega$ and ΔM. Retaining only the first order term, one gets, if we call :

$A_0 = A(a_0, e_0, i_0, \omega_0, M_0)$

$$\frac{di}{dt} = A_0 + \frac{\partial A_0}{\partial a_0}\Delta a + \frac{\partial A_0}{\partial e_0}\Delta e + \frac{\partial A_0}{\partial i_0}\Delta i + \frac{\partial A_0}{\partial \omega}\Delta \omega + \frac{\partial A_0}{\partial M}\Delta M \qquad (7.5)$$

In computing this expression, it is necessary to perform five multiplications between multi-trigonometric series with numerical coefficients - a basic operation that is easily programmed on present computers. The result of these multiplications is a series of the same type. So that the form of the right hand members is the same as described by 4.10 and can be integrated term by term. Actually the first part (A_0 in the right hand members) is the first order solution while the additional quantities produce after integration a new contribution to the solution. After this has been performed on all the unknowns, one may add Δa, Δe, ..., ΔM_1 to a_0, e_0...M_0 and proceed to a new iteration by a similar procedure. This procedure converges if the starting expressions are not too far from the solution. However, the concept of order introduced in analytical solutions does no more exist. The convergence must be checked numerically by testing that Δa, Δe,... decrease from one iteration to the next. One stops the procedure when two successive solutions appear to be sufficiently close one to another.

This example is an illustration of the general semi-numerical method. It has the great advantage that the operations on trigonometric series are very much simplified since, in comparison with analytical theories, operations on trigonometric series involve numbers instead of multiple litteral polynomials. The fact that not only the development of the right hand members of the equations are needed but also the development of their partial derivatives with respect to the unknowns is however an additional complication.

A more fundamental problem arises from the fact that the final result of the integration is a trajectory T_0 defined by a priori values of integration constants a_0, e_0 and i_0 (the other three constants, present in the arguments Ω_0, ω_0 and M_0 remain in litteral form and can take any values). This is in contrast with an analytical theory in which all parameters may vary at least within certain limits defining the validity of the perturbation technique used. In the case of a general semi-numerical method, a continuous set of trajectories may be obtained around T_0 by computing a mesh of trajectories defined by N sets of integration constants (a_k, e_k, i_k) k=1,N and interpolating between them. Another method is to express the solution in power series of a_k-a_0, e_k-e_0, i_k-i_0 around T_0. Each of the new solutions or each increment to T_0 are to be obtained by solving again from scratch the equation of motions. This complicated procedure is necessary in order to find the solution that satisfies for instance a certain number of observations. However, since the useful range of parameters is small, power series may be limited to order one or two.

7.2. SECULAR SEMI-NUMERICAL THEORIES

In some types of problems, the periods of some trigonometric arguments are much larger than the period during which a description of the orbit is desired. This is for instance the case of the motion of planets. The periods of the nodes and of the perihelia of planets are of the order of several tens or hundreds thousand years. In practice, one needs to know their orbits only for a few hundred years during which the arguments Ω and ω change only by a few degrees, so that one may write them as :

$$\bar{\Omega} = \Omega_0 + n_\Omega t$$

$$\bar{\omega} = \omega_0 + n_\omega t$$

where ω_0 and Ω_0 are values for t=0 and $n_\Omega t$+ or $n_\omega t$ are small angles. A general term of the theory is of the form :

$$\begin{matrix}\sin\\ \cos\end{matrix}\left(\sum j_k\lambda_k + h\Omega_0 + hn_\Omega t + l\omega_0 + ln_\omega t\right) \tag{7.6}$$

It can be developed in power series of t :

$$\begin{matrix}\sin\\ \cos\end{matrix}\left(\sum i_k\lambda_k + h\Omega_0 + l\omega_0\right) + \left(hn_\Omega + ln_\omega\right) t \,.\, \begin{matrix}\cos\\ -\sin\end{matrix}\left(\sum j_k\lambda_k + h\Omega_0 + l\omega_0 + \frac{\pi}{2}\right) + \ldots \tag{7.7}$$

The principle of secular theories is to accept time outside the trigonometric expressions; the terms of such series are called Poisson terms.

Because of the presence of constants in the arguments of 7.7, the strict separation between sine and cosine series desappears and the solution for all the six elements takes the form :

$$\sum\left(b_0+b_1t+b_2t^2+\ldots\right)\cos\left(\sum j_k\lambda_k\right) + \sum\left(c_0+c_1t+c_2t^2+\ldots\right)\sin\left(\sum j_k\lambda_k\right) \tag{7.8}$$

Having this form in mind, the semi-numerical approach to solve the equations of motion described in Section 7.1 can be applied to the present case. The advantage is that there are less arguments in the trigonometric series so that the total number of terms is drastically reduced. The disadvange is, of course, that the interval of validity of the solution is reduced to the interval of validity of the developments in time limited to a certain power.

7.3. NUMERICAL INTEGRATION

In many cases, even the construction of a semi-numerical theory is too difficult or time-consuming or is unsufficiently accurate for the needs. It is then necessary to adopt a completely different approach to solve the equations of motion. Let us write the equations in the following very general form :

$$\frac{d\mathbf{X}}{dt} = \mathbf{F}(\mathbf{X},t) \tag{7.9}$$

where $\mathbf{X}$ and $\mathbf{F}$ are vectors with N components. There are essentially two classes of numerical methods that are used to solve 7.9.

(i)- If we assume that $\mathbf{X}(t)$ is known for a given time t_0, there exist algorithms that permit to compute $\mathbf{X}(t_0+\Delta t)$ by approximating Taylor developments. Such methods are called single-step numerical methods.

(ii)- If, in addition to $\mathbf{X}(t_0)$ one knows the values of $\mathbf{X}$ for $t_0-\Delta t$, $t_0-2\Delta t$, etc..., very efficient multi-step numerical methods can be used.

7.4. SINGLE STEP METHODS

The simplest single step method is the development of $\mathbf{X}$ in Taylor series. This implies that one knows not only $d\mathbf{X}/dt$ but also its time derivatives. Since it is very rarely the case, several algorithms that somewhat imitate Taylor series developments without computing derivatives exist. They are known as various versions of Runge-Kutta methods.

The most commonly used is the fourth order symmetric Runge-Kutta algorithm. Let us define the vector $\mathbf{Y}$ of dimension N+1 including $x_1 x_2 \ldots x_N$ (vector $\mathbf{X}$) and $x_{N+1}=t$ so that $dx_{N+1}/dt=1$.

The symmetric differential equations are :

$$\frac{d\mathbf{Y}}{dt} = \mathbf{G}(\mathbf{X},t) = G(\mathbf{Y}) \tag{7.10}$$

The algorithm is :

$$\mathbf{Y}(t+\Delta t) = \mathbf{Y}(t) + \frac{\Delta t}{6}\left(\mathbf{F}_1 + 2\mathbf{F}_2 + 2\mathbf{F}_3 + \mathbf{F}_4\right) \tag{7.11}$$

with :

$$\left.\begin{aligned} \mathbf{F}_1 &= \mathbf{G}(\mathbf{Y}) \\ \mathbf{F}_2 &= \mathbf{G}(\mathbf{Y} + \frac{\mathbf{F}_1\,\Delta t}{2}) \\ \mathbf{F}_3 &= \mathbf{G}(\mathbf{Y} + \frac{\mathbf{F}_2\,\Delta t}{2}) \\ \mathbf{F}_4 &= \mathbf{G}(\mathbf{Y} + \mathbf{F}_3\,\Delta t) \end{aligned}\right] \tag{7.12}$$

The accuracy of this algorithm is similar to that of a fourth order Taylor development, the residual error being of the order of $(\Delta t)^5$. Actually, there are many other linear expressions that have approximatively the same accuracy, but they have more terms or more complicated coefficients.

7.5. MULTISTEP METHODS

This class of integration methods is based on approximate relations that exist between the values of a function at times separated by identical intervals. Let us consider a series of values of a function f(t) :

$$\ldots f(t-5\Delta t),\ f(t-4\Delta t) \ldots f(t) \ldots f(t+4\Delta t),\ f(t+5\Delta t) \ldots$$

that we shall call :

$$\ldots f_{-5},\ f_{-4} \ldots f_0 \ldots f_4,\ f_5 \ldots$$

Let us compute the first differences :

$$f^1_{n+\frac{1}{2}} = f_{n+1} - f_n$$

and so on. The 2k-th difference is :

$$f^{2k}_n = f^{2k-1}_{n+\frac{1}{2}} - f^{2k-1}_{n-\frac{1}{2}}$$

From these definitions, one can compute the differences in terms of successive values of the function. One obtains :

$$\left.\begin{aligned} f^{2k}_n &= \sum_{i=0}^{2k} \frac{(-1)^i (2k)!}{i!(2k-i)!} f_{n-k+i} \\ f^{2k+1}_{n+\frac{1}{2}} &= \sum_{i=0}^{2k+1} \frac{(-1)^{i+1}(2k+1)!}{i!(2k+1-i)!} f_{n-k+i} \end{aligned}\right] \qquad (7.13)$$

For continuity reasons, if Δt is sufficiently small, the differences decrease rapidly and can be neglected starting from some order 2k or 2k+1. In this case, if we set them equal to zero, this is equivalent to say that there exist linear relations between 2k or 2k+1 successive values of the functions, given by equating to zero the right-hand members of 7.13.

Differences are used to interpolate a function. For instance, using Bessel interpolation formula, one has, for $0 \le n \le 1$:

$$f(t+n\Delta t) = f_0 + n f^1_{1/2} + \frac{n(n-1)}{4}(f^2_0 + f^2_1) + \ldots \qquad (7.14)$$

Suppose one wishes to compute the integral of f between t_0 and $t_1 = t_0 + \Delta t$:

$$\int_{t_0}^{t_1} f(t)dt = \int_{t_0}^{t_0+\Delta t} f(t_0+n\Delta t)d(n\Delta t) = \int_0^1 f(t_0+n\Delta t)\Delta t dn$$

Hence :

$$\int_{t_0}^{t_0+\Delta t=t_1} f(t)dt = \Delta t \int_0^1 f(n)$$

Let us perform the second integral on the developed form 7.14. One gets :

$$\int_{t_0}^{t_1} f(t)dt = \Delta t\left[nf_0 + \frac{n^2}{2}f^1_{1/2} + \frac{2n^3-3n^2}{24}(f^2_0+f^2_1) + \ldots\right]_0^1$$

$$= \Delta t\left[f_0 + \frac{1}{2}f^1_{1/2} - \frac{1}{24}(f^2_0+f^2_1) + \ldots\right]$$

$$\int_{t_0}^{t_1} f(t)dt = \frac{\Delta t}{2}\left[f_0+f_1 - \frac{1}{12}(f^2_0+f^2_1) + \ldots\right] \qquad (7.15)$$

This formula shows that the integral of the function may be expressed in terms of the values of the function at equal intervals. Generally, one would obtain an expression of the following type :

$$\int_{t_0}^{t_1} f(t)dt = \Delta t \sum_{i=-k}^{+k} A_i f_i \qquad (7.16)$$

But if we admit that differences of order 2k are negligible, one can use expression 7.13 equated to zero with various values of n and successively eliminate f_k, f_{k-1}, etc..., so that 7.16 becomes either :

$$\int_{t_0}^{t_1} f(t)dt = \Delta t \sum_{i=-2k}^{0} B_i f_i \qquad (7.17)$$

or

$$\int_{t_0}^{t_1} f(t)dt = \Delta t \sum_{i=-2k+1}^{1} C_i f_i \qquad (7.18)$$

So, if one has the differential equation :

$$\frac{dx}{dt} = f(t)$$

and is x_0 is the value for $t=t_0$, then :

$$x(t_0+\Delta t) = x_0 + \int_{t_0}^{t_1} f(t)dt$$

can be expressed by one of the expressions 7.17 or 7.18, provided one knows sue cessive values of $f(t-k\Delta t)$, k=1,2,3....

Let us also remark that, for linearity reasons, these expressions can be written for all the components of a vector $\mathbf{F}$. In particular, 7.17 and 7.18 are generalized by:

$$\int_{t_0}^{t_1} \mathbf{F}(t)dt = \Delta t \sum_{i=-2k}^{0} B_i \mathbf{F}_i \tag{7.19}$$

or :

$$\int_{t_0}^{t_1} \mathbf{F}(t)dt) = \Delta t \sum_{i=-2k+1}^{1} C_i \mathbf{F}_i \tag{7.20}$$

so that a set of equations, like Lagrange equations, can be integrated step by step.

Repeating the proof with a double integration of 7.14, one also gets similar expressions for a double integration over time, allowing to integrate a set of second order differential equations like the equations of motion in rectangular coordinates.

There are many different forms of equations 7.17 or 7.18, depending on the order of differences that are neglected or what kind of interpolation formula one has derived them from. Let us give, as an example, the Adams-Moulton formulae to the fourth order, comparable in accuracy to the Runge-Kutta algorithm presented in 7.12 :

$$\mathbf{X}(t_0+\Delta t) = \mathbf{X}(t_0)+ \frac{\Delta t}{24}\left[55\mathbf{F}(\mathbf{X}(t_0))-59\mathbf{F}(\mathbf{X}(t_0-\Delta t)) + 37\mathbf{F}(\mathbf{X}(t_0-2\Delta t))-9\mathbf{F}(\mathbf{X}(t_0-3\Delta t))\right] \tag{7.21}$$

$$\mathbf{X}(t_0+\Delta t) = \mathbf{X}(t_0)+ \frac{\Delta t}{24}\left[9\mathbf{F}(\mathbf{X}(t_0+\Delta t))+19\mathbf{F}(\mathbf{X}(t_0)) - 5\mathbf{F}(\mathbf{X}(t-\Delta t))+\mathbf{F}(\mathbf{X}(t-2\Delta t))\right] \tag{7.22}$$

The procedure in using these equations or more generally 7.19 and 7.20 in order to integrate the differential equation :

$$\frac{d\mathbf{X}}{dt} = \mathbf{F}(\mathbf{X})$$

is the following :

(i)- Determine by some single step method the four successive values of $\mathbf{X}$ for $t_0-3\Delta t$, $t_0-2\Delta t$, $t_0-\Delta t$ and t_0 (or the 2k+1 values for $t_0-2k\Delta t$ to t_0).

(ii)- Compute $\mathbf{X}(t_0+\Delta t)$ using 7.21 or 7.19. These formulae are called "predictors", since they permit a first extrapolation of the vector $\mathbf{X}$

(iii)- Insert the predicted value $\mathbf{X}(t_0+\Delta t)$ just determined in the $\mathbf{F}_1=\mathbf{F}(\mathbf{X}(t_0+\Delta t))$ term of the right-hand member of 7.22 or 7.20, one gets an improved value of $\mathbf{X}(t_0+\Delta t)$. These formulae are called "correctors". Several iterations of the corrector algorithm may be useful.

(iv)- Return to step (ii) with the initial time value $t_1=t_0+\Delta t$, and so on ... The integration proceeds by steps of Δt. Procedures to modify the step (multiply or divide by 2 or another coefficient) may be used when necessary.

Similar procedures exist with predictor-corrector expressions for second order differential equations. Let us note that the choice of the best predictor and corrector formula for a given problem is a delicate task, because some of them may be more or less unstable.

7.6. DISCUSSION OF NUMERICAL INTEGRATION METHODS

Numerical integration methods present with respect to semi-numerical methods the same type of advantages and drawbacks as the latter as compared to analytical methods.

If the numerical integration algorithms are very simple and applicable to many different types of equations, the result of a calculation is a single orbit which does not generally fit to the physical conditions or observations. In addition, there is no hint about the properties of the orbit. Finally, the description ends at the time for which the integration stopped, while analytical and semi-numerical methods provide a slowly degrading representation for longer periods of time.

The main objective being to define an orbit that fits observations and often also to improve some of the parameters of the motion, it is necessary to built variational equations such as 7.5 for all parameters that are not perfectly known numerically: the initial conditions x_j, physical parameters p_i, etc... This means that one has to know the values of all the partial derivatives of all the variables with respect to any of the parameters throughout the complete interval of integration. Each of them has to be computed by numerical integration.

So, in addition to integrate :

$$\frac{d\mathbf{X}}{dt} = \mathbf{F}(\mathbf{X}, p_i)$$

where the p_i are the parameters, one has to integrate :

$$\frac{d}{dt}\left(\frac{\partial \mathbf{X}}{\partial p_i}\right) = \frac{\partial \mathbf{F}}{\partial p_i}(\mathbf{X}, p_i) \qquad 1 \leq i \leq n \tag{7.23}$$

One can either write explicitely $\partial\mathbf{F}/\partial p_i$ and integrate 7.23 or to integrate :

$$\frac{d\mathbf{X}}{dt} = \mathbf{F}(\mathbf{X},\ p_i + \Delta p_i)$$

and if $\mathbf{X}_j(t)$ is the result of this integration, deduce the approximate value :

$$\frac{\partial \mathbf{X}}{\partial p_i} = \left(\mathbf{X}_j(t) - \mathbf{X}(t)\right) / \Delta p_i \tag{7.24}$$

The partial derivatives with respect to a constant of integration - in the form of initial conditions of the integration algorithm - can only be computed through the procedure illustrated by 7.24.

In order to mitigate the drawbacks of semi-numerical and numerical methods, several authors have devised hybrid methods in which short period terms are first eliminated from the equations by an analytical procedure, then the remaining equations with longer periods are integrated numerically. This permits to obtain the general trend of the orbit over a very large interval of time with a much larger step than if the short periods were included. However, to eliminate short period terms is a job equivalent to the construction of an analytical secular theory in terms of work involved, so that in practice, these "averaging" methods have been applied only to simple problems.

GENERAL BIBLIOGRAPHY

Brouwer, D. and Clemence, G.M., 1961, "Methods of Celestial Mechanics", Academic Press, New-York.

Danby, J.M.A, 1970, "Fundamentals of Celestial Mechanics", The Macmillan Co. New-York.

Geyling, F.T. and Westerman, H.R., 1971, "Introduction to Orbital Mechanics", Addison-Wesley Publ.Co., London.

Kaula, W.M., 1966, "Theory of Satellite Geodesy", Blaisdell Co., Waltham (Maryland, USA).

Kovalevsky, J., 1967, "Introduction to Celestial Mechanics", D. Reidel Publ. Co., Dordrecht.

Levallois, J.J. and Kovalevsky, J., 1971, "Géodésie générale", Vol. 4, Ed. Eyrolles, Paris.

Milani, A., Nobili, A.M. and Farinella, P., 1987, "Non-gravitational Perturbations and Satellite Geodesy", Adam Hilger Ltd., Bristol.

Morando, B., 1970, "Mouvement d'un satellite artificiel de la Terre", Gordon and Breach, Paris.

Roy, A.E., 1978, "Orbital Motion", Adam Hilger Ltd., Bristol.

FOUR LECTURES ON SPECIAL AND GENERAL RELATIVITY

Erik W. Grafarend
Department of Geodetic Science, Stuttgart University
Keplerstr. 11, D-7000 Stuttgart,
Federal Republic of Germany

Lecture I

Flat spacetime, pseudo-Euclidean space, the Lorentz transformation

Summary

According to the fundamental postulate of special relativity, pseudo-orthonormal tetrads are introduced which span pseudo-Euclidean spacetime in inertial or Galilean frames of reference. The proper six-parameter *Lorentz transformation* between two pseudo-orthonormal *tetrads of Minkowski type* is derived and physically interpreted with respect to the velocity vector between two moving Minkowski tetrads. Six detailed *examples* are presented referring to simultaneity, length contraction, time dilatation, anholonomic proper time and proper length and accelerated systems of reference. Two *exercises* are left open for the reader referring to the Lorentz transformation in spinor notation and the Sagnac effect.

1. Flat spacetime, pseudo-Euclidean space, the Lorentz transformation

According to *Figure 1* we consider a *light flash* departing at the *origin* $O(t=0) = O'(t'=0)$ of *two inertial or Galilean frames* which are moving in a fixed direction by velocity $\underset{\sim}{v}$ or $\underset{\sim}{v}'$ with respect to each other. Then the spherical light front travels in vacuo according to the *fundamental postulate of special relativity* independent of the inertial frame of reference and of the direction such that for the velocity of light

$$c^2 = \frac{\Delta x^2 + \Delta y^2 + \Delta z^2}{\Delta t^2} \quad , \quad c'^2 = \frac{\Delta x'^2 + \Delta y'^2 + \Delta z'^2}{\Delta t'^2} \quad , \; c = c' \qquad 1(1)$$

holds. The $\{x,y,z,ct\}$- or $\{x',y',z',ct'\}$-coordinates are measured along the two *pseudo-orthonormal* or *Minkowski tetrads* $e = \{\underset{\sim}{e}_1, \underset{\sim}{e}_2, \underset{\sim}{e}_3, \underset{\sim}{e}_4\}$ and $e' = \{\underset{\sim}{e}_{1'}, \underset{\sim}{e}_{2'}, \underset{\sim}{e}_{3'}, \underset{\sim}{e}_{4'}\}$ characterized by the pseudo-scalar products

$$\begin{aligned} g(\underset{\sim}{e}_i, \underset{\sim}{e}_j) &= \langle \underset{\sim}{e}_i, \underset{\sim}{e}_j \rangle = \delta_{ij} = \begin{cases} 1 & \text{if } i = j = 1,2,3 \\ 0 & \text{otherwise} \end{cases} \\ g(\underset{\sim}{e}_4, \underset{\sim}{e}_j) &= \langle \underset{\sim}{e}_4, \underset{\sim}{e}_j \rangle = 0, g(\underset{\sim}{e}_4, \underset{\sim}{e}_4) = \langle \underset{\sim}{e}_4, \underset{\sim}{e}_4 \rangle = -1 \end{aligned} \qquad 1(2)$$

or

$$g(\underset{\sim}{e}_\mu,\underset{\sim}{e}_\nu) = \langle\underset{\sim}{e}_\mu,\underset{\sim}{e}_\nu\rangle = \delta^-_{\mu\nu} = \begin{bmatrix} +1 & \text{if } \mu = i,\ \nu = j \\ 0 & \text{if } \mu \neq \nu \\ -1 & \text{if } \mu = 4,\ \nu = 4 \end{bmatrix} \qquad 1(3)$$

The *Minkowski tetrads* span a four-dimensional *pseudo-Euclidean vector space* V^4 in which we represent a placement vector $\underset{\sim}{x},\underset{\sim}{x}'$, respectively by

$$\underset{\sim}{x} = x^1\underset{\sim}{e}_1 + x^2\underset{\sim}{e}_2 + x^3\underset{\sim}{e}_3 + x^4\underset{\sim}{e}_4\ , \quad \underset{\sim}{x}' = x^{1'}\underset{\sim}{e}_{1'} + x^{2'}\underset{\sim}{e}_{2'} + x^{3'}\underset{\sim}{e}_{3'} + x^{4'}\underset{\sim}{e}_{4'} \qquad 1(4)$$

$\underset{\sim}{x} \in V^4$, $\underset{\sim}{x}' \in V^4$ which we normalize by 1(3)

$$\begin{aligned} q(\underset{\sim}{x}) &= \underline{x}^*I_-\underline{x},\ q(\underset{\sim}{x}) = g_{\mu\nu}x^\mu x^\nu,\ q(\underset{\sim}{x}) = (x^1)^2 + (x^2)^2 + (x^3)^2 - (x^4)^2 \\ q(\underset{\sim}{x}') &= \underline{x}'^*I_-\underline{x}',q(\underset{\sim}{x}') = g_{\mu'\nu'}x^{\mu'}x^{\nu'},q(\underset{\sim}{x}') = (x^{1'})^2 + (x^{2'})^2 + (x^{3'})^2 - (x^{4'})^2 \end{aligned} \qquad 1(5)$$

where $\underline{x}$ is the 4×1 column array of coordinates $\{x^1,x^2,x^3,x^4\}$, $\underline{x}^*$ its *transpose*, I_- or $g_{\mu\nu} = \delta^-_{\mu\nu}$ the *pseudo-identity matrix*

$$I_- = \begin{bmatrix} 1 & 0 & 0 & 0 \\ 0 & 1 & 0 & 0 \\ 0 & 0 & 1 & 0 \\ 0 & 0 & 0 & -1 \end{bmatrix} \sim g_{\mu\nu} = \delta^-_{\mu\nu} = \begin{bmatrix} 1 & 0 & 0 & 0 \\ 0 & 1 & 0 & 0 \\ 0 & 0 & 1 & 0 \\ 0 & 0 & 0 & -1 \end{bmatrix} . \qquad 1(6)$$

Of fundamental importance are those transformations which in the sense of an *isometry* leave the quadratic form

$$\Delta x^2 + \Delta y^2 + \Delta z^2 - c^2\Delta t^2 = \Delta x'^2 + \Delta y'^2 + \Delta z'^2 - c^2\Delta t'^2 \qquad 1(7)$$

invariant. Those *isometric linear operators* A on a vector space V^4 which leave

$$g(x,x) = g(x',x') = g(Ax,Ax) \qquad 1(8)$$

invariant build up the *Lorentz group*. The invariance postulate is equivalent to

$$I_- = A^*I_-A \sim g_{\mu\nu} = a^{\mu'}_\mu\ a^{\nu'}_\nu\ g_{\mu'\nu'} \qquad 1(9)$$

where $A \sim a^{\mu'}_\mu$ is referred to as the *Lorentz matrix*.
Due to $\det I_- = \det A^* \det I_- \det A = (\det A)^2 \det I_-$, $\det A = \pm 1$ we select the *proper Lorentz transformation* by $\det A = +1$ and write explicitly the *pseudo-orthonormality conditions* 1(9)

$$\sum_{i'=1'}^{3'} a^{i'}_\mu\ a^{i'}_\nu - a^{4'}_\mu\ a^{4'}_\nu = \delta^-_{\mu\nu} = \begin{bmatrix} 1 & \text{if } \mu = \nu = 1,2,3 \\ 0 & \text{if } \mu \neq \nu \\ -1 & \text{if } \mu = \nu = 4 \end{bmatrix} \qquad 1(10)$$

So far we have partially experienced a four-dimensional vector space V^4 of vectors $\underset{\sim}{x}$ which we call

(i) *isotropic* or *lightlike* or *"null"* if $q(\underset{\sim}{x}) = 0$
(ii) *spacelike* if $q(\underset{\sim}{x}) > 0$
(iii) *timelike* if $q(\underset{\sim}{x}) < 0$.

In order to guarantee that positive, timelike vectors are transformed into positive, timelike vectors we have to postulate $a_4^{4'} > 0$ generating a special *Lorentz matrix* which we refer to as *orthochronous*. There are basically *two fundamental* special Lorentz transformations which we shall discuss now.

Case one: $\underset{\sim}{e}_4$ fixed

$$a_1^{4'} = a_2^{4'} = a_3^{4'} = 0 \quad \text{versus} \quad a_4^{4'} = 1$$

$$a_j^{i'} a_i^{j'} = \begin{bmatrix} 1 & \text{if } i = j = 1,2,3 \\ 0 & \text{if } i \neq j \end{bmatrix} \qquad 1(11)$$

Obviously 1(11) generates an orthogonal transformation we can represent by the unitary (orthogonal) matrix U, namely

$$A = \begin{bmatrix} U & 0 \\ 0 & 1 \end{bmatrix} \qquad 1(12)$$

Corollary: If A is a special Lorentz operator which leaves $\underset{\sim}{e}_4$ fixed, then the $\{\underset{\sim}{e}_{1'}, \underset{\sim}{e}_{2'}, \underset{\sim}{e}_{3'}\}$-subspace is formed by a *proper rotation*.

There are *three subgroups:*

U_{14}:

$$U_{14} = \begin{bmatrix} 1 & 0 & 0 & 0 \\ 0 & \cos\alpha & \sin\alpha & 0 \\ 0 & -\sin\alpha & \cos\alpha & 0 \\ 0 & 0 & 0 & 1 \end{bmatrix} \qquad 1(13)$$

U_{24}:

$$U_{24} = \begin{bmatrix} \cos\alpha & 0 & \sin\alpha & 0 \\ 0 & 1 & 0 & 0 \\ -\sin\alpha & 0 & \cos\alpha & 0 \\ 0 & 0 & 0 & 1 \end{bmatrix} \qquad 1(14)$$

U_{34}:

$$U_{34} = \begin{bmatrix} \cos\alpha & \sin\alpha & 0 & 0 \\ -\sin\alpha & \cos\alpha & 0 & 0 \\ 0 & 0 & 1 & 0 \\ 0 & 0 & 0 & 1 \end{bmatrix} \qquad 1(15)$$

Case two: $\underset{\sim}{e}_1$ and $\underset{\sim}{e}_2$ fixed
or $\underset{\sim}{e}_2$ and $\underset{\sim}{e}_3$ fixed
or $\underset{\sim}{e}_3$ and $\underset{\sim}{e}_1$ fixed

Example: $\underset{\sim}{e}_1$ and $\underset{\sim}{e}_2$ fixed

$$A = \begin{bmatrix} 1 & 0 & 0 & 0 \\ 0 & 1 & 0 & 0 \\ 0 & 0 & a_3^{3'} & a_4^{3'} \\ 0 & 0 & a_3^{4'} & a_4^{4'} \end{bmatrix} \qquad 1(16)$$

$$a_3^{3'} a_4^{3'} - a_3^{4'} a_4^{4'} = 0 \; ; \; (a_3^{3'})^2 - (a_3^{4'})^2 = 1 \; , \; (a_4^{3'})^2 - (a_4^{4'})^2 = -1 \qquad 1(17)$$

Obviously 1(17) generates a pseudo-orthogonal transformation we call a *proper hyperbolic rotation* since

$$a_3^{3'} = \cosh\alpha \quad , \; a_4^{3'} = a_3^{4'} = \sinh\alpha \quad , \; a_4^{4'} = \cosh\alpha \qquad 1(18)$$

satisfies the pseudo-orthonormality conditions 1(17).

Corollary: If A is a special Lorentz operator which leaves $\underset{\sim}{e}_1$ and $\underset{\sim}{e}_2$ fixed, then the $\{\underset{\sim}{e}_{3'},\underset{\sim}{e}_{4'}\}$ subspace is formed by a *proper hyperbolic rotation*.

In general, there are *three subgroups:*

A_{12}:

$$A_{12} = \begin{bmatrix} 1 & 0 & 0 & 0 \\ 0 & 1 & 0 & 0 \\ 0 & 0 & \cosh\alpha & \sinh\alpha \\ 0 & 0 & \sinh\alpha & \cosh\alpha \end{bmatrix} \qquad 1(19)$$

A_{13}:

$$A_{13} = \begin{bmatrix} 1 & 0 & 0 & 0 \\ 0 & \cosh\alpha & 0 & \sinh\alpha \\ 0 & 0 & 1 & 0 \\ 0 & \sinh\alpha & 0 & \cosh\alpha \end{bmatrix} \qquad 1(20)$$

A_{23}:

$$A_{23} = \begin{bmatrix} \cosh\alpha & 0 & 0 & \sinh\alpha \\ 0 & 1 & 0 & 0 \\ 0 & 0 & 1 & 0 \\ \sinh\alpha & 0 & 0 & \cosh\alpha \end{bmatrix} \qquad 1(21)$$

Summary:

A Lorentz matrix $A \sim a^{\mu'}_{\mu}$ *is a special Lorentz matrix if and only if* $\det A = +1$ *and* $A^{4'}_{4} > 0$. *The Lorentz group consists of 4×4 matrices* $A \sim a^{\mu'}_{\mu}$ *which fulfill*

$$\sum_{i'=1'}^{3'} a^{i'}_{\mu} a^{i'}_{\nu} - a^{4'}_{\mu} a^{4'}_{\nu} = \begin{cases} 1 & \text{if } \mu = \nu = 1,2,3 \\ 0 & \text{if } \mu \neq \nu \\ -1 & \text{if } \mu = \nu = 4 \end{cases} \qquad 1(22)$$

$$\det A = +1 \sim |a^{\mu'}_{\mu}| = +1 \quad \text{and} \quad a^{4'}_{4} > 0 \, . \qquad 1(23)$$

Any special Lorentz operator is the product of twodimensional and hyperbolic rotations being in general represented by

$$A = A_{12}(\alpha_6)A_{13}(\alpha_5)U_{14}(\alpha_4)A_{23}(\alpha_3)U_{24}(\alpha_2)U_{34}(\alpha_1) \qquad 1(24)$$

where six parameters $\alpha_1,\ldots,\alpha_6$ *appear.* $U_{i4}(\alpha)$, $i = 1,2,3$ *are proper orthogonal matrices,* $A_{ij}(\alpha)$ $(i \neq j;\ i,j = 1,2,3)$ *twodimensional hyperbolic rotation matrices.*

Finally we give a *physical representation* of the Lorentz matrix A. Let us refer according to *Figure 2* to *two inertial or Galilean frames* which are moving along the x-axis by velocity v or v' = -v with respect to each other. Once we pass over to the *differential Lorentz transformation*

$$dx^{\mu'} = a^{\mu'}_{\mu}\, dx^{\mu} \quad \text{if} \quad a^{\mu'}_{\mu} = \frac{\partial x^{\mu'}}{\partial x^{\mu}} \qquad 1(25)$$

we shall introduce the *velocity coordinates*

$$\begin{aligned} \frac{dx^{1'}}{dx^{4'}} &= \frac{dx^{1'}}{c\,dt'} = \frac{v'}{c} \Rightarrow dx^{1'} = \frac{v'}{c}\, dx^{4'}, \quad dx^{1} = 0 : x^{1} = \text{const. fixed} \\ \frac{dx^{1}}{dx^{4}} &= \frac{dx^{1}}{c\,dt} = \frac{v}{c} \Rightarrow dx^{1} = \frac{v}{c}\, dx^{4}, \quad dx^{1'} = 0 : x^{1'} = \text{const. fixed} \end{aligned} \qquad 1(26)$$

within 1(25):

$$dx^{1'} = a^{1'}_{4} dx^{4} = \frac{v'}{c}\, dx^{4'} \qquad 1(27)$$

$$dx^{4'} = a^{4'}_{4} dx^{4} \qquad 1(28)$$

$$\Rightarrow \quad \frac{dx^{1'}}{dx^{4'}} = \frac{a^{1'}_{4}}{a^{4'}_{4}} = \frac{v'}{c} = -\frac{v}{c}$$

$$a^{1'}_{4} + \frac{v}{c}\, a^{4'}_{4} = 0 \qquad 1(29)$$

In addition to 1(29) there are three conditions of pseudo-orthonormality:

$$(a^{1'}_{1})^2 - (a^{4'}_{1})^2 = 1, \quad (a^{1'}_{4})^2 - (a^{4'}_{4})^2 = -1, \quad a^{1'}_{1} a^{1'}_{4} - a^{4'}_{1} a^{4'}_{4} = 0 \qquad 1(30)$$

Only here we denote the unknown coefficients

$$a_1^{1'} = x_1 \ , \quad a_4^{1'} = x_2 \ , \quad a_1^{4'} = x_3 \ , \quad a_4^{4'} = x_4 \qquad 1(31)$$

in order to summarize the four equations with four unknowns $x_1,\ldots,x_4$.

$$\boxed{\begin{aligned} & x_2 + \frac{v}{c}\, x_4 = 0 \\ & x_1^2 - x_3^2 = +1 \\ & x_2^2 - x_4^2 = -1 \\ & x_1 x_2 - x_3 x_4 = 0 \end{aligned}} \qquad 1(32)$$

The *Ansatz*

$$\boxed{x_1 = x_4 = \frac{1}{\sqrt{1-v^2/c^2}} \ , \quad x_2 = x_3 = -\frac{v}{c}\,\frac{1}{\sqrt{1-v^2/c^2}}} \qquad 1(33)$$

fulfills 1(32), thus leading us to the *special Lorentz matrix*

$$\boxed{A = \begin{bmatrix} \frac{1}{\sqrt{1-v^2/c^2}} & 0 & 0 & -\frac{v}{c}\,\frac{1}{\sqrt{1-v^2/c^2}} \\ 0 & 1 & 0 & 0 \\ 0 & 0 & 1 & 0 \\ -\frac{v}{c}\,\frac{1}{\sqrt{1-v^2/c^2}} & 0 & 0 & \frac{1}{\sqrt{1-v^2/c^2}} \end{bmatrix}} \qquad 1(34)$$

The *general Lorentz transformation*, sometimes called the representation of the *Poincaré group*, solving 1(25) is the *inhomogeneous transformation*

$$\boxed{x^{\mu'} = a_{\mu}^{\mu'} x^{\mu} + a^{\mu'} \sim \underline{x}' = A\underline{x} + \underline{a} \ , \quad a^{\mu'} = x_o^{\mu'} \sim \underline{a} = \underline{x}_o} \qquad 1(35)$$

containing 6 + 4 = 10 parameters, which according to 1(34) reads for the *gauge* $a^{\mu'} = 0 \sim \underline{a} = 0$

$$\boxed{x' = \frac{x-vt}{\sqrt{1-v^2/c^2}} \ , \quad t' = \frac{t - \frac{v}{c^2}x}{\sqrt{1-v^2/c^2}} \ , \quad y' = y \ , \ z = z'.} \qquad 1(36)$$

When we come back to the general relative motion of two *inertial or Galilean frames* of Figure 1 the *proper Lorentz transformation* is generalized into

$$\boxed{\begin{aligned} & x^{a'} = r_a^{a'} \{x^a + v^a [\frac{v_b x^b}{\underset{\sim}{v}^2} (\frac{1}{\sqrt{1-\underset{\sim}{v}^2/c^2}} - 1) - \frac{t}{\sqrt{1-\underset{\sim}{v}^2/c^2}}]\} \\ & t' = \frac{t - v_b x^b/c^2}{\sqrt{1-\underset{\sim}{v}^2/c^2}} \end{aligned}} \qquad 1(37)$$

where $\underset{\sim}{v}^2 = v_a v^a = g_{ab}v^a v^b = \delta_{ab}v^a v^b$ and $v_b x^b = g_{ab}v^a x^b = \delta_{ab}v^a x^b$ and $r_a^{a'}$ denotes the proper rotation matrix of three parameters, which, in general, describes the orientation of *Minkowski tetrads* for fixed $\underset{\sim}{e}_4 = \underset{\sim}{e}_{4'}$ at initial epoch. In case that they are parallel, then $r_a^{a'} = \delta_a^{a'}$ holds.

At the end of our short review of flat spacetime, pseudo-Euclidean space and the Lorentz transformation we should give some references. The Lorentz transformation is named after *H.A. Lorentz: (i)* Versuch einer Theorie der elektrischen und optischen Erscheinungen in bewegten Körpern, Leiden 1895 and *(ii)* Electromagnetic phenomena in a system moving with any velocity smaller than that of light, Proceedings Acad. Sci. Amsterdam 6 (1904) 809-829. Special relativity was introduced by *A. Einstein:* Zur Elektrodynamik bewegter Körper, Ann. d. Phys. 17 (1905) 891-921. The concept of pseudo-Euclidean spacetime has been introduced by *H. Minkowski:* Raum und Zeit, Paper presented at 80. Versammlung Deutscher Naturforscher und Ärzte zu Cöln, 21. Sept. 1908. More general, but integrable relativistic transformations between *moving observers* like the *Gordon-Palacios* transformations are reviewed by D.J. Cashmore, Proc. Phys. Soc. 81 (1963) 181-185 published by the Institute of Physics and the Physical Society of London.

Figure 1: Light flash at the origin O(t=0) = O'(t'=0), two inertial or Galilean frames $\underset{\sim}{e}$, $\underset{\sim}{e}'$ moving with respect to each other by the velocity vector $\underset{\sim}{v}$, $\underset{\sim}{v}' = -\underset{\sim}{v}$

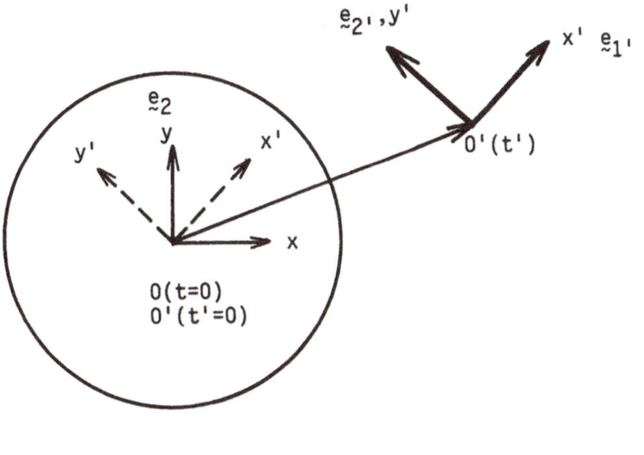

Figure 2: Two inertial or Galilean frames $\underset{\sim}{e}$,$\underset{\sim}{e}'$ moving with respect to each other by the velocity v, v' = -v along the x-axis

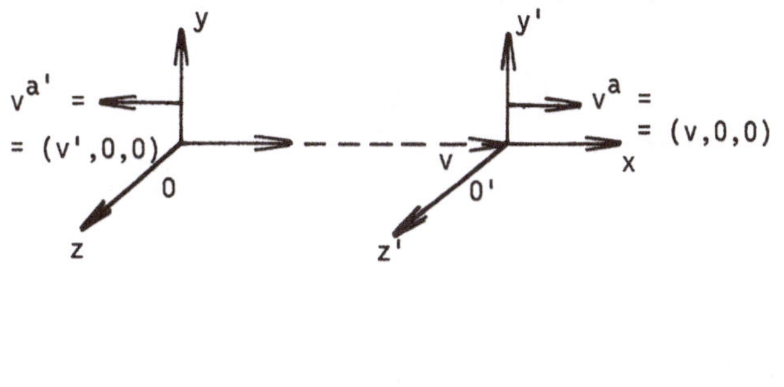

2. Examples

We shall work out six detailed examples. In the *first example* we reflect the notion of *simultaneity*. *Length contraction* and *time dilatation* are the topics of the *second* and *third example* before we move over to *spacetime diagrams* in the *fourth example*. Anholonomic *proper time* and *proper length* are discussed in the *fifth example*. Finally with the *sixth example* we deal with *accelerated systems of reference* being the necessary prerequisite of the *Sagnac effect*.

Example 2-1: Simultaneity

The statement "two events at places A and B are simultaneous" is a relative one, namely *dependent of the observer*. Two events at places A and B in the *inertial frame* $\{0,\underset{\sim}{e}\}$ are called simultaneous, if the *synchronized normal clocks* of the *inertial frame* $\{0,\underset{\sim}{e}\}$ at A and B, respectively, show the same time. For instance, two *light signals* at places A and B in the inertial frame $\{0,\underset{\sim}{e}\}$ arrive at the same time t if they have been transmitted from the mid of the junction line *simultaneously*. An observer *in another inertial frame*, say $\{0',\underset{\sim}{e}'\}$, defines simultaneity in his frame of reference, differently from the one in the $\{0,\underset{\sim}{e}\}$-frame.

If in the $\{0,\underset{\sim}{e}\}$-frame two events at places $A(-x_A,0,0)$ and $B(+x_A,0,0)$ are registered simultaneously, say at $t = t_A = t_B$, an observer in the $\{0',\underset{\sim}{e}'\}$-frame registers these events at the places at times

$$t'_A = \frac{t_A + vx_A/c^2}{\sqrt{1 - v^2/c^2}} \quad , \quad t'_B = \frac{t_B - vx_A/c^2}{\sqrt{1 - v^2/c^2}} \quad , \tag{2(1)}$$

thus by no means simultaneously:

$$t'_A - t'_B = \frac{2vx_A/c^2}{\sqrt{1 - v^2/c^2}} \tag{2(2)}$$

Example 2-2: Length contraction

Suppose that we have measured the length of a rod in the moving inertial frame $\{0',\underset{\sim}{e}'\}$, say $\ell' = x'_B - x'_A$ by taking the difference of position coordinates at places A and B on the x'-axis. According to the *proper Lorentz transformation* which connects coordinates in the $\{0,\underset{\sim}{e}\}$-inertial frame and in the $\{0',\underset{\sim}{e}'\}$-inertial frame which is moving by velocity v with respect to the $0,\underset{\sim}{e}$ - inertial frame, 1(36) leads to

$$x'_A = \frac{x_A - vt_A}{\sqrt{1 - v^2/c^2}} \quad , \quad x'_B = \frac{x_B - vt_B}{\sqrt{1 - v^2/c^2}} \tag{2(3)}$$

$$x'_B - x'_A = \frac{x_B - x_A}{\sqrt{1 - v^2/c^2}} \tag{2(4)}$$

$$\ell' = \frac{\ell}{\sqrt{1 - v^2/c^2}} \iff \ell = \ell'\sqrt{1 - v^2/c^2} \tag{2(5)}$$

Thus ℓ' suffers a *contraction:* $\ell < \ell'$! In the same way it can be shown that a volume V' given in the *inertial frame* $\{0', \underset{\sim}{e}'\}$ is contracted into $V = V'\ (1 - v^2/c^2)$ in the *inertial frame* $\{0, \underset{\sim}{e}\}$.

Example 2-3: Time dilatation

With respect to the *inertial frame* $\{0, \underset{\sim}{e}\}$ at a fixed point $(x^1, 0, 0)$ at times t_A and t_B two flash lights appear. The events are observed in the *inertial frame* $\{0', \underset{\sim}{e}'\}$ at times

$$t'_A = \frac{t_A - vx_1/c^2}{\sqrt{1 - v^2/c^2}} \quad , \quad t'_B = \frac{t_B - vx_1/c^2}{\sqrt{1 - v^2/c^2}} \tag{2(6)}$$

leading by subtraction to

$$t'_B - t'_A = \frac{t_B - t_A}{\sqrt{1 - v^2/c^2}} \iff t_B - t_A = (t'_B - t'_A)\sqrt{1 - v^2/c^2} \ . \tag{2(7)}$$

Thus $t_B - t_A$ suffers a *dilatation:* $t'_B - t'_A > t_B - t_A$!

Example 2-4: Spacetime diagrams

Let us illustrate a twodimensional vector space $\underset{\sim}{x} \in V^2$ characterized by a *"null"* quadratic form

$$q(\underset{\sim}{x}) = (x^1)^2 - (x^4)^2 = 0 \qquad x^1 = \pm x^4 , \tag{2(8)}$$

a vector space we have also identified as *isotropic* or *lightlike*. In contrast, the *anisotropic vectors* of a pseudo-Euclidean vector space V^2 with respect to the normal basis $\underset{\sim}{e}_1$ and $\underset{\sim}{e}_4$ are split into *four kinds:*

$$\begin{array}{lll} \text{(i)} & \underset{\sim}{x} = x^1\underset{\sim}{e}_1 + x^4\underset{\sim}{e}_4 : q(\underset{\sim}{x}) > 0 \ , \ x^1 > 0 \ , & \text{spacelike} \\ \text{(ii)} & \underset{\sim}{x} = x^1\underset{\sim}{e}_1 + x^4\underset{\sim}{e}_4 : q(\underset{\sim}{x}) > 0 \ , \ x^1 < 0 \ , & \text{spacelike} \\ \text{(iii)} & \underset{\sim}{x} = x^1\underset{\sim}{e}_1 + x^4\underset{\sim}{e}_4 : q(\underset{\sim}{x}) < 0 \ , \ x^1 > 0 \ , & \text{timelike} \\ \text{(iv)} & \underset{\sim}{x} = x^1\underset{\sim}{e}_1 + x^4\underset{\sim}{e}_4 : q(\underset{\sim}{x}) < 0 \ , \ x^1 < 0 \ , & \text{timelike} \end{array} \tag{2(9)}$$

In case of an *isotropic* vector space the "length" $\sqrt{q(\underset{\sim}{x})}$ of the vector $\underset{\sim}{x} \neq \underset{\sim}{0}$ is *null* and has therefore little in common with the daily experience of a length. For the *anisotropic* vectors of *first* and *second kind* $\sqrt{q(\underset{\sim}{x})}$ is a *positive number*. Here the notion of a length is reasonable since it agrees with our daily experience.

For the *anisotropic* vectors of *third* and *fourth kind* $\sqrt{q(\underset{\sim}{x})}$ is an *imaginary number*. In this case $i\sqrt{|q(\underset{\sim}{x})|}$ is complex. A diagrammatic representation is offered in *Figure 3*.

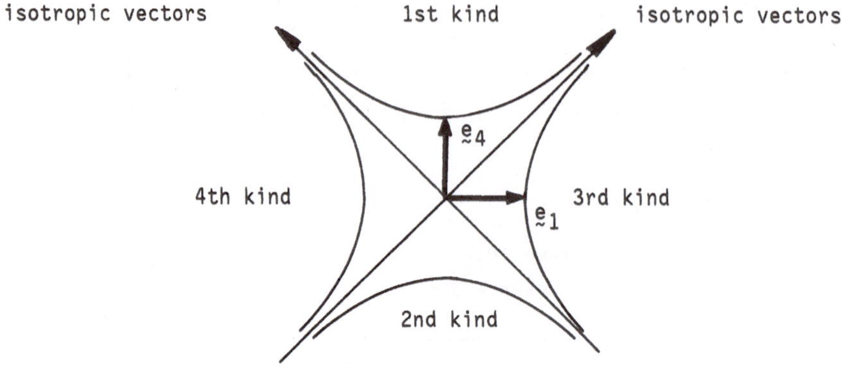

Figure 3: Pseudo-Euclidean twodimensional vector space; *isotropic* vector appear along the 45°- and 135°-straight lines; they split the plane in four angular parts corresponding to the four kinds of *anisotropic* vectors; the coordinates relating to $q(\underset{\sim}{x}) < 0$ or $q(\underset{\sim}{x}) > 0$ appear on a *hyperbola*.

Note that in the four-dimensional pseudo-Euclidean space V^4 spanned by $\{\underset{\sim}{e}_1, \underset{\sim}{e}_2, \underset{\sim}{e}_3, \underset{\sim}{e}_4\}$ the *null* quadratic form $q(\underset{\sim}{x}) = (x^1)^2 + (x^2)^2 + (x^3)^2 - (x^4)^2$ generates *isotropic vectors* on a *cone* separating future-pointing timelike vectors and past-pointing timelike vectors.

When we choose the angle between two *anisotropic* vectors of the same kind according to

$$\cosh\tau = \frac{q(\underset{\sim}{x}, \underset{\sim}{y})}{\sqrt{q(\underset{\sim}{x})}\sqrt{q(\underset{\sim}{y})}} \qquad 2(10)$$

we gain from

$$2q(\underset{\sim}{x}, \underset{\sim}{y}) = q(\underset{\sim}{x}) + q(\underset{\sim}{y}) - q(\underset{\sim}{x} - \underset{\sim}{y})$$

$$q(\underset{\sim}{x} - \underset{\sim}{y}) = q(\underset{\sim}{x}) + q(\underset{\sim}{y}) - 2\sqrt{q(\underset{\sim}{x})}\sqrt{q(\underset{\sim}{y})}\cosh\tau \qquad 2(11)$$

which is called "*hyperbolic cosine proposition*".

Example 2-5: Proper time, proper length

In an inertial frame $\{0, \underset{\sim}{e}\}$ the motion of a mass point is described by its *world line* $x^\mu(\lambda)$ where λ is an arbitrary parameter. We gain especially clear situations once we parameterize the worldline by *proper length* s or *proper time* τ being defined as following.

In the *pseudo-Euclidean* four-dimensional vector space V^4 we introduce the *infinitesimal metric*

$$q(\underset{\sim}{dx}) = ds^2 = dx^2 + dy^2 + dz^2 - c^2dt^2 = g_{\mu\nu}dx^\mu dx^\nu \qquad 2(12)$$

which we can rescale by

$$q(\tfrac{1}{c}\, i\underset{\sim}{dx}) = -\frac{1}{c^2}\, ds^2 = đ\tau^2 = dt^2 - (dx^2 + dy^2 + dz^2)/c^2 \qquad 2(13)$$

$$đ\tau^2 = dt^2\left(1 - \underset{\sim}{v}^2/c^2\right). \qquad 2(14)$$

Note that *proper time* τ is *non-integrable:*

$$\Delta\tau = \int_A^B \sqrt{1 - v^2/c^2}\, dt\ , \qquad \oint đ\tau \neq 0 \qquad 2(15)$$

The integral depends beside the worldline points A and B on the worldline connecting these two points: It is *path-dependent*. Here is the origin of many *paradoxes*!

Example 2-6: Accelerated systems of reference

As a *first example* we treat the transformation from Cartesian to polar coordinates being illustrated in *Figure 4*.

$$\begin{aligned} y^1 &= r = \sqrt{(x^1)^2 + (x^2)^2} \\ y^2 &= \lambda = \arctan \frac{x^2}{x^1} \end{aligned} \qquad 2(16)$$

versus

$$\begin{aligned} x^1 &= r\cos\lambda = y^1 \cos y^2 \\ x^2 &= r\sin\lambda = y^1 \sin y^2 \end{aligned} \qquad 2(17)$$

The squared worldline infinitesimal interval is

$$\begin{aligned} ds^2 &= g_{\mu\nu}dx^\mu dx^\nu = g_{\alpha\beta}dy^\alpha dy^\beta \\ ds^2 &= dx^2 + dy^2 + dz^2 - c^2dt^2 = dr^2 + r^2d\lambda^2 + dz^2 - c^2dt^2 \end{aligned} \qquad 2(18)$$

$$dx^\mu = \frac{\partial x^\mu}{\partial y^\alpha}\, dy^\alpha\ , \qquad dy^\alpha = \frac{\partial y^\alpha}{\partial x^\mu}\, dx^\mu \qquad 2(19)$$

$$dx = \frac{\partial x}{\partial r}\, dr + \frac{\partial x}{\partial \lambda}\, d\lambda = \cos\lambda\, dr - \sin\lambda\, r d\lambda$$

$$dy = \frac{\partial y}{\partial r}\, dr + \frac{\partial y}{\partial \lambda}\, d\lambda = \sin\lambda\, dr + \cos\lambda\, r d\lambda$$

$$g_{\mu\nu} = \begin{bmatrix} 1 & 0 & 0 & 0 \\ 0 & 1 & 0 & 0 \\ 0 & 0 & 1 & 0 \\ 0 & 0 & 0 & -1 \end{bmatrix} \quad , \quad g_{\alpha\beta} = \begin{bmatrix} 1 & 0 & 0 & 0 \\ 0 & r^2 & 0 & 0 \\ 0 & 0 & 1 & 0 \\ 0 & 0 & 0 & -1 \end{bmatrix} \qquad 2(20)$$

As a *second example* we generalize the above transformation into a *rotational* one being illustrated in *Figure 5*.

$$\begin{aligned} x^{1'} &= r = y^1 \\ x^{2'} &= \phi = \lambda - \omega t = y^2 - \frac{\omega}{c} y^4 \\ x^{3'} &= y^3 \\ x^{4'} &= y^4 \end{aligned} \qquad 2(21)$$

versus

$$\begin{aligned} y^1 &= r = x^{1'} \\ y^2 &= \lambda = \phi + \omega t = x^{2'} + \frac{\omega}{c} x^{4'} \\ y^3 &= x^{3'} \\ y^4 &= x^{4'} \end{aligned} \qquad 2(22)$$

The squared worldline infinitesimal interval is

$$\begin{aligned} ds^2 &= g_{\mu\nu} dx^\mu dx^\nu = g_{\alpha\beta} dx^\alpha dx^\beta = g_{\mu'\nu'} dx^{\mu'} dx^{\nu'} \\ ds^2 &= dx^2 + dy^2 + dz^2 - c^2 dt^2 = dr^2 + r^2 d\lambda^2 + dz^2 - c^2 dt^2 = \\ &= dr^2 + r^2 d\phi^2 + 2\omega r^2 d\phi dt + dz^2 - (c^2 - \omega^2 r^2) dt^2 \end{aligned} \qquad 2(23)$$

$$\begin{aligned} d\lambda &= \frac{\partial\lambda}{\partial\phi} d\phi + \frac{\partial\lambda}{\partial t} dt = d\phi + \omega dt \\ d\lambda^2 &= d\phi^2 + 2\omega d\phi dt + \omega^2 dt^2 \end{aligned} \qquad 2(24)$$

$$g_{\mu'\nu'} = \begin{bmatrix} 1 & 0 & 0 & 0 \\ 0 & r^2 & 0 & \frac{\omega}{c} r^2 \\ 0 & 0 & 1 & 0 \\ 0 & \frac{\omega}{c} r^2 & 0 & -1 + \frac{\omega^2}{c^2} r^2 \end{bmatrix} \qquad 2(25)$$

Note that the transformation is *not* time-orthogonal. Obviously reference points up to distances $r = \frac{c}{\omega}$ can be materialized: at larger distances the velocity exceeds c! We mention also that the worldline infinitesimal interval 2(23) of a rotating frame is *no longer pseudo-Euclidean!*

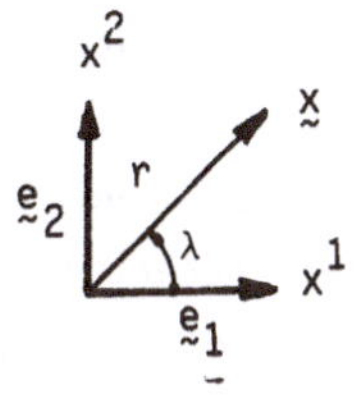

Figure 4: Polar coordinates

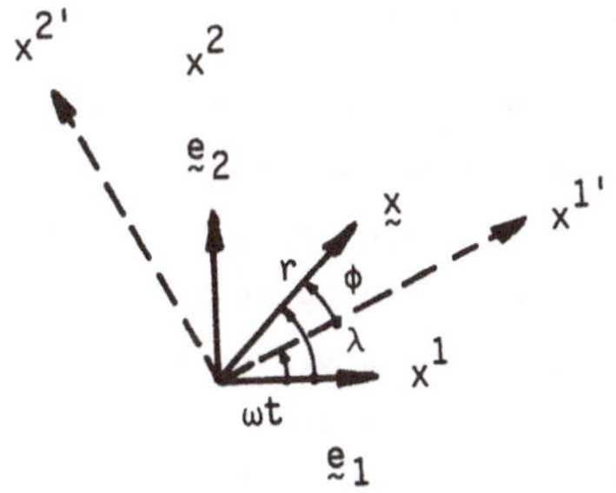

Figure 5: Polar coordinates, rotating frame

3. Exercises

Here we pose two problems in the form of exercises: *At first* we ask for a spinor notation of the Lorentz transformation. Secondly we ask for a derivation of the *relativistic Sagnac effect* which is an example concerning an accelerated system of reference.

3-1. Lorentz transformation in spinor notation

The starting idea for spinor notation of the Lorentz transformation is the array definition of the coordinates, namely

$$X := \begin{bmatrix} x^1 - x^4 & x^2 + ix^3 \\ -x^2 + ix^3 & x^1 + x^4 \end{bmatrix} , \quad X' := \begin{bmatrix} x^{1'} - x^{4'} & x^{2'} + ix^{3'} \\ -x^{2'} + ix^{3'} & x^{1'} + x^{4'} \end{bmatrix} \qquad 3(1)$$

such that

$$q(X) = \det X = (x^1)^2 + (x^2)^2 + (x^3)^2 - (x^4)^2 \qquad 3(2)$$

$$q(X') = \det X' = (x^{1'})^2 + (x^{2'})^2 + (x^{3'})^2 - (x^{4'})^2 \qquad 3(3)$$

The *vector Lorentz transformation* 1(35) is exchanged by the *spinor Lorentz transformation*

$$X' = AXA^* \qquad 3(4)$$

where A is a complex 2×2 (pseudo-) *unitary matrix* and A* the *Hermitean* of A (conjugate complex, transpose). *Find the complex matrix A!* A reference text for an alternative *spinor representation* of the Lorentz transformation is *R. Penrose and W. Rindler* (Spinors and Spacetime, Cambridge University Press, 2 volumes, Cambridge 1984).

3-2. The Sagnac effect

As an exercise with respect to *Example 2-6*, the worldline metric 2(23), 2(25) in a rotating frame, we outline the *Sagnac effect:* An observer is located on a rotating body. He sends two light rays along his own circular path, one *clockwise*, one *counterclockwise* according to *Figure 6*. If the observer is not rotating, both rays will arrive synchronously. Due to the rotation there will be different times of arrival. Use the values of ds^2 2(23), 2(25), together with the knowledge that $ds = 0$ for a light ray, to show that

$$\Delta t = \frac{\omega}{c^2 - \omega^2 r^2} \; 4\pi r^2$$

where πr^2 is the area of the circle. The interval Δt gives rise to a *phase shift* which can be measured by interferometric means. For a review of the Sagnac effect we refer to *E.J. Post* (Reviews of Modern Physics 39 (1967) 475-493, its application is reviewed in W.W. Chow et al. (Reviews of Modern Physics 57 (1985) 61-104).

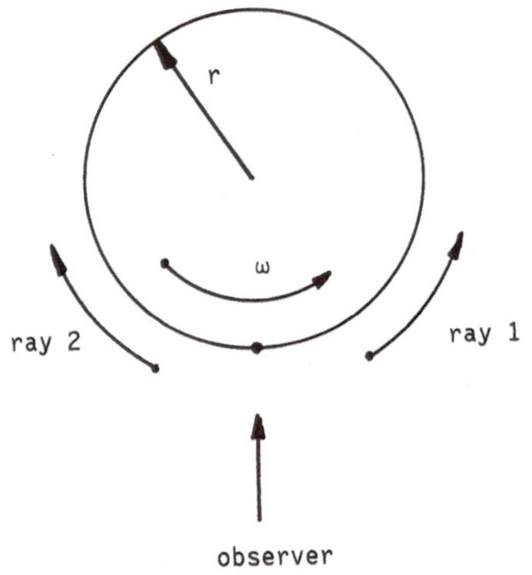

Figure 6: The Sagnac effect

Lecture II - IV

Curved spacetime, pseudo-Riemann space, the affine transformation

Summary

According to the fundamental postulate of general relativity that *(i)* inertial (passive) and gravitational (active) mass coincide (the weak equivalence theory) and *(ii)* all physical laws are invariant under any general or local affine coordinate transformation, Gaussian non-orthogonal, non-normalized tetrads are introduced which span pseudo-Riemannian spacetime. In their geometrical part they lead to the covariant derivative and the Riemann curvature tensor. Their physical counterpart expresses the threedimensional Newton equation of motion as a geodesic in spacetime. The Christoffel or connection symbols are derived from the metric tensor. The metric tensor is derived from a set of field equations, the *Einstein field equation* which balances the curvature invariants and the energy momentum tensor and is transferred from *Laplace-Poisson's* equation. Extended notes in geometry and physics motivate the relativistic transfer, namely on the basis of the Cartan formalism of Newton's dynamical equations. A variational formulation of the fundamental dynamical equations for gravitational interaction is finally given introducing the symplectic manifold. Two detailed *examples* present *(i)* the Schwarzschild solution of the Einstein equations, the field equations as well as the four geodesic equations, ("perihel motion") and *(ii)* weak gravitation, the first order perturbation theory of the Einstein equations and the PPN formalism. Two exercises are left to the reader. They refer to *(i)* the Kerr metric and *(ii)* the Schwarzschild metric in isotropic coordinates.

"The necessary mathematical apparatus for the general relativity theory is already completed in the absolute differential calculus. This rests on the researches of C.F. Gauß, B. Riemann and E.B. Christoffel on non-Euclidean manifolds, and was systematised by G. Ricci and T. Levi-Civita, who have already applied it to problems of theoretical physics ... Finally my thanks are due here to my friend, the mathematician M. Grossman, who not only helped me in the study of the extensive mathematical literature, but led me by his researches to the discovery of the field equations."

A. Einstein, quoted from "Foundations of General Relativity, Ann. d. Phys. 49 (1916) 769, engl. translation by C.W. Kilmister (General theory of relativity, Pergamon Press, Oxford 1973).

1. Curved spacetime, pseudo-Riemann space, the affine transformation

While the *Special Theory of Relativity* generalized Newton's theory of space and time *without gravitation*, the *General Theory of Relativity* extends Newton's theory of space and time *with gravitation*. General relativity is built on fundamental principles: There is the *weak equivalence principle* that inertial mass and gravitational mass (passive gravitational mass) coincide. While in special relativity physical laws are invariant under any Lorentz transformation, the *principle of general relativity* states that all physical laws are invariant under any general or affine coordinate transformation (*Principle of General Covariance*).

Geometrical part

Under the influence of gravitation *flat spacetime of pseudo-Euclidean type* (Minkowski space, Minkowski tetrads) has to be exchanged by *curved spacetime of pseudo-Riemann type* (Einstein space, Einstein tetrads). The *indefinite metric, affine space without torsion* is built up by the metric form

$$ds^2 = g_{\mu\nu}(x^\lambda)\, dx^\mu dx^\nu \,, \qquad \nabla_\lambda g_{\mu\nu} = 0 \qquad 1(1)$$

where the metric tensor is position-dependent, its *covariant derivative* ("delta operator") zero. The spacetime manifold is no longer flat, but curved. The curvature is described by the spacetime *Riemann curvature tensor*

$$R^\alpha{}_{\cdot\beta\mu\nu} = \frac{\partial}{\partial x^\nu}\Gamma^\alpha_{\mu\beta} - \frac{\partial}{\partial x^\mu}\Gamma^\alpha_{\nu\beta} + \Gamma^\alpha_{\nu\rho}\Gamma^\rho_{\mu\beta} - \Gamma^\alpha_{\mu\rho}\Gamma^\rho_{\nu\beta} \qquad 1(2)$$

where $\Gamma^\rho_{\mu\nu}$ are the spacetime *Christoffel symbols*, *symmetric connection symbols* which generate the covariant derivative, say of a vector $\underset{\sim}{A}$ by

$$\nabla_\mu A^\rho = \frac{\partial A^\rho}{\partial x^\mu} + \Gamma^\rho_{\mu\nu} A^\nu \qquad 1(3)$$

Physical part

The generalization of Newton's physical theory is twofold. *At first* the *Laplace-Poisson equation* which governs nonrelativistic gravitational field equations is generalized into

$$\Delta\phi = 4\pi g\rho \longrightarrow R^{\mu\nu} - \frac{1}{2} g^{\mu\nu} R = \frac{8\pi g}{c4} T^{\mu\nu} \qquad 1(4)$$

(Laplace-Poisson) (Einstein)

where $R^{\mu\nu} - \frac{1}{2} g^{\mu\nu}R = G^{\mu\nu}$ builds up as invariants of the *Riemann curvature tensor* the *Einstein tensor* $G^{\mu\nu}$ which contains *second order derivatives of the metric tensor*, e.g. namely $\partial^2 g_{\mu\nu}/\partial x^\alpha \partial x^\beta$. Thus the scalar *Newton gravitational potential* ϕ is replaced by the tensor-valued metric $g_{\mu\nu}$. Mass density ρ which causes the gravitational field ϕ is generalized by the *energy-momentum tensor* $T^{\mu\nu}$ which causes the tensor-valued gravitational field $g_{\mu\nu}$. *Secondly* the *Newton equation of mass point motion*

$$\frac{d^2x^i}{dt^2} + \mathrm{grad}\phi = 0 \longrightarrow \frac{d^2x^\mu}{d\lambda^2} + \Gamma^\mu_{\alpha\beta}\frac{dx^\alpha}{d\lambda}\frac{dx^\beta}{d\lambda} = 0 \qquad 1(5)$$

(Newton) (Einstein)

is generalized into the *four-dimensional geodesic equation* once we assume according to the postulate of weak equivalence that *active and passive masses coincide.*

Notes in geometry

A four-dimensional pseudo-Riemann space can be *locally* embedded into a maximal *ten-dimensional pseudo-Euclidean space.* A placement vector $\underset{\sim}{x}$ in a ten-dimensional pseudo-Euclidean space can therefore be represented by

$$\underset{\sim}{x} = \sum_{i=1}^{10} x^i(x^\mu)\underset{\sim}{e}_i \quad , \quad \mu = 1,2,3,4 \qquad 1(6)$$

where $\underset{\sim}{e}_i$ ($i = 1,2,\ldots,9,10$) characterize the pseudo-orthonormal 10-leg. (*Latin* indices refer to pseudo-Euclidean coordinates, while *Greek* indices are used for pseudo-Riemannian coordinates. Throughout we apply the summation convention over repeated indices.) The displacement vector $d\underset{\sim}{x}$ in the pseudo-orthonormal 10-leg is conventionally represented in the *Gaussian tetrad* $\underset{\approx}{g}_\mu$, namely by

$$d\underset{\sim}{x} = \frac{\partial \underset{\sim}{x}}{\partial x^\mu} dx^\mu = \frac{\partial x^i}{\partial x^\mu}\underset{\sim}{e}_i\, dx^\mu = \underset{\approx}{g}_\mu dx^\mu \qquad 1(7)$$

Note that in general the *Gaussian tetrad* $\underset{\approx}{g}_\mu$ is neither orthogonal nor pseudo-normal. The 4-leg $\underset{\approx}{g}_\mu$ ($\mu = 1,2,3,4$) spans the *tangent space* of the *pseudo-Riemannian manifold.* Its dual space is generated by $g^\nu = g^{\nu\mu}\underset{\approx}{g}_\mu$. 1(7) is called the formula of *derivational equations of the first kind.* The *derivational equations of the second kind* are gained by

$$d\underset{\approx}{g}_\mu = \nabla_\nu \underset{\approx}{g}_\mu dx^\nu = \frac{\partial}{\partial x^\nu}\underset{\approx}{g}_\mu dx^\nu - \Gamma^\alpha_{\mu\nu}\underset{\approx}{g}_\alpha dx^\nu \qquad 1(8)$$

where "d" represents the *Levi-Civita differential,* ∇_ν the *covariant derivative.* The *Leibniz-Newton* differential, in contrast, contains components *outside* the local tangent space: *For instance,* in a two-dimensional Riemannian manifold, also called the surface,

$$d\underset{\approx}{g}_\mu \Big|_{\text{Leibniz-Newton}} = \frac{\partial \underset{\approx}{g}_\mu}{\partial x^\nu} dx^\nu = \Gamma^\alpha_\mu \underset{\approx}{g}_\alpha + \Omega^3_\mu \underset{\approx}{g}_3 \;(\mu,\nu,\alpha = 1,2) \qquad 1(9)$$

enjoys components in the *tangent space* along g_1 and g_2, but another component along the *surface normal* g_3. Thus only the difference of vectors

$$dg_\mu \Big| - \Omega^3_\mu g_3 = dg_\mu \Big| = (\frac{\partial g_\mu}{\partial x^\nu} - \Gamma^\alpha_{\mu\nu}\, g_\alpha)dx^\nu \qquad 1(10)$$

Leibniz-Newton Levi-Civita

produces a vector field in the *tangent space*. Both derivational equations are vector-valued *1-differential forms*. Their *exterior covariant differential* leads to the 2-differential forms of type *Cartan torsion* $T^\alpha_{\mu\nu}\, dx^\mu \wedge dx^\nu$ and *Riemann curvature* $R^\alpha_{\mu\rho\nu} dx^\rho \wedge dx^\nu$, namely

$$\begin{aligned} d\underset{\sim}{x} \Rightarrow d(d\underset{\sim}{x}) &= \nabla_\nu g_\mu dx^\nu \wedge dx^\mu = \\ &= \frac{1}{2}(\Gamma^\alpha_{\mu\nu} - \Gamma^\alpha_{\nu\mu}) g_\alpha dx^\nu \wedge dx^\mu = T^\alpha_{\mu\nu} dx^\mu \wedge dx^\nu g_\alpha \end{aligned} \qquad 1(11)$$

$$\begin{aligned} dg_\mu \Rightarrow d(dg_\mu) &= d(\nabla_\nu g_\mu) dx^\nu = \\ &= \nabla_\rho \nabla_\nu g_\mu dx^\rho \wedge dx^\nu = \frac{1}{2}(\nabla_\rho \nabla_\nu - \nabla_\nu \nabla_\rho) g_\mu\, dx^\rho \wedge dx^\nu = \\ &= \frac{1}{2}\{\partial_\rho\, \Gamma^\alpha_{\nu\mu} - \partial_\nu\, \Gamma^\alpha_{\rho\mu} - \Gamma^\alpha_{\rho\beta}\, \Gamma^\beta_{\mu\nu} + \Gamma^\alpha_{\nu\beta}\, \Gamma^\beta_{\rho\mu}\}\, g_\alpha\, dx^\rho \wedge dx^\nu \\ &= \frac{1}{2} R^\alpha_{\mu\rho\nu}\, dx^\rho \wedge dx^\nu\, g_\alpha := \Sigma^\alpha_\mu\, g_\alpha \end{aligned} \qquad 1(12)$$

For a pseudo-Riemannian manifold *Cartan torsion* $T^\alpha_{\mu\nu} = 0$, while Riemann curvature $R^\alpha_{\mu\rho\nu} \neq 0$. The *Bianchi integrability conditions* are received once we postulate $d\Sigma = 0$: $\Gamma \wedge \Sigma = \Sigma \wedge \Gamma$. Finally, from $\nabla_\lambda g_{\mu\nu} = 0$ we obtain the *standard representation* of the *Christoffel or connection symbols*

$$\Gamma^\alpha_{\mu\nu} = \frac{1}{2} g^{\alpha\beta}\, \{\frac{\partial}{\partial x^\mu}\, g_{\beta\nu} + \frac{\partial}{\partial x^\nu}\, g_{\mu\beta} - \frac{\partial}{\partial x^\beta}\, g_{\mu\nu}\} \qquad 1(13)$$

Notes in physics

At first we want to bring the *dynamical Newton equation* 1(5) into the form of a geodesic following *E. Cartan's* line-of-thought of motivation. With respect to an orthonormal *inertial frame of reference* $\{0,\underset{\sim}{e}_j\}$ where the Latin indices run here 1,2,3 *Newton's equation* of motion reads

$$\underset{\sim}{e}_j \left\{\frac{d^2x^j}{dt^2} + \frac{\partial\phi}{\partial x^j}\right\} = 0 \tag{1(14)}$$

where ϕ is the gravitational potential. (We have chosen the negative sign in contrast to geodetic convention). $\frac{d^2x^j}{dt^2}$ denotes the coordinates of the acceleration vector in the frame $\{0,\underset{\sim}{e}_j\}$. Instead of using an *absolute Newton time* t we take reference to an "*affinely ticking*" Newton clock such that

$$\lambda = at + b \ , \quad t = a^{-1}\lambda - a^{-1}b \tag{1(15)}$$

thus $\{t(\lambda),x^j(\lambda)\}$ leading to

$$\frac{d^2t}{d\lambda^2} = 0 \ , \quad \frac{d^2x^j}{d\lambda^2} + \frac{\partial\phi}{\partial x^j}\left(\frac{dt}{d\lambda}\right)^2 = 0 \tag{1(16)}$$

1(16) is of the form of a general *spacetime geodesic*

$$\frac{d^2x^\mu}{d\lambda^2} + \Gamma^\mu_{\alpha\beta}\frac{dx^\alpha}{d\lambda}\frac{dx^\beta}{d\lambda} = 0 \qquad (\mu,\alpha,\beta = 1,2,3,4) \ . \tag{1(17)}$$

All *Christoffel symbols* are zero beside

$$\Gamma^j_{44} = \frac{\partial\phi}{\partial x^j} \ . \tag{1(18)}$$

Secondly we look into the Laplace-Poisson equation 1(4) governing Newton's gravitation. With reference to the *Riemann curvature tensor* of spacetime the only non-zero elements are

$$R^j_{4k4} = -R^j_{44k} = \frac{\partial^2\phi}{\partial x^j \partial x^k} \ . \tag{1(19)}$$

The *Ricci curvature* $R_{\alpha\beta} := R^\mu_{\cdot\alpha\mu\beta}$ leads to

$$R_{44} = \frac{\partial^2\phi}{\partial x^i \partial x^i} = 4\pi g\rho \tag{1(20)}$$

which is the *Laplace-Poisson equation*.

How to derive directly the spacetime geodesic? Thus *thirdly* we introduce the *Lagrangean*

$$L\{x^\mu(\lambda), x'^\mu(\lambda), \lambda\} = \frac{1}{2} g_{\mu\nu}(x^\lambda) \frac{dx^\mu}{d\lambda} \frac{dx^\nu}{d\lambda} - \phi(x^\lambda) \qquad 1(21)$$

where $\frac{dx^\mu}{d\lambda} = x'^\mu(\lambda)$ represents the four-velocity, here in an inertial frame of reference and with respect to an *"affinely ticking"* clock being parameterized by λ.

$$\frac{dx^i}{dt} = \frac{dx^i}{d\lambda}\frac{d\lambda}{dt} = a\frac{dx^i}{d\lambda} \qquad \frac{dx^i}{d\lambda} = \frac{dx^i}{dt}\frac{dt}{d\lambda} = a^{-1}\frac{dx^i}{dt} \qquad \frac{dx^4}{d\lambda} = a^{-1} \qquad 1(22)$$

$$g_{\mu\nu} = \begin{bmatrix} a^2 & 0 & 0 & 0 \\ 0 & a^2 & 0 & 0 \\ 0 & 0 & a^2 & 0 \\ 0 & 0 & 0 & 0 \end{bmatrix} \qquad 1(23)$$

$$L\{x^\mu, x'^\mu, \lambda\} = \text{EXTR} \Leftrightarrow \frac{d}{d\lambda}\left(\frac{\partial L}{\partial x'^\mu}\right) - \frac{\partial L}{\partial x^\mu} = 0 \qquad 1(24)$$

The Lagrangean has been decomposed into a *"geometrical part"* $\frac{1}{2} g_{\mu\nu}(x^\lambda)x'^\mu x'^\nu$ (kinetic energy) and the *"dynamical part"* $\phi(x^\lambda)$ (potential energy). The extremum postulate for the Lagrangean has led us to the *Euler-Lagrange equations* which we shall bring now into the familiar form of a *geodesic*.

$$\frac{\partial L}{\partial x'^\mu} = g_{\mu\nu}(x^\lambda)\frac{dx^\nu}{d\lambda}$$

$$\frac{d}{d\lambda}\left(\frac{\partial L}{\partial x'^\mu}\right) = g_{\mu\nu}(x^\lambda)\frac{d^2x^\nu}{d\lambda^2} + \frac{\partial g_{\mu\nu}}{\partial x^\lambda}\frac{dx^\lambda}{d\lambda}\frac{dx^\nu}{d\lambda} \qquad 1(25)$$

$$\frac{\partial L}{\partial x^\mu} = \frac{1}{2}\frac{\partial g_{\lambda\nu}}{\partial x^\mu}\frac{dx^\lambda}{d\lambda}\frac{dx^\nu}{d\lambda} - \frac{\partial \phi}{\partial x^\mu}$$

$$\Downarrow$$

$$\frac{d}{d\lambda}\left(\frac{\partial L}{\partial x'^{\mu}}\right) - \frac{\partial L}{\partial x^{\mu}} =$$
$$= g_{\mu\nu}(x^{\lambda})\,\frac{d^2x^{\nu}}{d\lambda^2} + \frac{\partial g_{\mu\nu}}{\partial x^{\lambda}}\frac{dx^{\lambda}}{d\lambda}\frac{dx^{\nu}}{d\lambda} - \frac{1}{2}\frac{\partial g_{\lambda\nu}}{\partial x^{\mu}}\frac{dx^{\lambda}}{d\lambda}\frac{dx^{\nu}}{d\lambda} + \frac{\partial\phi}{\partial x^{\mu}} = 0 \qquad 1(26)$$

As soon as we multiply 1(26) by the inverse of the metric matrix $g_{\mu\nu}$ we arrive at

$$\frac{d^2x^{\mu}}{d\lambda^2} + \Gamma^{\mu}_{\alpha\beta}\frac{dx^{\alpha}}{d\lambda}\frac{dx^{\beta}}{d\lambda} = 0 :$$
$$\frac{dx^{i}}{d\lambda^2} + \frac{\partial\phi}{\partial x^{i}}\frac{dx^4}{d\lambda}\frac{dx^4}{d\lambda} = 0\ , \quad \frac{d^2t}{d\lambda^2} = 0 \qquad 1(27)$$

It should be mentioned that $\Gamma^{j}_{44} = \partial_j\phi$ generates the non-vanishing *Christoffel symbols of the second kind*. For instance, if $\Gamma^{\mu}_{\alpha\beta} = 0$ globally, then the geodesic equation produces a straight line. We get a better geometric insight when we introduce the *four-momentum* p_μ *fourthly*.

$$p_\mu := \frac{\partial L}{\partial x'^{\mu}} = g_{\mu\nu}x'^{\nu}\ , \quad \frac{\partial p_\mu}{\partial x'^{\nu}} = \frac{\partial^2 L}{\partial x'^{\mu}\partial x'^{\nu}} \qquad 1(28)$$

integrability condition: $\dfrac{\partial p_\mu}{\partial x'^{\nu}} = \dfrac{\partial p_\nu}{\partial x'^{\mu}}$

$$\omega := p_\mu\, dx'^{\mu} = \frac{\partial L}{\partial x'^{\mu}}\, dx'^{\mu} \qquad 1(29)$$

$$\omega^2 := d\omega = dp_\mu \wedge dx'^{\mu} = \frac{\partial p_\mu}{\partial x'^{\nu}}\, dx'^{\nu} \wedge dx'^{\mu} =$$
$$= \frac{\partial^2 L}{\partial x'^{\mu}\partial x'^{\nu}}\, dx'^{\nu} \wedge dx'^{\mu} = 0 \qquad \Longrightarrow \qquad 1(30)$$

$$\omega = dL \qquad 1(31)$$

$$\eta := x'^{\mu}dp_\mu\ , \quad d\eta = dx'^{\mu} \wedge dp_\mu = -dp_\mu \wedge dx'^{\mu} =$$
$$= \frac{-\partial^2 L}{\partial x'^{\mu}\partial x'^{\nu}}\, dx'^{\nu} \wedge dx'^{\mu} = 0 \qquad \Longrightarrow \qquad 1(32)$$

$$\boxed{\eta = dH} \qquad 1(33)$$

$$\Downarrow$$

$$\begin{aligned} &d(L+H) = p_\mu\, dx'^\mu + x'^\mu\, dp_\mu \\ &L + H = p_\mu\, x'^\mu \qquad \frac{\partial H}{\partial p_\mu} = x'^\mu \end{aligned} \qquad 1(34)$$

$$\begin{aligned} &\frac{\partial L}{\partial x^\mu} + \frac{\partial H}{\partial x^\mu} = \frac{\partial p_\nu}{\partial x^\mu}\, x'^\nu + \frac{\partial H}{\partial p_\nu}\,\frac{\partial p_\nu}{\partial x^\mu} = 0 \\ &\frac{\partial H}{\partial x^\mu} = -\frac{\partial L}{\partial x^\mu} = -\frac{d}{d\lambda}\,\frac{\partial L}{\partial x'^\mu} = -p'_\mu \end{aligned} \qquad 1(35)$$

We have introduced the *symplectic manifold* being defined by the *closed* non-degenerate differential 2-form 1(30). The originator ω of ω^2 is the differential 1-form, the differential of the *Lagrangean*. Its *dual* differential 1-form η is the differential of the *Hamiltonian*. Quite naturally as a transformation between the *tangent space* $\partial L/\partial x'^\mu$ and the *co-tangent space* $\partial H/\partial p_\mu$ the *Legendre transformation* 1(34ii) comes up.

Fifthly we shall give the *geodesic* 1(5) another interpretation. Again we refer to the *four-momentum*

$$\boxed{p_\mu = m_o \frac{dx^\mu}{d\lambda}} \qquad 1(36)$$

where m_o refers to the *rest mass* of the material point under discussion and $dx^\mu/d\lambda$ is the *four-velocity* in the *Gaussian tetrad* $\underset{\sim}{g}_\mu$. The covariant directional derivative of Levi-Civita type applied to the *four-momentum* leads us to

$$\frac{d}{d\lambda}\, p_\mu \Big|_{\text{Levi-Civita}} = m_o \frac{d}{d\lambda}\, x'^\mu \Big| = m_o \left(\frac{d^2x^\mu}{d\lambda^2} + \Gamma^\mu_{\alpha\beta}\,\frac{dx^\alpha}{d\lambda}\,\frac{dx^\beta}{d\lambda}\right) = 0 \qquad 1(37)$$

The worldline parameter λ has not been specified so far. In case that a particle moves with light velocity, $ds = 0$, $d\tau = 0$ holds. In this case the *geodesic* reads

$$\boxed{\frac{d^2x^\mu}{d\mu^2} + \Gamma^\mu_{\alpha\beta}\,\frac{dx^\alpha}{d\mu}\,\frac{dx^\beta}{d\mu} = h(\mu)\,\frac{dx^\mu}{d\mu}} \qquad 1(38)$$

for a general affine parameter μ and $dx^\alpha/d\mu$ the coordinates of the *wave vector*. The standard form of the geodesic 1(5), 1(37) is reestablished if we solve the differential equation

$$\frac{d^2\mu}{d\lambda^2} + h(\mu)\left(\frac{d\mu}{d\lambda}\right)^2 = 0 \qquad 1(39)$$

Sixthly it is well understood in *Newton's mechanics* how to generalize the dynamics of a mass point to dynamics of *extended bodies*. Beside the dynamical equation governing *linear momentum*, the dynamical equations for *angular momentum* and *energy* are introduced. The motion of a *mass centre* has there been fundamental to establish an inertial frame of reference and to split translational and rotational degrees of freedom. Unfortunately there is no space to introduce general relativity for extended bodies. In short we refer to *J. Friedrich* ("Relativistic foundations of geodetic models", Mitteilungen aus den Geodätischen Instituten, Unviersität Bonn, Report 76, especially pages 56-65, Bonn 1988).

Seventhly we should give a short comment at least on the *Laplace-Poisson equation* being generalized into the *Einstein equation* 1(4). From the *Cartan-Newton dynamical equation* 1(16) we have learnt that also in non-relativistic physics *mass density produces curvature*, here R_{44}. It has been rather a speculative argument of *A. Einstein* to postulate 1(4). Here we refer to *A. Einstein: (i)* Die Grundlagen der allgemeinen Relativitätstheorie, Ann. d. Phys. 49 (1916) 769, *(ii)* Hamiltonsches Prinzip und allgemeine Relativitätstheorie, Sitzungsberichte der Preußischen Akademie der Wissenschaften (1916) 1111-1116. Another historical note for the establishment of a relativistic variational principle is *D. Hilbert:* Die Grundlagen der Physik, Königl. Ges. d. Wissenschaften, Nachr. Math.-Phys. Klasse (1915) 395-407, (1917) 53-76. In this context we must mention the review paper by *H. Rund and D. Lovelock* ("Variational principles in the general theory of relativity", Jahresberichte der Deutschen Mathematiker Vereinigung 74 (1972) 1-65) and the monography by *R. Hermann* ("Differential geometry and the calculus of variations", 440 pages, Academic Press, New York 1968).

2. Examples

Two examples are worked out in detail. In the *first example* we present the Schwarzschild solution of the Einstein field equations as well as of the four geodesic equations of motion (perihel motion). The *second example* deals with weak gravitation and the first order perturbation formalism with two special cases (mass and angular momentum in the stationary case; the central body with angular momentum). At the end we notice the PPN formalism.

Example 2-1: The Schwarzschild solution of the Einstein equations

In the class of *stationary, spherical symmetric* solutions *K. Schwarzschild* ("Über das Gravitationsfeld eines Massenpunktes nach der Einsteinschen Theorie", Sitzber. Deut. Akad. Wiss. Berlin, Kl. Math.-Phys. Tech. (1916) 189-196) presented one of the first solutions of the Einstein equations.

$$ds^2 = \frac{dr^2}{1-2c^{-2}gm/r} + r^2(\cos^2\phi d\lambda^2 + d\phi^2) - [1-2c^{-2}gm/r]c^2dt^2 \qquad 2(1)$$

$$x^1 = r\ ,\quad x^2 = \lambda\ ,\quad x^3 = \phi\ ,\quad x^4 = ct \qquad 2(2)$$

$$g_{\mu\nu} = \begin{bmatrix} [1-2c^{-2}gm/r]^{-1} & 0 & 0 & 0 \\ 0 & r^2\cos^2\phi & 0 & 0 \\ 0 & 0 & r^2 & 0 \\ 0 & 0 & 0 & -[1-2c^{-2}gm/r] \end{bmatrix} \qquad 2(3)$$

$$\frac{\partial}{\partial x^4} g_{\mu\nu} = 0 \quad \text{(stationarity postulate)} \qquad 2(4)$$

Obviously gm/r represents the *external* gravity field of a spherical symmetric mass in *Newton's theory of gravitation*. Note the special situation of the *Schwarzschild radius* $r_s = 2c^{-2}gm$, also referred to as the *"Schwarzschild horizon"* or *"Schwarzschild sphere"*. (A singularity only appears for r=0. Due to the diagonal structure of the metric tensor 2(3) the *Gaussian tetrad* is *orthogonal!*

In order to simplify the following manipulations, we represent the coordinates of the metric tensor by

$$g_{\mu\nu} = \text{diag}\ \{e^\sigma, r^2\cos^2\phi, r^2, -e^\nu\}\ ,\quad \sigma = \sigma(r)\ ,\quad \nu = \nu(r) \qquad 2(5)$$

In order to derive the four equations of the geodesic 1(5), 1(37) let us compute the *Christoffel symbols of the second kind* 1(13).

$$g^{\alpha\beta} = \text{diag}\ \{e^{-\sigma}\ ,\ \frac{1}{r^2\cos^2\phi}\ ,\ \frac{1}{r^2}\ ,\ -e^{-\nu}\} \qquad 2(6)$$

$$\det g_{\mu\nu} = -e^{\sigma+\nu}\, r^4\cos^2\phi\ , \qquad 2(7)$$

$$\Gamma^1_{11} = \{{}^{\ 1}_{1\cdot 1}\} = \frac{\sigma'}{2} \quad , \quad \sigma' = \frac{d\sigma}{dr}$$
$$\Gamma^1_{22} = \{{}^{\ 1}_{2\cdot 2}\} = - r\, e^{-\sigma} \cos^2\phi$$
$$\Gamma^1_{33} = \{{}^{\ 1}_{3\cdot 3}\} = - r\, e^{-\sigma}$$
$$\Gamma^1_{44} = \{{}^{\ 1}_{4\cdot 4}\} = \frac{\nu'}{2}\, e^{\nu-\sigma} \quad , \quad \nu' = \frac{d\nu}{dr}$$
$$\Gamma^2_{23} = \{{}^{\ 2}_{2\cdot 3}\} = \tan\phi \qquad 2(8)$$
$$\Gamma^2_{12} = \{{}^{\ 2}_{1\cdot 2}\} = \frac{1}{r}$$
$$\Gamma^3_{22} = \{{}^{\ 3}_{2\cdot 2}\} = - \sin\phi \cos\phi$$
$$\Gamma^3_{13} = \{{}^{\ 3}_{1\cdot 3}\} = \frac{1}{r}$$
$$\Gamma^4_{14} = \{{}^{\ 4}_{1\cdot 4}\} = \frac{\nu'}{2}$$

Next we compute the coordinates of the *Ricci tensor* $R_{\mu\nu} = R^{\alpha}_{\mu\alpha\nu}$ and the curvature scalar $R = R^{\mu}_{\cdot\mu}$.

$$R_{11} = \frac{\nu''}{2} + \frac{\nu'^2}{4} - \frac{\sigma'}{r} - \frac{\nu'\sigma'}{4}$$
$$R_{22} = e^{-\sigma} \cos^2\phi \left(1 + \frac{\nu'-\sigma'}{2}\, r\right) - \cos^2\phi$$
$$R_{33} = e^{-\sigma} \left(1 + \frac{\nu'-\sigma'}{2}\, r\right) - 1 \qquad 2(9)$$
$$R_{44} = -e^{\nu-\sigma} \left(\frac{\nu''}{2} - \frac{\sigma'\nu'}{4} + \frac{\nu'^2}{4} + \frac{\nu'}{r}\right)$$

All other unlisted terms vanish. We shall use only the gravitational field *outside* the central, spherical symmetric body, that is for *empty space*, containing no matter. $R_{\mu\nu} - \frac{1}{2} g_{\mu\nu} R = 0$, $T_{\mu\nu} = 0$. Note that

$$e^{\nu} = e^{-\sigma} = 1 - 2c^{-2}gm/r$$
$$\nu = - \sigma = \ln(1 - 2c^{-2}gm/r)$$
$$\nu' = - \sigma' = + \frac{2c^{-2}gm}{r^2} \; \frac{1}{1-2c^{-2}gm/r} \qquad 2(10)$$
$$\nu'' = - \sigma'' = - \frac{2c^{-2}gm}{r^3} \; \frac{(2-2c^{-2}gm/r)}{(1-2c^{-2}gm/r)^2}$$

hold. The four geodesic equations are obtained in the form 2(11) if we parameterize the geodesic by *proper time* τ:

$$\begin{aligned}
&\frac{d^2r}{d\tau^2} - \frac{\nu'}{2}\left(\frac{dr}{d\tau}\right)^2 - re^{\nu}\cos^2\phi\left(\frac{d\lambda}{d\tau}\right)^2 - re^{\nu}\left(\frac{d\phi}{d\tau}\right)^2 + \frac{\nu'}{2}e^{2\nu}\left(\frac{dx^4}{d\tau}\right)^2 = 0\\
&\frac{d^2\lambda}{d\tau^2} + 2\tan\phi\,\frac{d\phi}{d\tau}\frac{d\lambda}{d\tau} + \frac{2}{r}\frac{dr}{d\tau}\frac{d\lambda}{d\tau} = 0\\
&\frac{d^2\phi}{d\tau^2} + \frac{2}{r}\frac{dr}{d\tau}\frac{d\phi}{d\tau} - \sin\phi\cos\phi\left(\frac{d\lambda}{d\tau}\right)^2 = 0\\
&\frac{d^2x^4}{d\tau^2} + \nu'\frac{dx^4}{d\tau}\frac{dr}{d\tau} = 0
\end{aligned} \qquad 2(11)$$

The *fourth equation* 2(11iv) can be integrated directly leading to

$$\frac{dx^4}{d\tau} = Ce^{-\nu}, \quad x^4(\tau) = \frac{-1}{\nu'}Ce^{-\nu} + B \qquad 2(12)$$

We calibrate the solution of the *third equation* 2(11iii) by $\phi = 0$ generating a motion in the equatorial plane of the coordinate system. Using the solution $\phi = 0$ in the *second equation* 2(11ii), we find directly by integration

$$r^2\frac{d\lambda}{d\tau} = \frac{A}{m_o} \qquad 2(13)$$

where the normalized *angular momentum constant* D appears. In order to solve the *first equation* 2(11i) we go back to the metric form 2(1) $ds^2 = -c^2d\tau^2$:

$$\frac{1}{1-r_s/r}\left(\frac{dr}{d\tau}\right)^2 + r^2\left(\frac{d\lambda}{d\tau}\right)^2 - \left(1-\frac{r_s}{r}\right)\left(\frac{dx^4}{d\tau}\right)^2 + c^2 = 0 \qquad 2(14)$$

$$\left.\frac{1}{1-r_s/r}\left(\frac{dr}{d\tau}\right)^2 + \frac{A^2}{m_o^2r^2} - \frac{C^2}{1-r_s/r} + c^2 = 0\;\right] \Longrightarrow \qquad 2(15)$$

$$\frac{dr}{d\tau} = \frac{dr}{d\lambda}\frac{d\lambda}{d\tau} = \frac{A}{m_or^2}\frac{dr}{d\lambda} \qquad 2(16)$$

$$\left.\left(\frac{dr}{d\lambda}\right)^2 + r^4\frac{m_o^2}{A^2}(c^2-C^2) - r^3\frac{m_o^2}{A^2}c^2r_s + r^2 - rr_s = 0\right] \Longrightarrow \qquad 2(17)$$

$$r = \frac{1}{u} \qquad \frac{dr}{d\lambda} = -\frac{1}{u^2}\frac{du}{d\lambda} \qquad 2(18)$$

$$\left(\frac{du}{d\lambda}\right)^2 - r_su^3 + u^2 - \frac{m_o^2r_sc^2u}{A^2} + \frac{m_o^2}{A^2}(c^2-C^2) = 0 \qquad 2(19)$$

2(17) and 2(19) can be integrated directly leading to *elliptic integrals* of type

$$\lambda - \lambda_o = \int_{r_o}^{r} dr\,\frac{1}{P_4(r)} = \int_{u_o}^{u} du\,\frac{1}{P_3(u)} \qquad 2(20)$$

where the polynomials $P_4(r)$, $P_3(r)$ are of type $P_4 = a_o r^4 + a_1 r^3 + a_2 r^2 + a_3 r + a_4$, $P_3 = b_o u^3 + b_1 u^2 + b_2 u + b_3$. For more details see *F. Tricomi:* Elliptische Funktionen, Akad. Verlagsges. Geest/Pontig, Leipzig 1948. Here we apply perturbation theory in the following way. Let us differentiate 2(19) in order to receive an acceleration equation which can be compared to Newton's theory of gravitation.

$$\frac{d^2u}{d\lambda^2} + u - \frac{\kappa m_o^2 mc^4}{8\pi A^2} - \frac{3\kappa m\ c^2}{8\pi} u^2 = 0 \tag{2(21)}$$

The first three terms build up the corresponding equation of motion in Newton's mechanics. The term proportional to u^2 appears newly:

$$\Delta\Phi = -4\pi g\rho\ , \quad \frac{d^2u}{d\lambda^2} + u - \frac{gm_o^2 m}{A^2} = 0 \tag{2(22)}$$

Here we can relate the *Einstein gravitational constant* κ to the *Newton gravitational constant* g by $\kappa = \pi 8g/c^4$. The solution of the Newtonian equation of motion

$$u = a\cos(\lambda-\lambda_o) + \frac{gm_o^2 m}{A^2} \quad \text{(Newton)} \tag{2(23)}$$

is the *first order approximative solution* of the Einsteinian equation of motion:

$$v = b\cos\left\{\sqrt{1 - \frac{3r_s^2 m_o^2 c^2}{2A^2}}\,(\lambda-\lambda_o)\right\} + \frac{3r_s^3 m_o^4 c^4}{8A^4}\ , \quad v = u - \frac{r_s^2 m_o^2 c^2}{2A^2} \tag{2(24)}$$

This motion is known as the *motion of perihel*, since *after one revolution* the perihel has moved by

$$\Delta\lambda = \left(\frac{1}{\sqrt{1-3r_s^2 m_o^2 c^2/2A^2}} - 1\right) 2\pi \doteq \frac{3r_s^2 m_o^2 c^2 \pi}{2A^2} = \frac{3\pi r_s}{a(1-\varepsilon^2)} \tag{2(25)}$$

where

$$A^2 = \frac{a(1-\varepsilon^2) r_s m_o^2 c^2}{2} = a(1-\varepsilon^2) m_o^2 gm \tag{2(26)}$$

characterizes the *angular momentum*, a the *semi-major axis* of the ellipse and ε its *linear excentricity*.

Here is a list of perihel motion for planets and satellites:

Table 1: Perihel motion

Name	$\Delta\lambda$
Mercury	43,02"/century
Venus	8,6" /century
Earth	3,8" /century
Mars	1,35"/century
Moon	0,055"/century
Artificial satellite	1500"/century

Finally we go back to 2(1) in order to discuss length and time scale in a gravitational field:

$$d\underset{\sim}{x} = \mathbf{g}_r dr + \mathbf{g}_\lambda d\lambda + \mathbf{g}_\phi d\phi + \mathbf{g}_t dt \qquad 2(27)$$

$$\langle \mathbf{g}_r | \mathbf{g}_r \rangle = \frac{1}{1-r_s/r} \quad , \quad \langle \mathbf{g}_t | \mathbf{g}_t \rangle = -(1-r_s/r) \qquad 2(28)$$

We end up with a comment. The Schwarzschild metric was one of the first solutions of the Einstein equations. A vast literature exists nowadays for other solutions: We mention the monograph by *D. Kramer, H. Stephani, E. Herlt and M. MacCallum:* Exact solutions of Einstein's field equations, Cambridge University Press, 425 pages, Cambridge 1980. Analytical solutions for a *stationary, axisymmetric* field with *multipole moments* are given by *H. Quevedo* (Ph.D. Thesis, Cologne University, Cologne 1987). The review paper of *K.S. Thorne* on *multipole expansions* and *tensor spherical harmonics* (Reviews of Modern Physics 52 (1980) 299-339 must be mentioned here.

Example 2-2: Weak gravitation, first order perturbation theory of the Einstein equations, PPN

In the geodetic environment gravitation can be assumed to be *weak*. In this case spacetime differs little from flat *Minkowski space*. We set-up a perturbation scheme for the coordinates of the metric tensor of type

$$g_{\mu\nu}(x^\lambda) := g_{\mu\nu} + \delta g_{\mu\nu}(x^\lambda) = \bar{\delta}_{\mu\nu} + h_{\mu\nu}(x^\lambda) \qquad 2(29)$$

$$|g_{\mu\nu}(x^\lambda) - g_{\mu\nu}| \ll 1 \qquad 2(30)$$

The perturbation term $\delta g_{\mu\nu}(x^\lambda) = h_{\mu\nu}(x^\lambda)$ will be considered so small that products of them can be neglected. The *Christoffel symbols* $\Gamma^\lambda_{\mu\nu}$, the *Ricci curvature* $R_{\mu\nu}$ and the *scalar curvature invariant* R are given by

$$\Gamma^{\lambda}_{\mu\nu} = \frac{1}{2}\, g^{\lambda\rho}(h_{\rho\nu,\mu} + h_{\mu\rho,\nu} - h_{\mu\nu,\rho}) \qquad 2(31)$$

$$R_{\mu\nu} = \frac{1}{2}(\Box\, h_{\mu\nu} + h_{,\mu\nu} - h^{\sigma}_{\mu,\nu\sigma} - h^{\sigma}_{\nu,\mu\sigma} \qquad 2(32)$$

$$R = g^{\mu\nu}R_{\mu\nu} = \frac{1}{2}\,\Box\, h - g^{\lambda\rho}g^{\mu\nu}(h_{\lambda\mu} - \frac{1}{2}\, g_{\lambda\mu}h)_{\nu\rho} \quad . \qquad 2(33)$$

Here we have introduced the *d'Alembert operator* in *Minkowski space* $\Box$ and the metric invariant h by

$$\Box\, h_{\mu\nu} = g^{\lambda\rho}h_{\mu\nu,\lambda\rho} \quad , \quad g^{\lambda\rho} = \mathrm{diag}(+1,+1,+1,-1) \qquad 2(34)$$

$$h = g^{\rho\sigma}h_{\rho\sigma} \quad . \qquad 2(35)$$

In terms of the *reduced perturbation term*

$$k_{\mu\nu} = h_{\mu\nu} - \frac{1}{2}\, g_{\mu\nu}\, h \qquad 2(36)$$

we find for the *Einstein equation*

$$R_{\mu\nu} - \frac{1}{2}\, g_{\mu\nu}R = \kappa T_{\mu\nu}$$

$$\frac{1}{2}\,\Box\, k_{\mu\nu} - \frac{1}{2}\, g^{\lambda\rho}(k_{\lambda\mu,\nu\rho} + k_{\lambda\nu,\mu\rho}) + \frac{1}{2}\, g_{\mu\nu}g^{\lambda\rho}g^{\alpha\beta}k_{\lambda\alpha,\rho\beta} = \kappa T_{\mu\nu} \qquad 2(37)$$

$$g^{\lambda\nu}T_{\mu\nu,\lambda} = 0 \;, \quad g^{\lambda\nu} = \delta^{\lambda\nu}_{-} \qquad 2(38)$$

The last equation is the *conservation law of energy and momentum* in *special relativity*. The meaning of the perturbed Einstein field equations is therefore as following: The matter tensor is the source of the gravitational field in the linear approximation. The gravitational field is weak because of the relative smallness of the gravitational constant κ. Let us consider the *coordinate transformation*

$$\tilde{x}^{\mu} = x^{\mu} + \varepsilon\, \xi^{\mu}(x^{\nu}) \qquad 2(39)$$

with ε being of the order of $h_{\mu\nu}$ and $\xi(x^{\nu})$.

$$g_{\mu\nu}(x^{\lambda}) = g_{\mu\nu} + h_{\mu\nu} = \frac{\partial \tilde{x}^{\rho}}{\partial x^{\mu}}\frac{\partial \tilde{x}^{\sigma}}{\partial x^{\nu}}\, \tilde{g}_{\rho\sigma} = (\delta^{\rho}_{\mu} + \varepsilon\xi^{\rho}_{,\mu})(\delta^{\sigma}_{\nu} + \varepsilon\xi^{\sigma}_{,\nu})\tilde{g}_{\rho\sigma} \qquad 2(40)$$

$$\tilde{g}_{\rho\sigma}(\tilde{x}^{\mu}) := \delta^{-}_{\rho\sigma} + \tilde{h}_{\rho\sigma} \qquad 2(41)$$

$$h_{\mu\nu} = \tilde{h}_{\mu\nu} + \varepsilon(g_{\mu\rho}\xi^{\rho}_{,\nu} + g_{\rho\nu}\xi^{\rho}_{,\mu}) \qquad 2(42)$$

Retáining only terms of first order we arrive at *an Einstein field equation* which is much simpler

$$\frac{1}{2}\Box\tilde{k}_{\mu\nu} = \kappa T_{\mu\nu} \tag{2(43)}$$

Finally we shall consider two examples of the *perturbed Einstein field equation*, namely mass and angular momentum in the stationary case and for a central body with angular momentum.

Case 1 $T_{\mu\nu}$ diagonal

$$\frac{1}{c^2}T_{\mu\nu} = \mathrm{diag}(0,0,0,\rho c^2) \quad , \quad \partial\rho/\partial x^4 = 0 \tag{2(44)}$$

Consequently the *Einstein field equation* reduces to

$$\frac{1}{2}\Delta k_{\mu\nu} = \kappa T_{\mu\nu} \qquad \text{(no more "~" written)} \tag{2(45)}$$

where Δ is the *Laplace operator*. With the above energy-momentum or matter tensor we find

$$\frac{1}{2c^2}\Delta k_{44} = \kappa\rho c^2 \quad , \quad \Delta k_{\mu\nu} = 0 \qquad \text{if } \mu,\nu \neq 4 \tag{2(46)}$$

This solution corresponds to the field equation of Newton's theory of gravitation

$$\Delta\Phi = 4\pi g\rho \ , \tag{2(47)}$$

namely

$$k_{44} = \frac{\kappa c^4}{2\pi g}\Phi \quad , \quad k_{\mu\nu} = 0 \qquad \text{if } \mu,\nu \neq 4 \ . \tag{2(48)}$$

After some computations we find for the perturbation term

$$h_{11} = h_{22} = h_{33} = \frac{1}{c^2}h_{44} = \frac{\kappa c^2}{4\pi g}\Phi \quad , \quad h_{\mu\nu} = 0 \qquad \text{if } \mu \neq \nu \tag{2(49)}$$

Therefore, in the linear approximation the metric corresponding to the matter distribution is

$$ds^2 = \left(1 - \frac{\kappa c^2\Phi}{4\pi g}\right)(dx^2 + dy^2 + dz^2) - \left(1 + \frac{\kappa c^2\Phi}{4\pi g}\right)c^2dt^2 \tag{2(50)}$$

In case of a spherical symmetric mass distribution $\rho = \rho(r)$ the exterior solution of the *Laplace-Poisson equation* $\Delta\Phi = 4\pi g\rho$ is

$$\Phi = -\frac{gm}{r}\ , \quad m = \int d^3\underset{\sim}{x}\rho \tag{2(51)}$$

Thus we arrive at the metric of the *linearized isotropic Schwarzschild metric*

$$ds^2 = (1+2c^{-2}gm/r)(dx^2 + dy^2 + dz^2) - [1-2c^{-2}gm/r]c^2dt^2\ . \tag{2(52)}$$

Again we identify

$$2\frac{gm}{c^2} = \frac{\kappa c^2 gm}{4\pi g} \quad\Longrightarrow\quad \kappa = \frac{8\pi g}{c^4}\ . \tag{2(53)}$$

Case 2

We shall assume that a *body rotates* around the x^3-axis which is also the axis of symmetry of the matter distribution. Therefore the matter distribution and the gravitational field will be, in spite of the rotation, time-independent or *stationary*. The velocity v is assumed to be small compared with c, $v/c \ll 1$. Only retaining terms linear in v/c we arrive at the matter tensor

$$\frac{1}{c^2}T_{44} = \rho c^2\ , \quad T_{14} = \rho vc\sin\phi\ , \quad T_{24} = -\rho cv\cos\phi \tag{2(54)}$$

Because of the axial symmetry

$$\rho = \rho(\tilde{R},x^3)\ , \quad v = v(\tilde{R},x^3)\ , \quad \tilde{R}^2 = (x^1)^2 + (x^2)^2$$

holds. The linearized Einstein field equation reads now

$$\frac{1}{2c^2}\Delta k_{44} = \kappa\rho c^2\ , \quad \frac{1}{2}\Delta k_{14} = \kappa T_{14}\ , \quad \frac{1}{2}\Delta k_{24} = \kappa T_{24}\ ; \tag{2(55)}$$

all other $k_{\mu\nu}$ vanish. The exterior solution of these equations is determined by

$$k_{44} = \frac{4gm}{c^2 r} + \dots\ , \quad m = \int d^3\underset{\sim}{x}\rho \tag{2(56)}$$

$$k_{14} = \frac{\kappa}{2\pi}\int d^3\underset{\sim}{X}\,\frac{T_{14}}{R}\ , \quad k_{24} = \frac{\kappa}{2\pi}\int d^3\underset{\sim}{X}\,\frac{T_{24}}{R} \tag{2(57)}$$

Note that T_{14} and T_{24} generate a *dipole structure* as can be taken from the expansion

$$\frac{1}{R} = \frac{1}{r} + \frac{x_i X_i}{r^3} + \dots \tag{2(58)}$$

$$\int d^3\underset{\sim}{X}\,\frac{T_{14}}{R} = \frac{1}{r}\int d^3\underset{\sim}{X}\,T_{14} + \frac{x^i}{r^3}\int d^3\underset{\sim}{X}\,T_{14}X^i + \dots \tag{2(59)}$$

$$k_{14} = \frac{\kappa x^2}{2\pi r^3} \int d^3\underset{\sim}{X}\, T_{14} X^2_- = -\frac{\kappa x^2}{2\pi r^3} \int d^3\underset{\sim}{X}\, T^{14}\, X^2 \tag{2(60)}$$

In special relativity the x^3-component of *angular momentum* is given by

$$J^3 = \frac{1}{c} \int d^3\underset{\sim}{X}\, (X^1 T^{24} - X^2 T^{14}) \tag{2(61)}$$

thus leading to the first order *perturbation terms*

$$h_{11} = h_{22} = h_{33} = \frac{1}{c^2} h_{44} = \frac{2c^{-2}gm}{r} \tag{2(62)}$$

$$h_{14} = -\frac{2g}{c^3}\frac{x^2}{r^3} J^3 \,, \quad h_{24} = \frac{2g}{c^3}\frac{x^1}{r^3} J^3 \tag{2(63)}$$

Finally for the general case of angular momentum $\{J^1, J^2, J^3\}$ we arrive at

$$\delta g_{i4} = h_{i4} + \ldots = \frac{2g}{c^3 r^3}\, \delta_{ijk}\, (J^j x^k - J^k x^j) + \ldots \tag{2(64)}$$

($\delta_{ijk} = -\delta_{jik} = -\delta_{ikj}$, $\delta_{123} = +1$).

There is an interesting interpretation of the *Einstein equations* compared to the *Maxwell equations* of electromagnetism. We refer to case 2 and generalize slightly the matter tensor for a *rotating mass shell in stationary motion*. If we neglect stresses and products of source velocities the energy tensor becomes

$$T_{\mu\nu} = \begin{bmatrix} 0_3 & -c^2 v^i \\ -c^2 v^j & c^4 \rho \end{bmatrix} \tag{2(65)}$$

where 0_3 stands for the 3×3 zero matrix such that $k_{ij} = 0$ for $i = j = 1,2,3$. For slowly moving particles $ds = \sigma dt$. If we denote differentiation with respect to t by dots, the first three *geodesic equations of motion* become

$$\begin{aligned} \ddot{x}^i &= -\Gamma^i_{\mu\nu}\, \dot{x}^\mu \dot{x}^\nu = \\ &= -(k^i_{\mu,\nu} - \frac{1}{2} k_{\mu\nu,}{}^i - \frac{1}{4}\delta^i_{\cdot\mu} k_{,\nu} + \frac{1}{4}\delta^-_{\mu\nu} k_,{}^i)\dot{x}^\mu \dot{x}^\nu \end{aligned} \tag{2(66)}$$

Now let $\dot{x}^\mu = (v^i, 1)$ and neglect products of v's, we finally receive

$$\ddot{x}^i = -k^i_{4,j} v^j + k_{j4,}{}^i\, v^j + \frac{1}{4} k_{44}{}^i \tag{2(67)}$$

or in vectorial form

$$\ddot{\underset{\sim}{x}} = - \operatorname{grad}\Phi - \frac{1}{c} (\underset{\sim}{v} \wedge \operatorname{rot} \underset{\sim}{a}) \qquad 2(68)$$

where

$$\Phi = \frac{1}{4} k_{44} = -g \int d^3\underset{\sim}{x} \frac{\rho}{r} , \quad \underset{\sim}{a} = - \frac{c}{4} k^{i}_{\cdot 4} = \frac{g}{c} \int d^3\underset{\sim}{x} \frac{\rho \underset{\sim}{v}}{r} \qquad 2(69)$$

The formal similarity with Maxwell's theory is obvious! It must be noted that in the above problem the acceleration of *Coriolis type* is given by

$$\ddot{\underset{\sim}{x}} = - \frac{8}{3} \frac{gm}{c^2 R} \underset{\sim}{v} \wedge \underset{\sim}{\omega} = - \frac{1}{c} (\underset{\sim}{v} \wedge \underset{\sim}{h}) \qquad 2(70)$$

$$\underset{\sim}{h} := \frac{12}{5} \frac{R^2 gm}{e} \{(\underset{\sim}{\omega} \cdot \underset{\sim}{x}) \frac{\underset{\sim}{x}}{r^5} - \frac{1}{3} \frac{\underset{\sim}{\omega}}{r^3}\} \qquad 2(71)$$

For a more detailed discussion of the *"gravitomagnetic effect"* we refer to *B. Mashhoon, F.W. Hehl and D.S. Theiss:* "On the gravitational effects of rotating masses: The Thirring-Lense papers", General relativity and gravitation 16 (1984) 711-750 in which the classical papers of *H. Thirring* (1918) and *H. Thirring and J. Lense* (1918) have been translated into English and are being discussed with respect to the modern perspectives of the gravitational "magnetic" field, characterized by a vector potential $\underset{\sim}{a}$ ("magnetic dipole") corresponding to *angular momentum* $\underset{\sim}{h}$. The effect leads to the *Lense-Thirring precession* of the orbits of planets, moons and artificial satellites, namely a change in their *inclination*.

Based on the outlined post-Newton computation of orbits *A. Einstein, L. Infeld and B. Hoffmann* ("The gravitational equations and the problem of motion", Annals of Mathematics 39 (1938) 65-100) presented a detailed scientific program. Reference is very often taken to *EIH equations*. They are the basis of the *JPL empherides almanac* reviewed by *X.X. Newhall, E.M. Standish and J.G. Williams* ("DE 102: A numerically integrated ephemerides of the moon and planets spanning forty-four-centuries", Astronomy and Astrophysics 125 (1983) 150-167). For the computation of satellite orbits in post-Newton approximation we refer to *R. Wirrer, M. Soffel, H. Ruder and M. Schneider* ("Satellitenbahnen und Laufzeiten von Laserpulsen in nach-Newtonscher Näherung", Deutsche Geodätische Kommission, Report 48, p. 173-195, München 1986).

PPN

There are a number of alternative gravitation theories like the ones by *C. Brans and R.H. Dicke* ("Mach's principle and a relativistic theory of gravitation", Phys. Rev. 124 (1961) 925-935), *P. Jordan* ("Zum gegenwärtigen Stand der Diracschen kosmologischen Hypothesen", Z.Phys. 157 (1959) 112-121) or *W.T. Ni* ("Theoretical frameworks for testing relativistic gravity IV: A compendium of metric theories of gravity and their post-Newtonian limits", Astrophys. J. 176 (1972) 769-796.) There is no intention to discuss these theories here, but they give the framework for the understanding of the *parameterized post-Newtonian formalism* of metric theories of gravitation.

$$g_{ij} = \delta_{ij}(1 - 2\gamma\Phi)$$

$$g_{4j} = -\frac{7}{2}\Delta_1 V_j - \frac{1}{2}\Delta_2 W_j$$

$$g_{44} = -1 - 2\Phi - 2\beta\Phi^2 + 4\psi - \zeta A - \eta D$$

$$\Psi(\underset{\sim}{x},t) = \int d^3\underset{\sim}{x}' \frac{\rho_o(\underset{\sim}{x}',t)\psi(\underset{\sim}{x}',t)}{|\underset{\sim}{x}-\underset{\sim}{x}'|}$$

$$\psi = \beta_1 v^2 - \beta_2\Phi + \frac{1}{2}\beta_3\Pi + \frac{3}{2}\beta_4\, p/\rho_o\,, \quad \Pi = (\rho-\rho_o)/\rho_o$$

$$A(\underset{\sim}{x},t) = \int d^3\underset{\sim}{x}' \frac{\rho_o(\underset{\sim}{x}',t)\,\langle(\underset{\sim}{x}-\underset{\sim}{x}'),\underset{\sim}{v}(\underset{\sim}{x}',t)\rangle^2}{|\underset{\sim}{x}-\underset{\sim}{x}'|^3}$$

$$D(\underset{\sim}{x},t) = \int d^3\underset{\sim}{x}' \frac{[t_{jk}(\underset{\sim}{x}',t) - \frac{1}{3}\delta_{jk}t_{\ell\ell}(\underset{\sim}{x}',t)](x_j-x_j')(x_k-x_k')}{|\underset{\sim}{x}-\underset{\sim}{x}'|^3}$$

Table 2: PPN metric coefficients;
$\gamma,\beta,\beta_1,\beta_2,\beta_3,\beta_4,\zeta,\eta$ and Δ_1,Δ_2 unknown coefficients (PPN parameters),
$\gamma = 1, \beta = \beta_1 = \beta_2 = \beta_3 = \beta_4 = 1$, $\zeta = 0$, $\eta = 0$, $\Delta_1 = \Delta_2 = 1$
for Einstein's general relativity

For further details, namely the PPN stress-energy tensor and the PPN equations of motion we refer to *C.W. Misner, K.S. Thorne* and *J.A. Wheeler* ("Gravitation", W.H. Freeman, San Francisco 1973, namely § 39, pages 1066-1095).

3. Exercises

Here we only pose two problems as an exercise:

3-1. The Kerr metric

For a four dimensional treatment of the gravitational field of a rotating body the *Kerr metric* is of focal interest. (R.P. Kerr: "Gravitational field of a spinning mass as an example of algebraically special metrics", Phys. Rev. Letters 11 (1963) 237-238). The *Kerr metric* in Schwarzschild similar coordinates being introduced by *R.H. Boyer and R.W. Lindquist* ("Maximal analytic extension of the Kerr metric", J. Math. Phys. 8 (1967) 265-281) has the form

$$ds^2 = \frac{\Sigma}{\Delta}\, dr^2 + \Sigma d\phi^2 + \frac{\cos^2\phi}{\Sigma}\,[-acdt + (r^2 + a^2)d\lambda]^2 - $$
$$- \frac{\Delta}{\Sigma}\,(cdt - a\cos^2\phi\; d\lambda)^2 \tag{3(1)}$$

$$\Sigma := r^2 + a^2\sin^2\phi\ , \qquad \Delta = r^2 + a^2 - 2\,gm\,c^{-2}r \tag{3(2)}$$

The asymptotic angular momentum is represented in the "a-term". *Prove* that for $a = 0$ the *Kerr metric* in Schwarzschild similar coordinates approaches the *Schwarzschild metric*. For the general *Kerr metric* compute the Einstein field equations and the four geodesic equations as a *second* problem. *Thirdly* develop the *Kerr metric* in series of $1/r$ and show the off-diagonal element $g_{t\lambda} \sim 4gmc^{-1}\, a\cos^2\phi\; r^{-1}dt d\lambda$ where $J = gmc^{-1}a$ denotes the angular momentum (gravitomagnetic field).

3-2. Schwarzschild metric in isotropic coordinates

Use the *Schwarzschild metric* 2(1) and apply the transformation

$$\bar{r} \to r = \bar{r}\left(1 + \frac{gmc^{-2}}{2\bar{r}}\right)^2 \tag{3(3)}$$

in order to show that the Schwarzschild metric in *isotropic coordinates* reads

$$ds^2 = -\left(\frac{1-gmc^{-2}/2\bar{r}}{1+gmc^{-2}/2\bar{r}}\right)^2 c^2dt^2 +$$
$$+ \left(1 + \frac{gmc^{-2}}{2\bar{r}}\right)^4 [d\bar{r}^2 + \bar{r}^2(\cos^2\phi d\lambda^2 + d\phi^2)] \tag{3(4)}$$

Finally expand the metric in powers of gm/r to post-Newtonian accuracy, namely in order to arrive at

$$ds^2 = -\left[1 - 2\,\frac{gmc^{-2}}{r} + 2\,c^{-4}\left(\frac{gm}{r}\right)^2\right] c^2dt^2 +$$
$$+ \left[1 + 2\,\frac{gmc^{-2}}{r}\right][dr^2 + r^2(\cos^2\phi d\lambda^2 + d\phi^2)] \tag{3(5)}$$

$$ds^2 = -\left[1 - 2\,\frac{gmc^{-2}}{r} + 2\,c^{-4}\left(\frac{gm}{r}\right)^2\right] c^2dt^2 +$$
$$+ \left[1 + 2\,\frac{gmc^{-2}}{r}\right](dx^2 + dy^2 + dz^2) \tag{3(6)}$$

Literature

BOSECK, H.: Einführung in die Theorie der linearen Vektorräume. VEB Deutscher Verlag der Wissenschaften, Berlin 1965.

BOSECK, H.: Grundlagen der Darstellungstheorie. VEB Deutscher Verlag der Wissenschaften, Berlin 1973.

BOYER, R.H. and R.W. LINDQUIST: Maximal analytic extension of the Kerr metric. J. Math. Phys. 8 (1967) 265-281.

BRANS, C. and R.H. DICKE: Mach's principle and a relativistic theory of gravitation. Phys. Rev. 124 (1961) 925-935.

CASHMORE, D.C.: Integrable transformations between moving observers. Proc. Phys. Soc. 81 (1963) 181-185.

CHOW, W.W., J. GEA-BANACLOCHE, L.M. PEDROTTI, V.E. SANDERS, W. SCHLEICH and M.O. SCULLY: The ring laser gyro. Reviews of Modern Physics 57 (1985) 61-104.

EINSTEIN, A.: Zur Elektrodynamik bewegter Körper. Ann. d. Phys. 17 (1905) 891-921.

EINSTEIN, A.: Die Grundlagen der allgemeinen Relativitätstheorie. Ann. d. Phys. 49 (1916) 769.

EINSTEIN, A.: Hamiltonsches Prinzip und allgemeine Relativitätstheorie. Sitzungsberichte der Preuß. Akad. d. Wiss. (1916) 1111-1116.

EINSTEIN, A., L. INFELD and B. HOFFMANN: The gravitational equations and the problem of motion. Ann. of Math. 39 (1938) 65-100.

FRIEDRICH, J.: Relativistic foundations of geodetic models. Mitt. aus den Geod. Instituten, Universität Bonn, Report 76, 56-65, Bonn 1988.

HERMANN, R.: Differential geometry and the calculus of variations. Academic Press, New York 1968.

HILBERT, D.: Die Grundlagen der Physik. Königl. Ges. d. Wissenschaften, Nachr. Math.-Phys. Klasse (1951) 395-407, (1917) 53-76.

JORDAN, P.: Zum gegenwärtigen Stand der Diracschen kosmologischen Hypothesen. Z. Phys. 157 (1959) 112-121.

KERR, R.P.: Gravitational field of a spinning mass as an example of algebraically special metric. Phys. Rev. Letters 11 (1963) 237-238.

KILMISTER, C.W.: General theory of relativity. Pergamon Press, Oxford 1973.

KRAMER, D., H. STEPHANI, E. HERLT and M. MacCALLUM: Exact solutions of Einstein's field equations. Cambridge university Press, Cambridge 1980.

LORENTZ, H.A.: Versuch einer Theorie der elektrischen und optischen Erscheinungen in bewegten Körpern. Leiden 1895.

LORENTZ, H.A.: Electromagnetic phenomena in a system moving with any velocity smaller than that of light. Proceedings Acad. Sci. Amsterdam 6 (1904) 809-829.

MASHHOON, B., F.W. HEHL and D.S. THEISS: On the gravitational effects of rotating masses: The Thirring-Lense papers. General relativity and gravitation 16 (1984) 711-750.

MINKOWSKI, H.: Raum und Zeit. Paper presented at 80. Versammlung Deutscher Naturforscher und Ärzte zu Cöln, 21. Sept. 1908.

MISNER, C.W., K.S. THORNE and J.A. WHEELER: Gravitation. W.H. Freeman, San Francisco 1973.

NEWHALL, X.X., E.M. STANDISH and J.G. WILLIAMS: DE 102: A numerically integrated ephemerides of the moon and planets spanning forty-four-centuries. Astronomy and Astrophysics 125 (1983) 150-167.

NI, W.T.: Theoretical frameworks for testing relativistic gravity IV: A compendium of metric theories of gravity and their post-Newtonian limits. Astrophys. J. 176 (1972) 769-796.

NORDTVEDT, K.J. and C.M. WILL: Conservation laws and preferred frames in relativistic gravity. II. Experimental evidence to rule out preferred theories of gravity, Astr. J. 177 (1972) 775-792.

PENROSE, R. and W. RINDLER: Spinors and Spacetime. Cambridge University Press, 2 vol., Cambridge 1984.

POST, E.J.: Reviews of Modern Physics 39 (1967) 475-493.

QUEVEDO, H.: Statische und stationäre axialsymmetrische Lösungen der Einstein'schen Vakuumfeldgleichungen mit Multipolmomenten. Ph.D. Thesis, Cologne University, Cologne 1987.

RUND, H. and D. LOVELOCK: Variational principles in the general theory of relativity. Jahresberichte der Dt. Mathematiker-Vereinigung 74 (1972) 1-65.

SCHWARZSCHILD, K.: Über das Gravitationsfeld eines Massenpunktes nach der Einsteinschen Theorie. Sitzber. Deut. Akad. Wiss. Berlin, Kl. Math.-Phys. Tech. (1916) 189-196.

THORNE, K.S.: Multipole expansions of gravitational radiation. Reviews of Modern Physics 52 (1980) 299-339.

TRICOMI, F.: Elliptische Funktionen. Akad. Verlagsges. Geest/Portig, Leipzig 1948.

WILL, C.M. and K.J. NORDTVEDT: Conservation laws and preferred frames in relativistic gravity. I. Preferred-frame theories and an extended PPN formalism. Astr. J. 177 (1972) 757-774.

WIRRER, R., M. SOFFEL, H. RUDER and M. SCHNEIDER: Satellitenbahnen und Laufzeiten von Laserpulsen in nach-Newtonscher Näherung. Bayr. Komm. f. die Intern. Erdmessung, Astronomisch-Geodätische Arbeiten, Report 48, pp. 173-195, München 1986.

REFERENCE COORDINATE SYSTEMS: AN UPDATE

Ivan I. Mueller
Dept. of Geodetic Science and Surveying
Ohio State University, Columbus, Ohio 43210-1247

This paper is an updated version of the earlier work 'Reference Coordinate Systems for Earth Dynamics: A Preview', by the author published in the Proceedings of IAU Colloquium 56 on *Reference Coordinate Systems for Earth Dynamics*, Sept. 8-12, 1980, Warsaw, Poland, E.M. Gaposchkin and B. Kolaczek, eds., D. Reidel, 1981.

1. INTRODUCTION

Geodynamics has become the subject of intensive international research during the last decades, involving plate tectonics, both on the intraplate and interplate scale, i.e., the study of crustal movements, and the study of earth rotation and of other dynamic phenomena such as the tides. Interrelated are efforts improving our knowledge of the gravity and magnetic fields of the earth. A common requirement for all these investigations is the necessity of a well-defined coordinate system (or systems) to which all relevant observations can be referred and in which theories or models for the dynamic behavior of the earth can be formulated. In view of the unprecedented progress in the ability of geodetic observational systems to measure crustal movements and the rotation of the earth, as well as in the theory and model development, there is a great need for the definition, practical realization, and international acceptance of suitable coordinate system(s) to facilitate such work. Manifestation of this interest has been the numerous specialized symposia organized during the past decade or so, such as those held in Stresa (Markowitz and Guinot, 1968), Morioka (Melchior and Yumi, 1972; Yumi, 1971), Torun (Kołaczek and Weiffenbach, 1974), Columbus (Mueller, 1975b, 1978, 1985), Kiev (Fedorov, Smith and Bender, 1980), San Fernando (McCarthy and Pilkington, 1979), Warsaw (Gaposchkin and Kołaczek, 1981), and Coolfont (Wilkins and Babcock, 1987). There seems to be general agreement that only two basic coordinate systems are needed: a Conventional Inertial System (CIS), which in some 'prescribed way' is attached to extragalactic celestial radio sources, to serve as a reference for the motion of a Conventional Terrestrial System (CTS), which moves and rotates in some average sense with the earth and is also attached in some 'prescribed way' to a number of dedicated observatories operating on the earth's surface. In the latter, the geometry and dynamic behavior of the earth would be described in the relative sense, while in the former the movements of our planetary system (including the earth) and our galaxy could be monitored in the absolute sense. There also seems to be a need for certain interim systems to facilitate theoretical calculations in geodesy, astronomy, and geophysics as well as to aid the possible traditional decomposition of the transformations between the frames of the two basic systems. This scheme is shown in Fig. 1. The Earth Model block represents the current best knowledge of the geometry and dynamic behavior of the earth, partially deduced from the measurements made at the Dedicated Observatories. This model is continuously improving as more data of increasing accuracy becomes available, and it includes both local (L) and global (G) phenomena which have theoretical foundations based on physical reality and are mathematically describable. In the final and ideal situation, which may be achieved only after several iterations over an extended period of time, the global part of the model should be identical to the connection between the CIS and CTS frames. Departures (v) from the model (L′) observed at the observatories (j) or at other station (i) are of course most important since they represent new information based on which the model can be improved, after observational random and systematic errors have been taken into proper consideration. The model could eventually include the solid earth as well as the oceans and the atmosphere.

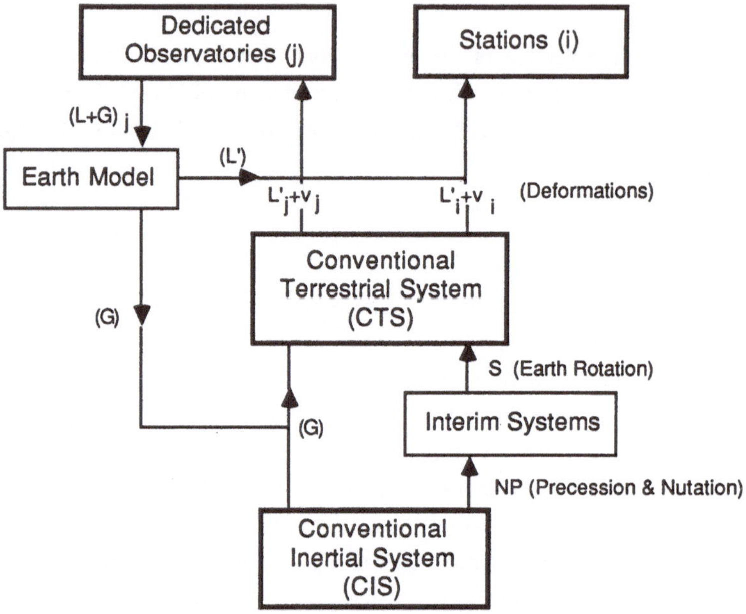

Fig. 1 Construction of conventional reference systems.

As we will see later, there is an understanding on how the two basic reference systems should be established; operational details are part of a recent international agreement. There are still, however, a number of open questions which have to be discussed further. These include the type of interim systems needed and their connections to be CIS and CTS, the type(s) of observatories, their number and distribution, whether all instruments need to be permanently located there or only installed at suitable regular intervals to repeat the measurements; how far the model development should go so as not to become impractical and unmanageable; and how independent observations should be referenced to the CTS, i.e., what kind of services need to be established for the user of the systems.

In order to clarify some of the conceptual aspects of various reference systems and frames, we propose to use specific terms suggested in (Kovalevsky and Mueller, 1981) that have been used somewhat inconsistently in the past:

The purpose of a *reference frame* is to provide the means to materialize a *reference system* so that it can be used for the quantitative description of positions and motions on the earth (terrestrial frames), or of celestial bodies, including the earth, in space (celestial frames). In both cases the definition is based on a general statement giving the rationale for an ideal case, i.e., for an *ideal reference system*. For example, one would have the concept of an ideal terrestrial system, through the statement that with respect to such a system the crust should have only deformations (i.e., no rotations or translations). The ideal concept for a celestial system is that of an inertial system so defined that in it the differential equations of motion may be written without including any rotational term. In both cases the term 'ideal' indicates the conceptual definition only, and no means are proposed to actually construct the system.

The actual construction implies the choice of a physical structure whose motions in the ideal reference system can be described by physical theories. This implies that the environment that acts upon the structure is modeled by a chosen set of parameters. Such a choice is not unique: there are many ways to model the motions or the deformations of the earth; there are also many celestial bodies that may be the basis of a dynamical definition of an inertial system (moon, planets, or artificial satellites). Even if the choice is based on sound scientific principles, there remains some degree of imperfection or arbitrariness. This is one of the reasons why it is suggested to use the term 'conventional' to characterize this choice. The other reason is related to the means, usually conventional, by which the reference frames are defined in practice.

At this stage, there are still two steps that are necessary to achieve the final materialization of the reference system so that one can refer coordinates of objects to them. First, one has to define in detail the model that is used in the relationship between the configuration of the basic structure and its coordinates. At this point, the coordinates are fully defined, but not necessarily accessible. Such a model is called a *conventional reference system*. The term 'system' thus includes the description of the physical environment as well as the theories used in the definition of the coordinates. For example, the FK4 (conventional) reference system is defined by the ecliptic as given by Newcomb's theory of the sun, the values of precession and obliquity, also given by Newcomb, and the Woolard theory of nutation. Once a reference system is chosen, it is still necessary to make it available to the users. The system usually is materialized for this purpose by a number of points, objects or coordinates to be used for referencing any other point, object or coordinate. Thus, in addition to the conventional choice of a system, it is necessary to construct a set of conventionally chosen (or arrived at) parameters (e.g., star positions or pole coordinates). The set of such parameters, materializing the system, define a *conventional reference frame*. For example, the FK4 catalogue of over 1500 star coordinates defines the FK4 frame, materializing the FK4 system.

Another example is the CTS for the deformable earth defined through the time varying coordinates of a number of terrestrial observatories whose positions are periodically reobserved by some international service. The frame of this CTS could then be derived from the changing coordinates through transformations containing rotational (and possibly translational) parameters. The service, as part of the system definition, would have to make the assumption that the progressive changes of the reference coordinates of the observatories, defining the frame, do not represent rotations (and translations) in a statistically significant sense.

2. CONVENTIONAL INERTIAL SYSTEMS (CIS) OF REFERENCE

2.1 Basic Considerations

The first law of Newton is as follows: 'Every body persists in its state of rest or uniform motion in a straight line unless it is compelled to change that state of forces impressed on it' (Newton, 1686). It should be obvious that the above *law of inertia* cannot hold in any arbitrary reference frame so that only certain specific reference frames are acceptable. In classical mechanics, reference frames in which the above law is valid are called *inertial frames*. Such 'privileged' frames move through space with a constant translational velocity but without rotational motion. Another privileged frame in classical mechanics is the *quasi-inertial*, which also moves without rotational motion, but its origin may have acceleration. Such a frame would be, for example, a nonrotating geocentric Cartesian coordinate system whose origin due to the earth orbit around the sun would move with a nonconstant velocity vector. Inertial reference frames thus are either at rest or are in a state of uniform rectilinear motion with respect to *absolute space*, a concept also mentioned by Newton and visualized as being observationally defined by the stars of invariable positions, a dogma in his time.

The refinement of classical mechanics through the theory of relativity requires changes in the above concepts. The theory of special relativity allows for privileged systems, such as the inertial frame but in the *space-time continuum* instead of the absolute space (Moritz, 1967). Transformation between inertial frames in the theory of special relativity are through the so-called Lorentz transformations, which leave all physical equations, including Newton's laws of motion, and the speed of light invariant. The special theory of relativity holds only in the absence of a gravitational field.

In the theory of general relativity, Einstein defined the inertial frames as 'freely falling coordinate systems' in accordance with the local gravitational field which arises from all matter of the universe. Thus the inertial frames lose their privileged status. Concerning the existence of inertial frames in the extended portions of the space-time continuum, Einstein (1956) states that 'there are finite regions, where, with respect to a suitably chosen space of reference, material particles move freely without acceleration, and in which the laws of special relativity hold with remarkable accuracy.' In other words, one can state (Weinberg, 1972) that

> 'At every space-time point in an arbitrary gravitational field, it is possible to choose a locally inertial coordinate system such that, within sufficiently small region of the point in question, the laws of nature take the same form as in unaccelerated Cartesian coordinate system in the absence of gravitation.'

(i.e., as in the theory of special relativity). Our sphere of interest, the area of the solar system, where the center of mass of the earth-moon system is 'falling' in an elliptic orbit around the sun, in a relatively weak gravitational field, seems to qualify as such a 'small region.' Thus we may assume that inertial or quasi-inertial frames of reference exist, and any violation of principles when using classical mechanics can be taken into account with small corrections appropriately applied to the observations and by an appropriate 'coordinate' time reference. The effects of special relativity for a system moving with the earth around the sun are of the order of 10^{-8}, while those of general relatively are 10^{-9} (Moritz, 1979). Since 10^{-8} on the earth's surface corresponds to about 6 cm, corrections at least for special relativity effects are needed when striving for such accuracies. Other than this, the problem, in the conceptual sense, need not be considered further.

Since the definition of the CIS may be based on dynamical properties of the solar system as well as on the kinematics of extragalactic sources, we are led to distinguish between two kinds of quasi-inertial systems (Fig. 2) (Kovalevsky and Mueller, 1981):

(a) *Conventional kinematical systems*, based on the assumption that the proper motions of some celestial bodies have known statistical properties. In the case of extragalactic sources, it is postulated that remote galaxies have no rotational component in their motions.

(b) *Conventional dynamical systems*, based on the theory of the motion of some bodies in the solar system (including artificial earth satellites) constructed in such a way that there remains no rotational term in the equations of motion.

If in the framework of Newtonian mechanics, both definitions are equivalent, this is not true in the theory of general relativity. A dynamical system of coordinates is a local reference that is locally tangent to the general space-time manifold. In contrast, the kinematical frame defined by the apparent directions of remote objects is a coordinate system that is subject to relativistic effects such as the geodesic precession. Even if this is being suitably corrected for, there remains a basic difference between the concepts, and this is another good reason to use the terminology '*quasi-inertial*' to characterize both kinematical and dynamical systems.

It is now well agreed that the best future CIS will be based on the position of extragalactic radio sources. But even is such a system is due to play a major role among conventional quasi-inertial systems, there may be great advantages, in some cases, to use a dynamical system. This is the case, for instance, when artificial satellites are used to monitor the earth rotation. This is why a certain hierarchy among these systems has been proposed in which the CIS, based on extragalactic radio sources is designated as a *primary system*, a role which used to be played by the FK4 System. Other systems, and in particular all the conventional dynamical systems, will have to be connected to the primary system in order to give consistent results (see later).

As mentioned, the actual availability of the systems is obtained through their realization in the form of reference frames. This materialization can be done in two different ways so that one can distinguish between two kinds of reference frames (Kovalevsky and Mueller, 1981):

(a) *Conventional kinematic frames*. The fiducial points are presently stars or extragalactic radio sources. In case of the latter, it is necessary to provide connection to stellar catalogues, so that the celestial system can be made available to optical instrumentation.

(b) *Conventional dynamic frames*. In such frames, one or several moving objects are used as the materialization of the system. The theory supporting the corresponding reference system provides the apparent ephemeris of the objects (satellites or planets) as a function of time and the observed successive positions are the fiducial points needed to refer the observations to the system.

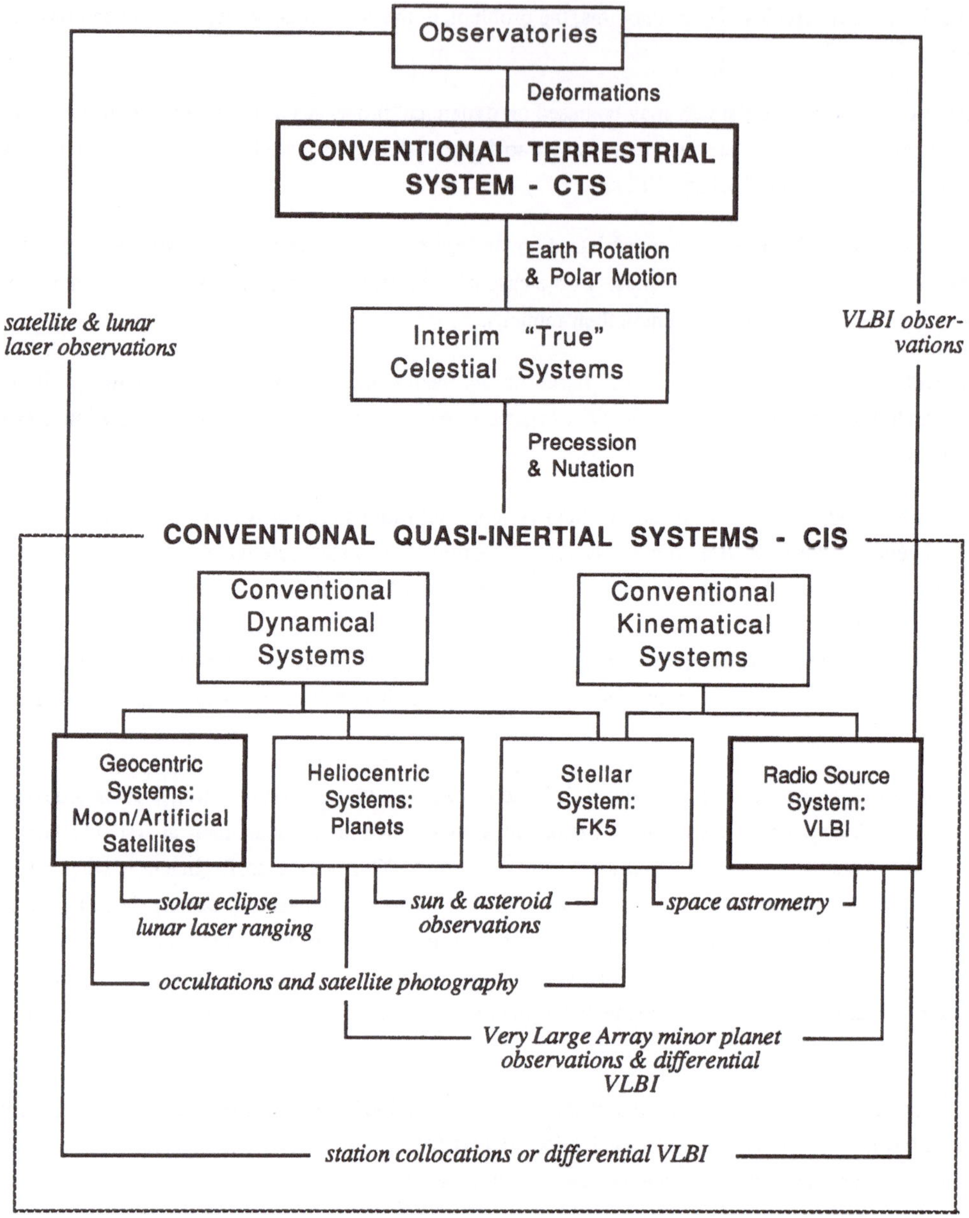

Fig. 2 Conventional terrestrial and quasi-inertial systems of reference with some possible connections.

It is to be noted that there is not necessarily a bi-univocal correspondence between the two types of frames and quasi-inertial systems. For instance, the FK4 or FK5 stellar systems are dynamical (due to the method of determination of the equinox), while their frame is stellar.

2.2 Inertial Systems in Practice

2.21 Extragalactic Radio Source System. This system is attached to radio sources which generally either are quasi-stellar objects (quasars) or galactic nuclei. Very long baseline interferometers rotating with the earth determine the declinations of these sources with respect to the instantaneous rotation axis of the earth (see Section 4.2), as well as their right ascension differences with respect to a selected source (3C273, NRAO 140, Persei (Algol), etc.) In addition, the observations also determine changes in the earth rotation vector with respect to a selected initial state, the baseline itself, and certain instrumental (clock) corrections. The frame of the Radio Source-CIS can be defined by the adopted true or mean coordinates of appropriately selected sources referred to some standard epoch. The mean coordinates naturally will depend on the model of the transformation from the true frame of date to the adopted mean standard. If, however, the reduction procedure is correct (see more on this later), there are no known reasons for nonradial relative motions of the sources, i.e., for the rotation of the frame. Thus such a frame could be considered inertial or at least quasi-inertial. The equatorial system of coordinates may be retained for convenience, but the frame could be attached to the sources in any other arbitrary way should this be necessary.

As far as the accuracy of the Radio Source-CIS is concerned, the question has meaning only in the sense of the formal precisions of the source positions in the catalogue. At the Torun meeting, this number was $0''.1$ (Moran, 1974); at Warsaw it was $0''.01$ (Purcell et al., 1980). Now the precision is of the order of $0''.001$ (Ma, 1989). The problem on this level is that the densification of such a catalogue is very difficult, since only a relatively few well-defined point-like radio sources have been observed. Others have structures such that identification of the center of the radiation with such accuracy may not be possible. This situation may change when the astrometric satellites (see below) are launched, or other new technological developments take place.

VLBI instrumentation has undergone considerable development since the initial efforts in the early 1960's. Table 1 describes the primary recording systems.

Table 1 VLBI Recording Systems (Ma, 1989)

System	*In Use*	*Basic Design*	*Sample Rate Megabit/s*	*Tape Time (min)*
Mark I	1967-78	Digital recording on computer tape	0.72	3
Mark II	1971-	Digital recording on various TV recorders	4	64-246
Mark III	1977-	Digital recording	112	13
Mark IIIA	1984-	Digital recording	112	164

VLBI networks, since they are composed of independent elements, vary with time and availability. Table 2 shows the stations which have contributed significantly to the current astrometric data base.

Table 2 VLBI Antennas Used for Astrometry (Ma, 1989)

Location	*Size*
Gilmore Creek, Alaska, USA	26 m
Goldstone Deep Space Station, California, USA	64
Hartebeesthoek Radio Observatory, So. Africa	26
Hat Creek Radio Observatory, California, USA	26
Harvard Radio Astronomy Station, Texas, USA	26
Haystack Observatory, Massachusetts, USA	37
Kashima Space Research Center, Japan	26
Kokee Tracking Station, Hawaii, USA	9
Madrid Deep Space Station, Spain	64
Maryland Point, Maryland, USA	26
Mojave Base Station, California, USA	12
National Radio Astronomy Observatory, West Virginia, USA	43
Onsala Space Observatory, Sweden	20
Owens Valley Radio Observatory, California, USA	40
Richmond, Florida, USA	18
Tidbinbilla Deep Space Station, Australia	64
Westford, Massachusetts, USA	18
Wettzell, Fed. Repub. Germany	20

Two connected element interferometer (CEI) instruments are also regularly used for astrometric measurements. The National Radio Astronomy Observatory interferometer in Green Bank, West Virginia, has a 35-km baseline and operates continuously as part of a program to monitor UT1. The Very Large Array (VLA) near Socorro, New Mexico, while primarily a mapping instrument is also used for differential and absolute astrometry. It consists of 27 25-m antennas laid out in a Y pattern with the longest arm 21 km.

There are at present several catalogs of extragalactic radio sources in the J2000.0 system. They vary considerably in number of sources, distribution of sources, and precision. See Table 3 for a summary (Ma, 1989).

An extragalactic reference frame which will serve as the initial system of the International Earth Rotation Service (IERS) was compiled on the basis of four individual catalogues from the NASA Goddard Space Flight Center, the Jet Propulsion Laboratory (JPL) and the U.S. National Geodetic Survey (NGS). The compilation was carried out at the IERS (Arias et al., 1988a) and includes 228 extragalactic, compact sources divided into primary, secondary and complementary sources depending upon geometrical and physical considerations as well as observational histories. Unfortunately, this reference frame contains no sources south of -45°, and of the 23 primary sources which define the directions of the axes, eight are in the Southern Hemisphere between the equator and -29°. This points up the fact that the distribution of well-observed radio sources and radio interferometry baselines is far from ideal for the purposes of a global reference frame.

Ma (1983) intercompared the catalogues of JPL and NASA, based on 45 overlapping sources and found an RMS difference of about 0.″005 in both right ascension and declination. A recent study by Arias et al. (1988b) intercompared JPL, NASA and NGS 1984-1986 catalogues based on 19-128 overlapping sources and found the directions of the axes of their respective reference frames consistent within 0.″003. This is considered a remarkable agreement on account of the diversity of observing strategies and data analysis.

Table 3 J2000.0 Catalogues of Extragalactic Compact Sources

Organization	*Instrument*	*Baseline Length (km)*	*No. of Sources*	*Uncertainties mas*	*Reference*
NRAO	CEI	35	36	20-40	Wade & Johnston, 1977
NRAO	CEI	35	16	10	Kaplan et al., 1982
JPL	Mark II	8000-11000	836	300	Morabito et al., 1982-86
NSF	VLA	<27	700	20-100	Perley, 1982
JPL	Mark II	8000-11000	117	1-5	Fanselow et al., 1984
NASA	Mark III	800-6000	85	0.3-13	Ma et al., 1986
NGS	Mark III	800-6000	26	0.5	Robertson et al., 1986
NASA	Mark III	800-11000	101	0.2-9	Ma, 1988
JPL	Mark III	8000-11000	128	0.5-7	Sovers et al., in press

The premier instrument for future radio astrometry will be the Very Long Baseline Array, currently under construction. It will consist of ten 25-m antennas spaced from Hawaii to Puerto Rico, each equipped with ten receivers from .33 GHz to 43 GHz.

Until the VLBA becomes fully operational in the mid-1990's, there are several ongoing programs which will continue to expand and refine the extragalactic catalogue. The NASA Crustal Dynamics Project has a VLBI survey program to expand its catalogue of unresolved sources to take advantage of improvements in sensitivity. The US Naval Observatory is starting an astrometric program using North American VLBI stations to densify the grid of optical/radio sources in the Northern Hemisphere. The JPL survey work will be further refined to support planetary spacecraft navigation using differential VLBI.

2.22 Stellar System. This system is attached to stars in the FK5 catalogue, i.e., the adopted right ascensions and declinations of the FK5 define the equator and the equinox and thus the frame of the Stellar-CIS. The FK5 is the fifth fundamental catalogue in a series which began with the FC in 1879 (Fricke and Gliese, 1978). In the fundamental catalogues the equator is determined from zenith distance (or distance difference) observations of the stars themselves, but the equinox determination also necessitates measurements of the sun or other members of the planetary system. It was always tacitly assumed that coordinate systems attached to the fundamental catalogues were quasi-inertial. However, as more and more observations became available for proper motions and on the various members of the planetary systems, certain small rotations were discovered, which required changes in the positions of the fundamental equator and equinox, in the proper motions and in the precessional constant (all intricately interwoven) when one fundamental catalogue replaces the other. This slow and painstaking process should lead to a quasi-inertial system. We hope that the FK5 is such a system.

When the FK4 was compiled, a small definitive correction to the declination of FK3 was applied, but there seemed to be no need to change the position of the equinox or the precessional constant (Fricke, 1974). The FK5 is a considerably different and improved catalogue. The main changes with respect to the FK4, regarding the issue of the coordinate systems, are as follows (Fricke, 1979a): 1) New value of general precession in longitude adopted by the IAU in 1976 was used (more on this later). 2) The centennial proper motions in right ascension were increased by 0.ˢ085/century to eliminate the motion of the FK4 equinox with respect to the dynamical equinox (the FK4 right ascensions are decreasing with time due to an error in the FK4 proper motions, see below). 3) Rotation of the FK4 equinox at 1950 by the amount of $0\overset{s}{.}035$ so that the FK5 and the dynamic equinoxes will be identical (the FK4 right ascensions at 1950 are too small). 4) Elimination of inhomogeneities of the FK4 system by means of absolute and quasi-absolute observations. 5) Determination of individual correction to positions and proper motions of FK4 stars. 6) Addition of 3130 fundamental stars to extend the visual magnitude to about 9.5, to be published in the *FK5 Extension.*

It should be mentioned that the above improvements were possible because of the availability and/or reanalysis of observations of the sun (1900-1970), of lunar occultations (1820-1977), of Mars (1941-1971), of minor planets (1850-1977), and the JPL DE-108 Ephemeris based on optical or radar observations of the sun, planets and some space probes (Mariner 9, Viking). All in all the number of these observations exceeds 350,000. In addition, more than 150 catalogues of star observations have become available since the completion of the FK4 (Fricke, 1979b).

One should also take note here of the work in progress at the Astronomisches Rechen-Institut Heidelberg providing FK5 coordinates of a few extragalactic radio sources with radio and optical positions and thus the connections between the Stellar-CIS and the Radio Source-CIS, though with somewhat limited accuracy ($\sim 0\overset{''}{.}1$). Improvement of this particular problem is expected from the Space Telescope (Van Altena, 1978) which could increase the number of radio stars, observable by VLBI, in the FK5 to about 50. Such missions (e.g., Hipparcos) could also contribute to the determination of the fundamental equator and equinox with increased accuracies, by observations of the minor planets. This, of course, would mean improved ties with the planetary-CIS (discussed below) which nowadays is based on the observations mentioned in connection with the establishment of the FK5 equator and equinox. The astrometric satellite Hipparcos is described to be able to measure relative positions of some 100,000 stars to a precision of $0\overset{''}{.}0015$ and annual proper motions to $0\overset{''}{.}002$ over a lifetime of 2.5 years (Barbieri and Bernacca, 1979). A second mission ten years later could improve this figure by a factor of 5. This compares well indeed with the precision of ground-based observations of $0\overset{''}{.}04$ at best, requiring something like 50 years to obtain proper motions of comparable precision ($0\overset{''}{.}002$).

As far as the accuracy of the FK5-CIS is concerned, the question again is meaningful only in the sense of how precise the star positions in the FK5 are. The 1535 'Basic FK5' stars have a mean precision of $0\overset{''}{.}02$ and 0.7 mas per year in proper motion. The 980 stars in the bright (magnitudes 5.5–7.0) 'FK5 Extension' are precise to about $0\overset{''}{.}03$, while the 2150 additional faint (magnitudes 6.5–9.5) stars to about $0\overset{''}{.}06$.

An important extension of the FK5 is the International Reference Star (IRS) catalogue which is almost completed and will include about one star per square degree. It will include the AGK3R stars in the northern hemisphere, the SRS (Southern Reference Stars) catalogue in the Southern Hemisphere (Zverev et al., 1986) and some additional stars to insure the homogeneity of the distribution on the celestial sphere (Smith, 1986). A special effort was made to obtain a homogeneous system of proper motions (Corbin, 1978).

Further extensions should be based on the IRS itself or on future larger and more homogeneous catalogues like the Hipparcos catalogue mentioned above (Froeschle and Kovalevsky, 1982).

2.23 Dynamical Systems. The dynamics expressed in the equations of motion define a number of nonrotating planes which could be the basis of reference frames. Considering the observable planes that could be the basis of such a Dynamic-CIS, there are the planetary (including the earth-moon barycenter) orbital planes, the equator, the lunar orbital plane, and the orbital planes of certain high flying, thus only slightly perturbed, artificial earth satellites (e.g., Lageos or GPS). Since all of these planes have relative rotations, it is possible to derive a mean plane for a given epoch from an observable apparent plane, or a nonobservable invariant plane could be adopted (Duncombe et al., 1974). At this point, the definition of the origin of the system becomes important also, because relativistic effects necessitate the distinction between proper and coordinate times. In the radio-source or stellar quasi-inertial systems, the question of origin can be settled through appropriate corrections for aberration and parallax, etc., but here it is also necessary that a uniform and unambiguous time scale referenced to a nonrotating frame of specified origin be established (coordinate time). The practical implications of a global coordinate time scale is not treated here, but the problem should not be ignored (cf. Ashby and Allan, 1978; Guinot, 1989). In more practical (observational) terms one can distinguish between Planetary, Lunar and (artificial) Satellite CIS's, each frame defined, in theory, by two of the above-mentioned planes, and in practice by the available ephemerides (see Fig. 2).

In the case of the *planetary systems*, the defining planes are the equator and the ecliptic, their intersection being the line of the equinoxes. In practical terms the frame of the Planetary-CIS is defined by the ephemerides of the centers of masses of the planets, including the barycenter of the earth-moon system. The ephemerides, such as the JPL DE-200, are based on observations of the sun, the planets, and space probes (see Table 4). Since most modern ephemerides are computed through the numerical integration of the orbital equations of motion, the degree of satisfaction that can be obtained depends only on the completeness of the modeling, including the astronomical constants, the determination of the starting conditions and, of course, on the type, accuracy and distribution of the observed data. In this sense each planetary ephemeris defines its own reference frame. These should agree with each other within the observational accuracies. Connection between the Planetary-CIS's and the Stellar-CIS's is through the determination of the equinox and the equator, as explained earlier (Fig. 2).

In the case of the *lunar system*, the main references are the orbital plane of the moon and the equator of the earth. In practice the Lunar-CIS frame is again defined by the lunar ephemeris, which nowadays is most accurately determined from lunar laser observations made from the surface of the earth to reflectors deposited on the lunar surface. For this reason, the adequacy of the definition also depends on how well the lunar rotation (librations) can be computed. Since the most frequently used lunar ephemerides are generally calculated through numerical integration, the above dependence on modeling (especially on the effect of tidal dissipation in the earth), and on initial conditions, apply here also. The identity of the coordinate frame, such defined, may be compared to the other frames to certain accuracies (Fig. 2). Lunar occultation of stars, or the earlier Markowitz moon-camera photography, provide a connection to the Stellar-CIS; differential VLBI observations between radio sources deposited on the moon and the extragalactic ones would tie to the Radio Source-CIS. The connection to the Planetary-CIS is through solar eclipse observations, and also through the planetary ephemeris used when calculating the lunar ephemeris. There are also some other looser connections stemming from the orientation of the earth when its nonspherical gravitational effects on the lunar motions are taken into consideration. Present observations reveal a residual rotation (or accelerations) of the order of a few seconds of arc per century squared. This seems to be the present stability (i.e., the accuracy) of this quasi-inertial frame. It is unlikely that without stronger connections to a frame of better stability, this rotation can be eliminated. As it is, the accuracy of this CIS should compare favorably with that defined by the FK5 but only over a period of, say, a decade (Kovalevsky, 1979).

Data types to which modern planetary and lunar ephemerides are adjusted are listed in Table 4 The post-fit rms residuals indicate the accuracy of the data. The values listed without brackets are the units of the original observations; those within brackets give the comparable values for comparison purposes.

Table 4 Data in Modern Lunar and Planetary Ephemerides (Williams and Standish, 1989)

Type of Observation	*Time Span*	*Post-Fit Rms (km)*	*Residuals*	*No. of Obs.*
Radar Ranging				
Mercury	1966-	1.5	[0″.002]	500
Venus	1965-	1.5	[0″.002]	1000
Mars	1967-	2.2	[0″.003]	40000
Mars Closure	1969-82	0.15	[0″.0002]	200
Spacecraft Ranging				
Ma9 Orbiter (Mars)	1972-73	0.040	[0″.0002]	600
Viking Lander (Mars)	1976-80	0.007	[0″.000003]	900
	1980-82	0.012	[0″.000006]	400
Spacecraft Tracking (Range, Doppler)				
Pion&Voy (Jup,Sat)	1973-80	[200, 400]	[0″.05]	20000
Lunar Laser Ranging	1969-70	0.00100	[0″.0005]	10
	1970-75	0.00030	[0″.00016]	1700
	1976-85	0.00015	[0″.00008]	3000
	1985-	0.00006	[0″.00003]	600
Radio Astrometry				
Jupiter, ..., Neptune	1983-	[100, ..., 600]	0″.03	10
Ring Occultation				
Uranus	1978-	[1500]	0″.1	14
Optical Transits (Manual)				
Sun, Mercury, Venus	1911-	[700]	1″.0	37000
Mars, ..., Neptune	1911-	[150, ..., 10000]	0″.5	18000
Optical Transits (Photoelectric)				
Mars, ..., Neptune	1982-	[100, ..., 6000]	0″.3	1000
Astrolabe				
Mars, ..., Uranus	1961-	[100, ..., 4000]	0″.3	1500
Astrometry				
Pluto	1914-	[15000]	0″.5	1600

Earlier ephemerides of the moon and planets, based upon optical observations, have inherited errors directly from the catalogues upon which they have been based. These errors amount to a number of tenths of an arcsecond in angular position and a number of tenths of an arcsecond per century in angular motion; i.e., errors comparable to those that are known to exist in the FK4 fundamental reference system. Modern ephemerides based upon ranging observations show at least an order of magnitude improvement over their optically based predecessors. Williams and Standish (1989) selected the most important data types and calculated how sensitive these data are to changes in certain ephemeris elements. The sensitivities, in turn, indicate how well each of these elements may be determined through the data fitting, keeping in mind that the statistics of the actual determinations are improved due to the large number of observations but also that there are correlations among the various parameters. The summary of their findings is quoted below:

> 'The lunar laser ranging data is sensitive to a change in the lunar mean anomaly and its rate at levels of 0".0006 and 0".02/cy respectively. The data is also sensitive to the rate of the lunar longitude with respect to inertial space at a level of 0".04/cy. This rate error is dominated by the uncertainties in the precessional rates of the lunar perigee; the precessional rates themselves are due to the perturbations which depend on the orbital elements and gravitational harmonics of the earth and moon. At times away from the data span, the uncertainty (1".0/cy^2) in the tidally induced acceleration in longitude becomes predominant.
>
> 'For the planets, the most important data are the ranges to the Viking landers on Mars. We show that these ranges have a remarkable sensitivity to a number of differential angles: the difference in heliocentric longitudes between earth and Mars at a level of 0".00001, each longitude with respect to the perihelion of Mars at a level of 0".00004 and each longitude with respect to the perihelion of the earth at a level of 0".0002. Further, the corresponding level for the inclination of Mars' orbit upon the ecliptic is about 0".0002.
>
> 'Radar ranging to Mercury and Venus determines the longitudes of these planets with respect to the longitude of the earth (and therefore to Mars). These sensitivities are on the order of 0".005 and 0".003 respectively, since the data are accurate to the level of 1.5 km. The sensitivities to the inclinations upon the ecliptic are two orders of magnitude worse than that for Mars.
>
> 'Solar perturbations upon the lunar orbit provide sensitivity to both the differential longitude between the heliocentric earth and the geocentric moon and to the inclination of the lunar orbit to the ecliptic; 0".001 and 0".007 respectively.
>
> 'Since the lunar ranges are taken from the spinning earth, sensitivities to the earth's orientation, coupled with the terrestrial coordinates of the observing station, allow determinations of
> (1) the mutual inclinations of the equator, the ecliptic and the lunar orbital plane (0".002);
> (2) the longitude of the earth and moon with respect to the dynamical equinox (0".005); and
> (3) a tie between the ephemeris frame and the terrestrial reference system (0".001 in longitude, comparable to 0.0001 seconds in UT0).
>
> 'Finally, the fact that the lunar retroreflectors and the Viking landers are situated on the surfaces of the bodies, the ranges are sensitive to the physical orientations of the bodies themselves. The lunar librations affect the LLR data; the spin rate, obliquity and equinox of Mars influence the Viking ranges.
>
> 'The analytical sensitivity analyses have been substantiated by numerical examples though the correspondence is not exact because of differences in numbers of observations, correlations, additional data and other perturbating forces. However, even when all of these factors are considered, it is seen that the dynamical reference system may be determined better than 0".01 in position with respect to the dynamical equinox. Further, the mean motions of earth and Mars with respect to inertial space may be determined as well as 0".003/cy during the times of the highly accurate ranging data; the uncertainty for Mars will grow to about 0".015/cy over the course of many decades away from the present data.'

In the case of *satellite systems*, the problem is compounded by additional modeling problems related to the force field in which the satellite moves and by the fact that nowadays there are no direct connections to other frames of reference. Modern satellite tracking techniques (laser, GPS, etc.) all basically observe ranges or range differences and contain no direct directional information. The main reference planes, the orbital plane of the satellite and the equator, intersect along the line of nodes, the initial orientation of which therefore must be defined more or less arbitrarily. In the 'old days' of satellite geodesy, when satellites were observed photographically in the background of stars, this direction could be determined with respect to the stellar frame, though not much better than a few tenths of a second of arc. The accumulation of errors in describing the motion of the node with respect to a selected zero point, even for the most suitable high flying and small heavy spherical satellites (Lageos), may prevent a Satellite-CIS from being accurate over a long period of time, say beyond several months. In any case, in observational terms such a frame would be defined by the satellite ephemeris made available to the users by organizations which provide for the continuous tracking of the satellite in question. A current example would be the Precise Ephemeris of the U.S. Navy Navigational Satellite (Transit) System. As far as the connections to other systems are concerned, the only accurate possibility seems to be indirectly through the tracking stations. If two observational systems occupy the same station, one observing the satellite, the other, say, the radio sources, either simultaneously or after a short time interval (during which the movement of the station can be modeled), the connection between the satellite and radio source frames can be established (see Fig. 2). In fact, the now classical disparity between the JPL and SAO frames came to light through just such an arrangement, when the SAO longitudes determined from satellite camera tracking (thus in the FK4 frame) differed by those determined by JPL space probe tracking (in the planetary frame) by an amount (about $0\overset{''}{.}7$ in the early 1970's) consistent with the FK4 equinox motion with respect to the dynamical equinox, mentioned earlier. Only through such continuously maintained connections can the lifetime of a Satellite-CIS be extended, thus its accuracy increased.

2.3 Conclusions

From the above discussion, the following conclusions can be drawn:

1. The most accurate, long-term CIS is the one attached to extragalactic radio sources. It is accessible through VLBI observations. Other systems can be accurately connected to it by station collocation, space astrometry, VLA and differential VLBI observations (Fig. 2). The number of primary radio sources must be increased, especially in the Southern Hemisphere, to achieve isotropy.

2. The CIS attached to the FK5 is somewhat less accurate. Direct access to it is through optical star observations, which by nature are generally less accurate than VLBI observations. Its main value is in defining the fundamental mean system of coordinates and thereby providing a direction (the FK5 equinox) for the time (UT1) definition, and for the possible orientation of the Radio Sources-CIS. The latter function, however, stems from more of a traditional requirement and not from theoretical needs.

3. Of the Dynamical-CIS's, the accuracy of the planetary (including lunar) system is better than the FK5, with a rotational stability of 3 mas per century. The satellite systems by themselves are suitable for medium-term to short-term work only. The rotational stability can be extended by connections to the Radio Source-CIS through accurate and continuous observations at collocated stations or differential VLBI. Lunar laser ranging provides the best connection between the lunar and planetary systems.

4. If a *dynamical system* is based on the motion of planets, the ecliptic plays a privileged role and, naturally, the ecliptic is used in the definition of coordinates. Since equatorial coordinates are preferred to ecliptic ones for obvious instrumental reasons, the intersection of the ecliptic and the equator, the vernal equinox, becomes the natural origin of right ascensions. When the dynamical system is geocentric, the natural reference plane is the Laplace plane whose position depends upon the relative magnitude of the perturbations. For the moon, the solar effects are dominant and, practically, the Laplace plane is the ecliptic and, again, the equinox is the natural origin of equatorial coordinates. In the case of artificial satellites the perturbations due to the earth flattening are predominant so that the Laplace plane is the equator. The equator is, therefore, the natural fundamental plane, but the origin may be arbitrary.

Similarly, the choice of the equinox in the stellar systems is justified by the fact that they are partially dynamical systems based upon planetary theories. However, in the construction of the corresponding stellar frame, the difficulty of maintaining the theoretical origin is so serious that one is led to distinguish between the dynamical equinox which defines the origin of the system and the catalogue equinox which is the origin of the frame. In practice, the actual origins of the stellar reference frames are purely conventional and are not the dynamical equinox.

5. The situation will become even more conspicuous for frames derived from conventional *kinematic systems*. Even if, for the sake of continuity, the origin and the fundamental plane of such a system should be close to the equinox and the equator, they should be conventional points defined only by the realization of the corresponding frame. Otherwise, it would be necessary to introduce a complex dynamical model to define the origin at the expense of introducing inaccuracies in the system and an uncertainty in its realization by the frame. In practice, the solution might be analogous to the present situation for the terrestrial reference frame (see Section 3.2). One would establish an international organization that would provide the coordinates of radio sources in the conventional kinematic frame, taking into account eventual changes in the number and position of the reference sources, due, for instance, to the disappearance or motion of quasars or better measurements, in such a way that the changes should not introduce a rotation (or translation) of the system in the average statistical sense. It is an almost unavoidable conclusion that for geodetic and geodynamic applications the most useful CIS is just such a system (Kovalevsky and Mueller, 1981; Guinot, 1986).

3. CONVENTIONAL TERRESTRIAL SYSTEMS (CTS) OF REFERENCE

As mentioned in the Introduction, the CTS is in some 'prescribed way' attached to observatories located on the surface of the earth. The connection between the CTS and CIS frames of common origin by tradition (to be preserved) is through the rotations (Mueller, 1969)

$$[\mathrm{C\vec{T}S}] = \mathbf{SNP}\ [\mathrm{C\vec{I}S}]$$

where **P** is the matrix of rotation for precession, **N** for nutation (to be discussed in Section 4), and **S** for earth rotation (including polar motion). Polar motion thus is defined as the angular separation of the third (Z) axis of the CTS and the axis of the earth for which the nutation (N) is computed (e.g., instantaneous rotation axis, Celestial Ephemeris Pole, Tisserand mean axis of the mantle (see Section 4)).

Geodynamic requirements for CTS may be discussed in terms of global or regional problems. The former are required for monitoring the earth's rotation, while the latter are mainly associated with crustal motion studies in which one is predominantly interested in strain or strain rate, quantities which are directly related to stress and rheology. Thus for these studies, global reference frames are not particularly important although it is desirable to relate regional studies to a global frame.

For the rotation studies one is interested in the variations of the earth's rotation rate and in the motions of the rotation axis both with respect to space (CIS) and to the crust (CTS). The problem therefore is threefold:

(i) To establish a geometric description of the crust, either through the coordinates of a number of points fixed to the crust, or through polyhedron(s) connecting these points whose side lengths and angles are directly estimable from observations using the new space techniques (laser ranging or VLBI). The latter is preferred because of its geometric clarity.

(ii) To establish the time-dependent behavior of the polyhedron due to, for example, crustal motion, surface loading or tides.

(iii) To relate the polyhedron to both the CIS and the CTS. For the global tectonic problems only the first two points are relevant although these may also be resolved through point (iii).

In the absence of deformation, the definition of the CTS is arbitrary. Its only requirement is that it rotate with the rigid earth, but common sense suggests that the third axis should be close to the mean position of the rotation axis and the first axis be near the origin of longitudes.

In the presence of deformations, particularly long-periodic or secular ones, the definition is more problematic, because of the inability to separate rotational (and translational) crustal motions of the crust from those of the CTS.

One geophysical requirement of the reference system is that other geophysical measurements can be related to it. One example is the gravity field. The reference frame generally used when giving values of the spherical-harmonic coefficients is tied to the mean axis of figure of the earth. This frame should be simply related with sufficient accuracy to the CTS as well as to the CIS in which, for example, satellite orbits are calculated. Another example is height measurements with respect to the geoid.

The vertical motions may require some special attention, because absolute motions with respect to the center of mass have an immediate geophysical interest and are realizable. Again, if the center of mass has significant motions with respect to the crust, such a motion will be absorbed in the future CTS, if defined as suggested above. At present there is no compelling evidence that the center of mass is displaced significantly, at least at the decade time scale.

Apart from the geometrical considerations, the configuration of observatories should be such that (i), there are stations on most of the major tectonic plates in sufficient number to provide the necessary statistical strength, and (ii) the stations lie on relatively stable parts of the plate so as to reduce the possibility that tectonic shifts in some stations will not overly influence, at least initially, the parameters defining the CTS frame.

Finally one should realize that the problem of the geometric origin of the CTS is linked to that of a geocentric ephemeris frame. The center of mass of the earth is directly accessible to dynamical methods and is the natural origin of a geocentric satellite-based dynamical system. But, as such, it is model dependent, and, unless the terrestrial reference frame is also constructed from the same satellites (as is the case in various earth models such as GEM, SAO, GRIM), there may be inconsistencies between the assumed origin of a kinematically obtained terrestrial system and the center of mass. A time-dependent error in the position of the center of mass, considered as the origin of a terrestrial frame, may introduce spurious apparent shifts in the position of stations that may then be interpreted as erroneous plate motions. To avoid this problem the parameters defining the CTS frame should include translational terms as well.

3.1 Brief History of the Past Decade

Until 1984 the internationally adopted Woolard series of nutation, based on a rigid earth model, was computed for the instantaneous rotation axis of the rigid earth, and the Z axis of the CTS was the Conventional International Origin (CIO), defined by the adopted astronomic latitudes of the five International Latitude Service (ILS) stations, located approximately on the 39°08' parallel. These were assumed to be motionless relative to each other, and without variations in their respective verticals (plumblines) relative to the earth. Thus, conceptually, polar motion was to be determined from latitude observations only at these ILS stations. This had been done for over 80 years, and the results are the best available *long-term* polar motions, properly, but not very accurately, determined. The first axis of the CTS was defined by the assigned astronomic longitudes of time observatories (around 50) participating in the work of the Bureau International de l'Heure (BIH).

Due to the fact that in most geodetic and astronomical applications accurate shorter-term variations of polar motion were needed, which were not available with sufficient accuracy from the ILS observations, and especially not fast enough, polar motion was also determined from latitude and/or time observations at a larger number of observatories participating in the work of the International Polar Motion Service (IPMS), as well as of the BIH. In the resulting calculations, the earlier definition of the CIO could not be maintained. The common denominator being the Woolard series of nutation, observationally the Z axis of the CTS was thus defined by the coordinates of the pole as published by the IPMS or by the BIH. Thus it was legitimate to speak of IPMS and BIH poles of the CTS (in addition to the CIO). The situation had become even more complicated because Doppler and laser satellite tracking, VLBI observations, and lunar laser ranging also can determine *variations* in the earth rotation vector (including polar motion), some of which were incorporated in the BIH computations. Further confusion arose due to the fact that the BIH had two systems: the BIH 1968 and the BIH 1979, the latter due to the incorporation of certain annual and semiannual variations of polar motion determined from the comparisons of astronomical (optical) results with those from Doppler and lunar laser observations (Feissel, 1980).

Though naturally every effort was made to keep the IPMS and BIH poles of the CTS as close as possible to the CIO, the situation was not satisfactory from the point of view of the geodynamic accuracy requirement of a few parts in 10^9. The accuracy of the pole position was estimated to be $0''.01$, and that of the UT1, 1 ms ($\sim 5 \times 10^{-8}$) for five-day averages (Guinot, 1978). These figures, of course, did not include biases from the definition problems mentioned.

From 1984 onward, the IAU 1980 (Wahr, 1981) series of nutation for the nonrigid earth gives the space position of the Celestial Ephemeris Pole (CEP) (see later). The pole of the CTS officially remained the same as before.

3.2 The New CTS

There seemed to be general agreement that the new CTS frame conceptually be defined similarly to the CIO-BIH system (Bender and Goad, 1979; Guinot, 1979; Kovalevsky, 1979; Mueller, 1975a), i.e., it should be attached to observatories located on the surface of the earth. The main difference in concept was that these can no longer be assumed motionless with respect to each other. Also they must be equipped with advanced geodetic instrumentation like VLBI or lasers, which are no longer referenced to the local plumblines. Thus the new transformation formula, again assuming a common origin, may have the form

$$[O\vec{B}S]_j = \vec{L}_j' + [C\vec{T}S]_j = \vec{L}_j' + \mathbf{SNP}\ [C\vec{T}S]_j \tag{1}$$

where $\vec{L}_j'$ is the vector of the 'j' observatory's movement on the deformable earth with respect to the CTS, computed from suitable models (see Fig. 1 and Section 4); **NP**, the nutation and precession matrices computed with the new 1976 IAU constants and the 1980 IAU series of nutation (see Section 4); and **S**, the rotation matrix between the CTS and the true frame for which the nutation is computed (see eqs. (7)-(9) in Section 4).

In the above equation the coordinates of the observatory 'j' $[C\vec{T}S]_j$, in the Conventional Terrestrial Reference Frame, are related to the coordinates determined by the technique 'o' $[O\vec{B}S]_j^o$, in its own reference frame, through the well-known transformation equation:

$$[C\vec{T}S]_j + \vec{\delta}^o + \mathbf{R}_1(\beta_1^o)\,\mathbf{R}_2(\beta_2^o)\,\mathbf{R}_3(\beta_3^o)[C\vec{T}S]_j + c[C\vec{T}S]_j + \vec{L}_j' = [O\vec{B}S]_j^o + \vec{v}_j \tag{2}$$

where $\vec{\delta}^o$ contains the three translation components between the CTS frame and that inherent in the technique 'o', $\vec{\beta}^o$ are the three (usually very small) rotations, and c a differential scale factor. $\vec{L}_j$ is the vector of deformation, not containing global rotations nor translations, and $\vec{v}_j$ is the residual vector.

Another set of equations derived in (Zhu and Mueller, 1983) relate other parameters in eq. (1), specifically the earth rotation parameters (ERP) in the matrix S determined by the technique 'o' in its own frame of reference, to those referring to the CTS frame:

$$\begin{aligned} x_p - \beta_2^o + \alpha_1^o \sin\theta + \alpha_2^o \cos\theta &= x_p^o + v_{x_p} \\ y_p - \beta_1^o - \alpha_1^o \cos\theta + \alpha_2^o \sin\theta &= y_p^o + v_{y_p} \\ \omega_d\,\mathrm{UT1} + \beta_3^o - \alpha_3^o &= \omega_d\,\mathrm{UT1}^o + v_{UT1} \end{aligned} \tag{3}$$

where x_p^o, y_p^o and $\mathrm{UT1}^o$ are the observed ERP's; x_p, y_p and UT1 are those referenced to the CTS; ω_d is the conversion factor, $\vec{v}$ is the residual vector of the observed ERP's and θ the sidereal time. Finally, $\vec{\alpha}^o$ are the small rotations between the Conventional Inertial Reference Frame and the Inertial Frame inherent in the technique 'o', i.e.,

$$[C\vec{I}S]^o = \mathbf{R}_1(\alpha_1)\,\mathbf{R}_2(\alpha_2)\,\mathbf{R}_3(\alpha_3)\,[C\vec{I}S] \tag{4}$$

For each ERP series 'k' of 1-1.2 years length (or longer), generated by the technique 'o', one can fit the following type of circular model:

$$
\begin{aligned}
&a_1^{o,k} + a_2^{o,k} \cos A + a_3^{o,k} \sin A + a_4^{o,k} \cos C + a_5^{o,k} \sin C = x_k^o + v_{x_k} \\
&a_6^{o,k} - a_2^{o,k} \sin A + a_3^{o,k} \cos A - a_4^{o,k} \sin C + a_5^{o,k} \cos C = y_k^o + v_{y_k}
\end{aligned}
\qquad (5)
$$

where A is the annual frequency and C the Chandler frequency, x_k^o, y_k^o are the observed ERP's in the series k, and v the residuals.

The coefficients $a^{o,k}$ allow the computation of the amplitude of the annual motion

$$\sqrt{(a_2^{o,k})^2 + (a_3^{o,k})^2} \;,$$

that of the Chandler motion

$$\sqrt{(a_4^{o,k})^2 + (a_5^{o,k})^2} \;,$$

as well as the coordinates of the center of the polhode $a_1^{o,k}$ and $a_6^{o,k}$.

If two techniques, 01 and 02, are collocated or tied together by local surveys at the station 'j', the following additional relationship holds

$$[O\vec{B}S]_j^{01} - [O\vec{B}S]_j^{02} = \vec{\Delta}_j + \vec{v}_{\Delta j} \qquad (6)$$

where $\vec{\Delta}_j$ is the coordinate difference vector from the local survey and $\vec{v}_{\Delta j}$ the residual vector.

Equations (2) - (6) can be used as observation equations by an international service to determine the parameters defining and maintaining the CTS and providing the relationship versus the terrestrial frame of each technique ($\vec{\delta}^{\,o}$, $\vec{\beta}^{\,o}$, c), versus the CIS (x_p, y_p, UT1), and the latter's relationship to the technique inertial frame ($\vec{\alpha}^{o}$):

$[C\vec{T}S]_j$ and $\vec{L}_j$	for each observatory 'j', to define the CTS
$\vec{\delta}^{\,o}$, $\vec{\beta}^{\,o}$, c, $\vec{\alpha}^{o}$ and $\vec{a}^{\,o,k}$	for each technique 'o' and ERP series 'k'
x_p, y_p, UT1	for the service to provide the S matrix in eqs. (1) and (9)

As far as the origin of the CTS is concerned, it could be centered at the center of mass of the earth, and its motion with respect to the stations can be monitored either through observations to satellites or the moon, or, probably more sensitively, from continuous global gravity observations at properly selected observatories (Mather et al., 1977). For the former method, the condition

$$\sum_D w_D \, \vec{\delta}_D^{\,o} = 0$$

could be imposed on the above adjustment. The summation would be extended to all the above dynamic techniques D with given relative weights w_D. A similar condition could also be imposed on the scale extended to techniques defining the best scales (probably VLBI):

$$\sum_{S} w_S \, c_S{}^o = 0$$

Other conditions between two independent techniques 01 and 02 may include the following obvious relations

$$a_1{}^{01} - a_1{}^{02} = \beta_2{}^{01} - \beta_2{}^{02}$$

$$a_6{}^{01} - a_6{}^{02} = \beta_1{}^{01} - \beta_1{}^{02}$$

The above method of determining ERP or some variation thereof needs to be initialized in a way to provide continuity. This could be done through the IPMS or BIH poles, and the BIH zero meridian, at the selected initial epoch (or averaged over a well-defined time interval, say 1 to 1.2 years), uncertainties in their definition mentioned elsewhere being mercifully ignored, i.e.,

$$\sum_{1\text{-}1.2\text{ yrs}} (x_p - x_p{}^{BIH}) = 0$$

$$\sum_{1\text{-}1.2\text{ yrs}} (y_p - y_p{}^{BIH}) = 0$$

etc.

It is probably not useless to point out that in the system described above, the most important information for the users of the service are the ERP's and the transformation parameters, but for scientists new knowledge about the behavior of the earth will come from the analysis of the residuals after the adjustment.

The IAU and IUGG recently made practical recommendations on the establishment of such a (or very similar) Conventional Terrestrial System, including the necessary plans for supporting observatories and services by establishing the International Earth Rotation Service, effective 1 January 1988 (Wilkins and Mueller, 1986). The goal of the service is the determination of the total transformation between the CTS and CIS. Thus the service will publish not only ERP determined from the repeated comparisons (the past situation), but also the models and parameters described above in eqs. (1) - (6), i.e., the parameters defining the whole system. (See Section 5.)

3.3 Reference Frame Ties

3.31 Ties Between the CIS Frames. 'Measurements are inherently simpler to make and generally more accurate in their "natural" frame and hence should always be reported as such. However, to benefit from the complementarity of the various techniques, knowledge of the frame interconnections (both the rotation and the time-variable offset) is essential' (Dickey, 1989). These are summarized in Figs. 2 and 3.

Recent activity in this area is indicated by the number of boxes and lines in Fig. 3, entitled Connections 1986 (the accuracy cutoff here is 0.″05); a similar figure in an earlier paper (Williams et al., 1983) had fewer boxes and connecting lines. For example, ten lines instead of fifteen connected the targets with the techniques, and radio stars were listed as prospects for the future. The *lunar planetary system*, integrated in a joint ephemeris, is by its nature unified by the dynamics (Williams and Standish, 1989). The *radio frame is tied to the ephemeris frame* in several ways; one is via differential VLBI measurements of planet-orbiting spacecraft and angularly nearby quasars (Newhall et al., 1986). Another is the determination of a pulsar's position in the ephemeris frame (via timing measurements) and the radio frame (via radio interferometry, see Backer et al., 1985). Very Large Array (VLA) observations of the outer planets (Jupiter, Saturn, Uranus and Neptune) or their satellite provide an additional tie between these two frames (Muhleman et al., 1985).

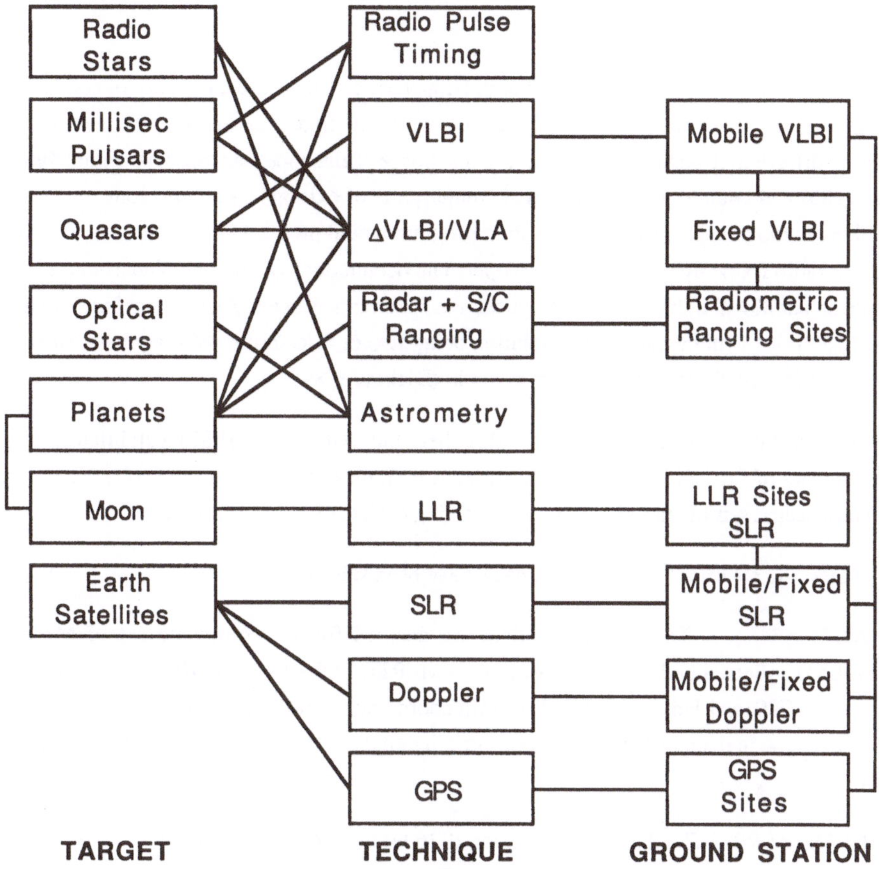

Fig. 3 Reference frame connections (Dickey, 1989).

As for an *optical-radio frame* tie, a preliminary link has been established between the FK5 optical frame and the JPL radio reference frame via the differential VLBI measurement of optically bright radio stars and angularly nearby quasars coupled with comparisons of their optical positions (see Lestrade et al., 1988), and also by the use of the optical positions of quasars (Purcell, 1979). The *optical and ephemeris frames* are tied by optical observations of the planets. Dickey (1989) also treats a few of the frame ties in greater detail; for example, for the connection between the radio and the ephemeris frames. For the other ties, the highlights are given with reference to a more detailed account.

Dickey (1989) also outlines the future with ongoing and planned efforts in several areas: Improved ephemeris-radio frame ties can be accomplished by VLBI observations of pulsars, additional VLA observations of the outer planets and satellites, and future differential VLBI experiments (such as that with orbiting spacecraft around Jupiter and Saturn). The millisecond pulsar PSR1937+214, having a period of

1.6 ms, has exceptionally low timing noise. Its position in the ephemeris frame can be measured to ~1 mas. This will allow a radio-planetary frame tie, limited only by the accuracy of an interferometric position measurement. Roughly, a factor of five improvement (down to 0."01) is expected here with the full implementation of VLBI observations. An initial experiment of this type has been executed by R. Linfield and C. Gwinn.

As already mentioned, for optical astrometry, Hipparcos will measure a network of stars over the entire sky with accuracies of ~2 mas (Kovalevsky, 1980), while the Space Telescope will measure small fields with similar differential accuracy. However, the Space Telescope can observe much fainter objects (Jeffreys, 1980) and could observe the optical counterparts of extragalactic radio sources, all but possibly one of which are too faint for Hipparcos. A joint program would produce an accurate stellar network linked to the quasar radio frame by the Space Telescope. The occultations of stars by planets and planetary rings can provide an additional link between the optical and ephemeris frames. Also, optical interferometry offers exciting possibilities with the potential resolution being two or three orders of magnitude finer than that of VLBI (Reasenberg, 1986). More details are given in (Dickey, 1989).

3.32 Ties Between the CTS Frames. Boucher and Altamimi (1987) established relationships between a number of Conventional Terrestrial Reference Frames based on colocated observation stations and eq. (2). The selected sets of station coordinates defining each CTS are as follows:

(i) CTS (VLBI). Three sets of station coordinates have been selected:

CTS(NGS) 87 R01. The coordinate data are derived from a composite set of Mark III VLBI observations collected under the aegis of project MERIT, POLARIS, and IRIS and conducted between September, 1980, and January, 1987. Westford coordinates were fixed to their initial values. The IRIS terrestrial frame is made more nearly geocentric by applying the BTS 1985 translations (Carter et al., 1987).

CTS(GSFC) 87 R01. The data acquired since 1976 by the NASA Crustal Dynamics Project and since 1980 by the NGS POLARIS/IRIS programs. The terrestrial frame is defined by the position of the Haystack 37-M antenna and the BIH Circular D values for 1980 October 17 (Ma et al., 1987).

CTS(JPL) 83 R05. The coordinate data are from the JPL Time and Earth Motion Precision Observations (TEMPO) project, using the DSN radio telescopes. The reference frame solution is tied to the BIH on 20 December 1979 (Eubanks et al., 1984).

(ii) CTS (Lunar Laser Ranging). The coordinate data are from the JPL solution: SSC(JPL) 87 M01 containing four stations, two at Fort Davis, one at Haleakala (Maui), and one at Grasse. The nominal planetary and lunar ephemeris DE121/LE65 was used in the reduction. The ephemeris uses the equator and equinox of B1950.0. It is on the dynamical equinox and has a zero point consistent with the FK5 catalogue (Newhall et al., 1987).

(iii) CTS (Satellite Laser Ranging). Two sets of station coordinates have been selected:

CTS(CSR) 86 L01. The solution is based on Lageos ephemeris from May, 1976, to September, 1986, using the model Lageos Long Arc 8511. The force model, referred to as the CSR 8511 system, adheres closely to the MERIT standards. The tectonic plate motion model AM1-2 of Minster and Jordan (1978) was used and the epoch of the derived station coordinates is 1983 January 1. The GM value is 398600.4404 km^3/s^2 (Schutz et al., 1987).

CTS(DGFI) 87 L01. The solution is computed from Lageos observations covering the period 1980 to end 1984 and based on five yearly solutions. By the rates of change of the yearly solutions, the station coordinates then were related to the same reference epoch 1984.0. The reference frame was defined by the three coordinates (longitude, latitude of Yaragadee (7090) and latitude of Wettzell (7834)) which were held fixed in the five solutions. The GM value is 3.98600448 E + 14 $m^3\ s^{-2}$, initial ERP series were from homogeneous BIH series and other constants from MERIT Standards (Reigber et al., 1987).

(iv) CTS (Doppler). Station coordinates are from DMA Doppler project SSC(DMA) 77 D01 solution, and other Doppler campaigns containing more than 100 station positions. They are determined in the NSWC9Z2 datum by point positioning using Precise Ephemerides.

Three comparisons have been performed to get an idea about the consistency of different solutions and relations between these solutions related to a same technique. Table 5 summarizes these different comparisons.

Table 5 Transformation Parameters Between Different CTS Frames (Boucher and Altamimi, 1987) (the uncertainties are given in the second line)

CTS	δ_1 *m*	δ_2 *m*	δ_3 *m*	$(c\text{-}1)10^{-6}$	β_1	β_2	β_3	*CS*/RMS*
NGS 87R01 – GSFC87R01	1.697 0.006	-0.998 0.006	0.339 0.007	0".003 0.001	-0".001 0.000	-0".001 0.000	-0".003 0.000	12 1 cm
CSR 86L01 – DGFI87L01	-0.007 0.023	-0.026 0.023	0.074 0.023	0.015 0.003	0.013 0.001	-0.009 0.001	0.125 0.001	37 12 cm
CSR 86L01 – CSR 85L07	-0.080 0.024	0.040 0.023	0.080 0.022	0.015 0.003	0.004 0.001	-0.003 0.001	0.009 0.001	35 11 cm

*Number of collocated stations.

The first comparison is between two VLBI solutions CTS(NGS) 87 R01 and CTS(GSFC) 87 R01 containing 12 colocated stations. Note the 1 cm of rms issued from this comparison. The origin difference between the two solutions is due to the arbitrary choice of the VLBI origin in the definition of the terrestrial frame.

The second comparison is between two SLR solutions CTS(CSR) 86 L01 and CTS(DGFI) 87 L01 containing 37 collocated sites. In this case the rms is about 12 cm. Note also a rotation of 125 mas about the Z axis between the two solutions.

The last comparison is between the two last SLR solutions of CSR of 85 and 86 giving an RMS of about 11 cm. Note here that the scale factor has been decreased of about 1.5×10^{-8} from 85 to 86 solution.

The slightly larger scatter (10 cm level) of SLR data is mainly explained by the mixture of good third generation stations (4 cm level) with some older ones (20 to 50 cm).

A combination of all above data has also been performed incorporating 51 collocated sites and making use of the plate tectonic absolute motion model AMO-2 derived from the global RM-2 model (Minster and Jordan, 1978).

Table 6 lists the transformation parameters of the individual systems with respect to a global one whose origin is constrained to that of JPL 87M01 (LLR) and CSR 86L01 (SLR), the scale to CSR 86L01 (SLR), and the orientation to NGS 87R01 (VLBI). Some conclusions about the origin, scale and orientation of the individual CTS's with respect to the global one: Knowing that the origin of the adjusted system is from SLR and LLR, the origin of all VLBI solutions remains arbitrary. The level of consistency of the scale factor is 10^{-8} for the different solutions. Some variations for VLBI and LLR solutions are due to a relativistic bias in the definition of the terrestrial system (Hellings, 1986; Boucher, 1986). The orientation of the individual terrestrial systems is usually realized through BIH values. The differences in orientation of the different solutions are arbitrary and of some mas level.

Tables similar to Table 6 are published in the Annual Reports of the BIH (or IERS) giving the transformation parameters for all CTS techniques participating in the work of the Service, as per Section 3.2.

Table 6 Transformation Parameters from the Individual 1984.0 CTS Systems to the 'Global' CTS (Boucher and Altamimi, 1987) (the uncertainties are given on the second line)

CTS	δ_1 *m*	δ_2 *m*	δ_3 *m*	$(c-1)10^{-6}$	β_1	β_2	β_3
NGS 87 R01	-0.009	-0.111	-0.112	0.023	0".000	0".000	0".000
	0.035	0.036	0.035	0.004	0.000	0.000	0.000
GSFC 87 R01	-1.696	0.862	-0.463	0.020	0.001	0.000	0.003
	0.029	0.034	0.032	0.004	0.001	0.001	0.001
JPL 83 R05	-0.062	0.234	0.140	0.015	0.001	0.011	0.000
	0.032	0.036	0.035	0.005	0.002	0.002	0.001
JPL 87 M01	0.000	0.000	0.000	0.020	-0.004	0.009	0.004
	0.000	0.000	0.000	0.017	0.005	0.005	0.005
CSR 86 L01	0.000	0.000	0.000	0.000	0.003	0.005	0.008
	0.000	0.000	0.000	0.000	0.002	0.001	0.002
DGFI 87 L01	-0.015	0.021	-0.053	-0.015	-0.010	0.014	-0.115
	0.041	0.041	0.040	0.006	0.002	0.002	0.002
DMA 77 D01	0.302	0.096	4.645	-0.605	-0.030	-0.005	0.797
	0.219	0.206	0.195	0.026	0.009	0.009	0.006

4. MODELING THE DEFORMABLE EARTH

In this section we will try to highlight the modeling problems associated with the components of transformation between the CIS and CTS mentioned in Section 3.

4.1 Precession (P)

At the XVIth General Assembly in Grenoble in 1976, the IAU adopted a new speed of general precession in longitude of 5029."0966 per Julian century at the epoch J2000.0 (JED 2451545.0) effective 1984. This value when referred to the beginning of the Besselian year B1900.0 is 5026."767 per tropical century, which may be compared to the previously adopted value of 5025."64 per tropical century at B1900.0. The change was calculated by Fricke (1977) from proper motions of stars in the systems GC, FK3, N30 and FK4. From the results, a correction to Newcomb's lunisolar precession in longitude was recommended. This value combined with a correction to Newcomb's planetary precession, due to the improved 1976 IAU values of planetary masses, resulted in the above new precessional constant. Expressions to compute the effect of precession from one epoch to another were developed by Lieske et al. (1977); and the usual equatorial parameters, z, θ, ζ_0, to be used in the precession matrix (Mueller, 1969),

$$\mathbf{P} = \mathbf{R}_3(-z)\, \mathbf{R}_2(\theta)\, \mathbf{R}_3(-\zeta_0) \tag{7}$$

to and from the epoch J2000.0 were computed by Lieske (1979). This transformation is the currently adopted one between the CIS (say, the FK5 at J2000.0) and an interim 'Mean Equator and Equinox Frame' of some date (see Section 4.52).

4.2 Nutation (N)

The nutation story is much more complex. First of all, the nutation matrix is (Mueller, 1969)

$$\mathbf{N} = \mathbf{R}_1\,(-\varepsilon - \Delta\varepsilon)\, \mathbf{R}_3(-\Delta\psi)\, \mathbf{R}_1(\varepsilon) \tag{8}$$

where ε is the obliquity of the ecliptic, $\Delta\varepsilon$ is the nutation in obliquity, and $\Delta\psi$ the nutation in longitude, computed from a certain theory of nutation. This matrix allows transformation from the aforementioned interim mean frame of date to the (also) interim true frame of the same date. This part is clear and without controversy. The complexities are in the agreement reached on the theory of nutation when computing the above parameters. Kinoshita et al. (1979) give an historical review:

> 'In astronomical ephemerides, nutation has been computed until now by the formulae which were given by Woolard (1953). The coefficients of the formulae are calculated assuming that the Earth is rigid. However, it has been found in recent analyses of observations ... that some coefficients of actual nutations are in better agreement with values calculated by the nonrigid Earth theory.
>
> 'Moreover, Woolard (1953) gave the nutation of the axis of rotation. Therefore, a small and nearly diurnal variation appears in the latitude and time observations, which is the so-called dynamical variation of latitude and time, or Oppolzer terms. In the global reduction of latitude and time observations, such as polar motion or time services, the Oppolzer terms have been until now removed from the data at each station (cf. BIH Rapport Annuel 1977, p. A-3) or counted out as a part of the nonpolar common z and τ-terms (IPMS Annual Report 1974, p. 11). On the other hand, Atkinson (1973) pointed out that if the (forced) nutation of the axis of figure is calculated instead of rotation axis, such a complicated treatment becomes unnecessary.
>
> 'Considering these situations, the IAU investigated the treatment of nutations, together with the system of astronomical constants which should be used in new ephemerides, and set up the "Working Group of IAU Commission 4, on Precession, Planetary Ephemeris, Units, and Time-Scales." The results by the Working Group are given in the report of Joint Meeting of Commissions 4, 8, and 31, in Grenoble, 1976 (Duncombe et al., 1976). In the report, the proposal by Atkinson is adopted, and the formula for computing the (forced) nutation of figure axis is shown clearly and in detail, by using the equation-numbers given by Woolard (1953).

However, the amendments of coefficients taking account of the nonrigidity of the Earth have not been adopted. In regard to this problem, it was noted that there should be a possibility of making further amendments in Kiev Symposium ... .

'At the IAU Symposium No. 78 in Kiev in 1977, the problem with the nonrigid values of nutation was discussed, and a series of new values were recommended which seemed to be based on Molodenskij's nonrigid theory. In the Symposium, however, it was recommended that the axis for which the nutation should be computed was the axis of rotation. This recommendation reversed the resolution given at Grenoble.

'In accordance with the resolution at the Kiev Symposium, an "IAU Working Group on Nutation under Commission 4" was set up and is investigating these two problems, in order to prepare a fully documented proposal for the next IAU General Assembly in Montreal in 1979. In the second draft of the Working group circulated on Nov. 16, 1978, the following conclusions are reported: (1) as for the axis to be referred, the Grenoble resolution is still valid, and (2) as for the coefficients of nutation series, the value in which the nonrigidity of the Earth is taken into account should be adopted as a working standard of astronomical observations. In the draft, a table of nutation series is given, and the numerical values in the table are based on the rigid theory by Kinoshita (1977), with use of IAU (1976) System of Astronomical Constants, and are modified by Molodenskij's nonrigid theory (Molodenskij, 1961).'

As we understand it, the Kinoshita theory above is for the nutation of the axis of maximum moment of inertia of the 'mean shape of the elastic mantle' (briefly, 'mean axis of figure of the mantle'). To add to the history, after the above-quoted Working Group Report was circulated, a new proposal was made by J.M. Wahr and M.L. Smith of CIRES that it would be preferable to adapt the nonrigid earth results of Wahr (1979) for the earth model 1066A developed by Gilbert and Dziewonski (1975). This model is a rotating, elliptically stratified linearly elastic and oceanless earth with a fluid outer core and a solid inner core. The nutations are computed for the 'Tisserand mean figure axis of the surface', which is also a mean mantle fixed axis (Wahr, 1979). The IAU in Montreal in 1979 considered both proposals and opted for the Kinoshita et al. (1979) series. A few months later in December, 1979, the IUGG in Canberra, in Resolution No. 9 addressed to the IAU, requested reconsideration in favor of the Wahr model. The IAU subsequently adopted Wahr's series as the IAU 1980 Theory of Nutation.

It should be pointed out that regardless of the fact that in geodetic or geodynamic applications we are only concerned with the total transformation SNP, it is of scientific importance to understand clearly the definition of the interim true equator and equinox frame of date, more specifically, the exact definition and the desirability (from the observability point of view) of the axis for which the nutation is computed.

In order to simplify the discussion, let us start with the rigid model. The motion of each of the axes, i.e., the axis of figure (F) (maximum moment of inertia), of the angular momentum (H), and the instantaneous rotation axis (I) are described by differential equations. If we want to refer to one of these axes we have to consider the complete solution of the differential equations, i.e., the free solution and the forced solution components. Confusion can arise if one refers to only one solution component (forced or free), but still calls it axis of figure, instantaneous rotation axis, etc. It is mandatory to point out which solution component one refers to. Neglecting to do so has been the reason for the by now classical confusing controversy about the Atkinson papers, though Atkinson (1975, p. 381) clearly states: 'Accordingly, when we speak of computing the nutations for either axis, we mean here computing the *forced motion* only, excluding the appropriate fraction of the noncomputable Chandlerian wobble.' Unfortunately, he, and others as well, then continue to use the term 'axis of figure' sometimes in the sense of the axis of maximum moment of inertia and at other times in the sense of the forced motion of the axis of figure.

A remark concerning the 'Eulerian pole of rotation' (E_0) as given by Woolard seems in order also. Quoting once again Atkinson (1976):

> 'The wording of the resolution on nutation, and the notes on it, which have been circulated by the Working Group, avoid all explicit mention of the axis of figure, even though they specify that the coefficients which Woolard gives for that axis shall be inserted, and they refer to the "Eulerian pole of rotation" although this cannot ever, in principle, coincide with the celestial pole and really has no more direct connection with the observations than is shown for it in [his] Fig. 2, i.e., none at all.'

The difference between the Eulerian pole of rotation (E_0) and the pole which Atkinson talks about is due to a homogeneous solution component. (E_0) is obtained from the complete solution of (I) by subtracting the periodic diurnal body-fixed motions of (I).

Consequently, the point E_0 has no periodic motion with respect to the crust, but it does have such a motion is space which is exactly the free nutation. Although this spatial motion is conceptually insignificant considering the observation technique (fundamental observations at both culminations), one gets another point, which is called the (true) Celestial Pole (C) in (Leick and Mueller, 1979), by subtracting the forced body-fixed motions of (H) from the complete nutation set of (H). The thus obtained axis (C) has no periodic diurnal spatial motion because the homogeneous solution of the angular momentum (H) is constant (zero). Equivalently, one can say that the nutations of (C) correspond to the forced solution of the axis of figure (rigid case, of course). This is the pole which Atkinson talks about and which is called (mistakenly) the 'mean axis of figure'. There is no doubt that this is the point to which the astronomical observations as well as lunar laser ranging refer, and the nutation should be adopted for this point. As for terminology, the IAU in 1979 named this (C) pole appropriately the *Celestial Ephemeris Pole* because its motion characteristics, i.e., no periodic diurnal motion relative to crust or space, have always been associated with the concept of the celestial pole. It would be preferred that the word 'figure' be dropped entirely for several reasons. First, one intuitively associates the axis of figure with the one for which the moment of inertia is maximum. This is true for the (C) only if the free solution (Chandler) is zero. But this is, generally, not the case. Second, the conceptual definition of (C) can easily be extended to elastic models or models with liquid core (the IAU 1980 case). Moreover, in order to emphasize that the observations take place on the earth surface, it would be useful to denote the actual pole accessible to the fundamental observation techniques by another designation, e.g., (CO), similarly to UT0. The 'O' would indicate that the nutations of this pole can in principle be determined only from observations because of the lack of a perfect earth model. Any nutation set based on a model is only an approximation to the nutations of the (CO). In this sense the rigid earth nutations of (I), (H) or (F) are all equivalent. Each of these nutations defines its own pole which has a diurnal motion around the (CO). The purpose of the measuring efforts is to find the corrections to the adopted set of nutations in order to get those of the (CO), the only pole which is observable.

Some have suggested the term 'zero excitation figure axis' for what is called above the (CO). The term 'zero excitation' would not reduce the confusion. The spatial motion of this axis is computed by adding

Atkinson's terms to Woolard's series, but this is equivalent to the *forced* motion of the axis of figure (rigid case). The observed motion of the (CO) relative to the crust only appears as a motion of zero excitation (free motion) at the first sight. Since the conceptual observation time of one position determination is one day, the observed position of the (CO) will always include effects due to oceans, atmospheric mass redistribution, etc., i.e., the geophysical nutations. These motions are better known as the annual polar motion and the subharmonics. Therefore, the zero-excitation pole is not directly observable. On the other hand, the concept of the (CO) can still be used in this case since it is by definition the pole which has no periodic diurnal motions relative to the crust or to space.

There is also the common offset of both the rotation axis and the (CO) caused by the tidal deformation (McClure, 1973). This is an offset of (I) and (CO) relative to (H) for the perfectly elastic model as compared with the rigid model. We have to remember, again, that the observations refer to the (CO). Therefore, any nutation correction which is derived from observations (based on an adopted set of nutations) will automatically give the corrections to the (CO). Consequently, there is no need for a special consideration of this possible separation, at least not for those harmonic motions whose amplitudes are derived from observations. In fact, the analysis of the observed fortnightly term seems to contradict somewhat the predicted amplitude for the perfectly elastic model.

From the above discussion, it also seems clear that ideas advocating the adoption of nutations for the axis of angular momentum violate the concept of observability. It is true that the direction of (H) in space is the same for the rigid, elastic, or any other reasonable earth model. But this property is not of much interest to the astronomer or geodesist who tries to determine the orientation of the earth. It is *conceptually simpler* to refer to an axis which is observable. More on this in Section 4.4.

Returning now to the problem of the IAU 1979/1980 adopted sets of nutations, from the geodetic point of view there seemed to be little difference whether the Kinoshita series was retained or the Wahr set was adopted. Using more and more realistic earth models is certainly appealing. On the other hand, severely model-dependent developments are liable to change as models improve (see Section 4.51). Improving the nutation series from geodetic observations leads to earth model improvements, one of the main goals in geophysics. This is one of the dichotomies between geodesy and geophysics.

4.3 Earth Rotation (S)

The two components of the **S** matrix (Mueller, 1969),

$$\mathbf{S} = \mathbf{R}_2(-x_p)\,\mathbf{R}_1(-y_p)\,\mathbf{R}_3(\mathrm{GAST}) \tag{9}$$

are the rotational angle of the first (X) axis of the CTS with respect to the first axis of the interim true equator and equinox frame of date, measured in the equator of the Celestial Ephemeris Pole (or whatever is defined in the **N** matrix), also known as the Greenwich Apparent Sidereal Time (GAST), and the polar motion coordinates (x_p, y_p) referred to the same pole and the Z axis of the CTS.

In this connection it should be mentioned that some authors prefer a different 'true' frame, which would have 'no rotation' about the Z axis (Guinot, 1979; Murray, 1979; Kinoshita et al., 1979). It is in such an interim frame where, for example, a nutational theory can be conveniently developed, or satellite orbits calculated (Kozai, 1974). Such a frame can be obtained from the CIS by a modified **NP** transformation, where

$$\mathbf{N} = \mathbf{R}_1 (-\Delta\varepsilon \cos M + \Delta\psi \sin \varepsilon \sin M)\, \mathbf{R}_2 (\Delta\psi \sin \varepsilon \cos M + \Delta\varepsilon \sin M) \qquad (10)$$

and

$$\mathbf{P} = \mathbf{R}_3 (-z + M)\, \mathbf{R}_2(\theta)\, \mathbf{R}_3 (-\zeta_0) \qquad (11)$$

where M is the precession in right ascension.

In this case the rotation of CTS about the Z axis (ϕ) is the Apparent Sidereal Time from which the general precession and nutation in right ascension are removed. What is left, thus, is the rotational angle of the X axis of the CTS directly with respect to that of the CIS. Such a definition of the sidereal angle would, of course, necessitate the redefinition of UT1, a possibility for controversy. It should be noted also that the above transformation is independent of the ecliptic, a preference of many astronomers. More on this in Section 4.5.

Here there is not very much modeling that can be considered really useful. Of course, the rotation rate of the earth could be modeled as constant and possibly in the UTC scale. This would then mean that observed departures could immediately be referenced to that scale, a current practice. If one really wanted to go overboard, polar motion could also be modeled with the Chandlerian cycle of, say, 428 days and a circular movement of radius 0".15, centered at the Z axis of the CTS. More complex models may be developed (e.g., Markowitz, 1976, 1979), but since there are no valid physical concepts yet for the excitation of the amplitude of the Chandler motion, such modeling would not serve much purpose.

Reviewing Sections 4.1–4.3, one could conclude that if the phenomena of precession, nutation and polar motion as well as the concepts of the ecliptic and the vernal equinox can be disconnected from the realization of a reference frame, and be regarded as simply describing various aspects of the Earth's complicated motions, then a great simplification will have been achieved. Of course all of the above phenomena and concepts are basic, and a knowledge of them is absolutely necessary. This knowledge will continue to be supplied by the classical, dynamical observations, radio astrometry and pulsar observations. However, it is now possible to consider these items in their proper context and to define a reference frame which is independent of them. Such independence will benefit not only the reference frame, but also aid in the study of the very phenomena from which the concept of a reference frame will have been freed. Essentially, observations will have been decoupled from the observing platform. As a result of this, the accuracy of the reference frame will become primarily dependent upon the precision and accuracy of the underlying measurements, and will have a minimal, noncritical dependence upon any companion theories. More on this in Section 4.53.

4.4 Deformations ($\vec{L}'$)

The deformations which reasonably can be modeled at the present state of the art are those due to the tidal phenomena and to tectonic plate movements.

4.41 Tidal Deformations. Tides are generated by the same forces which cause nutation; thus models developed for the latter should be useful for the former. One would think that for earth tides it may not be necessary to use the theories based on the very sophisticated earth models: the amplitude of the phenomena being only around 30 cm, an accuracy of 3% should be adequate for centimeter work. This should be compared, for example, with the accuracy of the Wahr nutation model claimed to be at the 0.3% level. However, the tides and nutations differ in one important respect. The nutations hardly depend upon the elasticity and are affected only slightly by the liquid core (this is one reason why modern theories such as those of Wahr and Kinoshita give only slightly different results). Thus, except perhaps for the largest terms, one can depend upon theory when dealing with nutation. The tides, on the other hand, depend intimately upon the internal properties of the earth, and one must use tidal theories with caution (Newton, 1974). Additional problems are handling the transformation of the potential into physical displacements and on the calculations of regional (ocean loading) or local tidal deformations.

As far as the transformation of the tidal potential into displacement is concerned, the traditional way to do this is through the Love numbers for the solid effect and through 'load' numbers for ocean loading. These numbers, however, are spherical approximations which, for the purely elastic earth, are global constants. For more sophistication, elliptic terms can be added, but they will change the results by 1-2% only. A liquid core model produces resonance effects, which will result in a frequency dependency. The actual numbers representative for a given location can be determined only through in situ observations, such as gravity, tilt, deflections, which are all sensitive to certain Love number combinations and frequencies. Difficulties in this regard include the frequency dependence of the Love number. For example, the Love number h for radial (vertical) displacement can be determined locally from combined gravity and tilt meter observations by the analysis of the O_1 tidal component, but the real radial motion of geodetic interest is influenced by the M_2 and other semidiurnal tidal components.

Tidal loading effects have been successfully computed by Goad (1979) using the 1° square Schwiderski (1978) M2 ocean tide model. Global results show agreement with gravimetrically observed deformation on the 0.5 μgal (5×10^{-10}) level. From this it would seem that with good quality ocean tide models and with proper attention to the frequency dependence, this problem is manageable.

Suitable equations for displacement, gravity change, deflection change, tilt and strain calculations due to tides may be found in (Melchior, 1978; Vanicek, 1980) and in (Wahr, 1979) for the elliptic case.

As a conclusion one can reasonably state that the global and regional station movements due to tides can be estimated today within centimeters. Local effects, however, can be sizable and unpredictable, and therefore they are best determined from in situ observations. Thus most of the tidal effect in fact can and should be removed from the observations.

4.42 Plate Tectonic Mass Transfer. The concept that the earth lithosphere is made up of a relatively small number of plates which are in motion with respect to each other is the central theme of global plate tectonics. The theory implies the transfer of masses as the plates move with velocities determined from geologic evidence (see, e.g., Solomon and Sleep, 1974; Kaula, 1975; or Minster and Jordan, 1978). Material rises from the asthenosphere and cools to generate new oceanic lithosphere, and the lithospheric slabs descend to displace asthenospheric material (see, e.g., Chapple and Tullis, 1977). A good example of how such a theory can be used to estimate the vertical motions of observatories located on the lithosphere (in terms of changes in geoid undulations) is given in (Larden, 1980), based on specific models constructed in (Mather and Larden, 1978). The results indicate that changes in the geoid can reach 150 mm/century. Horizontal displacements can be estimated from the plate velocity models mentioned directly with certain possible amendments (Bender, 1974).

4.43 Other Deformations. If one wants to carry the modeling further, it is possible to estimate seasonal deformations due to variations in air mass and groundwater storage, for which global datasets are available (Van Hylckama, 1956; Stolz and Larden, 1979; Larden, 1980). A more esoteric effect would be the expansion of the earth (e.g., Dicke, 1969; Newton, 1968). The rate of possible expansion is estimated to be 10 - 100 mm/century.

One could continue with other modeling possibilities, but there is a real question on the usefulness of modeling phenomena of this level of magnitudes and uncertainties. As a general philosophy, one could accept the criteria that modeling should be attempted only if reliable and global data is available related to the phenomena is question, and if the magnitudes reach the centimeter per year level or so.

One last item which should be brought up is the fact that the issue of referencing observations and/or geodynamic phenomena is not exhausted by the establishment of reference frames of the Cartesian types discussed in this paper. An outstanding issue is still the geoid as a reference surface. Though it is true that three-dimensional advanced geodetic observational techniques do not need the geoid as a reference, there are still others, such as spirit leveling, which are used in the determination of crustal deformations in the local scale. In addition, the geoid is needed to reference gravity observations on a global scale (one should remember that a 1 cm error in the geoid corresponds to a 3 μgal error in the gravity reduction, which is (or soon will be) the accuracy of modern gravimeters). Further, in connection with the use of satellite altimetry for the determination of the departures of sea surface topography from the equipotential geoid (a topic of great oceanographic interest), there is a requirement for a geoid of at least 10 cm accuracy. The determination of such a geoid globally, or even over large areas, is a very difficult problem, which, however, is not the subject of the present paper.

4.44 Current (1988) Practice. Some of the above effects can be modelled with good accuracy. A review of current models can be found in MERIT Standards (Melbourne, 1983). Two models are of particular interest for terrestrial frames (Boucher, 1987):

(i) The solid earth tide correction for ground station positions. Especially important is the vertical component

$$\Delta h = -0.121 \left(\frac{3}{2}\sin^2\phi - \frac{1}{2}\right) \text{meters},$$

the permanent tidal deformation, where ϕ is the latitude of the station.

(ii) Tectonic plate motion correction for the horizontal components of ground station position: The usual ones, such as the series of Minster-Jordan models, are defined through a set of angular velocity vectors $\bar{\Omega}_p$, one for each plate, and expressed in the terrestrial system, so that the velocity of a point of coordinate $\bar{X}$ is

$$\dot{\bar{X}} = \bar{\Omega}_p \wedge \bar{X}$$

Two absolute motion models are usually used: AMO-2, derived from the RM-2 model by applying a 'no global rotation' condition; AM1-2, which minimizes the motion of a set of hot spots, also derived from RM-2 (Minster and Jordan, 1978).

AMO-2 depends only on the adopted plate boundaries, whereas AM1-2 depends on the selection of the hot spots which are more subject to uncertainties. On the other hand, AMO-2 corresponds to the type of law of evolution one wants to give to terrestrial frames and has been consequently adopted by MERIT Standards (Update 1, December 1985). Nevertheless, AM1-2 leads to a system linked to the mantle which is needed to express a geopotential model without secular variations due to a residual rotation of the system. It is therefore favoured by groups which perform dynamical analysis of satellite tracking data.

4.5 Recent Developments

4.51 Expected Changes in the Adopted Series of Nutation. Recent analysis of modern highly accurate observations (e.g., VLBI) indicates significant departures from the IAU 1980 nutation series. None of the existing theories based on various Earth models can adequately explain these departures from Wahr's model. Apparently more efforts are required both in theory and in observations to arrive at a resolution. In the interim, the corrections in Table 7, based on (Herring et al., 1986), and further analysis are being recommended until such time when adequate theoretical coefficients can be determined. See also (Carter, 1988; Sovers and Edwards, 1988; and Kinoshita, 1988).

Assuming that the CTS is to be maintained unchanged, corrections to the nutation terms in longitude ($\delta\Delta\psi$) and obliquity ($\delta\Delta\varepsilon$) would theoretically change the polar motion components and GAST, utilized in the transformation equation (1), i.e., in the matrix S, as follows (Zhu and Mueller, 1983):

$$\begin{aligned}\Delta x_p &= \delta\Delta\varepsilon \sin\theta + \delta\Delta\psi \sin\varepsilon \cos\theta \\ \Delta y_p &= -\delta\Delta\varepsilon \cos\theta + \delta\Delta\psi \sin\varepsilon \sin\theta \\ \Delta(\mathrm{GAST}) &= \delta\Delta\psi \cos\varepsilon\end{aligned} \tag{12}$$

Table 7 Corrections to the Long-Period Terms of the IAU 1980 Nutation Series

Period (years)	*Δε In Phase (mas)*	*Δε Out of Phase (mas)*	*Δψ In Phase (mas)*	*Δψ Out of Phase (mas)*
18.6	2.15	1.81	-5.55	3.37
9.3	-0.24	0.0	1.20	0.0
1.0	2.08	-0.24	5.23	-0.61
0.5	-0.41	-0.47	1.02	-1.18
0.037	0.32	0.0	-0.81	0.0

Δ precession constant -2.7 mas/yr

where θ is the sidereal time. As it is seen, the theoretical effects on polar motion are diurnal terms ($\delta\Delta\psi$ and $\delta\Delta\varepsilon$ being long periodic).

4.52 Expected Change in the Constant of Precession. Modern, LLR and VLBI, observations also indicate a possible correction of $-0''.2$ to $-0''.3$/Julian century to the IAU 1976 constant of precession. This correction is uncertain due to the relatively short time span of available observations.

Williams and Melbourne (1982) and Zhu and Mueller (1983) investigated the effects of such a change. The effect on polar motion is a diurnal periodic term with an amplitude increasing linearly in time; on the GAST it is a linear term.

4.53 Intermediate Reference Frame Issues. The complete transformation from the CIS to the terrestrial frame CTS is given by eq. (1). In geodetic applications generally only the complete transformation **SNP** is needed. Changes in the 'intermediate' reference frame defined by the **NP** transformation must either by 'absorbed' in the **S** matrix by changing appropriately x_p, y_p and GAST (UT1), or the CTS must change its orientation. There are seven options to choose from, and they are a matter of preference (Zhu and Mueller, 1983). One of these which would neither change the CTS orientation nor the UT1 is probably preferred by geodesists. It would however change the definition of the Greenwich Mean Sidereal Time by referring it to a point on the equator insensitive to precession. As mentioned in Section 4.3, a similar option has been advocated by Guinot (1979) during the past decade but for different reasons. A recent proposal by Capitaine and Guinot (1988) is based on the observation that the classical definition of GAST representing the rotation of the Earth (i.e., CTS) is not satisfactory mainly for two reasons:

(i) It is referred to the true equinox of date which is an inadequate and unnecessary intermediate reference point because modern observations of the CTS's orientation in space (especially VLBI) are practically insensitive to the orientation of the ecliptic and consequently to the position of the equinox.

(ii) The presently adopted expression converting GAST to UT1 (Aoki et al., 1982) neglects some cross-terms between precession and nutation which are of the order of $0''.001$ and should now be considered.

The definition advocated would thus be better adapted to the new methods of observation and would provide an accuracy of the order of $0''.0001$. It would also result in a new definition of Universal Time which would remain valid even if the adopted model for the **NP** transformation is revised (see also (Capitaine et al., 1986)). The proposal is not without its critics. See (Aoki and Kinoshita, 1983; Aoki, 1988).

Related to the above issue is the definition of the third axis of the intermediate frame as defined by the transformation model **NP**, specifically, by the adopted theory of nutation (see Section 4.2). This pole, the Celestial Ephemeris Pole (CEP), conceptually has no diurnal motion with respect to an Earth-fixed or a space-fixed reference frame. Some of the modern observational techniques, however, are not very sensitive to this axis and, in fact, on the level of $0''.001$ accuracy, define a variety of technique dependent conventional poles. Capitaine et al. (1985) and Capitaine (1986) point out that clarification of this issue is necessary in order to intercompare and interpret polar motion coordinates determined at the level of 0".001 accuracy, by means of a variety of techniques ranging from VLBI to superconducting gravimetry.

5. THE INTERNATIONAL EARTH ROTATION SERVICE

5.1 The MERIT-COTES Programs

The acronyms MERIT and COTES refer to two international programs that were started independently, but which developed together. MERIT refers to an international program to monitor the earth's rotation and intercompare the techniques of observation and analysis with a view to making recommendations about the form of a new international service. On the other hand, the objective of the COTES program program was to provide a basis for recommendations on the establishment and maintenance of a new conventional terrestrial reference system for the specification of positions on or near the earth's surface. The two programs were linked when it became clear that the observational campaign planned for MERIT and the new earth rotation service would provide results that could be used for COTES. In particular, in order to determine the earth rotation parameters to high accuracy, it is necessary to establish the positions of the observing sites (or 'stations') in a worldwide network that provides a suitable basis for a new terrestrial reference system. The observational data and results that have been obtained in the course of these programs have been collected together for further analysis and for use in current and future scientific studies and practical applications.

Project MERIT was conceived in 1978 at IAU Symposium No. 82 on "Time and the Earth's Rotation." The Symposium recommended the appointment of a "working group to promote a comparative evaluation of the techniques for the determination of the rotation of the earth and to make recommendations for a new international program of observation and analysis in order to provide high quality data for practical applications and fundamental geophysical studies." Two years later, in 1980, the participants in IAU Colloquium No. 56 on "Reference Coordinate Systems for Earth Dynamics" recommended the setting up of a working group "to prepare a proposal for the establishment and maintenance of a Conventional Terrestrial Reference System." Information discussions at the First MERIT Workshop in 1981 were followed eventually by the merging of the two groups and the production of a Joint Summary Report (Wilkins and Mueller, 1986). This report describes briefly the development of the programs of observation and analysis and gives recommendations for new terrestrial and celestial reference systems and for the setting up of a new International Earth Rotation Service (IERS); this report also includes references to earlier reports that describe the techniques used, the organizational arrangements and the programs of the activities, and that give the principal results and references to relevant papers.

The MERIT and COTES programs have been very successful in stimulating the use and development of new techniques of observations using laser ranging and radio interferometry; they also led to improvements in the results from optical astrometry and the Doppler (radio) tracking of satellites, which were in regular use before 1978. Coordinators were appointed for each technique and for certain associated activities, such as the operation of a Coordinating Center for the combination and dissemination of results, the preparation of MERIT Standards, and the collocation of equipment of different techniques.

The quantities measured by each of the techniques that were used in the programs are as follows:

Doppler tracking of satellites: The Doppler shifts (range-rates) in the radio transmissions from Transit navigation satellites.

Satellite laser ranging: The time for pulses of laser light to travel to and from geodetic satellites carrying retroreflectors.

Lunar laser ranging: Time of flight for pulses of laser light to travel to and from retroreflectors on the surface of the moon.

Optical astrometry: Directions to stars measured with respect to local reference frames.

Connected-element radio interferometry, and
Very long baseline radio interferometry: Differences between the travel times of the radio emission from quasars to two or more radio telescopes.

Organizational arrangements for the regular transmission and processing of data already existed for optical astrometry and Doppler tracking, but for the other techniques it was necessary to set up both operational centers and analysis centers. The operational centers coordinated the observations, collected the observational data, computed earth rotation parameters on a rapid-service basis from 'quick-look data', and distributed the observational data (perhaps after some processing) to the analysis centers, which determined both earth rotation parameters and station coordinates from all the available data.

There were several designated periods when all stations were requested to make observations and send them as quickly as possible to the operational centers. The first was the MERIT Short Campaign from 1 August to 31 October 1980. This was primarily a test of the technical and organizational arrangements, but it also produced much valuable data and showed clearly the potential of the new techniques. The MERIT Main Campaign covered the 14-month period from 1 September 1983 to 31 October 1984 and included the first COTES Intensive Campaign, which ran from 1 April until 30 June 1984. The data were analyzed independently at two or more analysis centers for each technique, and many excellent series of earth rotation parameters and sets of station coordinates were obtained. These data are still being studied to determine, for example, the systematic differences between the reference systems of the various techniques. The results have established beyond doubt the very close correlation between the short-period variations in the length of day and in the angular momentum of the atmosphere. The pole of rotation has been shown to move much more smoothly than had earlier been thought, but there is still controversy about the sources of excitation of the 14-month term in the motion.

5.2 The International Earth Rotation Service

By the end of the MERIT Main Campaign it had become clear that laser ranging and radio interferometry were able to provide more precise estimates of polar motion, universal time and length of day than could optical astrometry and the Doppler tracking of satellites, which were the prime contributors to the international services in 1978. This conclusion has since been substantiated by the more detailed analyses of the data that have been reported at the MERIT Workshop and Conference held at Columbus, Ohio, on 29 July - 2 August 1985 (Mueller, ed., 1985). The accuracy of the regular determination of the coordinates of the poles by SLR and VLBI is about 5 cm, compared with 30 cm by optical astrometry and Doppler tracking, while for UT and excess length of day the accuracy is about 0.2 ms and 0.06 ms, compared with 1 ms and 0.2 ms.

It must be realized, however that other factors besides precision had to be taken into account before recommendations about the future international services could be formulated. Perhaps the most important factor was whether it is reasonable to expect that the organizations concerned are likely to continue to make and process observations at an appropriate level and to make the results available to the international community without restriction. The MERIT Main Campaign was a period of special activity, and it cannot be assumed that any technique would provide results of the same high quality (as judged by the combination of precision, accuracy, frequency, reliability and promptness) on a long-term basis.

The International Latitude Service was initially set up a a set of five dedicated stations, but it was eventually replaced by the International Polar Motion Service which relied on receiving data from a much larger number of instruments which provided local services and data for other scientific purposes as their prime justification. It is to be expected that any new International Earth Rotation Service will also have to depend largely on the use of observations and results that are obtained for other national and international programs.

In particular it must be recognized that an important application of the Service will be the establishment and maintenance of the new conventional terrestrial reference system. The permanent stations used for monitoring earth rotation will comprise a primary geodetic network of large scale and high precision that will be densified, partly by the use of mobile systems using the same techniques, but mainly by the use of other geodetic techniques, such as the use in radio interferometric mode of signals for navigation satellites.

The choice of the techniques to be used in the new service depends on the subjective evaluation of many factors and not merely on a comparison of the potential quality of the determination of each rotation parameters. Although it is conceivable that a single VLBI network could provide an adequate international earth rotation service, the general conclusions of the discussions in the MERIT and COTES working groups is that the new service should be based on both laser ranging and VLBI and should also utilize any other appropriate data that are made available to it.

The three recommendations given in Appendix 1 were adopted at a joint meeting of the MERIT Steering Committee and the COTES Working Group that was held at Columbus, Ohio, on 3 August 1985. Earlier drafts had been subject to critical review at the MERIT Workshop on 30 July and by interested participants in the Conference on Earth Rotation and Reference Systems held 31 July to 2 August. The joint meeting also adopted a draft resolution for consideration by a Joint Meeting of the IAU Commissions 19 and 31 on 22 November 1985 during the XIXth General Assembly of the IAU at New Delhi. Amended versions of this resolution were adopted by the Joint Meeting and subsequently by the Union on 28 November 1985. A further recommendation concerning the assignment of responsibility within the IAU for matters relating to the celestial and terrestrial reference systems was adopted by the MERIT/COTES meeting on 3 August and served to stimulate a discussion within the IAU, but no decision was announced.

The final version of the IAU resolution on the MERIT/COTES program and recommendations is given in organization and interim arrangements. As a consequence the MERIT and COTES Working Groups were replaced by a Provisional Directing Board for the new International Earth Rotation Service which was to come into operation on 1 January 1988. The IAU resolution was endorsed by the Executive Committee of the International Association of Geodesy in March, 1986 (Mueller and Wilkins, 1986). The recommendations of the Provisional Directing Board were considered and adopted by the IUGG during its XIXth General Assembly in Vancouver, B.C., in August, 1987 (Appendix 3).

With this last action, after ten years of preparation the new International Earth Rotation Service became a reality.

Organization of the Service. For each technique of observation (VLBI, SLR and LLR), prospective host organizations were invited to submit proposals for participation in one or more of the following ways:

- as a coordinating center,
- as an observing station or a network of stations,
- as a data collection (and distribution) center for quick-look and/or full-rate observational data. Such a center could, if appropriate, also process the data to form normal point data for use in analyses, or the task could be carried out by separate centers,
- as a quick-look operational center that would provide rapid service results,
- as a full-rate analysis center that would determine ERP's, station coordinates and other parameters to a regular schedule.

Several of these activities might be carried out by one center, and the actual organization would differ according to the number of observing stations and networks and to the nature of the processing required. There will be nod need for associate analysis centers in the formal structure, although it is expected that many groups will wish to analyze data provided by the Service. Offers of the deployment of mobile systems for use in improving the terrestrial reference system would be welcomed.

The principal tasks of the Central Bureau are specified in Recommendation B in Appendix 1, and some of them would be carried out by sub-bureaus. There is a need also for separate centers for relevant data from other fields, such as data on atmospheric angular momentum (AAM) and appropriate geodetic data (e.g., GPS results). The former might prove to be useful in predicting the variations in the rate of rotation of the earth, while the latter would be useful in the establishment and maintenance of the terrestrial reference system.

Kovalevsky and Mueller in their 1980 review of the Warsaw Conference listed a number of actions required to assure that the reference system issue be resolved "early and that the uniformity is assured by means of international agreements." There were the following:

Re CTS:

1 Selection of observatories whose catalogue will define the CTS.

2. Initiation of measurements at these observatories.

3. Recommendation on the observational and computational maintenance of the CTS (e.g., permanent versus temporary and repeated station occupations, constraints to be used).

4. Decision on how far and which way the earth deformation should be modeled initially.

5. Plans and recommendations for the establishment of new international service(s) to provide users with the appropriate information regarding the use of the CTS frame.

Re CIS:

6. Selection of extragalactic radio sources whose catalogue will define the CIS.

7. Improvement of the positions of these sources to a few milliseconds (arc).

8. Final decision on the IAU series of nutation and to assure that it describes the motion of the Celestial Ephemeris Pole.

9. Early completion of the FK5 and revision of astronomical equations due to the changed equinox (e.g., transformation between sidereal and Universal times).

10. Extension of the stellar catalogues (FK5 and later Hipparcos) to higher magnitudes.

11. Connection of the FK5, and later Hipparcos, reference frames to the CIS frame.

Eight years later it is gratifying to note that significant progress has been made on all items.

REFERENCES

Aoki, S., 1988, Relation between the celestial reference system and the terrestrial reference system of a rigid Earth, *Celes. Mechan.*, in press.

Aoki, S. and Kinoshita, H., 1983, *Celes. Mechan.*, **29**, 335.

Aoki, S., Guinot, B., Kaplan, G., Kinoshita, H., McCarthy, D. and Seidelmann, P., 1982, *Astron. Astrophys.*, **105**, 359.

Arias, E., Feissel, M. and Lestrade, F., 1988a, in BIH Annual Report for 1987, Paris.

Arias, E., Lestrade, F. and Feissel, M., 1988b, in Wilkins and Babcock (eds.), Reidel.

Ashby, N. and Allan, D., 1979, *Radio Science*, **14**, 649.

Atkinson, R.d'E., 1973, *Astron. J.*, **78**, 147.

Atkinson, R.d'E., 1975, *Monthly Notices Roy. Astron. Soc.*, **71**, 381.

Atkinson, R.d'E., 1976, On the Earth's Axes of Rotation and Figure, pres. XVI Gen. Assembly of IUGG, Grenoble.

Backer, D., Fomalont, E., Goss, W., Taylor, J., and Weisberg, J., 1985, *Astron. J.*, **90**, 2275.

Barbieri, C. and Bernacca, P. (eds.), 1979, *European Satellite Astrometry*, Ist. di Astronomia, Univ. di Padova, Italy.

Bender, P. and Goad, C., 1979, in *The Use of Artificial Satellites for Geodesy and Geodynamics, Vol. II*, G. Veis and E. Livieratos (eds.), National Technical Univ., Athens.

Bender, P., 1974, in Kolaczek and Weiffenbach (eds.), 85.

Boucher, C. and Altamimi, Z. , 1987, Intercomparison of VLBI, LLR, SLR, and GPS Derived Baselines on a Global Basis, IGN No. 27.450, France.

Boucher, C., 1986, GRGS Tech. Rep. No. 3, IGN, France.

Boucher, C., 1987, IGN/SGN No. 27.459.

Bureau International de l'Heure, 1985, BIH Annual Rep. for 1984, Paris.

Bureau International de l'Heure, 1986, BIH Annual Rep. for 1985, Paris.
Bureau International de l'Heure, 1987, BIH Annual Rep. for 1986, Paris.
Bureau International de l'Heure, 1988, BIH Annual Rep. for 1987, Paris.
Capitaine, N. and Guinot, B., 1988, in Wilkins and Babcock (eds.), Reidel , 33.
Capitaine, N., 1986, *Astron. and Astrophys.*, ***162***, 323.
Capitaine, N., Guinot, B. and Souchay, J., 1986, *Celes. Mechan.*, **39**, 283.
Capitaine, N., Williams, J. and Seidelmann, P., 1985, *Astron. Astrophys.*, **146**, 381.
Carter W., 1988, in BIH Annual Report for 1987, Paris, p. D-105.
Carter, W., Robertson, D. and Fallon, F., 1987, in BIH Annual Rep. for 1986, Paris, p. D-19.
Chapple, W. and Tullis, T., 1977, *J. Geophys. Res.*, **82**, 1967.
Corbin, T., 1978, in F. Prochazka and R. Tucker (eds.).
Dicke, R., 1969, *J. Geophys. Res.*, **74**, 5895.
Dickey, J., 1989, in Kovalevsky, Mueller and Kolaczek (eds.), Kluwer.
Duncombe, R., Fricke, W., Seidelmann, P., and Wilkins, G., 1976, *Trans. IAU,* **XVIB**, 52.
Duncombe, R., Seidelmann, P., and Van Flandern, T., 1974, in Kolaczek and Weiffenbach (eds.), 223.
Eichhorn, H. and Leacock, R. (eds.), 1986, *Astrometric Techniques*, Reidel.
Einstein, A., 1956, *The Meaning of Relativity*, Princeton Univ. Press, Princeton, New Jersey.
Eubanks, T., Steppe, J. and Spieth, M., 1985, in BIH Annual Rep. for 1984, Paris, p. D-19.
Fanselow, J. et al., 1984, *Astron. J.*, ***89***, 987.
Fedorov, E., Smith, M., and Bender, P. (eds.), 1980, *Nutation and the Earth's Rotation,* IAU Symp. 78, Reidel.
Feissel, M., 1980, *Bull. Geodesique,* **54**, 81.
Fricke, W. and Gliese, W., 1978, in Prochazka and Tucker, 421.
Fricke, W., 1974, in Kolaczek and Weiffenbach (eds.), 201.
Fricke, W., 1977, *Veröffentlichungen Astron. Rechen-Inst. Heidelberg,* **28**, Verl. G. Braun, Karlsruhe.
Fricke, W., 1979a, Progress Rept. on Preparation of FK5, pres. at Commission IV, IAU XVII Gen. Assembly, Montreal.
Fricke, W., 1979b, in Barbieri and Bernacca (eds.), 175.
Froeschle, M. and Kovalevsky J., 1982, *Astron. and Astrophys.*, ***116***, 89.
Gaposchkin, E. and Kolaczek, B. (eds.), 1981, *Reference Coordinate Systems for Earth Dynamics*, Reidel.
Gilbert, F. and Dziewonski, A., 1975, *Phil. Trans. R. Soc. London,* **A278**, 187.
Goad, C., 1979, Gravimetric Tidal Loading Computed from Integrated Green's Functions, NOAA Tech. Memorandum NOS NGS 22, NOS/NOAA, Rockville, Maryland.
Guinot, B., 1978, in Mueller (ed.), 1978, 13.
Guinot, B., 1979, in McCarthy and Pilkington (eds.), 7.
Guinot, B., 1981, in Gaposchkin and Kolaczek (eds.).
Guinot, B., 1986, in Eichhorn and Leacock (eds.), Reidel.
Guinot, B., 1989, in Kovalevsky, Mueller and Kolaczek (eds.), Kluwer.
Hellings, R., 1986, *Astron. J.*, **91**, 650.
Herring, T., Gwinn, C. and Shapiro, I., 1986, *J. Geophys. Res.*, ***91***, 4745.
Jefferys, W., 1980, *Celestial Mech.*, ***22***, 175.
Kaplan, G. et al., 1982, *Astron. J.*, ***87***, 570.
Kaula, W., 1975, *J. Geophys. Res.*, **80**, 244.
Kinoshita, H., 1977, *Celes. Mech.*, **15**, 227.
Kinoshita, H., 1988, in BIH Annual Report for 1987, Paris, p. D-103.
Kinoshita, H., Nakajima, K., Kubo, Y., Nakagawa, I., Sasao, T. and Yokoyama, K., 1979, *Publ. Int. Lat. Obs., of Mizusawa,* **XII**, 71.
Kolaczek, B. and Weiffenbach, G. (eds.), 1974, *On Reference Coordinate Systems for Earth Dynamics*, IAU Colloq. 26, Smithsonian Astrophys. Obs., Cambridge, Mass.
Kovalevsky, J. and Mueller, I., 1981, in Gaposchkin and Kolaczek (eds.).
Kovalevsky, J., 1979, in McCarthy and Pilkington (eds.), 151.
Kovalevsky, J., 1980, *Celestial Mech.*, ***22***, 153.
Kovalevsky, J., 1985, *Bull. Astronomique,* ***10***, 87.
Kovalevsky, J., 1989, in Kovalevsky, Mueller and Kolaczek (eds.).
Kovalevsky, J., Mueller, I. and Kolaczek, B.(eds.), 1989, *Reference Frames*, Kluwer Publ.
Kozai, Y., 1974, in Kolaczek and Weiffenbach (eds.), 235.
Larden, D., 1980, Some Geophysical Effects on Geodetic Levelling Networks, *Proc. 2nd Int. Symp. on Problems Related to the Redefinition of North American Vertical Geodetic Networks,* Canadian Inst. of Surveying, Ottawa.
Leick, A. and Mueller, I., 1979, *manus. geodaetica,* **4**, 149.
Lestrade, J., Requieme, Y., Rapaport, M., and Preston, R., 1988, in Wilkins and Babcock (eds.).
Lieske, J., 1979, *Astron. Astrophys.*, ***73***, 282.
Lieske, J., Lederle, T., Fricke, W. and Morando, B., 1977, *Astron. Astrophys.*, **58**, 1.

Ma, C., 1983, *EOS*, ***64***, 674.
Ma, C., 1988, in Wilkins and Babcock (eds.).
Ma, C., 1989, in Kovalevsky, Mueller and Kolaczek (eds.).
Ma, C., Clark, T., Ryan, J., Herring, T., Shapiro, I., Corey, B., Hinteregger, H., Rogers, A., Whitney, A., Knight, C., Lundquist, G., Shaffer, D., Vandenburg, N., Pigg, J., Schupler, B. and Ronnang, B., 1986, *Astron., J.* ***92***, 1020.
Ma, C., Himwich, W., Mallama, A. and Kao, M., 1987, in BIH Annual Rep. for 1986, Paris, p. D-11.
Markowitz, W. and Guinot, B. (eds.), 1968, *Continental Drift, Secular Motion of the Pole, and Rotation of the Earth*, IAU Symp. 32, Reidel.
Markowitz, W., 1976, Comparison of ILS, IPMS, BIH and Doppler Polar Motions with Theoretical, Rep. to IAU Comm. 19 and 31, IAU Gen. Assembly, Grenoble.
Markowitz, W., 1979, Independent Polar Motions, Optical and Doppler; Chandler Uncertainties, Rep. to IAU Comm. 19 and 31, IAU Gen. Assembly, Montreal.
Mather, R. and Larden, D., 1978, *Uniserv G* **29**, Univ. of New South Wales, Sidney, Australia.
Mather, R. et al., 1977, Uniserv G 26, Univ. of New So. Wales, Australia.
Mather, R., Masters, E. and Coleman, R., 1977, *Uniserv G* **26**, Univ. of New South Wales, Sidney, Australia.
McCarthy, D. and Pilkington, J. (eds.), 1979, *Time and the Earth's Rotation*, IAU Symp. 82, Reidel.
McClure, P., 1973, Diurnal Polar Motion, GSFC Rep. X-592-73-259, Goddard Space Flight Center, Greenbelt, Maryland.
Melbourne, W. (ed.), 1983, Project MERIT Standards, US Naval Obs. Circular, No. 167.
Melchior, P. and Yumi, S. (eds.), 1972, *Rotation of the Earth*, IAU Symp. 48, Reidel.
Melchior, P., 1978, *The Tides of the Planet Earth*, Pergamon Press, Oxford.
Minster, J. and Jordan, T., 1978, *J. Geophys. Res.*, **83**, 5331.
Molodenskij, M., 1961, *Comm. Obs. R. Belgique, 188 S. Geoph.* **58**, 25.
Morabito, D., Preston, R., Linfield, R., Slade, M., Jauncey, D., 1986, *Astron. J.*, ***92***, 546.
Moran, J., 1974, in Kolaczek and Weiffenbach (eds.), 269.
Moritz, H. and Mueller, I., 1987, *Earth Rotation: Theory and Observation*, Ungar Publ., New York.
Moritz, H., 1967, *Dept. of Geod, Sci,. Rep.* **92**, Ohio State Univ., Columbus.
Moritz, H., 1979, *Dept. of Geod, Sci,. Rep.* **294**, Ohio State Univ., Columbus.
Mueller, I., 1969, *Spherical and Practical Astronomy As Applied to Geodesy*, Ungar Publ. Co., New York.
Mueller, I., 1975a, *Geophys. Surveys*, **2**, 243.
Mueller, I. (ed.), 1975b, *Dept. of Geod, Sci,. Rep.* **231**, Ohio State Univ., Columbus.
Mueller, I. (ed.), 1978, *Dept. of Geod, Sci,. Rep.* **280**, Ohio State Univ., Columbus.
Mueller, I., 1981, in Gaposchkin and Kolaczek (eds.).
Mueller, I., 1985, *Bull. Geod.*, ***59***, 181.
Mueller, I. (ed.), 1985, *Proc. Int. Conf. on Earth Rotation and the Terrestrial Reference Frame,"* publ. Dept. of Geodetic Sci. and Surveying, Ohio State Univ.
Mueller, I. and Wilkins, G., 1986, *Adv. Space Res.*, ***9***, 5.
Muhleman, D., Berge, G., Rudy, D., Niell, A., Linfield, R. and Standish, E., 1985, *Celestial Mech.*, ***37***, 329.
Murray, C., 1979, in McCarthy and Pilkington (eds.), 165.
Newhall, X, Preston, R. and Esposito, P., 1986, in Eichhorn and Leacock (eds.), 789.
Newhall, X, Williams, J. and Dickey, J., 1987, in BIH Ann. Rep. for 1986, Paris, p. D-29.
Newton, I., 1686, *Philosophiae Naturalis Principia Mathematica*, Univ. of California Press, 1966.
Newton, R., 1968, *J. Geophys. Res.*, **73**, 3765.
Newton, R., 1974, in Kolaczek and Weiffenbach (eds.), 181.
Perley, R., 1982, *Astron. J.*, ***87***, 859.
Prochazka, F. and Tucker, R. (eds.), 1978, *Modern Astrometry*, IAU Colloq. 48, Univ. Obs. Vienna.
Purcell, G., Cohen, E., Fanselow, J., Rogdstad, D., Skjerve, L., Spitzmesser, D. and Thomas, J., 1978, in Prochazka and Tucker (eds.), 185.
Purcell, G., Jr., Fanselow, J., Thomas, J., Cohen, E., Rogstad, D., Sovers, O., Skjerve, L., and Spitzmesser, D., 1980, Radio Interferometry Techniques for Geodesy, NASA Conference Publ. 2115, p. 165, NASA Scientific & Tech. Information Office, Washington, D.C.
Reasenberg, R., 1986, in Eichhorn and Leacock (eds.), 789.
Reigber, C., Schwintzer, P., Mueller, H. and Massmann, F., 1987, in BIH Ann. Rep. for 1986, Paris, p. D-39.
Robertson, D., Fallon, F., Carter, W., 1986, *Astron. J.*, ***91***, 1456.
Schutz, B., Tapley, D. and Eanes, R., 1987, in BIH Ann. Rep. for 1986, Paris, p. D-33.
Schwan, H., 1986, in Eichhorn and Leacock (eds.).
Schwan, H., 1987, *Mapping the Sky*, Reidel.
Schwiderski, E., 1978, Global Ocean Tides, Part 1; A Detailed Hydrodynamical Interpolation Model, U.S. Naval Surface Weapons Center TR-3866, Dahlgren, Virginia.

Seidelmann, P., 1982, *Celes. Mech.*, **27**, 79.
Smith, C., 1986, in Eichhorn and Leacock (eds.).
Solomon, S. and Sleep, N., 1974, *J. Geophys. Res*, **79**, 2557.
Sovers O. and Edwards, C., 1988, in BIH Annual Report for 1987, Paris, p. D-109.
Sovers, O., Edwards, C., Jacobs, C., Lanyi, G., Liewer, K., Treuhaft, R., *Astron. J.*, in press.
Stolz, A. and Larden, D., 1979, *J. Geophys. Res*, **84**, 6185.
Van Altena, W., 1978, in Prochazka and Tucker (eds.), 561.
Van Hylckama, T., 1956, *Climatology*, **9**, 59.
Vanicek, P., 1980, Tidal Corrections to Geodetic Quantities, NOAA Tech. Rep. NOS 83 NGS 14, NOS/NOAA, Rockville, Maryland.
Wade, C. and Johnston, K., 1977, *Astron. J.*, **82**, 791.
Wahr, J., 1979, *The Tidal Motions of a Rotating, Elliptical, Elastic and Oceanless Earth,* PhD diss., Dept. of Physics, Univ. of Colorado, Boulder.
Wahr, J., 1981, *Geophys. J. R. Astr. Soc.*, **64**, 705.
Weinberg, S., 1972, *Gravitation and Cosmology: Principles and Applications of the General Theory of Relativity*, Wiley & Sons, New York.
Wilkins, G. and Babcock, A., 1988, *The Earth's Rotation and Reference Frames for Geodesy and Geodynamics*, Reidel.
Wilkins, G. and Mueller, I., 1986, *EOS, Trans. Am. Geophys. Union*, **67**, 601.
Williams, J. and Melbourne, W., 1982, in *High-Precision Earth Rotation and Earth-Moon Dynamics,* O. Calame (ed.), Reidel Publ., Dordrecht, 293.
Williams, J. and Standish, E., 1989, in Kovalevsky, Mueller and Kolaczek (eds.).
Williams, J., Dickey, J., Melbourne, W. and Standish, E., 1983, in Proc. of IAG Symposia, IUGG XVIIIth Gen. Assembly, Hamburg, FRG, publ. Dept. of Geodetic Sci. and Surveying, Ohio State Univ.
Woolard, E., 1953, *Astronomical Papers Prepared for the Use of the American Ephemeris and Nautical Almanac, XV*, Part I, U.S. Govt. Printing Office.
Yumi S. (ed.), 1971, *Extra Collection of Papers Contributed to the IAU Symp. No. 48, Rotation of the Earth*, International Latitude Obs., Mizusawa, Japan.
Zhu, S.Y. and Mueller, I., 1983, *Bull. Geodes.*, **57**, 29.
Zverev, M., Polozhentsev, D., Stepanova, E., Khrutskaya, E., Yagudin, L., and Polozhentsev, A., 1986, in Eichhorn and Leacock (eds.).

APPENDIX 1
PRINCIPAL RECOMMENDATIONS OF THE MERIT AND COTES WORKING GROUPS

A. Technical Recommendation on Concepts

The IAU/IUGG MERIT and COTES Joint Working Groups recommend that the following concepts be incorporated in the operation of an international earth orientation service:

(1) The Conventional Terrestrial Reference System (CTRS) be defined by a set of designated reference stations, theories and constants chosen so that there is no net rotation or translation between the reference frame and the surface of the earth. The frame is to be realized by a set of positions and motions for the designated reference stations.

(2) The Conventional Celestial Reference System (CCRS) be defined by a set of designated extragalactic radio sources, theories and constants chosen so that there is no net rotation between the reference frame and the set of radio sources. The frame is to be defined by the positions and motions of the designated radio sources. The origin of the frame is to be the barycenter of the solar system.

(3) This international service should provide the information necessary to define the Conventional Terrestrial Reference System and the Conventional Celestial Reference System and relate them as well as their frames to each other and to other reference systems used in the determination of the earth rotation parameters. The information should include, but not be limited to, pole positions, universal time, precession, nutation, dynamical equinox, positions of the designated reference stations and radio sources, and crustal deformation parameters.

B. Recommendation for the Organization of a New International Earth Rotation Service

The IAU/IUGG MERIT and Cotes Joint Working Groups recommend that IAU and IUGG establish a new international service within FAGS for monitoring the rotation of the earth and for the maintenance of the Conventional Terrestrial Reference System to replace both the International Polar Motion Service (IPMS) and the Bureau International de l'Heure (BIH) as from 1 January 1988.

The new service will be known as the International Earth Rotation Service (IERS) and will consist of a Directing Board, a Central Bureau, coordinating centers and observatories. The Central Bureau, the centers and the observatories will be hosted by national organizations.

The Directing Board will exercise organizational, scientific and technical control over the activities and functions of the Service including such modifications to the organizational structure and participation in the Service as are appropriate to maintain an efficient and reliable service while taking full advantage of advances in technology and theory. The voting membership of the Directing Board will consist of one representative each of the IAU, the IUGG, the Central Bureau, and each of the coordinating centers. Additional nonvoting members may be appointed to advise the Board on complex technical and scientific issues.

The Central Bureau will combine the various types of data collected by the Service to derive and disseminate to the user community the earth rotation parameters in appropriate forms, such as predictions, quick-look and refined solutions, and other information relating to the rotation of the earth and the associated reference systems. The Central Bureau will conduct research and analysis to develop improved methods of processing and interpreting the data submitted. The Central Bureau may include sub-bureaus that carry out some of the specific tasks of the Central Bureau.

Coordinating centers will be designated for each of the primary techniques of observation to be utilized by the Service as well as for other major activities which the Directing Board may deem appropriate. Initially, there will be three centers for (1) very long baseline interferometry (VLBI), (2) satellite laser ranging (SLR), and (3) lunar laser ranging (LLR). Additional coordinating centers may be designated for the improvement of the determination of the earth rotation parameters and the maintenance of the conventional reference system by other techniques and to ensure that relevant data on the atmosphere, oceans and seismic events are available.

The coordinating centers will be on the same level as the Central Bureau in the organizational structure o the Service and will be responsible for developing and organizing the activities by each technique to meet the objectives of the Service. Associated with the coordinating centers there may be network centers for subsets of observatories that may, for reasons of geometry or system compatibility, work more efficiently as autonomous units. There may also be associated analysis centers to process the observational data regularly or for special applications and studies. These centers may submit their results directly to the Central Bureau.

National Committees for the International Unions for Astronomy and for Geodesy and Geophysics will be invited to propose before 1 January 1987 national organizations and observatories that will be willing to host the Central Bureau or one of the centers and/or to provide observational data for use by the Service.

It is essential that the new service have redundancy throughout the organizational structure to insure the uninterrupted timely production of consistent, accurate, properly documented earth orientation and reference frame parameters, even in the event that one of the host national organizations should terminate its participation. A widespread distribution of observatories that regularly make high precision observations by one, or preferably more, modern space techniques by fixed and/or mobile equipment will be needed for this purpose, and national organizations are urged to provide appropriate resources.

APPENDIX 2
RESOLUTION OF INTERNATIONAL ASTRONOMICAL UNION (1985)

The following resolution was adopted by the XIXth General Assembly of the International Astronomical

The International Astronomical Union

recognizing the highly significant improvement in the determination of the orientation of the earth in space as a consequence of the MERIT/COTES program of observation and analysis, and

recognizing the importance for scientific research and operational purposes of regular earth orientation monitoring and of the establishment and maintenance of a new Conventional Terrestrial Reference Frame,

thanks all the organizations and individuals who have contributed to the development and implementation of the MERIT and COTES programs and to the operations of the International Polar Motion Service and the Bureau International de l'Heure,

endorses the final report and recommendations of the MERIT and COTES Joint Working Groups;

decides

(1) to establish in consultation with IUGG a new International Earth Rotation Service within the Federation of Astronomical and Geophysical Services (FAGS) for monitoring earth orientation and for the maintenance of the Conventional Terrestrial Reference Frame; the new Service is to replace both the IPMS and the BIH as from 1 January 1988,

(2) to extend the MERIT/COTES program of observation, analysis, intercomparison and distribution of results until the new service is in operation,

(3) to recommend that an optical astrometric network be maintained for the rapid determination of UT1 for so long as this is recognized to be useful,

(4) to set up a Provisional Directing Board to submit recommendations on the terms of reference, structure and composition of the new service, and to serve as the Steering Committee for the extended MERIT/COTES program,

invites National Committees for the International Unions for Astronomy and for Geodesy and Geophysics to submit proposals for the hosting of individual components of the new service by national organizations and observatories, and

urges the participants in Project MERIT to continue to determine high precision data on earth rotation and reference systems and to make the results available to the BIH until the new service is in operation.

APPENDIX 3
RESOLUTION 1 OF THE INTERNATIONAL UNION OF GEODESY AND GEOPHYSICS, XIX GENERAL ASSEMBLY VANCOUVER, 21 AUGUST 1987

The International Union of Geodesy and Geophysics

Noting that the improved determination of the Earth's orientation parameters resulting from the MERIT and COTES programmes of observation and analysis is highly significant,

considering the importance for scientific research and operational purposes of regularly monitoring the Earth's orientation and of establishing and maintaining a new conventional terrestrial frame of reference,

approving the replacement of the International Polar Motion Service (IPMS) and of the Bureau International de l'Heure (BIH) by the International Earth Rotation Service (IERS) which will be responsible both for earth rotation and for the associated conventional frames of reference, and

recognizing that organizations in many countries have indicated their willingness to participate in such a new service,

endorses the recommendations of its Provisional Directing Board on the terms of reference, structure and composition of the new service,

decides to establish, in cooperation with the International Astronomical Union, the International Earth Rotation Service within the Federation of Astronomical and Geophysical Data Analysis Services (FAGS) a from 1 January 1988,

thanks all organizations and individuals who have helped to develop and implement the MERIT and COTES programmes, all who have operated IPMS and BIH in the past and all who have indicated their willingness to participate in the new Service.

GRAVITY FIELD RECOVERY FROM SATELLITE TRACKING DATA

Christoph Reigber

Deutsches Geodätisches Forschungsinstitut (DGFI), Abt. I
8000 München 22, Marstallplatz 8, F.R.G.

1. Introduction

Near Earth orbiting satellites have become one of the basic tools for solid-Earth and ocean physic studies, that is investigation of processes and forces which act upon the solid, the viscous and fluid parts of the planet Earth and the related physical and geometric properties of our planet.

Manifestations of such forces and processes which are derivable from near Earth satellite observations are irregular variations in the orientation of the Earth in space (including both changes of the direction of the rotation axis and variations in the rotational speed), vertical and horizontal changes in the Earth's surface geometry (continents and oceans), the anomalous gravity field, the solid Earth and ocean tides, and the anomalous magnetic field. These phenomena cannot be regarded as single features in isolation: they have rather to be considered in the light of how they react with one another and with the atmosphere and hydrosphere, in order to obtain a significant overall picture both of the processes taking place on and inside the Earth and of the forces which drive these processes.

Satellite geodesy with its broad spectrum of existing observing systems or systems under development for: (i) target tracking from ground to space, satellite to ground or satellite to satellite, (ii) remote sensing of the ocean and ice surfaces with satellite-borne altimeters and (iii) measuring gravity gradient tensor components with satellite-borne gradiometer sensors, has in the last 10 years already contributed significantly to the improved understanding of the Earth and ocean dynamics and will continue to contribute to the explosive increase in our understanding of these phenomena with improved observational techniques and satellite missions.

Those aspects of the overall complex solid Earth and ocean physics which have profited most from the rapid development in satellite geodesy are

(i) the global Earth gravity field;

(ii) precise positions of terrestrial points;

(iii) the orientation of the Earth.

All these topics are more or less non-static and therefore have to be treated as functions of time.

We are concerned here in this lecture with the global gravity field determination from satellite tracking data, that is with the determination of coefficients of mathematical base functions by which the anomalous field is represented.

In the course of time a variety of measurement systems and data reduction strategies have been conceived and implemented to derive from tracking information a better and better picture of the global gravity field of the Earth. Quality and amount of gravity parameters that could be extracted from tracking information have been increased according to improvements in technology and physical modelling. Nevertheless the spatial detail with which the gravity field can be resolved from tracking data is limited and will remain limited, primarily because of the fact that only medium to high altitude orbits can be observed reasonably dense from a manageable global ground station network. Specifying the resolution in terms of (i) the shortest wavelength λ which is still discernable or (ii) by the maximum resolvable degree n of a representation in terms of spherical harmonics or (iii) by the size s of rectangular blocks on the sphere, for which average gravity values are available, it is with the classification in table 1 the long-to-medium wavelength structure of the field which is derivable from ground tracking data. We are dealing here with this portion of the field.

WAVELENGTH RANGES				
	long	medium	short	very short
λ km	>8000	>1000	>200	<200
n	<5	<36	<200	>200
s	>10°	>5°	>1°	<1°

Table 1: Typical wavelength ranges in which the gravity field is represented

Adapted to the summer school objectives, only the basic principles and procedures for gravity field recovery from tracking data are described in the following sections. For more intensive studies of the subject the reader will have to contact the comprehensive literature.

2. Principles of Gravity Parameter Determination

The determination of Earth gravity parameters from tracking data has to be seen in the context of what is called a dynamic analysis procedure in satellite geodesy.

Dynamic methods for estimating orbital, kinematic and dynamic model parameters from satellite tracking data start from the basic vector equation

$$\mathbf{M} := \vec{\varrho} + \mathbf{X}_T - \mathbf{X}_S \equiv \mathbf{O} \tag{1}$$

expressed in a reference frame $S(X_1, X_2, X_3)$ which the analyser considers to be most appropriate for the data reduction process. This equation connects (fig.1) the motion $\mathbf{X}_T(t)$ of the topocentre T (which can be a tracking station on the solid parts of the Earth surface or an altimeter footprint in the oceans) with the geocentric motion $\mathbf{X}_s(t)$ of the satellite through the topocentric position vector $\vec{\varrho}(t)$.

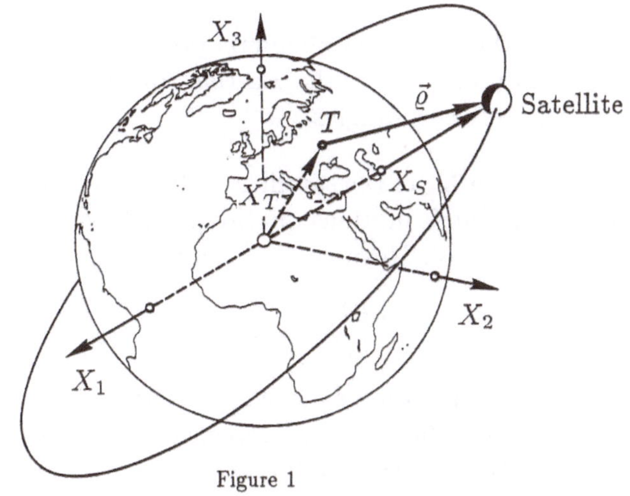

Figure 1

Single elements of the vector $\vec{\varrho}$ result from the measurement and data preprocessing process. Such elements can be ranges ϱ provided by tracking systems operating in the optical or radio-frequency spectrum, range-rates or Doppler counts obtainable with micro-wave tracking systems and directional information $(\vec{\varrho}/\varrho)$ in the form of right ascension and declination values as they were obtained in the sixties by camera tracking of bright satellites. The most precise tracking data are currently laser ranges to retroreflector equipped satellites, the most prominent ones of which are presently LAGEOS, STARLETTE and AJISAI (table 2). All such measurements, after they have passed the preprocessing step, provide the observables t, q:

t ... time of observations

q ... ranges, range-rates, angular information

The vector $\mathbf{X}_S$ results from the theory of the satellite's orbital motion in space and the vector $\mathbf{X}_T$ from the theory of the motion of the topocentre on the deformable Earth and the motion of the Earth in space. These theories are based on models which are described by a great number of parameters.

Supposing equation (1) is formulated in the quasi-inertial frame S_I whose axes at the basic epoch J2000.0 coincide with the mean equatorial frame (s. Prof. Mueller's lecture). For the description of the vector $\mathbf{X}_T$ for a crust-fixed tracking station in this system at time t one needs parameters describing

- the geocentric position of the station in a conventional terrestrial reference system S_{T_0} defined at epoch T_O
- the motion of the Earth crust at the location of the station, which means the global motion of the lithospheric plates in the time interval $t - T_O$, tidal deformations, ocean loading effects
- the motion of the celestial ephemeris pole (CEP) with respect to the terrestrial system pole (CTP), the Earth's rotation around this axis, and the nutation and precession angles.

The geocentric position $\mathbf{X}_S$ of the satellite in the inertial system S_I results from an integration of the satellite's motion equation, which when using the Newton-Euler formulation reads (s. Prof. Kovalevsky's lecture)

$$\frac{m_S d^2\mathbf{r}_I}{dt^2} = \mathbf{F}_I(t; \mathbf{r}_I, \frac{d\mathbf{r}_I}{dt}) \tag{2}$$

with $\mathbf{F}_I$ the composite set of forces which accelerate the satellite as seen from S_I. For a satellite orbiting in the near Earth space these forces result from interactions of the spacecraft with the primary body, other natural celestial bodies and the environmental field to which the satellite is exposed along its path. The latter non-gravitational forces, also often referred to as surface forces, are a function of the satellite size, mass and orientation. For the precise computation of geodetic satellite orbits the total acceleration vector $\mathbf{F}_I/m_S$ has to include the components

$$\mathbf{F}_I/m_S = \mathbf{a} = \mathbf{a}_K + \mathbf{a}_R + \mathbf{a}_B + \mathbf{a}_E + \mathbf{a}_O + \mathbf{a}_D + \mathbf{a}_S + \mathbf{a}_A \tag{3}$$

the designations of which are

$\lvert\mathbf{a}_K\rvert \cdots$	Central force term acceleration
$\lvert\mathbf{a}_R\rvert \cdots$	anomalous geopotential acceleration
$\lvert\mathbf{a}_B\rvert \cdots$	third body attractions, including sun, moon and major planets
$\lvert\mathbf{a}_E\rvert \cdots$	solid Earth tides acceleration

$\|\mathbf{a}_O\| \cdots$	ocean tides acceleration
$\|\mathbf{a}_D\| \cdots$	atmospheric drag acceleration
$\|\mathbf{a}_S\| \cdots$	solar radiation pressure acceleration
$\|\mathbf{a}_A\| \cdots$	Earth albedo pressure acceleration

Further perturbing accelerations have to be modelled if extreme orbit precision is required. All these force field components are described by models which contain a diversity of parameters.

For some geodetic satellites orbiting at different altitudes, the magnitude of the various accelerations is given in table 2. It is obvious from this table that the Earth geopotential U, conveniently expressed in S_{T_O} by a spherical harmonic expansion as (Heiskanen and Moritz, 1967)

$$U = \frac{GM}{r}[1 + \sum_{l=2}^{\infty} \sum_{m=0}^{l} (\frac{a_E}{r})^l P_{lm}(\sin\varphi)(C_{lm} \cos m\lambda + S_{lm} \sin m\lambda)] \tag{4}$$

with G the gravitational constant, M the mass and a_E the mean equatorial radius of the Earth and the C_{lm}, S_{lm} the Stokes coefficients, produces the largest accelerations on these satellites

$$\mathbf{a}_K + \mathbf{a}_R = \nabla U_o + \nabla(U - U_o) \tag{5}$$

with $U_o = GM/r$ the zero degree term, corresponding to the potential of a spherically symmetric Earth. Expressions to compute the other accelerations can be found in several textbooks on satellite dynamics and satellite geodesy (e.g. Kaula, 1966, Milani et al., 1987).

Now, given initial epoch (t_o) values for the spacecraft position $\mathbf{r}_I(t_o)$ and velocity $\dot{\mathbf{r}}_I$ and all the forces $\mathbf{F}_I$ which accelerate the satellite, also expressed in the frame S_I, for precision orbit computation the differential equation (2) is numerically integrated directly in rectangular coordinates, leading to the satellite's position vector $\mathbf{r}_I(t)$ and velocity vector $\dot{\mathbf{r}}_I(t)$ in S_I for times $t > t_O$.

Source of acceleration	Designation in equ.(3)		Magnitude of acceleration [m/s^2]			
			STARLETTE	**AJISAI**	**LAGEOS**	**GPS**
		semimajor axis (km)	**7337**	**7869**	**12266**	**26559**
		area/mass ratio (m^2/kg)	**9,6·10^{-4}**	**5,3·10^{-3}**	**6,9·10^{-4}**	**2,0·10^{-2}**
Kepler Term	$\mid a_K \mid$		7,4	6,4	2,6	0,6
C_{20}	$\mid a_R \mid$		$8 \cdot 10^{-3}$	$6 \cdot 10^{-3}$	$2 \cdot 10^{-3}$	$5 \cdot 10^{-5}$
other harmonics			$1 \cdot 10^{-4}$	$9 \cdot 10^{-5}$	$5 \cdot 10^{-6}$	$3 \cdot 10^{-7}$
Third body perturbations	$\mid a_B \mid$		$1 \cdot 10^{-6}$	$1 - 2 \cdot 10^{-6}$	$2 \cdot 10^{-6}$	$5 \cdot 10^{-6}$
Earth tides	$\mid a_E \mid$		$2 \cdot 10^{-7}$	$1 - 2 \cdot 10^{-7}$	$3 \cdot 10^{-8}$	$1 \cdot 10^{-9}$
Ocean tides	$\mid a_O \mid$		$3 \cdot 10^{-8}$	$2 \cdot 10^{-8}$	$2 \cdot 10^{-9}$	$1 \cdot 10^{-10}$
Atmospheric drag	$\mid a_D \mid$		$1 - 2 \cdot 10^{-10}$	$1 - 2 \cdot 10^{-10}$	$3 \cdot 10^{-12}$ (empirical)	0
Solar radiation pressure	$\mid a_S \mid$		$5 \cdot 10^{-9}$	$5 \cdot 10^{-8}$	$4 \cdot 10^{-9}$	$1 \cdot 10^{-7}$
Albedo pressure	$\mid a_A \mid$		$5 \cdot 10^{-10}$	$8 \cdot 10^{-9}$	$7 \cdot 10^{-11}$	$1 \cdot 10^{-9}$

Table 2: Accelerations on some geodetic satellites in m/sec^2

Only if the measurement is perfect and all models are described perfectly by their parameters, the balance equation (1) will be fulfilled at the observation epoch t. In the real world case with the observations having errors and the theories being incomplete or inaccurate this will never happen, but discrepancies

$$\mathbf{M}(t, \mathbf{P}) - \mathbf{M}(t, \mathbf{P}_R) = \Delta\mathbf{M} \neq \mathbf{O} \tag{6}$$

$\mathbf{P}$ ··· set of real parameters
$\mathbf{P}_R$ ··· set of a priori parameters

will occur, which will show a wide spectrum of amplitudes and periods, the magnitude of which will largely depend on the satellite's orbital characteristics.

For low-Earth orbiting satellites the main contributors to these discrepancies will be mismodelling effects in the Earth gravity and in the surface forces.

2.1. Linear Observation Equations

If $\Delta\mathbf{M}$ is small the a priori non-linear problem (6) can be approximated by a linear one. Abundant data can be used to set up a system of linear observation equations, which when combined with a stochastic model, is solved by applying least squares adjustment techniques.

By this the misclosure $\Delta\mathbf{M}$ is distributed onto all the components of equation (1).

To derive the linear observation equations the observation functional q (range ϱ, range-rate $\dot{\varrho}$, right-ascension α, declination δ) is expanded in a Taylor series about the computed value

$$q(t, \mathbf{P}_R) = q_c,$$

where $\mathbf{P}_R$ is the vector of the initial or reference values of the model parameters mentioned above, so that

$$\mathbf{P} = \mathbf{P}_R + \Delta\mathbf{P},$$

where $\Delta\mathbf{P}$ is the sought-for adjustment. If $\mathbf{P}_R$ is sufficiently close to $\mathbf{P}$, terms of order higher than the first in the expansion can be neglected and we get in a general form

$$q(t, \mathbf{P}) = q_c(t, \mathbf{P}_R) + \sum_i \frac{\partial q}{\partial P_i} \Delta P_i. \tag{7}$$

As one can easily realize from equation (1) ground-to-satellite tracking observations are in their most general form a function of the instantaneous position and velocity of the spacecraft and the position of the topocentre.The instantaneous state of the motion of the satellite is derived from an integration of the motion equations (2) and is therefore a function of the state vector components at epoch and the parameter subsets describing the various force field components. The instantaneous position of the topocentre, when described in the same system as the geocentric satellite position, is primarily a function of the three-dimensional coordinates in the conventional terrestrial reference frame and the Earth orientation parameters between these two frames. Considering e.g. only the solid Earth gravity, tidal and surface forces effects, one can therefore describe the dependency of the observables in the form

$$q(t, \mathbf{P}) = q(\mathbf{X}_S(t; \mathbf{Y}_S^O, \beta, \tau, \epsilon), \mathbf{X}_T(t; \mathbf{S}_T, \mathbf{o})) \tag{8}$$

where

q: the observed quantity (e.g. range)

t: time of observation

$\mathbf{P}$: the set of all parameters to which the observable is sensitive

$\mathbf{X}_S/\mathbf{X}_T$: the satellite/the topocentre position coordinates at the observation epoch t in the frame S_I

$\mathbf{S}_T$: the station position coordinates in the terrestrial frame S_{T_O}

$\mathbf{Y}_S^o$: the satellite state vector containing the coordinates of position and velocity at epoch in the frame S_I

β: the set of parameters representing the geopotential

τ: the set of tidal model parameters

ϵ: the set of surface force model parameters

$\mathbf{o}$: the set of Earth orientation parameters $(x_p, y_p, \Delta UT1)$ at the observation epoch t

With (8) equation (7) can be rewritten in the form

$$\Delta q = q(t, \mathbf{P}) - q_c(t, \mathbf{P}_R) = \left[\frac{\partial q}{\partial \mathbf{X}_S}, \frac{\partial q}{\partial \mathbf{X}_T}\right]_{\mathbf{P}_R} \sum_i \left(\frac{\partial}{\partial P_i}\right)_{\mathbf{P}_R} \begin{bmatrix} \mathbf{X}_S \\ \mathbf{X}_T \end{bmatrix} \Delta P_i. \qquad (9)$$

Separating the parameter set P_i into

- arc-dependent parameters (internal parameters)
  $\mathbf{P}_I \ldots$ state vector at epoch
  $\mathbf{P}_\epsilon \ldots$ drag and solar radiation pressure modelling parameters
- and arc-independent parameters (external parameters)
  $\mathbf{P}_S \ldots$ coordinates of tracking stations
  $\mathbf{P}_O \ldots$ earth rotation parameters
  $\mathbf{P}_\beta \ldots$ earth gravity field parameters
  $\mathbf{P}_\tau \ldots$ tidal parameters

the previous equation reads in matrix form

$$\Delta q = [\,\mathbf{I}\;\mathbf{E}\;\mathbf{S}\;\mathbf{R}\;\mathbf{G}\;\mathbf{T}\,] \begin{pmatrix} \Delta\mathbf{P}_I \\ \Delta\mathbf{P}_\epsilon \\ \Delta\mathbf{P}_S \\ \Delta\mathbf{P}_o \\ \Delta\mathbf{P}_\beta \\ \Delta\mathbf{P}_\tau \end{pmatrix} \qquad (10)$$

Writing the observation equation in this form, we implicitly assume that the observation and the time are not affected by biases or that possible biases are eliminated in a previous processing step.

The vectors **I** to **T** in equation (10) give the partial derivatives of the observable with respect to the internal and external parameters to be determined. All the internal and most of the external parameters are concerned with the dynamics of the satellite motion. Their partial derivatives are determined according to the chain rule as e.g. for the components of **G**

$$\frac{\partial q}{\partial \beta} = \frac{\partial q}{\partial \mathbf{X}_S}\frac{\partial \mathbf{X}_S}{\partial \beta} \tag{11}$$

where $\mathbf{X}_S$ is the vector describing the satellite position and velocity in the inertial system. The partials $\partial q/\partial \mathbf{X}_S$ are computed directly from the expressions relating the observed quantity with the satellite's position and velocity. The partial derivatives $\partial \mathbf{X}/\partial \beta$ are called the variational partials and are obtained by numerical integration of the variational equations. For these variational equations one obtains with the motion equations having the form

$$\begin{aligned} \ddot{\mathbf{x}} &= f(\mathbf{x}, \dot{\mathbf{x}}, \mathbf{p}) \\ \frac{\partial \ddot{\mathbf{x}}}{\partial \mathbf{p}} &= \frac{\partial f}{\partial \mathbf{p}} = \frac{\partial f}{\partial \mathbf{x}}(\frac{\partial \mathbf{x}}{\partial \mathbf{p}}) + \frac{\partial f}{\partial \dot{\mathbf{x}}}(\frac{\partial \dot{\mathbf{x}}}{\partial \mathbf{p}}) + \frac{\partial f}{\partial \mathbf{p}}\ /expl. \end{aligned} \tag{12}$$

or if the variation with time for the parameters is small

$$\frac{d}{dt^2}(\frac{\partial \mathbf{x}}{\partial \mathbf{p}}) = \frac{\partial f}{\partial \mathbf{x}}(\frac{\partial \mathbf{x}}{\partial \mathbf{p}}) + \frac{\partial f}{\partial \dot{\mathbf{x}}}(\frac{\partial \dot{\mathbf{x}}}{\partial \mathbf{p}}) + \frac{\partial f}{\partial \mathbf{p}}\ /expl. \tag{13}$$

For the state at epoch parameters $\mathbf{P}_I$ the last term cancels. For the surface model parameters $\mathbf{P}_\epsilon$ the last term is dominant. Details on the construction of these variational equations are given, for example, in Cappellari et al. (1976).

Using many observations from a single satellite arc, a complete system of observation equations of the type (10) can be set up. Independent of how many observations are used, such a system will be considerably underdetermined, simply because for example the number of orientation parameters is three times the number of observation instants and because of the theoretically infinite number of geopotential parameters. To convert the system into an overdetermined one, which can be solved by least squares adjustment techniques, model approximations have to be introduced. This is achieved for example by expressing the Earth orientation parameters by a simple function of time (e.g. a polygon function using only values at finite time intervals), by reducing the number of geopotential parameters and by replacing parameters by a priori known information, according to the respective purpose of the computation.

The estimation process is therefore in most cases splitted into two steps:

(a) the differential orbit improvement process, in which only the internal parameters are determined from single arc tracking data (and sometimes a few external parameters)

(b) the geodetic parameter estimation process, in which the external parameters are determined from measurements to many data arcs of the same satellite as for example for the determination of kinematic parameters, or many satellites as in the case of gravity model determinations.

The way the stepwise processing in the orbit and the geodetic parameter estimation process is done, is described in more detail in section 6.

In any case, the question which parameters have to be held at a priori "known" (approximate) values and for which parameters robust estimates can be obtained from a single or multiple-satellite, multi-arc solution is largely a function of the tracking data type (range, velocity, angles), their quality, their distribution along an orbital arc and primarily a function of the spectrum of perturbations in the observed orbit induced by the gravitational and non-gravitational forces. Before discussing in more detail the estimation process, it is important to have a closer look at the orbit perturbations.

3. Gravity Induced Linear Orbit Perturbations

As shown in table 2, of all perturbing forces acting on a geodetic satellite the force produced by the anomalous part of the geopotential, i.e.

$$\mathbf{a}_R = \nabla(U - U_O), \tag{14}$$

is dominant for orbital altitudes up to 20000 km. Thus the deviation of a geodetic satellite's motion from the central force motion (the Kepler orbit) will primarily be produced by the anomalous geopotential, a fact of course fortunate for gravity field studies from tracking data. Two other aspects are immediately visible from an inspection of the spherical harmonic expansion of the disturbing potential (s. equation (4))

$$R = (U - U_O) = \frac{GM}{r}[\sum_{l=2}^{\infty}\sum_{m=0}^{l}(\frac{a_E}{r})^l \bar{P}_{lm}(\sin\varphi)(\bar{C}_{lm}\cos m\lambda + \bar{S}_{lm}\sin m\lambda)] \tag{15}$$

where

GM	$\cdots$	gravitational constant times mass of Earth and atmosphere
r, φ, λ	$\cdots$	spherical coordinates for geocentric position in radius, latitude and longitude
a_E	$\cdots$	mean equatorial radius of the Earth
$\bar{C}_{lm}, \bar{S}_{lm}$	$\cdots$	fully normalized harmonic coefficients ($m = 0$ *zonals*; $m > 0$ tesserals)
l, m	$\cdots$	degree, order of development
$\bar{C}_{20}$	$\approx$	-5.10^{-4}
$O(\bar{C}_{lm}, \bar{S}_{lm})$	$\approx$	$10^{-5}/l^2$ (Kaula's rule)

From all harmonic terms $\bar{C}_{20}$ will lead to the largest perturbations. With increasing distance from the Earth the disturbing potential gets smaller and the equipotential surfaces (figure 2) become smoother (because of the $(a_E/r)^l$factor), and so will the orbit perturbations.

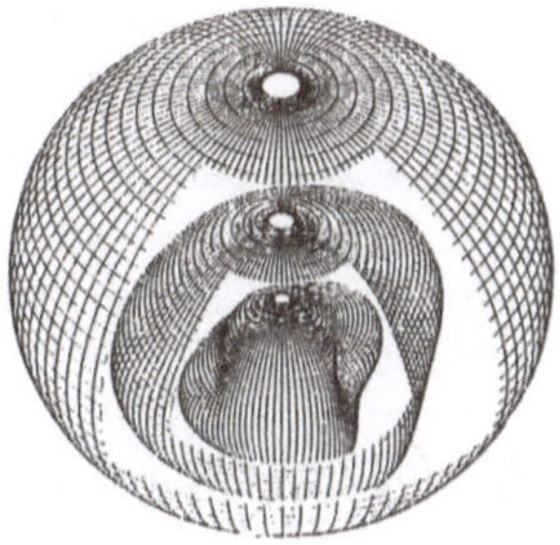

Figure 2: Smoothing of equipotential surfaces with increasing altitude.

A detailed picture of the spectrum of the gravity induced orbit perturbations can be obtained from the first order solution of the Lagrange Planetary Equations (LPE) (s. Prof. Kovalevsky's lecture).

Considering only Earth gravity, the perturbing acceleration $\mathbf{a}_R$ is represented by the gradient of the disturbing potential R

$$\mathbf{a}_R = \nabla R.$$

Then the perturbation equations read

$$\begin{aligned}
\frac{da_I}{dt} &= \frac{2}{na_I} A_M, \\
\frac{de_I}{dt} &= \frac{1-e_I^2}{na_I^2 e_I} A_M - \frac{(1-e^2)^{1/2}}{na_I^2 e_I} A_\omega, \\
\frac{di_I}{dt} &= \frac{\cos i_I A_\omega - A_\Omega}{na_I^2(1-e_I^2)^{1/2}\sin i_I}, \\
\frac{d\omega_I}{dt} &= -\frac{\cos i_I A_i}{na_I^2(1-e_I^2)^{1/2}\sin i_I} + \frac{(1-e_I^2)^{1/2}}{na_I^2 e_I} A_e, \\
\frac{d\Omega_I}{dt} &= \frac{A_i}{na_I^2(1-e_I^2)^{1/2}\sin i_I}, \\
\frac{dM_I}{dt} &= n - \frac{1-e_I^2}{na_I^2 e_I} A_e - \frac{2}{na_I} A_a.
\end{aligned} \tag{16}$$

with

$$A_{E_i} = \mathbf{a}_R \cdot \frac{\partial \mathbf{r}_I}{\partial E_i} = \nabla R_I \cdot \frac{\partial \mathbf{r}_I}{\partial E_i} = \frac{\partial R_I}{\partial E_i} \tag{17}$$

n ... the mean motion

and E_i the Kepler elements : $a, e, i, \omega, \Omega, M$: semimajor axis, eccentricity, inclination, argument of perigee, right ascension of node and mean anomaly respectively. With Kaula's development of the geopotential disturbing function R in terms of Keplerian elements (Kaula, 1966), the partials A_{E_i} (17) can easily be obtained and introduced into the right-hand sides of equations (16).

This development results in

$$R = \sum_{l=2}^{\infty} \sum_{m=0}^{l} \sum_{p=0}^{l} \sum_{q=-\infty}^{\infty} R_{lmpq} \tag{18}$$

where the individual terms belonging to an $(lmpq)$ index combination are

$$R_{lmpq} = \frac{\mu a_E^l}{a^{l+1}} \bar{J}_{lm} \bar{F}_{lmp}(i) G_{lpq}(e) \left\{ {\cos \atop \sin} \right\}_{l-m \text{ odd}}^{l-m \text{ even}} \psi_{lmpq}. \tag{19}$$

and

$$\psi_{lmpq} = (l-2p)\omega + (l-2p+q)M + m(\Omega - \theta - \lambda_{lm}) \tag{20}$$

with

- $\bar{F}_{lmp}$(i) the normalized inclination function (Kaula, 1966; Allan, 1965; Kostelecky, 1985), a polynomial in $\sin i$ and summation over p is finite.
- G_{lpq}(e) the eccentricity function (Kaula, 1966), a polynomial in e and of the order $e^{|q|}$. Terms with $q = 0$ are generally dominant and for close Earth satellites with e small, one normally needs only to sum over a few additional ± 1, ± 2 terms.
- $\bar{J}_{lm}$, λ_{lm} the harmonic amplitudes and phase angles ($\bar{C}_{lm} = \bar{J}_{lm} \cos m\lambda_{lm}$; $\bar{S}_{lm} = \bar{J}_{lm} \sin \lambda_{lm}$)
- θ the right ascension of Greenwich
- μ=GM

Substituting the development (19) into the right-hand sides of the perturbation equations (16) and differentiating with respect to the E_i leads to

$$\frac{dE^i}{dt} = \sum_{lmpq} \bar{J}_{lm} K^i_{lmpq} \left\{ {-\sin \atop \cos} \right\}_o^e \psi_{lmpq} \tag{21}$$

$$\text{for} i = 1, 2, 3 (a, e, i)$$

$$\frac{dE^i}{dt} = \sum_{lmpq} \bar{J}_{lm} K^i_{lmpq} \left\{ {\cos \atop \sin} \right\}_o^e \psi_{lmpq} \tag{22}$$

$$\text{for} i = 4, 5 (\omega, \Omega)$$

$$\frac{dE^6}{dt} = n + \sum_{lmpq} \bar{J}_{lm} K^6_{lmpq} \{{}^{\cos}_{\sin}\}^e_o \psi_{lmpq} \tag{23}$$

with the abbreviations

$$\sum_{lmpq} = \sum_{l=2}^{\infty} \sum_{m=0}^{l} \sum_{p=0}^{l} \sum_{q=-\infty}^{\infty}$$

$e = l - m$ even; $o = l - m$ odd

and the K factors as

$$K^1_{lmpq} = \frac{2\mu a_e^l}{na^{l+2}} \bar{F}_{lmp} G_{lpq}(l - 2p + q) \tag{24}$$

$$K^2_{lmpq} = \frac{\mu a_e^l}{a^{l+3} ne} \bar{F}_{lmp} G_{lpq}(1-e^2)^{1/2}[(1-e^2)^{1/2} (l-2p+q) - (l-2p)] \tag{25}$$

$$K^3_{lmpq} = \frac{\mu a_e^l \bar{F}_{lmp} G_{lpq}}{na^{l+3}(1-e^2)^{1/2} \sin i}[(l-2p)\cos i - m] \tag{26}$$

$$K^4_{lmpq} = \frac{\mu a_e^l}{na^{l+3}}[\frac{(1-e^2)^{1/2}}{e} \bar{F}_{lmp} \partial G_{lpq}/\partial e - \frac{\operatorname{cotg} i}{(1-e^2)^{1/2}} \cdot \partial \bar{F}_{lmp}/\partial i \; G_{lpq}] \tag{27}$$

$$K^5_{lmpq} = \frac{\mu a_e^l \; \partial \bar{F}_{lm}/\partial i \; G_{lpq}}{na^{l+3}(1-e^2)^{1/2} \sin i} \tag{28}$$

$$K^6_{lmpq} = \frac{\mu a_e^l}{na^{l+3}} \bar{F}_{lmp}[2(l+1)\, G_{lpq} - \frac{1-e^2}{e} \partial G_{lpq}/\partial e] \tag{29}$$

(30)

To get approximate insight into the orbit perturbations it is sufficient to assume that the total perturbation is obtained by linear superposition of the $lmpq$-term effects, i.e.

$$E^i(t) \approx E^i_o + \sum_{lmpq} \Delta E^i_{lmpq} \tag{31}$$

with E^i_o the constant elements of the Keplerian orbit.

In this approximation the integrals to be solved read with e.g. expressions (22)

$$\int^t \frac{dE^i_{lmpq}}{dt'} dt' = \bar{J}_{lm} \int^t K^i_{lmpq} \{{}^{\cos}_{\sin}\} \psi_{lmpq} dt' \tag{32}$$

Both K^i_{lmpq} and ψ_{lmpq} are computed with the elements of the Kepler orbit. Thus the only time dependency in the integrand is through ψ_{lmpq} and here only through the arguments M and θ.

With

$$\psi_{lmpq} := (l-2p)\omega + (l-2p+q)nt + m(\Omega - (\dot{\theta}t + \theta_0)) \tag{33}$$

$$= (l-2p+q)n - m\dot{\theta})t + (l-2p)\omega + m(\Omega - \theta_0)) \tag{34}$$

it becomes evident that two classes of perturbations result: (1) secular perturbations (section 3.1) if $(l-2p+q)n - m\dot{\theta}$ vanishes, (2) periodic perturbations (section 3.2) if not.

3.1. Secular Perturbations

An $(lmpq)$ index combination contributes to the total perturbation of an element a secular portion if in equation (34) the frequency $(l-2p+q)n - m\dot{\theta}$ vanishes. This is only possible if either

$$I := \frac{n}{\dot{\theta}} = \frac{m}{l-2p+q} \qquad \text{or} \qquad \begin{aligned} l-2p+q &= 0 \\ m &= 0. \end{aligned}$$

The first case, with I an integer, would correspond to an exact commensurability between the mean motion of the satellite and the earth rotation rate. (s. section 3.2)

In the second case, with $m = 0$, the secular perturbation will result from the zonal harmonics. $l-2p$ has to become zero. With p being summed from 0 to l, $l-2p+q=0$ and $l-2p=0$ this is for $q=0$ and l even. Secular perturbations therefore can only result from even zonal harmonics.

Going through the perturbation equations (21)-(23) with the index combinations in question ($m=0$; $l-2p+q=0$; $l=2p$) one finds

$$(\frac{da}{dt})_{lopo} = (\frac{de}{dt})_{lopo} = (\frac{di}{dt})_{lopo} = 0 \tag{35}$$

and

$$(\frac{d\omega}{dt})_{lopo} = \bar{C}_{l0}\, K^4_{lopo} \tag{36}$$

$$(\frac{d\Omega}{dt})_{lopo} = \bar{C}_{l0}\, K^5_{lopo} \tag{37}$$

$$(\frac{dM}{dt})_{lopo} = \bar{C}_{l0}\, K^6_{lopo}. \tag{38}$$

Thus, the semimajor axis, the eccentricity and inclination of the orbit undergo no secular change. The nodal line and the apsidal line precess with a rate which is proportional to all even zonal harmonics. The maximum contribution of course will come from $\bar{C}_{20}$. Also the mean motion is changed according to the rate given by (38).

For a typical geodetic satellite such as STARLETTE ($i = 49.8$ deg.) the secular rates in the perigee and node are

$$\dot{\omega} = 3.30°/\text{day} \qquad \dot{\Omega} = -3.95°/\text{day} \tag{39}$$

3.2. Periodic Perturbations

First-order periodic perturbations are obtained from equation (32) by assuming a, e, i are constant and ω, Ω, M have only linear rates. Substituting

$$dt = \dot{\psi}_{lmpq}^{-1}\, d\psi_{lmpq}$$

shows that the integrals can still be solved in closed form if

$$\dot{\psi}_{lmpq} = (l-2p)\dot{\omega} + (l-2p+q)\dot{M} + m(\dot{\Omega} - \dot{\theta}) \tag{40}$$

is constant. Under this condition the first-order perturbations for a specific $lmpq$-term are given by

$$\Delta E^i_{lmpq} = \frac{\bar{J}_{lm}\, K^i_{lmpq}}{\dot{\psi}_{lmpq}} \{{\cos \atop \sin}\}^e_o\, \psi_{lmpq} \quad \text{for } i = 1,2,3 \tag{41}$$

$$\Delta E^i_{lmpq} = \frac{\bar{J}_{lm}\, K^i_{lmpq}}{\dot{\psi}_{lmpq}} \{{\sin \atop -\cos}\}^e_o\, \psi_{lmpq} \quad \text{for } i = 4,5 \tag{42}$$

$$\Delta E^i_{lmpq} = \bar{J}_{lm} \left[\frac{K^i_{lmpq}}{\dot{\psi}_{lmpq}} - \frac{3n\, K^1_{lmpq}}{2a\, \dot{\psi}^2_{lmpq}}\right] \{{\sin \atop -\cos}\}^e_o\, \psi_{lmpq} \quad \text{for } i = 6 \tag{43}$$

With equations (41) to (43) it is possible to compute the periodic perturbation contribution of each $lmpq$-combination in each orbital element. Summing over pq results in the combined effect of an lm-pair

$$\Delta E^i_{lm} = \sum_{pq} \Delta E^i_{lmpq}. \tag{44}$$

Of more interest in orbit analysis is often the corresponding effect in the satellite's along-track (λ), cross-track (β) and radial position (r) component.

Upper bounds for the lm-pair perturbations in the orbit position components result from

$$\begin{aligned} \Delta\lambda^2_{lm} &= \sum_{p,q} (\Delta\Omega_{lmpq} \cdot \cos I + \Delta\omega_{lmpq} + \Delta M_{lmpq})^2 \cdot a^2(1+e)^2; \\ \Delta\beta^2_{lm} &= \sum_{p,q} [\Delta I_{lmpq} \cdot a(1+e)]^2 + \sum_{p,q} [\Delta\Omega_{lmpq} \sin I \cdot a(1+e)]^2; \\ \Delta r^2_{lm} &= \sum_{p,q} [(1+e)\cdot \Delta a_{lmpq} - a \cdot \Delta e_{lmpq}]^2. \end{aligned} \tag{45}$$

This leads to what we call the maximum orbit position perturbation due to a $(\bar{C}_{lm}, \bar{S}_{lm})$-pair

$$\Delta O_{lm} = (\Delta\lambda^2_{lm} + \Delta\beta^2_{lm} + \Delta r^2_{lm})^{1/2}. \tag{46}$$

For the medium altitude STARLETTE orbit these ΔO_{lm} values for l, m up to 70 are shown in figure 3 (threshold 1 cm).

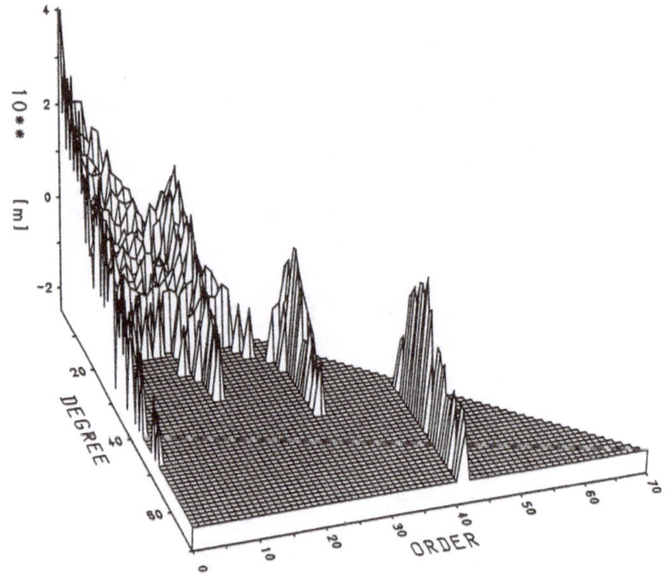

Figure 3: Maximum orbit position perturbations due to geopotential STARLETTE (altitude ~ 950 km)

According to equations (41)-(43), with the K^i_{lmpq} being functions of a, e, i, the periodic perturbation amplitudes are controled primarily by the semimajor axis a, the size of the C_{lm}, S_{lm} terms and also by $\dot{\psi}$ and i.

The frequencies of the perturbations are all contained in

$$\dot{\psi}_{lmpq} = (l-2p)\dot{\omega} + (l-2p+q)\dot{M} + m(\dot{\Omega} - \dot{\theta}) \tag{47}$$

and the period of the perturbation which an $(lmpq)$ index combination is contributing to the total perturbation is

$$P_{lmpq} = \frac{2\pi}{|\dot{\psi}_{lmpq}|}. \tag{48}$$

A natural classification of the frequencies or periods results according to m, q and $k = (l-2p+q)$. The basic periods and their designations are given in table 3.

m	$l-2p$	$l-2p+q$ $=k$	Condition	Period P_{lmpq}	Designation (BasicPeriod)
0		$\neq 0$		$\frac{2\pi}{\lvert k\bar{n}\rvert}$	short period (satellite revolution)
0	$\neq 0$	0	$q \neq 0$	$\frac{2\pi}{\lvert (l-2p)\dot{\omega}\rvert}$	long period (apsidal line revolution)
$\neq 0$		$\neq 0$		$\frac{2\pi}{\lvert k\bar{n}-m\dot{\theta}\rvert}$	short period (satellite revolution)
$\neq 0$		0		$\frac{2\pi}{\lvert m(\dot{\Omega}-\dot{\theta})\rvert} \approx \frac{2\pi}{m\dot{\theta}}$	m-daily (siderial day)
$\neq 0$		>0	$\bar{n} \approx m\dot{\theta}/k$	$\frac{2\pi}{\lvert k\bar{n}-m\dot{\theta}\rvert}$	shallow resonant

Table 3: Characterisation of perturbations of zonal and tesseral terms

As shown by Klosko and Wagner, 1982, the basic periods for the short period, m-daily and shallow resonance perturbations will be modulated by a $2\pi/\mid q\dot{\omega}\mid$ period. This simply results when rewriting $\dot{\psi}_{lmpq}$ in the form

$$\dot{\psi}_{lmpq} = -q\dot{\omega} + k(\dot{M} + \dot{\omega}) + m(\dot{\Omega} - \dot{\theta}) \tag{49}$$

We have thus a secondary slow frequency. It is important to take this into account for the determination of tesserals from orbit perturbations. Orbits should be distributed over a complete apsidal period.

Zonal Harmonics

Zonal harmonics of odd order are primarily computed from the long-period perturbations. As shown for STARLETTE these effects are pronounced even for high degree terms. Short period perturbation information can be used in addition for the even and odd zonal term determination.

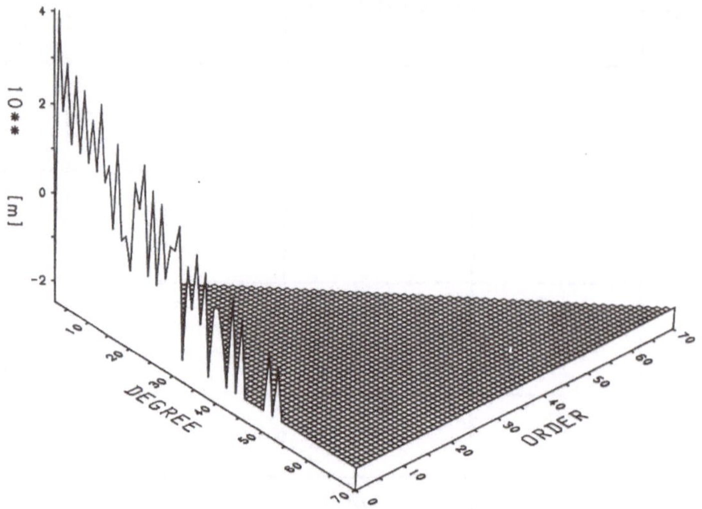

Figure 4: Zonal long period orbit perturbations due to geopotential
STARLETTE (altitude ~ 950 km)

Tesseral Harmonics

Low to medium degree and order tesseral harmonics are primarily determined from the large m-daily perturbations on the satellite motion, also from short period perturbations. For STARLETTE the perturbation spectrum for the m-daily and short-periods is shown on the next two figures.

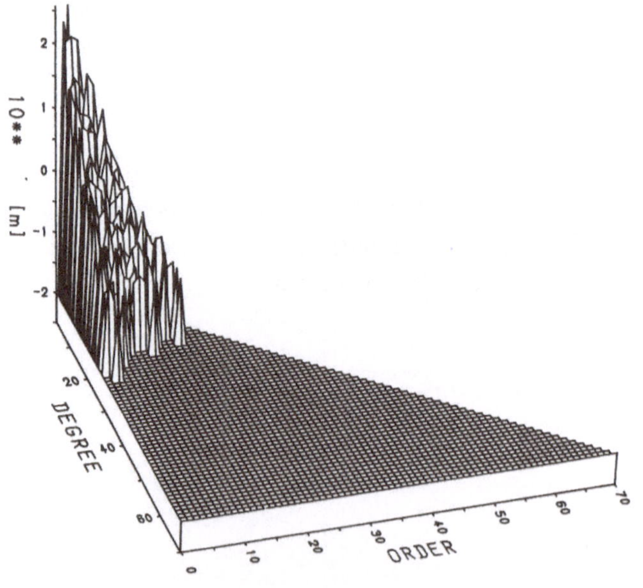

Figure 5: M-daily orbit perturbations due to geopotential
STARLETTE (altitude ~ 950 km)

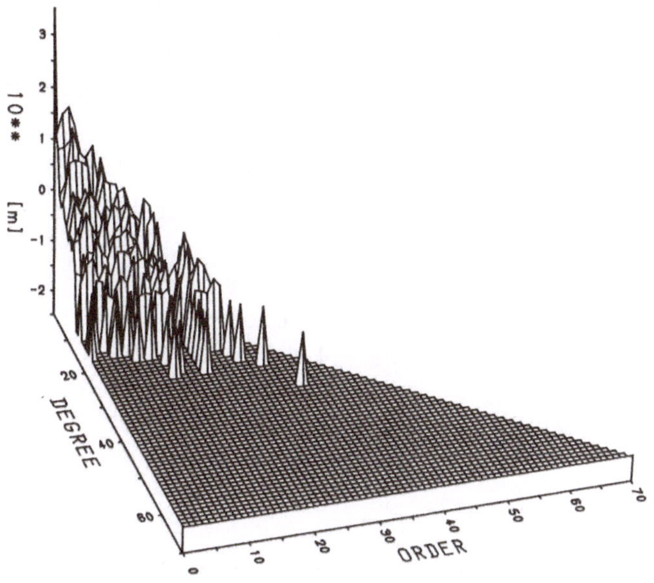

Figure 6: Short period orbit perturbations due to geopotential
STARLETTE (altitude ~ 950 km)

Resonant Harmonics

Resonance is a common effect experienced by almost all near Earth orbiting objects. As is evident from equations (41)-(43), large perturbation will occur if the argument $\dot{\psi}_{lmpq}$ is small or will become singular if

$$\dot{\psi}_{lmpq} = (l-2p)\dot{\omega} + (l-2p+q)\dot{M} + m(\dot{\Omega} - \dot{\theta}) = 0. \tag{50}$$

For $q = 0$ this is the case if

$$\alpha(\dot{\omega} + \dot{M}) = \beta(\dot{\theta} - \dot{\Omega}), \tag{51}$$

where α and β are some pair of mutually prime integers. To interpret (51), the exact condition for commensurability is that the satellite performs β nodal periods while the Earth rotates α times relative to the precessing satellite orbit plane. After this interval the path of the satellite relative to the Earth repeats exactly, which is the physical reason for the resonance effects. Alternatively the approximate condition is that the mean motion of the satellite is β/α times the angular velocity of the Earth.

For most satellites, the exact resonance condition is not met, but its approximation yields substantial perturbations nevertheless. While deep resonance can be thought of as an event, shallow resonance typically is a stable periodic condition.

The shallow resonance for close orbits is always a problem for orbit determination because the minimum such period is about two days.

A slowly varying argument in the vicinity of an α/β commensurability is obtained if

$$l - 2p = \alpha\gamma; \qquad m = \beta\gamma; \qquad \gamma = 1, 2, 3 \ldots.$$

For STARLETTE performing 13.82 revolutions per day the resonance periods are

γ=1 resonance				Period
	$l - 2p = 1,$	$q = 0,$	$m = 14$	2.8 days
γ=2 resonance				Period
	$l - 2p = 2,$	$q = 0,$	$m = 28$	1.4 days
γ=3 resonance				Period
	$l - 2p = 3,$	$q = 0,$	$m = 41$	16.6 days
γ=4 resonance				Period
	$l - 2p = 4,$	$q = 0,$	$m = 55$	2.4 days

Figure 7 shows the corresponding resonant perturbations. It becomes quite clear from this picture that the $\gamma = 3$ resonance, although not dominant, produces quite pronounced perturbations which can be used to determine harmonics of order 41 up to quite high degrees. .

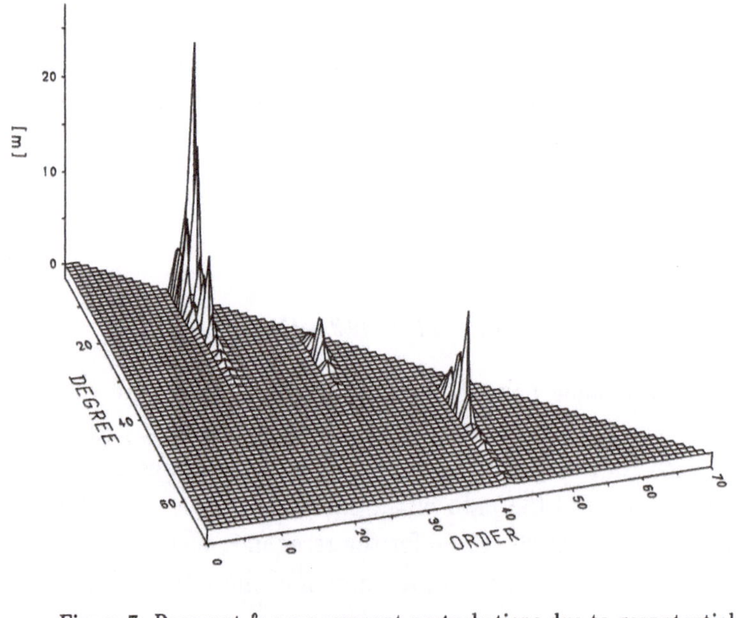

Figure 7: Resonant & near resonant perturbations due to geopotential STARLETTE (altitude ~ 950 km)

4. Adjustment Procedures

4.1. Single Arc Solution

As shown in section 2, in the differential orbit correction process we are faced with linear observation equations of the form

$$\underset{(m\times n)}{\mathbf{A}}\;\underset{(n\times 1)}{\mathbf{X}} = \underset{(m\times 1)}{\mathbf{L}} \tag{52}$$

where the relation between the m observables $\mathbf{L}$ and the n unknown parameters $\mathbf{X}$ is given through the $m \times n$ matrix $\mathbf{A}$ of the partial derivatives. In general because of errors in the observations we have to replace $\mathbf{L}$ by $\mathbf{l}+\mathbf{r}$, where $\mathbf{l}$ is the vector of observed quantities $(q - q_c)_{t_i}$ and $\mathbf{r}$ is the vector of residuals. Equation (52) then becomes

$$\underset{(m\times n)}{\mathbf{A}}\;\underset{(n\times 1)}{\mathbf{X}} - \underset{(m\times 1)}{\mathbf{l}} = \underset{(m\times 1)}{\mathbf{r}} \tag{53}$$

with $m > n$. The least squares estimate $\hat{\mathbf{X}}$ of $\mathbf{X}$ is obtained under the condition

$$\mathbf{r}^T\mathbf{P}_l\mathbf{r} = minimum \tag{54}$$

with the weight matrix

$$\mathbf{P}_l = \sigma_O^2\mathbf{C}_l^{-1}, \tag{55}$$

where σ_0 is the a priori standard error of unit weight, $\mathbf{C}_l$ the covariance matrix of the observations and the following relationship holds between the weight matrix $\mathbf{P}_l$, the covariance matrix $\mathbf{C}_l$ and the weight coefficient matrix $\mathbf{Q}_l$

$$\mathbf{P}_l^{-1} = \mathbf{C}_l/\sigma_O^2 = \mathbf{Q}_l. \tag{56}$$

In the case of uncorrelated observations, as is normally assumed for the tracking data, the weight matrix $\mathbf{P}_l$ is given by

$$\mathbf{P}_l = \sigma_0^2 \begin{pmatrix} \sigma_1^{-2} & & 0 \\ & \sigma_2^{-2} & \\ 0 & & \sigma_n^{-2} \end{pmatrix} \tag{57}$$

with σ_i^2 the variance of the ith observed quantity.

Solving the extremal problem leads to

$$\mathbf{A}^T\mathbf{P}_l(\mathbf{AX} - \mathbf{l}) = 0 \tag{58}$$

or if we let

$$\mathbf{N} = \mathbf{A}^T\mathbf{P}_l\mathbf{A}\;;\;\mathbf{b} = \mathbf{A}^T\mathbf{P}_l\mathbf{l} \tag{59}$$

$$\mathbf{NX} = \mathbf{b}, \tag{60}$$

which are the least squares normal equations.

From the above equation one gets the least squares estimate of $\mathbf{X}$

$$\hat{\mathbf{X}} = (\mathbf{A}^T\mathbf{P}_l\mathbf{A})^{-1}\mathbf{A}^T\mathbf{P}_l\mathbf{l} = \mathbf{N}^{-1}\mathbf{b}. \tag{61}$$

These are primarily the corrections for the internal parameters and perhaps for a few external parameters.

Applying the law of error propagation, the weight coefficient matrix of the solution $\hat{\mathbf{X}}$ is given by

$$\mathbf{Q}_{\hat{x}} = (\mathbf{A}^T\mathbf{P}_l\mathbf{A})^{-1} = \mathbf{N}^{-1} \tag{62}$$

and the covariance matrix $\mathbf{C}_{\hat{x}}$ by

$$\mathbf{C}_{\hat{x}} = \hat{\sigma}_O^2\mathbf{Q}_{\hat{x}} \tag{63}$$

with the a posteriori variance of unit weight

$$\hat{\sigma}_O^2 = \hat{\mathbf{V}}^T\mathbf{P}_l\hat{\mathbf{V}}/(m-n) = (\mathbf{l}^T\mathbf{P}_l\mathbf{l} - \mathbf{l}^T\mathbf{P}_l\mathbf{A}\hat{\mathbf{X}}). \tag{64}$$

From the diagonal elements of the covariance matrix (62) we get the estimated standard deviation for each unknown $\hat{\mathbf{X}}_i$

$$\hat{\sigma}_{x_i} = (C_{\hat{x}}^{ii})^{1/2}, \tag{65}$$

and from the off-diagonal elements the correlation between the ith and j^{th} unknown

$$K_{ij} = \frac{C_{\hat{x}}^{ij}}{(C_{\hat{x}}^{ii}\, C_{\hat{x}}^{jj})^{1/2}}. \tag{66}$$

4.2. Solution from Combined Normal Equations

As described in section 2, the solution vector $\mathbf{X}$ contains

- arc-dependent parameters, called $\mathbf{X}_I$
- arc-independent parameters, called $\mathbf{X}_E$.

Splitting equation (60) into partitioned matrices, we thus finish for the Kth arc with the normal equation system

$$\begin{pmatrix} \mathbf{N}_{II} & \mathbf{N}_{IE}^T \\ \mathbf{N}_{IE} & \mathbf{N}_{EE} \end{pmatrix}^K \begin{pmatrix} \mathbf{X}_I \\ \mathbf{X}_E \end{pmatrix}^K = \begin{pmatrix} \mathbf{b}_I \\ \mathbf{b}_E \end{pmatrix}^K \begin{matrix} \}r \\ \}m-r \end{matrix}. \tag{67}$$

To reduce the number of unknowns, the corrections to the arc-dependent parameters are eliminated after forming the normal equations for the Kth arc.

Decomposing equation (67), one gets

$$\mathbf{X}_I^K = \mathbf{N}_{II}^{-1}(\mathbf{b}_I - \mathbf{N}_{IE}^T\mathbf{X}_E)^K \tag{68}$$

and with this result the reduced normal equations

$$\mathbf{N}_K^*\mathbf{X}_E = \mathbf{b}_K^* \tag{69}$$

with

$$\begin{aligned}\mathbf{N}_K^* &= (\mathbf{N}_{EE} - \mathbf{N}_{IE}\mathbf{N}_{II}^{-1}\mathbf{N}_{IE}^T)^K \\ \mathbf{b}_K^* &= (\mathbf{b}_E - \mathbf{N}_{IE}\mathbf{N}_{II}^{-1}\mathbf{b}_I)^K.\end{aligned} \tag{70}$$

Combining the reduced normal equations for all arcs ($K = 1, \ldots N$) yields

$$\sum_{K=1}^{N} \mathbf{N}_K^* \, \mathbf{X}_E = \sum_{K=1}^{N} \mathbf{b}_K^* \tag{71}$$

from which we get the solution

$$\hat{\mathbf{X}}_E = \sum_{K=1}^{N} \mathbf{N}_K^{*\;-1} \sum_{K=1}^{N} \mathbf{b}_K^* \tag{72}$$

for e.g. the spherical harmonics and the station coordinates.

With the solution (72) the partial solution $\mathbf{X}_I^K$ of the arc-dependent parameters of the Kth system can be computed from equation (68).

According to equation (62) the weight coefficient matrix of the solution $\hat{\mathbf{X}}_E$ will be

$$\mathbf{Q}_{\hat{X}_E} = \sum_{K=1}^{N} \mathbf{N}_K^{*\;-1} \tag{73}$$

Without regrouping all subsystems (68), a good approximation for the a posteriori variance of the weight unit $\hat{\sigma}_O^2$ can be computed according to equation (64) from

$$\sigma_0^2 = [\, \mathbf{l}^T\mathbf{P}_l\,\mathbf{l} - \hat{\mathbf{X}}_E^T \sum_{K=1}^{N} \mathbf{b}_E^K - \sum_{K=1}^{N} \hat{\mathbf{X}}_I^{T^K}\,\mathbf{b}_I^K \,] / (M - N) \tag{74}$$

M = number cf equations
N = number of unknowns.

Finally, with equations (73), (74) the covariance matrix for the arc-independent parameters is computed

$$\mathbf{C}_{\hat{x}_E} = \sigma_O^2\mathbf{Q}_{\hat{x}_E}, \tag{75}$$

from which all error estimates can be obtained.

4.3. Constraint Equations

In both the orbit and geodetic parameter estimation process it may be necessary to impose constraints on particular unknowns. Such constraints can be imposed through condition equations of special importance and added to equations (53) with appropriate weights. If the constraints are absolute very high weights have to be applied.

4.4. Light Constraint Solutions

To permit stable solutions for high degree and order satellite fields, most analysis groups have introduced constraints in the form of a priori weights for the geopotential unknowns into their systems. This is accomplished as follows:

Equation (53) implies the assumption that we have no a priori information on the unknowns. For the gravity unknowns there is no justification for this assumption because since 1966 (Kaula, 1966) we know that the degree variances of the $\bar{C}_{lm}$, $\bar{S}_{lm}$ coefficients follow the approximate rule

$$\sigma_l^2 \approx 10^{-10}/l^4. \tag{76}$$

Adapting this value to what is known from the power spectrum of more recent gravity solutions, gives

$$\sigma_l^2 \approx 5 \cdot 10^{-11}/l^4.$$

This information can be used in the form that we enlarge the model (53) by additional equations of the form

$$\begin{aligned} \bar{C}_{lm} &= 0 \pm 7 \cdot 10^{-6}/l^2 \\ & \qquad\qquad\qquad l = m, \ldots N_{max} \\ \bar{S}_{lm} &= 0 \pm 7 \cdot 10^{-6}/l^2, \end{aligned} \tag{77}$$

which means that we take the parameters as direct observed quantities with a zero mean and a standard deviation of $\pm 7 \cdot 10^{-6}/l^2$.

The combination of the systems then becomes

$$\begin{pmatrix} \mathbf{A} \\ \mathbf{I} \end{pmatrix} \mathbf{X} - \begin{pmatrix} \mathbf{l} \\ \mathbf{O} \end{pmatrix} = \begin{pmatrix} \mathbf{r} \\ \mathbf{V} \end{pmatrix} \quad \mathbf{V} \equiv \mathbf{X} \tag{78}$$

and the weight matrices read, if $\mathbf{l}$ and $\mathbf{X}$ are uncorrelated,

$$\mathbf{P} = \begin{pmatrix} \mathbf{P}_l & o \\ o & \mathbf{P}_X \end{pmatrix}. \tag{79}$$

The least squares normal equations obtained under the condition

$$\left(\mathbf{r}^T \mathbf{V}^T\right) \mathbf{P} \begin{pmatrix} \mathbf{r} \\ \mathbf{V} \end{pmatrix} = \mathbf{r}^T \mathbf{P}_l \mathbf{r} + \mathbf{V}^T \mathbf{P}_X \mathbf{V} = \mathbf{r}^T \mathbf{P}_l \mathbf{r} + \mathbf{X}^T \mathbf{P}_X \mathbf{X} = min$$

are given by

$$\begin{aligned} \mathbf{N}\mathbf{X} &= \mathbf{b} \\ \mathbf{N} = \mathbf{A}^T \mathbf{P}_l \mathbf{A} + \mathbf{P}_X \quad &; \quad \mathbf{b} = \mathbf{A}^T \mathbf{P}_l \mathbf{l}. \end{aligned} \tag{80}$$

From the above equation we obtain for the solution vector and the error covariance matrix

$$\hat{\mathbf{X}} = (\mathbf{A}^T\mathbf{P}_l\mathbf{A} + \mathbf{P}_X)^{-1}\mathbf{A}\mathbf{P}_l\mathbf{l} \quad (81)$$

$$\mathbf{C}_{\hat{X}} = \sigma_O^2(\mathbf{A}^T\mathbf{P}_l\mathbf{A} + \mathbf{P}_X)^{-1}. \quad (82)$$

Considering the collocation definition of $\mathbf{P}$

$$\mathbf{P} = \sigma_O^2\begin{pmatrix}\mathbf{C}_l & O \\ O & \mathbf{C}_X\end{pmatrix}^{-1} \quad (83)$$

where the a priori variance factor σ_O^2 is set to unity and using the matrix identity

$$\mathbf{C}_X\mathbf{A}^T(\mathbf{A}\mathbf{C}_X\mathbf{A}^T + \mathbf{C}_l)^{-1} = (\mathbf{A}^T\mathbf{C}_l^{-1}\mathbf{A} + \mathbf{C}_X^{-1})^{-1}\mathbf{A}^T\mathbf{C}_l^{-1} \quad (84)$$

it is easy to show that equations (81) and (82) are identical to the signal and error covariance equations (85) and (86) of the least squares collocation method (Moritz 1973) when no systematic parameters are estimated.

$$\hat{\mathbf{X}} = \mathbf{C}_X\mathbf{A}^T(\mathbf{A}\mathbf{C}_X\mathbf{A}^T + \mathbf{C}_l)^{-1}\mathbf{l} \quad (85)$$

$$\mathbf{C}_{\hat{X}} = \mathbf{C}_X - \mathbf{C}_X\mathbf{A}^T(\mathbf{A}\mathbf{C}_X\mathbf{A}^T + \mathbf{C}_l)^{-1}\mathbf{A}\mathbf{C}_X. \quad (86)$$

One of the main advantages of using the informations on the statistical behaviour of the coefficients by introducing the additional equations (77) is the better condition of the normal equations. This leads to stable solutions and accuracy estimates even when solving for high degrees N_{max}. Truncation errors in the solution are greatly retarded by this procedure. It is a disadvantage though that such solutions have the tendency to reduce the absolute values of the harmonics because of the expected zero mean.

4.5. Parameters Considered for Adjustment

The number and type of solution parameters vary considerably in the various steps of an orbit and geodetic parameter estimation process. Giving what is done in the DGFI Orbit and Geodetic Parameter Estimation System DOGS, the following results

- In the data fine screening step a large number of internal and a few external parameters have to be estimated

  $\mathbf{X}_I$: corrections to epoch position and velocity of spacecraft, C_R, and C_D m-daily scaling values for solar radiation and drag, $a, \dot{a}$ values for unmodelled along-track acceleration, pass dependent range and time biases for laser data, pass or station dependent time biases for optical data, pass dependent frequency offsets and drifts for Doppler data, centre of mass correction.

$\mathbf{X}_E$: station position corrections if a not well coordinated station appears in the data arc.

Especially for Doppler arcs this can result in more than 1000 unknowns per arc.

- In the definitive orbit determination step preceding the generation of the satellite normal equations only internal parameters are adjusted

  $\mathbf{X}_I$: the epoch state vector corrections, C_R, C_D daily or multi-day values and/or $a, \dot{a}$ (along track accel.)

- In the following satellite normal equation generation step for a large number of parameters partials are computed, independent of whether they will finally be solved for or perhaps be fixed in the final solution by absolute constraints.

  $\mathbf{X}_I$: the epoch state vector corrections C_R^i, C_D^i values, $a, \dot{a}$ (LAGEOS)

  $\mathbf{X}_E$: positions of stations participating in arc, m-day pole position and LOD values, solid Earth and ocean tidal terms, Love and Shida numbers, all potential coefficients to which tracking data are sensitive. This can result in up to 1500 and even more external unknowns in extreme cases for a single arc data set.

5. Tracking Data

5.1. Existing Data

The tracking data types almost exclusively used for the computation of "satellite-only" solutions are: laser ranges, Doppler and right ascension/declination values. There are almost ninety objects which have been tracked by cameras, lasers and Doppler systems, but only about fifty to sixty with a reasonably dense coverage.

Routine tracking of the early near Earth satellites was performed with cameras. All data analysed for the first SAO gravity model in 1966 was of this type. After the invention of lasers and the launch of the first retroreflector equipped spacecraft BEB in 1964 SAO, GSFC and CNES groups started to build and operate laser systems on an irregular basis. In 1968 already six retroreflector equipped satellites were launched by France and the United States. Observations accurate to about 5 - 10 metres were made during the late 60's, but the number of data points was very small as compared to present standards. The first larger amount of laser tracking data became available in the course of the ISAGEX campaign in 1971, when for the first time more laser than optical data were acquired. Although laser accuracies were still at the 1 - 5 metre level, the data from that campaign contributed significantly to the early US and European satellite gravity models.

After this campaign most of the agencies stopped camera tracking and intensified laser tracking. Five additional retroreflector satellites were launched after that period: GEOS 3, STARLETTE, SEASAT, LAGEOS, AJISAI. Tracking accuracies have improved since the mid-seventies by almost one order of magnitude and have now reached for most of the systems the few centimetre level. Tracking has since 1980 completely concentrated on LAGEOS, STARLETTE and recently also on AJISAI.

Doppler tracking data, although acquired since the early sixties, have for various reasons only recently become available for gravity field modelling. OSCAR, SEASAT and NOVA satellite Doppler data are included in the new generation of GSFC, CSR and DGFI/GRGS models. So, in summary, there is a large amount of geodetic quality tracking data available since the beginning of the space age. But the number of tracked geodetically relevant objects, the density of tracking, the measurement types and measurement qualities have been very different over the last 30 years.

5.2. Data Selection

The first task in a gravity model computation from tracking data is the selection of the appropriate observation material for the processing. The main criteria for this selection are:

- magnitude of drag perturbations in the orbit
- sensitivity of satellite orbit to the various field constituents
- distribution in inclination
- the distribution of orbital arcs over the satellite's apsidal period
- the amount of data in an orbital arc, the quality of data and distribution along the arc

The list of satellites which have been selected by the TOPEX team for the GEM-T1 model determination in view of these criteria and in view of specific TOPEX requirements is given in table 4. Data from these satellites and a few additional ones are also processed by the DGFI/GRGS team for the next model of the GRIM series.

SATELLITE NAME	SATELL. ID NO.	SEMI MAJOR AXIS	ECC	INCL. (DEG.)	DATA* TYPE
ANNA-1B	620601	7501.	.0082	50.12	O
BE-B	640841	7354.	.0135	79.69	O
BE-C	650321	7507.	.0257	41.19	L,O
COURIER-1B	600131	7469.	.0161	28.31	O
D1-C	670111	7341.	.0532	39.97	L,O
D1-D	670141	7622.	.0848	39.46	L,O
GEOS-1	650891	8075.	.0719	59.39	L,O
GEOS-2	680021	7711.	.0330	105.79	L,O
GEOS-3	750271	7226.	.0008	114.98	L
LAGEOS	760391	12273.	.0038	109.85	L
OSCAR	670921	7440.	.0029	89.27	D
PEOLE	701091	7006.	.0164	15.01	L,O
SEASAT	780641	7170.	.0021	108.02	D,L
STARLETTE	750101	7331.	.0204	49.80	L
TELESTAR-1	620291	9669.	.2429	44.79	O
VANGUARD-2RB	590012	8496.	.1832	32.92	O
VANGUARD-2	590011	8298.	.1641	32.89	O

*D=Doppler L=Laser O=Optical

Table 4: Satellites used for GEM-T1 model determination (after Marsh et al., 1987)

6. Processing Steps

After selection of some hundred arcs and a lengthy, time consuming data screening process, in which all spurious data points and biases are removed, the finally accepted arcs are precision-processed. For most satellites data are processed in five to seven-day orbital arcs. This arc length is adequate to provide good resolution in all geopotential coefficients including the zonal and resonant terms which are generally derived from their long-period orbital perturbations. For most satellites the resonant periods are less than one week.

Although for a better resolution of the zonal and resonant terms one would have the tendency to enlarge the arc length, this is not advisable for most of the satellites because of the increasing mismodelling effects from atmospheric drag in the observation residuals.

In general terms the processing steps are as follows:

Common for All Arcs

- Establish an initial dynamic model which allows to predict with highest accuracy the satellite motion around the Earth under the attraction exerted on it by conservative and non-conservative forces of different origin. This involves all forces mentioned in section 2.

- Establish an initial tracking station position set materializing the mean terrestrial reference frame at an epoch T_O.
- Establish a transformation model which provides with highest accuracy the connection between the quasi-inertial frame, in which the satellite motion is represented and the mean terrestrial frame at epoch T_O. This involves the currently adopted theory for the orientation of the Earth in inertial space, that is precession and nutation and the observationally determined variations of the celestial ephemeris pole (CEP) with respect to the aforementioned terrestrial frame, that is polar motion and UT1 variations.
- Set up a measurement correction model, which takes into account all geometric corrections on both end points of the ground-to-satellite measurements and all atmospheric refraction corrections of the measured signal.

For the Kth Arc

- Combine preprocessed tracking data from all stations which contributed to the observation of the considered orbital arc in the time interval Δt into a single arc tracking data file.
- Combine this file with the parameter files describing the dynamic, transformation and measurement models and with a file containing the initial state vector of the spacecraft (S/C) at the beginning of the orbital arc and the tracking station position file into the single arc orbit determination (OD) input file.
- Generate with the initial state vector and the aforementioned dynamic and transformation files an initial S/C orbit ephemeris file over the period Δt through numerical integration of the motion equations.
- Compute with the station file, the measurement, the measurement correction, the transformation and the orbit ephemeris file at each observation instant initial, predicted (or computed) observables $q_c(t_i)$.
- Reduce each preprocessed measurement with the correction file data to obtain the corrected observables $q(t_i)$.
- Compute the residual $\Delta q(t_i)$ between the predicted and corrected observables, eliminate residuals above a certain threshold and with the accepted residuals set up the linear observation equation for each observation instant by computing the partial derivatives of the observable with respect to the

solution parameters. These solution parameters are in the OD case corrections to the initial state vector components and Lagrange parameters for the surface forces.

- Form the weighted observation equation system with all data from the considered arc.
- Compute the solution parameters as well as their precision estimates in the least squares sense by minimizing the "weighted sum of the squared observation residuals".
- Add the solution parameters to the initial parameters to get an improved set of solution parameters.
- Repeat the last seven steps in the sense of a differential correction process until convergence is reached.
- Compute, after the definitive orbit has been reached, the partials for all internal and all external parameters (including the geopotential) considered necessary for adjustment and set up the satellite normal equation system.
- To eliminate the internal parameters, in which we are not interested in gravity modelling work, from the reduced normal equations of the Kth arc.

Combination

- Combine all n reduced satellite normals for the m satellites into the combined normal equation system SI.
- Add normals for constraints to the system SI. This results in the final satellite normal equation system SII.
- Combine system SII with normals resulting from an analysis of altimeter and/or gravity anomalies. This results in the normal equation system $SIII$ for a combination solution.
- Invert the normal equation systems to get the solutions and error estimates.
- Calibrate and quality assess the solutions.

In a generalized form the processing procedure is visualized in figure 8:

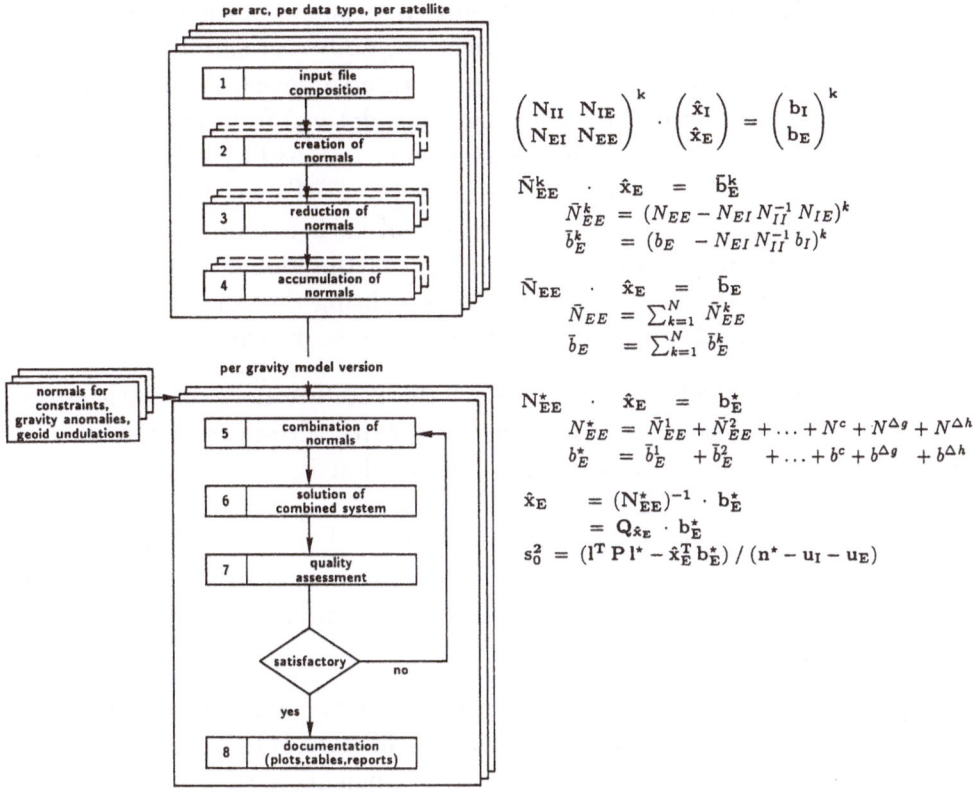

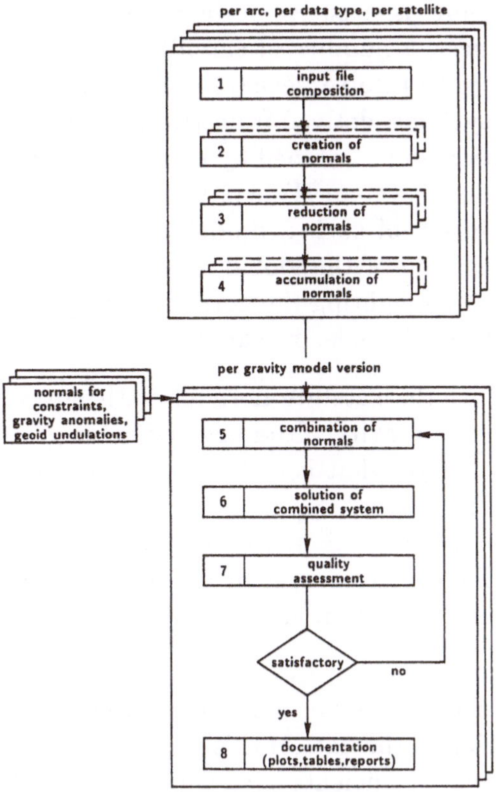

$$\begin{pmatrix} \mathbf{N_{II}} & \mathbf{N_{IE}} \\ \mathbf{N_{EI}} & \mathbf{N_{EE}} \end{pmatrix}^k \cdot \begin{pmatrix} \hat{\mathbf{x}}_\mathbf{I} \\ \hat{\mathbf{x}}_\mathbf{E} \end{pmatrix} = \begin{pmatrix} \mathbf{b_I} \\ \mathbf{b_E} \end{pmatrix}^k$$

$$\bar{\mathbf{N}}^k_{\mathbf{EE}} \cdot \hat{\mathbf{x}}_\mathbf{E} = \bar{\mathbf{b}}^k_\mathbf{E}$$

$$\bar{N}^k_{EE} = (N_{EE} - N_{EI}\, N_{II}^{-1}\, N_{IE})^k$$

$$\bar{b}^k_E = (b_E - N_{EI}\, N_{II}^{-1}\, b_I)^k$$

$$\bar{\mathbf{N}}_{\mathbf{EE}} \cdot \hat{\mathbf{x}}_\mathbf{E} = \bar{\mathbf{b}}_\mathbf{E}$$

$$\bar{N}_{EE} = \sum_{k=1}^{N} \bar{N}^k_{EE}$$

$$\bar{b}_E = \sum_{k=1}^{N} \bar{b}^k_E$$

$$\mathbf{N}^*_{\mathbf{EE}} \cdot \hat{\mathbf{x}}_\mathbf{E} = \mathbf{b}^*_\mathbf{E}$$

$$N^*_{EE} = \bar{N}^1_{EE} + \bar{N}^2_{EE} + \ldots + N^c + N^{\Delta g} + N^{\Delta h}$$

$$b^*_E = \bar{b}^1_E + \bar{b}^2_E + \ldots + b^c + b^{\Delta g} + b^{\Delta h}$$

$$\hat{\mathbf{x}}_\mathbf{E} = (\mathbf{N}^*_{\mathbf{EE}})^{-1} \cdot \mathbf{b}^*_\mathbf{E} = \mathbf{Q}_{\hat{\mathbf{x}}_\mathbf{E}} \cdot \mathbf{b}^*_\mathbf{E}$$

$$\mathbf{s}_0^2 = (\mathbf{l}^\mathbf{T}\, \mathbf{P}\, \mathbf{l}^* - \hat{\mathbf{x}}^\mathbf{T}_\mathbf{E}\, \mathbf{b}^*_\mathbf{E}) \,/\, (\mathbf{n}^* - \mathbf{u_I} - \mathbf{u_E})$$

Figure 8: Gravity field processing chain

Depending on the amount of information which has been used for the creation of the combined normal equation systems, the following classification is normally used for the estimated set of gravity harmonics:

- **Super-Tailored Model:** These are models derived in the extreme case from tracking data over a limited observation period and from a single satellite only. Since only a very limited number of coefficients can be computed in this case, the effect of unadjusted coefficients and mismodelled effects will creep into the solution vector. Through this aliasing with the solved-for parameters the accuracy and self-consistency of the solution is destroyed. In the case of using such models for OD purposes, aliasing can be used to our advantage so that we can restrict the highest degree and order of solved-for geopotential coefficients by including the higher terms effects in part in the values determined. Geophysically these solved-for coefficients have of course no meaning.

- **Satellite-only Model:** No single satellite is sensitive in distinctly different ways to each harmonic and we thus need a number of satellites in various orbit configurations in order to be able to resolve decisively between those harmonics. Gravity normal equation systems, derived from tracking of a set of satellites well distributed in inclination and combined by balanced weighting to a common normal equation system (system II) lead to what is called a "satellite-only" solution. The resolution of such a model is limited but is a geophysically meaningful representation of the global field as derivable from multi-satellite tracking data. Highly weighting the normals of a specific satellite leads to a tailoring to the data of this specific satellite.

- **General-Purpose Model:** Combining the satellite normals with normal equation systems derived from e.g. surface gravity, altimetry and other gravity related information (system III) leads to what is called general purpose or combination models. These models are "best" approximations of the real geopotential with a spatial resolution which is dictated by the resolution of the terrestrial data sources, the computer resources of the analyser and the application needs. Models up to degree and order 50 are derivable with present day computers and data sources.

- **Tailored Model:** Highly weighting the normals of a specific satellite in a system II or system III leads to a tailoring to the data of this specific satellite. The geophysical meaning of the potential coefficients is restricted.

7. Special Topics

There are a few topics which impact in addition to what has already been said the recovery of gravity field harmonics and therefore need to be mentioned.

- **Reference Frames**

  Besides the many dynamical and measurement models which have to be carefully selected and tested before going into a global gravity model solution, it is of great importance to put great efforts into the

  - appropriate definition and initial practical realization of a terrestrial reference frame, materialized by a set of tracking station coordinates (SSC) at a specified epoch and their motions,
  - and the determination of a homogeneous series of Earth rotation parameters (ERP) referred to the axes of the realized terrestrial frame.

  The inclusion of motions, either computed or taken from a geophysical model, is important because in gravity modelling we deal with data sets spanning over a period of almost 30 years now.

Station coordinates and Earth rotation parameters are presently by one to two orders of magnitude better determined with LAGEOS than with any of the other lower orbiting laser and Doppler satellites. Thus deriving the station coordinates and Earth rotation parameters from a multi-year laser tracking record of LAGEOS leads to the most precise and consistent realization of a conventional terrestrial reference frame and its link to the true of date equator and equinox frame within the multi-year analysis period. Doppler and camera stations can be precisely referred to this terrestrial frame if survey ties or other good connections to the designated stations exist. The polar motion series of BIH back to 1962 can be referenced to the computed series through a transformation adjustment with the pole path data in the interval common to both series. In this way precise, homogeneous and consistent initial station and Earth rotation parameter sets are obtained, allowing the fixation of a great number of parameters in the global Earth model adjustment process.

Since satellite laser ranging, primarily used in the new models, provides no directional information, the absolute orientation of the coordinate axes of any of the aforementioned systems cannot be determined by the measurements. One has to fix the zero point of the terrestrial longitude preferably by the longitude of a colocated laser/VLBI site and the terrestrial equator by fixing two latitudes or by introducing special conditions on the z-axis. Linking the SLR terrestrial observing site(s) through the SLR/VLBI connection(s) would allow to bring the LAGEOS satellite inertial frame close to the VLBI inertial frame if in the satellite analysis $UT1$ values are used which do not exhibit offsets or drifts with respect to inertial techniques $UT1$ values. This last aspect is important in connection with the determination of even zonal harmonics, resulting from secular perturbations of the node.

The definition of the polar axis of the terrestrial system has, as we know from the spherical harmonic expression of the field, implications on the size of the terms of degree 2 and order 1. If the conventional terrestrial pole is placed into the center of a multi-year pole path, then the z-axis origin of the terrestrial system will be very near to the mean of the non-diurnal part of the figure axis motion over the same period. With respect to the so defined terrestrial z-axis the $\bar{C}_{2,1}, \bar{S}_{2,1}$ coefficients will become zero. The deformation effect in the 2,1-terms will almost cancel out, when analysing data over a number of Chandler periods.

- **Parametrization of Surface Forces**

Most of the satellite orbits used for gravity field studies are significantly perturbed by the atmospheric drag. To minimize the atmospheric drag effect in the observation residuals, one tends to parametrize the poorly known drag acceleration as much as possible. Solutions for several scale parameters C_D over specified time intervals (e.g. 1 day) over the arc length are common practice (s. Marsh et al., 1987). We found this not to be uncritical, if not properly handled (e.g. by introducing strong a priori information), in particular when it comes to separating between real range and time biases and apparent biases.

- **Weighting of Satellite Normals**
When computing the combined normal equation system (71), it is a common procedure to apply weight factors t_K to the individual arc systems. The balanced size of the weight factors is obtained from an inspection of the normal equation diagonal terms and by experimenting with the normals. There is a lengthy discussion on this subject in (Marsh et al., 1987). The size of the t_K factors will of course depend on what has been used as measurement errors in forming the observation equations. We are using sigmas composed of the measurement error plus a model error, which accounts for model uncertainties in the generation of the observation residuals. Different modelling accuracies of different satellite orbits are by this automatically taken into account in the satellite normals.

8. Global Gravity Field Models

To the lecturer's knowledge there are four major software systems which allow extensive global gravity field modelling from tracking data and for combination solutions. These are

- GEODYN II - NASA/GSFC
- UTOPIA - CSR/Austin
- GIN/DYNAMO - GRGS/CNES
- DOGS - DGFI/Munich

These groups (in a joint German-French effort on the European side) have contributed significant advancements to the modelling of the global Earth gravity field and are presently all involved in the production of a new generation of improved models in support of the upcoming altimeter missions TOPEX and ERS-1.

8.1. Recent Gravity Field Models

Some most recent examples of gravity field models for the long and medium wavelength structures of the field which are available in the open literature are given in table 5.

Model	Complete Harmonics	Field Resol. (km)	Data used	References
GEM9	20	1000	ST	Lerch et al., 1977
GEM10B	36	550	ST+SG+SA	Lerch et al., 1978
GEM-l2	20	1000	ST	Lerch et al., 1983
PGS-1331*	36	550	GEM10B +ST+SA	Marsh et al., 1985
PGS-S4*	36	550	GEM10B +ST+SA	Lerch et al., 1982
GRIM3	36	550	ST+SG+SA	Reigber et al., 1983
GRIM3B*	36	550	GRIM3+ST	Reigber et al. 1984
GRIM3-L1	36	550	ST+SG+SA	Reigber et al., 1985
GEM-T1	36	550	ST	Marsh et al., 1987

ST...Satellite Tracking; SG...Surface Gravity; SA...Satellite Altimetry
* Tailored Gravity Models

Table 5: Description of recent Earth gravity field models

These models and their resolution are characterized by the types of data which went into the solutions, the relative weighting of the heterogeneous data sources and the analysis approach. As explained in the next section, the GEM-T1 model belongs to a new class of gravity models which outperform all the older models. Insofar the other models do no longer have a real computational value and are presented here only for completeness.

The Goddard Space Flight Center models GEM9 and GEM-L2 belong to the class of "satellite-only" models, whereas the GSFC GEM10B and the DGFI/GRGS models GRIM3 and GRIM3-L1 are "combination" solutions. In contrast to this, models like the PGS-1331, PGS-S4, GRIM3B were aiming at an optimal representation of the perturbation behaviour of a specific satellite's orbit by highly weighting the tracking data of this satellite in a combination solution: STARLETTE (PGS-1331), SEASAT (PGS-S4) and LAGEOS (GRIM3B).

A number of drawbacks of the pre-GEM-T1 models can be mentioned. The successive GEM and GRIM solutions did not represent real new iterations of the previous solution, but were for a long time based on the GEM9 and GRIM3 normals respectively. Although the solutions following GEM9 and GRIM3 benefitted from additional and higher quality data, improvements in data processing, improved surface force modelling, as well as from continuous upgradings of the analysis software, the inclusion of the "old" GEM9 and GRIM3 normals respectively, and by this the introduction of inconsistencies in the reduction of old and newer data, did not allow real big improvements for the comprehensive fields. The major improvement was for the longest wavelength portion of models, achieved through the inclusion of a great deal of precise LAGEOS SLR observations.

8.2. New Gravity Model Developments

The situation with regard to reference systems realization, dynamic, transformation and measurements model developments and also with regard to the observational material changed a great deal over the last few years due to the tracking and analysis efforts we have seen for projects like MERIT, CDP and Wegener/MEDLAS. In view of the pressing need for better gravity field models for the upcoming altimetric missions TOPEX/POSEIDON and ERS-1 more money has been made available in the United States and Europe. Groups in the USA (GSFC/EG & G, CSR) in Germany (DGFI) and France (GRGS) are at present thoroughly reanalysing historical and new data and are in the process of producing, or have already produced as in the case of the GSFC/EG & G team, a new iteration of their models.

In 1987 the GSFC/EG & G group published the first TOPEX model GEM-T1 (Marsh et al., 1987). This model was derived exclusively from satellite tracking data acquired on the 17 different satellites given in table 4. In all, almost 800,000 observations were used, half of which were from third generation laser systems. The GEM-T1 model provides a simultaneous solution for:

- a gravity model in spherical harmonics complete to degree and order 36;
- a subset of 66 ocean tidal coefficients for the long wavelength components of 12 major tides. This adjustment was made in the presence of 550 other ocean tidal coefficients representing 32 major and minor tides; and
- 5-day averaged Earth rotation and polar motion parameters for the 1980 period onwards.

All details on the computation, calibration and validation of this model can be taken from the referenced report. As already mentioned, GEM-T1 outperforms all older models when used in orbit computation including the tailored models.

The same group is preparing a new "satellite-only" model GEM-T2, which will be based on tracking data for 31 satellites and will give all 50 × 50 terms. It will be released in the third quarter of 1988. A combination solution, GEM-T3, is in preparation and will include altimetry and surface gravity information.

The group at the Center for Space Research in Austin is involved in the development of a TOPEX gravity field (TGF). A few preliminary versions of this field have been derived in the meantime. To the lecturer's knowledge the last version is the model PTGF 3a, based on tracking data of 12 satellites and direct altimeter and crossover data from SEASAT.

The DGFI in Munich and the GRGS in Toulouse are jointly preparing a new gravity model which is planned to be ready in 1989. This model is characterized by the following features:

- **Tracking Data**
  21 satellites (laser, Doppler, camera)
- **Resolution**
  complete to 50 × 50
- **Reference Systems**
  - inertial: J2000
  - terrestrial: CTP of SSC(DGFI)L03:
    zero mean of pole from 1980-86 LAGEOS analysis: Laser station positions from SSC(DGFI)L03 solution at epoch 1984.0 plus rates. All other stations transformed into this system.
- **Transformation Model**
  - IAU Precession/Nutation
  - Homogeneous polar motion series for 1962-1987 including laser derived values for 1980-1986 period.
  - Homogeneous $UT1 - UTC$ series 1962-1987 with VLBI values for 1984-1987.
- **Third Body Effects:** All major planets
- **Earth Tides:** Wahr model
- **Ocean Tides:** Schwiderski extended
- **Tidal Deformations:** MERIT standard values
- **Atmosphere Model:** DTM
- **Indirect Radiation:** Albedo, IR modelled

So, in summary, on both sides of the Atlantic quite some effort is put into gravity field improvements. The results obtained so far are promising as concerns the use of such models for the data reduction of altimetric missions, but also for geodynamic studies with low orbiting objects such as STARLETTE. Additional tracking data from dense satellites (LAGEOSII, STELLA) and from microwave tracking with DORIS and PRARE are likely to support the gravity model generation in the early nineties.

References

Allan, R.R.: On the Motion of Nearly Synchronous Satellites, Proc. Roy. Soc. A288, 1965

Cappellari, J.O., Velez, C.E., Fuchs, A.J.: Mathematical Theory of the Goddard Trajectory Determination System, GSFC-X-582-76-77, 1976

Heiskanen, W.A., Moritz, A.: Physical Geodesy, Freeman and Comp., San Francisco, 1967

Kaula, W.M.: Theory of Satellite Geodesy, Blaisdell Publ. Comp., Waltham, Mass., 1966

Klosko, S.M., Wagner, C.A.: Spherical Harmonic Representation of the Gravity Field from Dynamic Satellite Data, Planet. Space Sci., Vol. 30, No. 1, 1982

Kostelecky, J.: Recurrence Relation for the Normalized Inclination Function, Bull. Astron. Inst. CS., 36, 1985

Marsh, J.G., Lerch, F.J., Putney, B.H., Christodoulidis, D.C., Felsentreger, T.L., Sanchez, B.V., Smith, D.E., Klosko, S.M., Martin, T.V., Pavlis, E.C., Robbins, J.W., Williamson, R.G., Colombo, O.L., Chandler, N.L., Rachlin, K.E., Patel, G.B., Bhati, S., and Chinn, D.S.: An Improved Model of the Earth's Gravitational Field: *GEM-T1*, NASA Techn. Memorandum 4019, 1987

FUNDAMENTALS OF ORBIT DETERMINATION

B. D. Tapley
Center for Space Research
The University of Texas at Austin
Austin, Texas 78712 USA

Introduction

For the satellite orbit determination problem, the satellite's state, $X(t)$, at a general time, t, is the n-dimensional column vector which contains, as its elements, the components of the satellite position and velocity as well as any constant, but unknown, parameters which appear in the dynamic force model or the measurement model. If at some time, t_o, the state X_o of a satellite is known and if the differential equations which govern the satellite's motion are known, these equations can be integrated to determine the state of the satellite at any subsequent time $t \geq t_o$. However, for earth-orbiting satellites, the initial state is never known exactly. Moreover, certain physical constants required to define the differential equations of motion are known only approximately. Errors in these constants will cause the actual motion to deviate from the predicted motion. Consequently, to determine the position of the satellite at some time $t > t_o$, it is necessary that observation of the satellite's motion be used to obtain a better estimate of the satellite's trajectory. The observational data, which will be subject to both random and systematic errors, will usually consist of measurements of such quantities as range, range-rate, azimuth, elevation or some other observable quantity. The state variables (position, velocity, unknown model parameters, etc.) will not be observed directly, but rather the observable will usually be some nonlinear function of the state variables. In this discussion, the problem of determining the best estimate of the state of the satellite, at some epoch t_k, from observations influenced by random errors, using a mathematical force model which is not exact, is referred to as the problem of orbit determination.

To illustrate some of the basic ideas involved in the orbit determination process, consider the motion of a vehicle as shown in Figure 1. In the mission design, an initial state X_o^* is selected such that the vehicle will follow a specified nominal (or design) trajectory. At mission initiation, the *true initial state*, X_o, will differ from the *nominal initial state*, X_o^*, and consequently, the true vehicle trajectory will differ from the nominal trajectory. To determine an estimate of the true motion, tracking information must be used. Figure 1 shows a tracking station at a location, $\bar{r}_s$. The range (linear distance along the line of sight), R, and the angular orientations (A, E) of the line of sight to the vehicle can be measured. Measurement of the range-rate, $\dot{R}$, is also frequently used as an observation of the vehicle's motion. The tracking information, or observations, R, A and E, depend on the true vehicle motion and the position of the tracking station. Alternately, the calculated values of the observations will depend on the nominal or reference state of the vehicle, X^*. The difference between these two quantities provides the information which is used to obtain an improved estimate of the vehicle's motion. Other observations such as range rate, azimuth rate, etc., depend on both the velocity and position of the satellite and tracking station. The observations, made at different times, can be related through the equations of motion.

The Orbit Determination Problem

From Newton's law, the vector differential equations of motion can be expressed as

$$\ddot{\bar{r}} = -\mu \frac{\bar{r}}{r^3} + \frac{\bar{P}}{m}(\bar{r}, \dot{\bar{r}}, t) \tag{1}$$

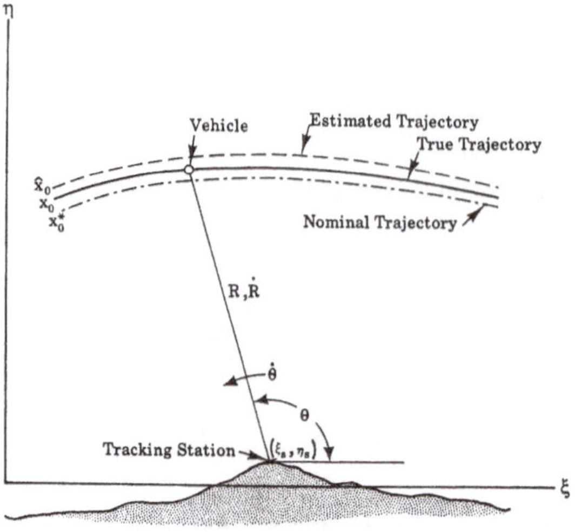

Figure 1. Uniform Gravity Field Trajectory

where $\bar{P}(\bar{r},\dot{\bar{r}},t)$ represents the acceleration due to forces other than the central body force. Appendix A gives a description of the major forces (or accelerations) which influence the motion of a satellite. Eqs. (1) can be expressed in first order form as follows:

$$\dot{\bar{r}} = \bar{v}$$

$$\dot{\bar{v}} = -\mu \frac{\bar{r}}{r^3} + \bar{R}(\bar{r},\dot{\bar{r}},t) \tag{2}$$

If the vectors $\bar{\xi}$ and $f(\bar{\xi},t)$ are defined as follows:

$$\bar{\xi} = \begin{bmatrix} x \\ y \\ z \\ u \\ v \\ w \end{bmatrix} \qquad f(\bar{\xi},t) = \begin{bmatrix} u \\ v \\ w \\ -\mu x/r^3+P_x \\ -\mu y/r^3+P_y \\ -\mu y/r^3+P_z \end{bmatrix}$$

then, Eqs. (2) can be written as

$$\dot{\bar{\xi}} = f(\bar{\xi},t) \qquad \bar{\xi}(t_o) = \bar{\xi}_o \tag{3}$$

where x, y, z and u, v, w are the components of position and velocity, respectively, with respect to a non-rotating geocentric coordinate system.

Let an augmented state vector $X(t)$ and a force vector $F(t)$ be defined as

$$X = \begin{bmatrix} \xi \\ \cdots \\ \alpha \end{bmatrix} \qquad F(X,t) = \begin{bmatrix} f(\bar{\xi},\alpha,t) \\ \cdots \\ 0 \end{bmatrix} \tag{4}$$

where α is a q-vector of unknown model constants which satisfy the relation $\dot{\alpha}=0$. Then Eqs. (3) can be combined with the condition, $\dot{\alpha}=0$, to obtain the differential equations of state as follows:

$$\dot{X} = F(X,t) \qquad X(t_o) = X_o \tag{5}$$

Eqs. (5) represent a system of n nonlinear first order ordinary differential equations.

Assume that observations have been made at times $t_1, \ldots, t_l$ and that for each t_i, a $p \times 1$ vector of observations, Y_i, has been obtained, where

$$Y_i = G(X_i, t_i) + \varepsilon_i \qquad i = 1, \ldots, l. \tag{6}$$

That is, the actual observation, Y_i, is assumed to be a nonlinear function of the true observation, $G(X_i, t_i)$, and the random measurement noise, ε_i.

Now by noting that the solution to Eq. (5) can be expressed as

$$X(t_i) = \Theta(X_o, t_o, t_i) \tag{7}$$

it follows that

$$Y_i = G(\Theta_i(X_o, t_o, t_i), t_i) + \varepsilon_i = \tilde{G}_i(X_o, t_o, t_i) + \varepsilon_i$$

Note that $\tilde{G}_i$ is an implicit relationship. For a general system of differential equations, an explicit relationship usually cannot be determined. The solution implied by Eq. (7) is usually obtained by numerically integrating Eqs. (5). Let the m-vectors Y, $\tilde{G}$ and ε, where $m = l \times p$, be defined as

$$Y = \begin{bmatrix} Y_1 \\ \vdots \\ Y_l \end{bmatrix}, \quad \tilde{G} = \begin{bmatrix} \tilde{G}_1(X_o, t_o, t_1) \\ \vdots \\ \tilde{G}_l(X_o, t_o, t_l) \end{bmatrix}, \quad \varepsilon = \begin{bmatrix} \varepsilon_1 \\ \vdots \\ \varepsilon_l \end{bmatrix} \tag{8}$$

and the data set or collection of observations can be defined as

$$Y = \tilde{G}(X_o, t_o) + \varepsilon \tag{9}$$

Examination of Eqs. (9) indicates that the relation represents a system of m algebraic equations in terms of the n unknown components of the state and the m unknown components of the observation error. If $\varepsilon_i = 0$, $i = 1, 2, \ldots, l$, then any n of Eqs. (9) that are independent can be used to determine X_o or the state at any other time X_k, through Eqs. (7).

If $\varepsilon_i \neq 0$, then some best estimate must be obtained where "best" is used to select one estimate or solution from the many possible solutions. One criterion which has wide acceptance in practice is to minimize the sum of the square of the residual observation errors. For generality, let the observation state relation be expressed in terms of the state at an arbitrary epoch t_k (e.g., the reference epoch used in Eq. (9) is t_o). Then,

$$Y_i = \tilde{G}_i(X_k, t_k, t_i) + \varepsilon_i \qquad i = 1, \ldots, l$$

and let $J(X_k^*)$ be defined as

$$J(X_k^*) = \sum_{i=1}^{l} \varepsilon_i^{*T} \varepsilon_i^* = \sum_{i=1}^{l} [Y_i - \tilde{G}_i(X_k^*, t_k, t_i)]^T [Y_i - \tilde{G}_i(X_k^*, t_k, t_i)] \tag{10}$$

where X_k^* is a specified value of the state at t_k and ε_i^* is the difference between the actual observation, Y_i, and the calculated observation, $G(\tilde{X}_k^*, t_k, t_i)$.

Now, let $\hat{X}_k$ be the value of X_k^* which minimizes $J(X_k^*)$. Then, it is necessary that

$$\left.\frac{\partial J}{\partial X_k^*}\right|_{\hat{X}_k} = 0 \qquad \delta X_k^T \left[\frac{\partial^2 J}{\partial X_k^* \partial X_k^*} \right]_{\hat{X}_k} \delta X_k \geq 0 . \tag{11}$$

for arbitrary ∂X_k. From the first of Eqs. (20),

$$\left.\frac{\partial J}{\partial X_k^*}\right|_{\hat{X}_k} = -\sum_{i=1}^{l} \left[Y_i - \bar{G}_i(\hat{X}_k,t_k,t_i) \right]^T \frac{\partial \bar{G}_i}{\partial \hat{X}_k} (\hat{X}_k,t_k,t_i) = 0 \tag{12}$$

Eq. (12) forms a system of n nonlinear algebraic equations involving the unknown n-vector, $\hat{X}_k$. Eqs. (12) must be solved iteratively by a numerical procedure, such as the Newton-Raphson iteration procedure, for solving nonlinear algebraic equations.

Linearization of the Orbit Determination Process

If a reasonable reference trajectory is available and if X, the true trajectory, and X^*, the reference trajectory, remain sufficiently close throughout the time interval of interest, the trajectory for the actual motion can be expanded in a Taylor's series about the reference trajectory at each point in time to obtain a set of linear differential equations with time dependent coefficients. A linear relation between the observation deviation and the state deviation can be obtained by a similar expansion procedure. Then, the nonlinear orbit determination problem in which the complete state vector is to be estimated can be replaced by a linear orbit determination problem in which the deviation from some reference trajectory is to be determined [*Lawson and Hanson*, 1964; *Tapley*, 1973; *Gelb*, 1974; *Bierman*, 1977].

To conduct this linearization procedure, let

$$x(t) = X(t) - X^*(t) \qquad y(t) = Y(t) - Y^*(t) \tag{13}$$

where $X^*(t)$ is a specified reference trajectory and $Y^*(t)$ is the value of the observation, calculated by using $X^*(t)$. Then, substituting into Eqs. (5) and (6) and expanding in a Taylor's series leads to

$$\begin{aligned} \dot{X}(t) &= F(X,t) = F(X^*,t) + \left[\frac{\partial F}{\partial X} \right]^* x + \cdots \\ Y_i &= G(X_i,t_i) + \varepsilon_i = G(X_i^*,t_i) + \left[\frac{\partial G}{\partial X} \right]^* x_i + \cdots + \varepsilon_i \end{aligned} \tag{14}$$

If the terms of order higher than the first are neglected and if the condition $\dot{X}^* = F(X^*,t)$ and $Y_i^* = G(X_i^*,t_i)$ are used, Eqs. (14) can be written as

$$\begin{aligned} \dot{x} &= A(t)x \\ y_i &= \tilde{H}_i x_i + \varepsilon_i \qquad (i = 1, \ldots, l) \end{aligned} \tag{15}$$

where

$$A(t) = \frac{\partial F}{\partial X}(X^*,t) \qquad \tilde{H} = \frac{\partial G}{\partial X}(X^*,t) \tag{16}$$

Hence, the original nonlinear estimation problem is replaced by the linear estimation problem described in Eqs. (15).

The first of Eqs. (15) represents a system of linear differential equations with time dependent coefficients. The general solution for this system can be expressed as

$$x(t) = \Phi(t,t_k) x_k \tag{17}$$

where $x(t)$ is the value of x at a specific time t, and where the $n \times n$ state transition matrix $\Phi(t,t_k)$ satisfies the differential equation:

$$\dot{\Phi}(t,t_k) = A(t)\Phi(t,t_k) \quad , \qquad \Phi(t_k,t_k) = I \tag{18}$$

Using Eq. (16), the second of Eqs. (14) may be written in terms of the state at t_o as

$$y_i = \tilde{H}_i \, \Phi(t_i,t_o)x_o + \varepsilon_i \quad , \qquad i = 1, \ldots, m \tag{19}$$

If the following definitions are used

$$y = \begin{bmatrix} y_1 \\ \vdots \\ y_m \end{bmatrix} \qquad H = \begin{bmatrix} \tilde{H}_1\Phi(t_1,t_o) \\ \vdots \\ \tilde{H}_m\Phi(t_l,t_o) \end{bmatrix} \qquad \varepsilon = \begin{bmatrix} \varepsilon_1 \\ \vdots \\ \varepsilon_m \end{bmatrix} \tag{20}$$

and if the subscript on x_o is dropped, then Eqs. (19) can be expressed as follows:

$$y = Hx + \varepsilon \tag{21}$$

where y is an $m \times 1$ vector, x is an $n \times 1$ vector, ε is an $m \times 1$ vector, H is an $m \times n$ mapping matrix where m is the total number of observations. For the cases of interest, the essential condition $m > n$ is satisfied.

The Least Squares Solution

The least squares solution selects the estimate of x as the value which minimizes the sum of the squares of the calculated observation residuals. That is, x is selected to minimize the following performance index:

$$J(x) = \varepsilon^T \varepsilon \tag{22}$$

With the observations, y, and a specified value of x, say x°, the value for the observation error, ε, can be computed from Eq. (21). An intuitive measure of the quality of x can be obtained by squaring all of the observation errors and adding them together. If the calculated value of *epsilong*, ε°, is substituted into Eq. (22), the following expression is obtained:

$$J(x) = \varepsilon^{\circ T} \varepsilon^\circ = (y - Hx^\circ)^T (y - Hx^\circ) \tag{23}$$

Note that Eq. (23) is a quadratic function of x and as a consequence, the expression will have a unique minima when

$$\left. \frac{\partial J}{\partial x^\circ} \right|_{x^\circ = \hat{x}} = 0$$

Hence, from Eq. (23),

$$\left. \frac{\partial J}{\partial x} \right|_{\hat{x}} = 0 = -2H^T (y - H\hat{x}) \tag{24}$$

which requires that

$$H^T (y - H\hat{x}) = 0 \tag{25}$$

or

$$(H^T H)\hat{x} = H^T y \tag{26}$$

Note that if $m \geq n$, the matrix $H^T H$ is an $n \times n$ positive-definite matrix (H is rank n), and the solution for the estimate of x, denoted as $\hat{x}$, can be obtained as follows

$$\hat{x} = (H^T H)^{-1} H^T y \tag{27}$$

Equation (27) is the well-known least squares solution for the best estimate of x given the linear observation state relationship expressed by the Eq. (21).

The Minimum Norm Solution

Equation (46) is used if $m > n$ and H has rank n. Consider the case where H is of rank $< n$. In other words, there are more unknowns than linearly independent observations. In this case, one can use the minimum norm criterion to determine $\hat{x}$. One could choose to solve any m independent observation equations exactly by setting $\varepsilon = 0$; however, this leads to an infinite number of solutions for $\hat{x}$. Generally, a nominal or initial guess for x exists. Recall that the differential equations have been linearized and $x = X - X^*$. Therefore, one can choose x to minimize the deviations between X and X^*, subject to the constraint that $\varepsilon = 0$, i.e., $y = Hx$. Hence, the performance index becomes

$$J(x,\lambda) = (1/2)x^T x + \lambda(y - Hx) \tag{28}$$

where the constraint has been adjoined with an $m \times 1$ vector of Lagrange multipliers. The necessary conditions for a minimum of $J(x,\lambda)$ is that the first variation with respect to x and λ vanish. This leads to

$$\delta J(x,\lambda) = (1/2)[\delta x^T x + x^T \delta x] - \lambda H \delta x + \delta\lambda(y - Hx) = 0 \tag{29}$$

since $\delta J(x,\lambda)$ must vanish for arbitrary δx and $\delta\lambda$, and noting that $\delta x^T x = x^T \delta x$, it follows that

$$\hat{x} - H^T \lambda^T = 0 \tag{30}$$

and

$$y - H\hat{x} = 0 \tag{31}$$

Substituting the expression for $\hat{x}$ from Eq. (30) into Eq. (31) yields

$$y = HH^T \lambda^T$$

solving for λ

$$\lambda^T = (HH^T)^{-1} y \tag{32}$$

Substituting Eq. (32) into (30) yields

$$\hat{x} = H^T (HH^T)^{-1} y \tag{33}$$

where HH^T is an $m \times m$ matrix of rank m. Eq. (33) is the solution for x of minimum length. The quantities $H^T (HH^T)^{-1}$ in Eq. (33) and $(H^T H)^{-1} H^T$ in Eq. (27) are referred to as the pseudo inverse of H for the equation $H\hat{x} = y$. In summary,

$$\begin{aligned} \hat{x} &= (H^T H)^{-1} H^T y \ , && \text{if } m > n \\ \hat{x} &= H^{-1} y \ , && \text{if } m = n \\ \hat{x} &= H^T (HH^T)^{-1} y \ , && \text{if } m < n \end{aligned} \tag{34}$$

Weighted Least Squares Solution

Equations (1.34) have no means of preferentially ordering one observation with respect to another. A more general expression can be obtained by considering the following formulation. Given a sequence of observations $y_1, y_2, \ldots, y_l$ related through the state transition matrix to the state at some general time, x_k, and an associated weight w_i for each of the observations, one can write

$$\begin{aligned} y_1 &= H_1 x_k + \varepsilon_1 \quad ; \quad w_1 \\ y_2 &= H_2 x_k + \varepsilon_2 \quad ; \quad w_2 \\ &\vdots \\ y_l &= H_l x_k + \varepsilon_l \quad ; \quad w_l \end{aligned} \tag{35}$$

where

$$H_i = \tilde{H}_i \Phi(t_i, t_k)$$

In Eqs. (35) the weights, w_i, are normalized to range between zero and one. Following the previously discussed procedures, Eq. (35) can be expressed as

$$y = Hx_k + \varepsilon \quad ; \quad W \tag{36}$$

where W is an $l \times l$ diagonal matrix. One can then pose the weighted least squares problem as follows. Given the linear observation state relationship expressed by (37), find the estimate of x_k to minimize the weighted sum of the squares of the calculated observation errors.

$$J(x_k^*) = \varepsilon^{*T} W \varepsilon^* = \sum_{i=1}^{l} \varepsilon_i^{*T} w_i \varepsilon_i^* \tag{37}$$

where $e^* = y - H\,x_k^*$.

Applying the condition that the gradient of J with respect to x_k^* must vanish at the minimum of $J(x_k)$ leads to the following condition

$$\left.\frac{\partial J}{\partial x_k}\right|_{\hat{x}_k} = 0 = -2H^T W (y - H\hat{x}_k) \tag{38}$$

where the value, $\hat{x}_k$ has been substituted for x_k when the condition that the derivative vanish is satisfied. This expression can be rearranged to obtain the normal equations analogous to Eq. (26) in the least squares formulation

$$(H^T W H)\hat{x}_k = H^T W y. \tag{39}$$

If the normal matrix $H^T W H$ is positive definite, it will have an inverse and the solution to (39) is

$$\hat{x}_k = (H^T W H)^{-1} H^T W y \tag{40}$$

The $n \times n$ matrix P_k is symmetric, as can be seen from the definition. Furthermore, if it exists, it must be positive definite, since it was computed as the inverse of $H^T W H$. The parameter observability is related to the rank of this matrix. If all the parameters in x_k are observable, then P_k will be full rank and P_k will have an inverse. The number of observations must be greater than or equal to the number of parameters being estimated, if P_k is to be invertible. Furthermore, P_k is related to the accuracy of the estimate, x_k. In general, the larger the magnitude of the elements of the matrix, P_k, the less accurate the estimate. Since the weighting matrix, W, usually results from an initial judgment on the accuracy of the observations followed by a normalization procedure to scale the weights to values between zero and one, this interpretation is not strictly valid in the statistical sense, and some caution should be used when attempting to infer the accuracy of an estimate from the magnitude of P_k as obtained in the weighted least squares estimate. In the next section, it will be shown that P_k is the variance-covariance matrix associated with the estimate, $\hat{x}_k$.

The Minimum Variance Estimate

The minimum variance criterion is widely used in developing solutions to estimation problems because of the simplicity in its use [*Liebelt*, 1967; *Tapley*, 1973]. It has the advantage that the complete statistical description of the random errors in the problem is not required. Rather, only the first and second moments of the probability density function of the observation errors are required. This information is expressed in the mean and covariance matrix associated with the random error.

If it is assumed that the observation error ε_i is random with zero mean and specified covariance, i.e.,

$$E[\varepsilon_i] = 0 \qquad E[\varepsilon_i \varepsilon_i^T] = R_i \tag{41}$$

then the state estimation problem can be formulated as follows:

Given: The system of state-propagation equations and observation state equations

$$x_i = \Phi(t_i, t_k) x_k$$

$$y_i = \tilde{H}_i x_i + \varepsilon_i \qquad i = 1, \dots, l \tag{42}$$

where $E[\varepsilon_i] = 0$; $E[\varepsilon_i \varepsilon_j^T] = R_i \delta_{ij}$, where δ_{ij} is the Kroneker delta defined by

$$\delta_{ij} = \begin{cases} 1\ , & i = j \\ 0\ , & i \neq j \end{cases}$$

Find: The best linear unbiased minimum variance estimate, $\hat{x}_k$, of the state x_k.

The solution to this problem proceeds as follows. Using the state transition matrix, reduce Eqs. (42) to the following form

$$y = H x_k + \varepsilon \tag{43}$$

where

$$E[\varepsilon] = \begin{bmatrix} E[\varepsilon_1] \\ E[\varepsilon_2] \\ \vdots \\ E[\varepsilon_l] \end{bmatrix} = 0 \qquad E[\varepsilon\varepsilon^T] = \begin{bmatrix} R_1 & & 0 \\ & R_2 & \\ & & \\ 0 & & R_l \end{bmatrix} = R \tag{44}$$

Generally, $R_1 = R_2 = \cdots = R_l$, but this is not a necessary restriction in the following argument. Furthermore, the more general case of time-correlated observation errors where $E[\varepsilon_i \varepsilon_j^T] = R_{ij}$ and where the off-diagonal zero's in the definition of the R matrix will be replaced by non-zero quantities can be treated within the framework of the following discussion.

From the problem statement, the estimate is to be the best linear, unbiased minimum variance estimate. The consequences of each of these requirements are examined in the following steps.

(1) *Linear*: The requirement of a linear estimate implies that the estimate is to be made up of a linear combination of the observations, i.e.,

$$\hat{x}_k = M y. \tag{45}$$

The $(n \times m)$ matrix M is unspecified and is to be selected to obtain the best estimate.

(2) *Unbiased*: If the estimate is unbiased, then

$$E[\hat{x}] = x. \tag{46}$$

Substituting, Eqs. (45) and (43) into Eq. (46) leads to the following requirement

$$E[\hat{x}_k] = E[My] = E[MHx_k + M\varepsilon_k] = x_k.$$

But, since $E[\varepsilon_k] = 0$, this reduces to

$$MHx_k = x_k$$

from which the following constraint on M is obtained

$$MH = I. \tag{47}$$

That is, if the estimate is to be unbiased, the linear mapping matrix M must satisfy Eq. (47).

(3) *Minimum Variance*: The task of selecting the particular $n \times m$ matrix M which minimizes the covariance matrix can be developed using the following formal arguments. The minimization of an $n \times n$ matrix requires specifying the particular character of the minimum sought. In the discussion here, the value of M that minimizes the trace of the covariance matrix will be sought. A more rigorous, but more complicated, development can be framed in the content of finding the matrix, M, that minimizes the trace of the covariance matrix, P_k. The formal development proceeds as follows.

If the estimate is unbiased, then the covariance matrix can be expressed as

$$P_k = E[(\hat{x}_k - E[\hat{x}_k])(\hat{x}_k - E[\hat{x}_k])^T] = E[(\hat{x}_k - x_k)(\hat{x}_k - x_k)^T] \tag{48}$$

Hence, the problem statement requires the $\hat{x}_k$ be selected to minimize Eq. (48) while satisfying Eqs. (46) and (47). Substituting Eq. (45) into Eq. (48) leads to the following result.

$$\begin{aligned} P_k &= E[(My - x_k)(My - x_k)^T] \\ &= E[\{M(Hx_k + \varepsilon) - x_k\}\{M(Hx_k + \varepsilon) - x_k\}^T] \\ &= E[M\varepsilon\varepsilon^T M^T] \end{aligned}$$

since $MH = I$. It follows then from Eq. (44) that the covariance matrix can be written as

$$P_k = MRM^T \tag{49}$$

where M is to be selected to satisfy Eq. (47). To involve the constraint imposed by Eq. (47) and to keep the constrained relation for P_k symmetric, Eq. (47) is adjoined to Eq. (49) in the following form

$$P_k = MRM^T + \Lambda^T[I - MH]^T + (I - MH)\Lambda \tag{50}$$

where Λ is a $n \times n$ matrix of unspecified Lagrange multipliers. Then, for a minimum of P_k, it is necessary that the first variation with respect to M and Λ vanish. From this requirement, it follows that

$$\Lambda^T = (H^T R^{-1} H)^{-1} \tag{51}$$

and

$$M = (H^T R^{-1} H)^{-1} H^T R^{-1}. \tag{52}$$

Substitution of Eq. (52) into Eq. (49) leads to the following expression for the covariance matrix.

$$P_k = (H^T R^{-1} H)^{-1}. \tag{53}$$

With Eqs. (52) and (45), the best linear unbiased minimum variance estimate of x_k is given as

$$\hat{x}_k = (H^T R^{-1} H)^{-1} H^T R^{-1} y. \tag{54}$$

Note that computation of the estimate, $\hat{x}_k$, requires inverting the $n \times n$ normal matrix and for a large dimension system, the computation of this inverse is difficult. Further discussion of the numerical techniques for computing the solution is given in a subsequent section on this topic. The solution given by Eq. (54) will agree with the weighted least squares solution if the weighting matrix, W, used in the least squares approach is equal to the inverse of the observation noise covariance matrix, i.e., if $W = R^{-1}$. Also, the minimum variance estimate given in Eq. (54) will agree with the maximum likelihood estimate, if the observation errors are assumed to be distributed normally with zero mean and covariance R.

Propagation of the Estimate

After determining an estimate at a time, t_j, it is often required that an estimate of the state be determined at some future time. If the estimate at a time t_j is obtained by using Eq. (54), the estimate at a time prior to the next observation can be determined as follows. Let

$$\begin{aligned} \hat{x}_j &= E[x_j \mid y_1, y_2, \ldots, y_j] \\ \bar{x}_k &= E[x_k \mid y_1, y_2, \ldots, y_j]. \end{aligned} \tag{55}$$

That is, $\bar{x}_k$, is the best estimate of x_k for $t_k > t_j$ based on the observations available up to t_j. From the first of Eqs. (42), it follows that

$$E[x_k \mid y_1, \ldots, y_j] = \Phi(t_k, t_j) E[x_j \mid y_1, \ldots, y_j].$$

In view of Eqs. (55), the *a priori* estimate of x_k, i.e., $\bar{x}_k$, is given by the following expression

$$\bar{x}_k = \Phi(t_k, t_j)\hat{x}_j. \tag{56}$$

This expression can be generalized to include the effects of process noise in the dynamics. This topic will be considered in a subsequent discussion.

The expression for propagating the covariance matrix can be obtained as follows.

$$\bar{P}_k \equiv E[(x_k - \bar{x}_k)(x_k - \bar{x}_k)^T \mid y_1, \ldots, y_j]. \tag{57}$$

In view of Eq. (48), Eq. (49) becomes

$$\bar{P}_k = E[\Phi(t_k, t_j)(x_j - \hat{x}_j)(x_j - \hat{x}_j)^T \Phi^T(t_k, t_j) \mid y_1, \ldots, y_j].$$

Since the state transition matrix is deterministic, it follows that

$$\bar{P}_k = \Phi(t_k, t_j) P_j \Phi^T(t_k, t_j) \tag{58}$$

where $P_j = E[(x_j - \hat{x}_j)(x_j - \hat{x}_j)^T \mid y_1, \ldots, y_j]$. Equations (56) and (58) can be used to determine the estimate of the state and its associated covariance matrix at t_k based on an estimate at t_j.

Minimum Variance Estimate With A Priori Information

If an estimate and the associated covariance matrix are obtained at a time t_j, and an additional observation or observation sequence is obtained at a time t_k, the estimate and the observation can be combined in a straightforward manner to obtain the new estimate $\hat{x}_k$. The estimate $\hat{x}_j$ and P_j are propagated forward to t_k, using Eqs. (56) and (58) given by

$$\bar{x}_k = \Phi(t_k,t_j)\hat{x}_j \ , \quad \bar{P}_k = \Phi(t_k,t_j)P_j\Phi^T(t_k,t_j). \tag{59}$$

The problem to be considered can be stated as follows:

Given $\bar{x}_k$, $\bar{P}_k$ and $y_k = \tilde{H}_k x_k + \varepsilon_k$, where $E[\varepsilon_k] = 0$, $E[\varepsilon_k \varepsilon_j^T] = R_k \delta_{kj}$ and $E[(x_j - \hat{x}_j)\varepsilon_k^T] = 0$, find the best linear minimum variance unbiased estimate of x_k.

The solution to the problem can be obtained by reducing it to the previously solved problem. To this end, note that if $\hat{x}_j$ is unbiased, $\bar{x}_k$ will be unbiased since $E[\bar{x}_k] = \Phi(t_k,t_j)E[\hat{x}_j]$. Hence, $\bar{x}_k$ can be interpreted as an observation and the following relations will hold

$$\begin{aligned} y_k &= H_k x_k + \varepsilon_k \\ \bar{x}_k &= x_k + \eta_k \end{aligned} \tag{60}$$

where

$$E[\varepsilon_k] = 0 \, , \quad E[\varepsilon_k \varepsilon_k^T] = R_k \ , \quad E[\eta_k] = 0$$

$$E[\eta_k \varepsilon_k^T] = 0 \quad \text{and} \quad E[\eta_k \eta_k^T] = \bar{P}_k .$$

Now, if the following definitions are used

$$y = \begin{bmatrix} y_k \\ \cdots \\ \bar{x}_k \end{bmatrix} \qquad H = \begin{bmatrix} H_k \\ \cdots \\ I \end{bmatrix} \qquad \varepsilon = \begin{bmatrix} \varepsilon_k \\ \cdots \\ \eta_k \end{bmatrix} \qquad R = \begin{bmatrix} R_k & 0 \\ \cdots & \\ 0 & \bar{P}_k \end{bmatrix} \tag{61}$$

Equations (60) can be expressed as $y = Hx_k + \varepsilon$, and the solution obtained in the previous section can be applied to obtain the following estimate for $\hat{x}_k$.

$$\hat{x}_k = (H^T R^{-1} H)^{-1} H^T R^{-1} y.$$

In view of the definitions in Eq. (61),

$$\hat{x}_k = \left\{ [H_k^T : I] \begin{bmatrix} R_k^{-1} & 0 \\ \cdots & \cdots \\ 0 & \bar{P}_k^{-1} \end{bmatrix} \begin{bmatrix} H_k \\ \cdots \\ I \end{bmatrix} \right\}^{-1} \left\{ [H_k^T : I] \begin{bmatrix} R_k^{-1} & 0 \\ \cdots & \cdots \\ 0 & \bar{P}_k^{-1} \end{bmatrix} \begin{bmatrix} y_k \\ \cdots \\ \bar{x}_k \end{bmatrix} \right\}$$

or in expanded form,

$$\hat{x}_k = (H_k^T R_k^{-1} H_k + \bar{P}_k^{-1})^{-1} (H_k^T R_k^{-1} y_k + \bar{P}_k^{-1} \bar{x}_k). \tag{62}$$

The covariance associated with $\hat{x}$ is easily shown to be

$$P_k = E[(\hat{x}_k - x_k)(\hat{x}_k - x_k)^T] = (H_k^T R_k^{-1} H_k + \bar{P}_k^{-1})^{-1} \tag{63}$$

The following remarks are pertinent to Eq. (62).

1. The vector y_k may be only a single observation or it may include an entire batch of observations.

2. The *a priori* estimate, $\bar{x}_k$, may represent the estimate based on *a priori* initial conditions or the estimate based on the reduction of a previous batch of data.

3. As in the previous solution, the $n \times n$ normal matrix must be inverted and if the dimension n is large, this inversion can lead to computational problems. Alternate solution techniques which avoid the inversion of the normal matrix have been developed and are discussed in a subsequent section.

The Sequential Estimation Algorithm

In this section, an alternate approach to the batch processor is discussed in which each observation is processed as soon as it is received. An advantage of this approach, which is referred to as the sequential processing algorithm, is that the matrix to be inverted will be of the same dimension as the observation vector. Hence, if the observations are processed individually, only scalar divisions will be required to obtain the estimate of x_k. The algorithm was developed originally by *Swerling* [1958], but the treatment which has received a more popular acclaim is that due to the work of *Kalman* [1960] and *Kalman and Bucy* [1961]. In fact, the sequential estimation algorithm discussed here is often referred to as the Kalman-Bucy filter.

Recall that, if there is no dynamic model error other than errors in the dynamic model parameters, an estimate $\hat{x}_j$ and a covariance matrix P_j can be propagated forward to a time t_k by the relations,

$$\begin{aligned} \bar{x}_k &= \Phi(t_k,t_j)\hat{x}_j \\ \bar{P}_k &= \Phi(t_k,t_j)P_j\Phi^T(t_k,t_j) \end{aligned} \tag{64}$$

and with an additional observation at t_k,

$$y_k = H_k x_k + \varepsilon_k \tag{65}$$

where $E[\varepsilon_k] = 0$ and $E[\varepsilon_k \varepsilon_j^T] = R_k \delta_{kj}$, the best estimate of x_k is obtained in Eq. (54) as

$$\hat{x}_k = (H_k^T R_k^{-1} H_k + \bar{P}_k^{-1})^{-1}(H_k^T R_k^{-1} y_k + \bar{P}_k^{-1}\bar{x}_k) \tag{66}$$

The primary computational problems associated with this expression lie in computing the $(n \times n)$ matrix inverse. Recall that in the minimum variance derivation (Eq. (62)), it is shown that the quantity to be inverted is the covariance matrix P_k associated with estimate $\hat{x}_k$. That is,

$$P_k = (H_k^T R_k^{-1} H_k + \bar{P}_k^{-1})^{-1}. \tag{67}$$

It can be shown that Eq. (67) can be rearranged algebraically to obtain the following expression [*Tapley*, 1973]:

$$P_k = \bar{P}_k - \bar{P}_k H_k^T [H_k \bar{P}_k H_k^T + R_k]^{-1} H_k \bar{P}_k. \tag{68}$$

Note that Eq. (68) is an alternate way of computing the inverse in Eq. (67). In Eq. (68), the matrix to be inverted is of dimension $p \times p$, e.g., the same as the observation error covariance matrix. If only a scalar observation is involved, the inverse is obtained through only one scalar division. The identity in Eq. (68) is referred to in the literature as the Schurr identity or the inside out rule. If the weighting matrix, K_k, is defined as

$$K_k \equiv \bar{P}_k H_k^T [H_k \bar{P}_k H_k^T + R_k]^{-1} \tag{69}$$

then Eq. (68) can be expressed in the compact form

$$P_k = [I - K_k H_k]\bar{P}_k. \tag{70}$$

Now, if Eq. (67) is substituted into Eq. (66), the sequential form for computing x_k can be obtained as

$$\hat{x}_k = \bar{x}_k + K_k[y_k - H_k \bar{x}_k]. \tag{71}$$

Equation (71), along with Eqs. (64), (69) and (70), can be used in a recursive fashion to compute the estimate of $\hat{x}_k$, incorporating the observation y_k.

As pointed out previously, the only inverse required is that used in calculating the weighting matrix K_k, and the matrix to be inverted will be the size of the observation noise covariance matrix. The estimate of the state of the nonlinear system is given by $\hat{X}_k = X_k^* + \hat{x}_k$.

One disadvantage of both the batch and sequential algorithm lies in the fact that, if the true state and the reference state are not close together, then the linearization assumption leading to Eqs. (15) may not be valid and the estimation process may diverge. A second unfavorable characteristic of the sequential estimation algorithm is that the state estimate covariance matrix will approach zero as the number of observations becomes large. Examination of the estimation algorithm shows that, as $P_k \to 0$, the estimation procedure will become insensitive to the observations, and the estimate will diverge due either to errors introduced in the linearization procedure, computational errors or errors due to an incomplete mathematical model. Two modifications which improve these deficiencies are discussed in the following sections.

The Extended Sequential Estimation Algorithm

In order to minimize the effects of errors due to the neglect of higher order terms in the linearization procedure leading to Eqs. (15), the extended form of the sequential estimation algorithm is used. The primary difference between the sequential and the extended sequential algorithm is that the fixed reference trajectory used for the sequential algorithm is updated after each observation to reflect the best estimate of the true trajectory. For example, after processing the kth observation, the reference trajectory at time t_k is modified to be

$$\hat{X}_k = X_k^* + \hat{x}_k \tag{72}$$

The integration for the reference trajectory and the state transition matrix is reinitialized at each observation epoch, and the equations are integrated forward to t_{k+1}. The estimate for $\hat{x}_{k+1}$ is then computed from

$$\hat{x}_{k+1} = K_{k+1} y_{k+1} \tag{73}$$

where K_{k+1} and y_{k+1} are computed based on the new reference orbit. Note that updating the reference orbit at t_k results in $\bar{x}_{k+1} = 0$. Then, the reference orbit is updated at time t_{k+1} to incorporate $\hat{x}_{k+1}$ and the process proceeds to t_{k+2}.

The advantage of the extended sequential algorithm is that convergence to the correct estimate will be more rapid because errors introduced in the linearization process are reduced. A major disadvantage of the extended sequential algorithm is that the differential equations for the reference trajectory must be reinitialized after each observation is processed. This necessitates restarting the numerical integration procedure after each observation.

State Noise Compensation Algorithm

In addition to the effects of the nonlinearities, the effects of errors in the dynamic model can lead to divergence in the estimate. As pointed out previously, for a sufficiently large number of observations the value of the covariance matrix $\bar{P}_k$ will asymptotically approach zero and the estimation algorithm will be insensitive to any further observations, a condition which can lead to filter divergence. One approach to eliminating this divergence is to 1) recognize that the linearized equations for propagating the estimate of the state are in error, and 2) to compensate for this effect with the addition of process noise to the state dynamics.

The state dynamics of a linear system under the influence of state or process noise are described by

$$\dot{x}(t) = A(t)\, x(t) + B(t)\, u(t) \tag{74}$$

where $A(t)$ and $B(t)$ are known functions of time. The function $u(t)$ is a random process with

$$\begin{aligned} &E[u(t)] = 0 \\ &E[u(t)u^T(\tau)] = Q(t)\delta(t-\tau) \end{aligned} \tag{75}$$

where $\delta(t-\tau)$ is the Dirac Delta function. The solution of Eq. (74) can be obtained by the method of variation of parameters as follows.

$$x(t) = \Phi(t,t_o)x_o + \int_{t_o}^{t} \Phi(t,\tau)B(\tau)u(\tau)d\tau \tag{76}$$

Equation (76) is the general solution for the inhomogeneous system of linear equations given in Eq. (74).

The equations for propagating the state estimate $\hat{x}(t_{k-1})$ to the next observation time, t_k, are obtained by recalling that

$$\bar{x}(t) = E[x(t)|\ y_{k-1}] \quad \text{for} \quad t \geq t_{k-1} \tag{77}$$

Differentiating Eq. (77) after substituting (74) leads to

$$\dot{\bar{x}}(t) = A(t)\,\bar{x}(t) \tag{78}$$

with initial conditions $\bar{x}(t_{k-1}) = \hat{x}(t_{k-1})$. In Eq. (78), the assumption has been made that the state noise $u(t)$ has zero mean and is independent of the observations, i.e.,

$$E[u(t)|\ y_{k-1}] = E[u(t)] = 0$$

Hence, the equation for propagating the state estimate is the same as without process noise, i.e.,

$$\bar{x}(t) = \Phi(t,t_{k-1})\hat{x}_{k-1} \tag{79}$$

The equation for propagation of the estimation error covariance matrix is obtained by using the definition for $\bar{P}(t)$, given by

$$\bar{P}(t) = E[(x(t)-\bar{x}(t))(x(t)-\bar{x}(t))^T |\ y_{k-1}] \qquad t \geq t_{k-1} \tag{80}$$

On substituting Eq. (76) and Eq. (79) into (80) and applying the expected value operator, the following result is obtained.

$$\bar{P}(t) = \Phi(t,t_{k-1})P_{k-1}\Phi^T(t,t_{k-1}) + \int_{t_{k-1}}^{t} \Phi(t,\tau)B(\tau)Q(\tau)B^T(\tau)\Phi^T(t,\tau)d\tau \tag{81}$$

Equations (78) and (81) are the equations for propagating the estimate of the state and the covariance for a continuous dynamic system. As an alternative, Eq. (81) can be differentiated to obtain the matrix differential equation

$$\dot{\bar{P}}(t) = A(t)\bar{P}(t) + \bar{P}(t)A^T(t) + B(t)Q(t)B^T(t);\ \bar{P}(t_o) = P_o \tag{82}$$

Batch and Sequential Estimation Compared

As described in previous sections, two general categories of estimators are used, the batch processor and the sequential processor, both with distinct advantages and disadvantages. The batch formulation provides an estimate of the state at some chosen epoch using an entire batch or set of data. This estimate and its associated covariance matrix can then be mapped to other times. The sequential processor, on the other hand, provides an estimate of the state at each observation time which also can be mapped to another time.

In the sequential formulation without process noise, a mathematical equivalence can be shown between the batch and sequential algorithms, i.e., given the same data set both algorithms produce the same estimates when the estimates are mapped to the same times. In the extended form of the sequential algorithm where the reference orbit is updated at each observation time, the equivalence is not so clear but numerical experiments have shown a very close agreement.

Normally, the batch and sequential algorithm will need to be iterated to convergence, while the extended sequential will essentially converge in a single iteration. On the other hand, the extended sequential requires restarting a numerical integrator at each observation whereas more efficient integrators can be applied to the single reference trajectory of the batch filter. In general, the sequential processor is used in onboard applications and when it is appropriate to incorporate some representation of the state noise. This implementation provides a means of compensating for various error sources in the processing of ground-based or onboard data. As indicated previously, inclusion of process noise in the batch algorithm substantially complicates the solution of the normal equations by increasing the dimensions of the normal matrix from n (the number of state parameters) to m (the number of observations).

Error Sources

In the application of an estimation procedure to a satellite or trajectory problem, measurements are obtained by various ground-based or onboard instruments. For example, a ground-based ranging system may make the measurements shown in Figure 2 with time measured in minutes since the first measurement. Based on a mathematical model of the dynamical system and the measurement system, a predicted or computed measurement could be generated and compared with the actual measurement. If, in fact, the models are quite accurate, the difference (or residual) between the actual and predicted (or computed) measurements (O-C) will simply exhibit the random component in the measurement system as in Figure 3. On the other hand, as is usually the case, the model has some inaccuracies associated with it, and the residual pattern will exhibit the character shown in Figure 4. These residuals are used by the estimators in improving the state used to predict the measurement.

In the ideal case, the nonzero difference between the actual measurement and the predicted value should be simply due to the noise and biasing in making the measurement. In practice, however, the mathematical models which describe the satellite force environment and those which describe the instrument performing some measurement are not completely accurate, or certain approximations are made for the benefit of computer storage and/or computer execution time which introduce some discrepancy or error in the data processing. It is frequently necessary to ascribe the source of an error to a phenomena in the physical world or to an approximation made in the model of the real world. Knowledge of various parameters in the mathematical models, such as the mass of the earth, have been obtained through various experiments or through use of many measurements and are only approximately known. These error sources are dependent on the satellite under consideration, e.g., the altitude

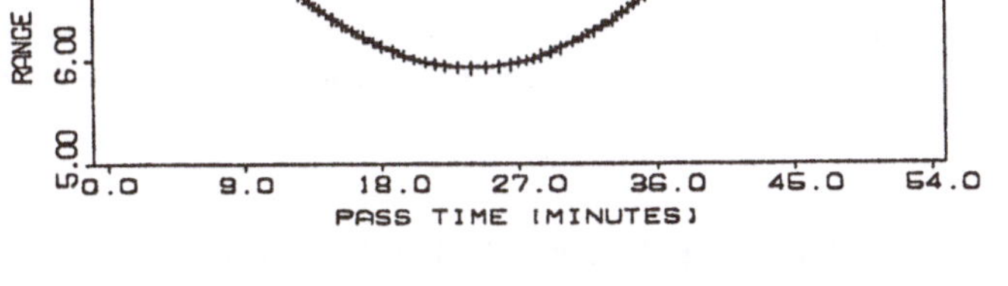

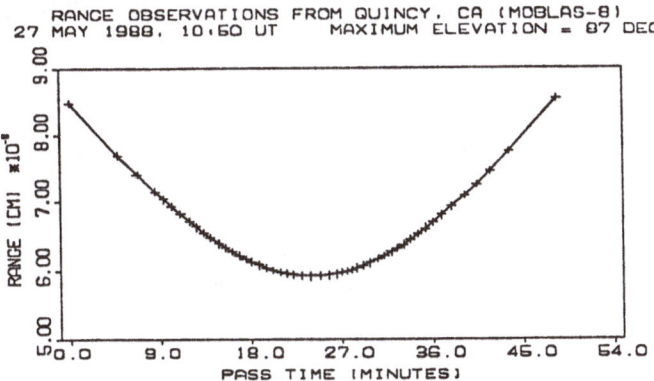

Figure 2. Range vs. Time From Ground-Based Station

and inclination, and on the measurement systems. Many of these error sources have distinct signatures in the data, while others may be very similar thus producing aliasing between these components and making it difficult or impossible to separate their effects into individual components.

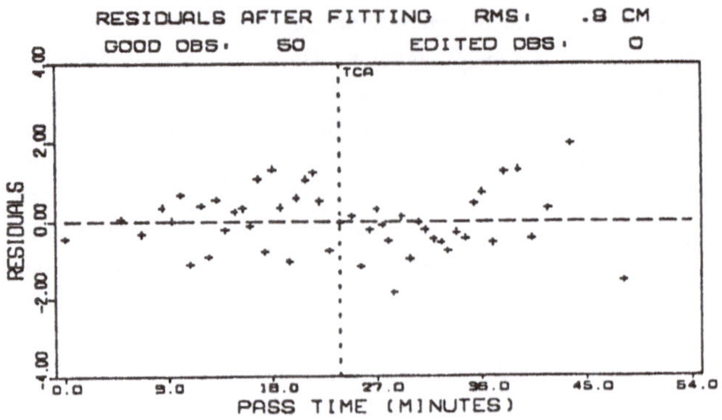

Figure 3. O-C, Random Component

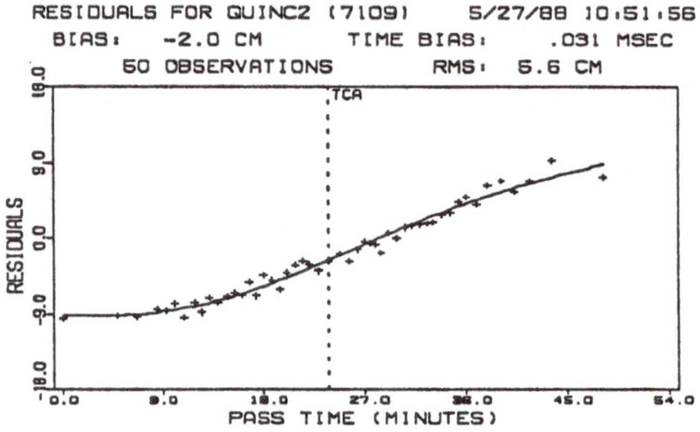

Figure 4. O-C, Random and Systematic Component

Solution Methods for the Orbit Determination Problem

As shown in Eq. (27), the solution to the least squares estimation problem is usually represented in the normal equation form

$$H^T H\hat{x} = H^T y \tag{83}$$

which can be expressed as

$$M\hat{x} = N \tag{84}$$

where computational problems are encountered in forming and in inverting the normal equation matrix $M = H^T H$. An orthogonal transformation approach can be used to reduce the solution to the form

$$R\hat{x} = b \tag{85}$$

where R is upper triangular and $\hat{x}$ can be obtained by backward substitution. The orthogonal transformation approach has the advantage that increased numerical precision is achieved for short word length systems [*Lawson and Hanson*, 1963]. Using the orthogonal transformation approach, accuracy can be achieved with a single-precision computation which is comparable to double-precision accuracy with the normal equation approach.

The normal equation approach has several operational and conceptual advantages which have led to the widespread adoption of this technique for most operational orbit determination systems.

Cholesky Decomposition

The conventional solution of the linear system

$$M\hat{x} = N \tag{86}$$

is expressed as

$$\hat{x} = M^{-1}N \tag{87}$$

where the operation M^{-1} implies that the inverse of the $(n \times n)$ matrix M is computed and then post multiplied by the column vector N. Since the matrix M is symmetric and positive definite, an alternate approach based on the Cholesky decomposition is usually used in practice. The solution by Cholesky decomposition is more efficient and, in most cases, more accurate. The Cholesky decomposition is applicable only if M is symmetric and positive definite, a condition satisfied for the case considered here.

Let M be a symmetric positive definite matrix, and let R be an upper triangular matrix defined by the relation

$$R^T R = M \tag{88}$$

Eq. (86) can be expressed as

$$R^T R\hat{x} = N \tag{89}$$

Then, let

$$z = R\hat{x} \tag{90}$$

and Eq. (89) can be written as

$$R^T z = N \tag{91}$$

where R^T is lower triangular. The components of z can be solved using a forward recursion relation. Then, Eq. (90) can be solved using a backward recursion to obtain the elements of $\hat{x}$.

The expressions for the Cholesky decomposition of M is obtained by expanding Eq. (88) and solving term by term for the elements of R, e.g., R_{ij}, in terms of the elements of M, e.g., M_{ij}.

Least Squares Solution via Orthogonal Transformation

An alternate approach which avoids some of the numerical problems encountered in the normal equation approach described above is described in the following discussions. The method obtains the solution by applying successive orthogonal transformations to the information array. Consider the quadratic performance index

$$J = \tfrac{1}{2} \| W^{1/2}(Hx - y)\|^2 = \tfrac{1}{2} (Hx - y)^T W (Hx - y) \tag{92}$$

The solution to the least squares estimation problem (as well as the minimum variance and the maximum likelihood estimation problem, under certain restrictions) is obtained by finding the value $\hat{x}$ which minimizes Eq. (92). To achieve the minimum value of Eq. (92) let Q be an $m \times m$ orthogonal matrix. Hence Q will satisfy the conditions

$$QQ^T = Q^T Q = I \tag{93}$$

Note that the matrix Q^{-1} satisfies the conditions

$$QQ^{-1} = Q^{-1}Q = I \tag{94}$$

Hence, $Q^{-1} = Q^T$. It follows then that (92) can be expressed as

$$J = \tfrac{1}{2} \left\| QW^{\frac{1}{2}} (Hx - y) \right\|^2 \tag{95}$$

Now, if Q is selected such that

$$QW^{\frac{1}{2}}H = \begin{bmatrix} R \\ O \end{bmatrix} \qquad QW^{\frac{1}{2}}y = \begin{bmatrix} b \\ e \end{bmatrix} \tag{96}$$

where R is a $n \times n$ upper-triangular matrix
O is a $(m-n) \times n$ null matrix
b is a $n \times 1$ column vector
e is a $(m-n) \times 1$ column vector

Eq. (95) can be written as

$$J(x) = \tfrac{1}{2} \| Rx - b \|^2 + \tfrac{1}{2} \| e \|^2 \tag{97}$$

The value of x which minimizes (97) is obviously obtained by the solution

$$R\hat{x} = b \tag{98}$$

and the minimum value of the performance index becomes

$$J(\hat{x}) = \tfrac{1}{2} \| e \|^2 \tag{99}$$

That is, e is the residual error vector.

The procedure is direct and for mechanization requires only that a convenient computational procedure for computing Q be obtained. One such procedure can be developed based on the Given's rotation. Let x be a 2×1 vector having components $x^T = [x_1 \; x_2]$ and let G be a 2×2 orthogonal matrix associated with the plane rotation through the angle θ. Then select G such that

$$Gx = x' = \begin{bmatrix} x_1' \\ 0 \end{bmatrix} \tag{100}$$

To this end, consider the transformation

$$\begin{bmatrix} x_1' \\ x_2' \end{bmatrix} = \begin{bmatrix} \cos\theta & \sin\theta \\ -\sin\theta & \cos\theta \end{bmatrix} \begin{bmatrix} x_1 \\ x_2 \end{bmatrix} \tag{101}$$

or

$$x_1' = \cos\theta x_1 + \sin\theta x_2$$

$$x_2' = -\sin\theta x_1 + \cos\theta x_2 \tag{102}$$

Eqs. (102) represent a system of two equations in three unknowns, e.g., x_1', x_2' and θ. The possible solutions to these equations are:

Specify θ: 2 eqs, 2 unknown x_1', x_2'
Specify x_2': 2 eqs, 2 unknown x_1', θ
Specify x_1': 2 eqs, 2 unknown x_2', θ

The Given's rotation is defined by selecting the rotation θ such that $x_2' = 0$. That is, let

$$x_1' = \cos\theta x_1 + \sin\theta x_2$$

$$0 = -\sin\theta x_1 + \cos\theta x_2 \tag{103}$$

From the second of Eqs. (103), it follows that

$$\tan\theta = \frac{x_2}{x_1}$$

$$\sin\theta = \frac{x_2}{\sqrt{x_1^2+x_2^2}} \qquad \cos\theta = \frac{x_1}{\sqrt{x_1^2+x_2^2}} \tag{104}$$

Also, from the first equation

$$x_1' = \frac{x_1^2}{\sqrt{x_1^2+x_2^2}} + \frac{x_2^2}{\sqrt{x_1^2+x_2^2}} = \sqrt{x_1^2+x_2^2} \tag{105}$$

Now consider the application of this transformation to two general row vectors, e.g.

$$G\begin{bmatrix} h_{ii} & h_{ii+1} & \cdots & h_{in} \\ h_{ki} & h_{ki+1} & \cdots & h_{kn} \end{bmatrix} = \begin{bmatrix} h_{ii}' & h_{ii+1}' & \cdots & h_{in}' \\ 0 & h_{ki+1}' & \cdots & h_{kn}' \end{bmatrix} \tag{106}$$

For any two general row vectors, h_k and h_i, the transformation is applied to the first column so as to null h_{ki}. Then the transformation is applied to each remaining column to obtain the transformed matrix. Hence,

$$\begin{bmatrix} \cos\theta & \sin\theta \\ -\sin\theta & \cos\theta \end{bmatrix} \begin{bmatrix} h_{ii} \\ h_{ki} \end{bmatrix} = \begin{bmatrix} h_{ii}' \\ 0 \end{bmatrix}$$

or

$$\sin\theta = h_{ki}/\sqrt{h_{ii}^2 + h_{ki}^2} = h_{ki}/h_{ii}'$$
$$\cos\theta = h_{ii}/\sqrt{h_{ii}^2 + h_{ki}^2} = h_{ii}/h_{ii}' \tag{107}$$
$$h_{ii}' = \sqrt{h_{ii}^2 + h_{ki}^2}$$

Then for all other columns

$$h_{ij}' = h_{ij}\cos\theta + h_{kj}\sin\theta$$
$$j = i+1, \ldots, n \tag{108}$$
$$h_{kj}' = -h_{ij}\sin\theta + h_{kj}\cos\theta$$

By using this transformation repetitively as k goes from $i+1$ to m, a complete column of an $m+n$ matrix can be reduced to a non-zero first element with zeros for the remaining terms.

Then by applying the transformation to successive columns, the matrix can be reduced to an upper triangular $n\times n$ matrix with a lower $(m-n)\times n$ null matrix.

As an example, the transformation to null the fourth element in the third column is shown as follows

$$\begin{bmatrix} 1 & & & & & & & & \\ & 1 & & & & & & & \\ & & C^{3,4} & S^{3,4} & & & & & \\ & & -S^{3,4} & C^{3,4} & & & & & \\ & & & & 1 & & & & \\ & & & & & 1 & & & \\ & & & & & & 1 & & \\ & & & & & & & \ddots & \\ & & & & & & & & 1 \end{bmatrix} \begin{bmatrix} h_{11}^1 & h_{12}^1 & h_{13}^1 & \cdots & h_{1n}^1 & y_1^1 \\ 0 & h_{22}^2 & h_{23}^2 & \cdots & h_{2n}^2 & y_2^2 \\ 0 & 0 & h_{33}^2 & \cdots & h_{3n}^2 & y_3^2 \\ 0 & 0 & h_{43}^2 & \cdots & h_{4n}^2 & y_4^2 \\ 0 & 0 & h_{53}^2 & \cdots & h_{5n}^2 & y_5^2 \\ 0 & 0 & h_{63}^2 & \cdots & h_{6n}^2 & y_6^2 \\ \vdots & \vdots & \vdots & & \vdots & \vdots \\ 0 & 0 & h_{m3}^2 & \cdots & h_{mn}^2 & y_m^2 \end{bmatrix} = \begin{bmatrix} h_{11}^1 & h_{12}^1 & h_{13}^1 & \cdots & h_{1n}^1 & y_1^1 \\ 0 & h_{22}^2 & h_{23}^2 & \cdots & h_{2n}^2 & y_2^2 \\ 0 & 0 & h_{33}^{2'} & \cdots & h_{3n}^{2'} & y_3^{2'} \\ 0 & 0 & 0 & \cdots & h_{4n}^{3'} & y_4^{3'} \\ 0 & 0 & h_{53}^2 & \cdots & h_{5n}^2 & y_5^2 \\ 0 & 0 & h_{63}^2 & \cdots & h_{6n}^2 & y_6^2 \\ \vdots & \vdots & \vdots & & \vdots & \vdots \\ 0 & 0 & h_{m3}^2 & \cdots & h_{mn}^2 & y_m^2 \end{bmatrix} \quad (109)$$

By using the transformation

$$Q^{3,5} = \begin{bmatrix} 1 & & & & & & & & \\ & 1 & & & & & & & \\ & & C^{3,5} & 0 & S^{3,5} & & & & \\ & & 0 & 1 & 0 & & & & \\ & & -S^{3,5} & 0 & C^{3,5} & & & & \\ & & & & & 1 & & & \\ & & & & & & 1 & & \\ & & & & & & & 1 & \\ & & & & & & & & \ddots \\ & & & & & & & & & 1 \end{bmatrix} \quad (110)$$

The fifth row will be transformed so that the term h_{53}^2 will be zero.

After application of this algorithm, the $m \times (n+1)$ matrix will appear as

$$Q\,[H \,\vdots\, y] = \begin{bmatrix} R & b \\ O & e \end{bmatrix} \quad (111)$$

which is the required form for solution of the least squares estimation problem as given by Eq. (98).

Once the array has been reduced to the form given by Eq. (111), subsequent observations can be included by considering the following array:

$$\begin{bmatrix} R & b \\ H^{m+1} & y^{m+1} \\ 0 & e \end{bmatrix} = \begin{bmatrix} R_{11} & R_{12} & \cdots & R_{1n} & b_1 \\ 0 & R_{22} & \cdots & R_{2n} & b_2 \\ 0 & 0 & \cdots & R_{3n} & b_3 \\ \vdots & & & & \\ 0 & 0 & \cdots & R_{nn} & b_n \\ H_{m+1,1} & H_{m+1,2} & \cdots & H_{m+1,n} & y_{m+1} \\ 0 & 0 & \cdots & 0 & e^2 \end{bmatrix} \quad (112)$$

where $e^2 = \Sigma\, e_i^2$. Then by application of a Given's rotation to the first and last row, $H_{m+1,1}$ can be zeroed. Successive applications moving down the main diagonal can be used to null the $m+1$ row and reduce the array to upper triangular form. Note that the augment matrix

$$\begin{bmatrix} R & b \\ 0 & e \\ H^{m+1} & y^{m+1} \end{bmatrix}$$

is rearranged to obtain the relation given in (112) by reordering to insert the last row as the $n+1^{\text{st}}$ row.

Alternate approaches based on the Householder transformation can be used to obtain a solution equivalent to the expression shown in Eq. (111). From this expression,

$$\hat{x} = R^{-1} b$$
$$P = R^{-1} R^{-T} \tag{113}$$
$$J(\hat{x}) = \| e \|^2$$

where the solution for $\hat{x}$ can be obtained through a backward recursion.

APPENDIX A

THE PRIMARY FORCES ON A NEAR-EARTH SATELLITE

The forces which affect the motion of a near-earth satellite can be separated into gravitational forces and surfaces forces.

Gravitational Perturbations

The gravitational perturbations include the gravitational effects of the earth, solid earth tides, lunar, solar and planetary perturbations, ocean tide and general relativistic forces.

Gravitational Potential for the Earth

The potential function for the earth is generally expressed in terms of a spherical harmonic expansion and is referred to as the geopotential

$$U = \frac{GM}{r} \left\{ 1 - \sum_{l=2}^{\infty} J_l \left[\frac{a_e}{r} \right]^l P_l(\sin\phi) + \sum_{l=2}^{\infty} \sum_{m=1}^{l} \left[\frac{a_e}{r} \right]^l P_{lm}(\sin\phi)[C_{lm} \cos m\lambda + S_{lm} \sin m\lambda] \right\} \tag{A.1}$$

assuming the origin of the spherical coordinates coincides with the center of mass of the earth $J_1 = C_{11} = S_{11} = 0$. The terms in Eq. (A.1) are:

$P_l(\sin\phi)$: Degree l Legendre polynomials

$P_{lm}(\sin\phi)$: Associated functions of Legendre of degree l and order m

a_e : Mean radius of the earth

$J_l = -C_{lo}$: Zonal harmonic coefficient

C_{lm}, S_{lm}: Tesseral harmonic coefficient if $l \neq m$; sectoral or sectoral harmonic coefficient if $l = m$.

The harmonic coefficients are functions of the Earth's mass distribution and are obtained from observing the motions of artificial satellites. In practice, the summations over l in Eq. (A.1) are taken to a finite number, N, instead of ∞ where N may be chosen as a function of the individual satellite orbit. By far, the dominant perturbation of the departure of the earth's gravity field from sphericity is produced by the oblateness, represented in the geopotential by J_2.

Solid Earth Tides

The earth, which is not perfectly rigid, deforms under the solar and lunar gravitational attractions. These deformations, associated with the redistribution of mass, can be conveniently defined using the Love numbers introduced by A. Love in 1909. The deformation can be expressed as a change to the external geopotential by the following expression

$$\Delta U(r) = \sum_{l=2}^{\infty} k_l \left[\frac{R_e}{r} \right]^{2l+1} V_l(r) \tag{A.2}$$

where $\Delta U(r)$ is the change in potential at position $\bar{r}$, k_l are Love numbers of degree l, and V_l is the disturbing tidal potential of degree l.

The luni-solar tidal potential $V_l(r)$ is of the following form

$$V_l(r) = GM_i \left[\frac{1}{\rho_i} - \frac{\bar{r}_i \cdot \bar{r}}{r_i^3} \right] \tag{A.3}$$

where M_i is the mass of the i^{θ} disturbing body, refers to the moon or the sun; ρ_i is the distance between the ith disturbing body and a particle on the solid earth; and $\bar{r}_i$ is the position vector of the i^{θ} disturbing body.

The changes in geopotential caused by the luni-solar tides can be expressed in terms of time dependent geopotential coefficients; that is, $\Delta\bar{C}_{lm}$ and $\Delta\bar{S}_{lm}$ which can be expressed as follows [*Sanchez*, 1974],

$$\Delta\bar{C}_{lm} = \frac{k_l R_e^{n+1}}{GM} q_{lm} \left[\frac{4(l+2)(l-m)!}{(l+m)!} \right]^{1/2}$$

$$\Delta\bar{S}_{lm} = \frac{k_l R_e^{n+1}}{GM} u_{lm} \left[\frac{4(l+2)(l-m)!}{(l+m)!} \right]^{1/2}$$

and

$$q_{lm} = \sum_{j=2}^{3} \frac{GM}{r_{1j}^{l+1}} \frac{2(l-m)!}{(l+m)!} \bar{P}_{lm}(\cos\theta_j)\cos m\psi_j$$

$$u_{lm} = \sum_{j=2}^{3} \frac{GM_j}{r_{1j}^{l+1}} \frac{2(l-m)!}{(l+m)!} \bar{P}_{lm}(\cos\theta_j)\sin m\psi_j$$

where the index $j = 1,2,3$ denotes earth, moon and sun, respectively; (θ,ψ) are colatitude and longitude of the disturbing body; $\Delta\bar{C}_{lm}$ and $\Delta\bar{S}_{lm}$ are the time-varying geopotential coefficients affected by the luni-solar tidal effect. Furthermore, a parameter δ, which represents tidal lag angle associated with the response of the solid earth to the tidal forces, can be accounted for the time delay caused by the earth's inelasticity.

N-Body

The perturbing forces of the sun, moon and other planets, namely, Mercury, Venus, Mars, Jupiter, Saturn, Uranus, Neptune and Pluto, can be approximated with a sufficient accuracy as point masses. Expressing in a geocentric coordinate system, the central body and the N-body forces can be expressed as

$$\bar{F} = (1 + \frac{m}{M})\nabla U + \sum_i GM_i \left[\frac{\nabla U}{GM} - \frac{\bar{\Delta}_i}{\Delta_i^3} \right] \tag{A.4}$$

where $\bar{F}$ is the force acting on the satellite due to N-body attraction; m and M are masses of the satellite and the earth, respectively; $\bar{\Delta}_i$ is the position vector between the satellite and the perturbing mass, M_i; and ∇U refers to the gradient of the geopotential. Since m is small compared to M, the m/M term in Eq. (A.4) can be dropped.

The values of $\bar{\Delta}_i$ can be obtained using planetary ephemerides, for example, the Jet Propulsion Laboratory Development Ephemeris-200 (JPL DE-200).

Ocean Tides

The dynamical contribution of ocean tides due to the gravitational attraction of the sun and moon can be formulated in terms of time-varying geopotential coefficient corrections. The disturbing ocean tide potential, ΔU, can be expressed as follows

$$\Delta U = 4\pi G \rho_w R_e \sum_{\mu} \sum_{l=0}^{\infty} \sum_{m=0}^{l} \sum_{+}^{-} \frac{1+k_l'}{2l+1} \left[\frac{R_e}{r} \right]^{n+1} \bar{P}_{lm} (\sin\phi) \cdot$$

$$\bar{C}^{\pm}_{\mu lm} \sin(\bar{\eta}_{\mu} \cdot \bar{\beta}(t) \pm m\lambda \pm \bar{\varepsilon}^{\pm}_{\mu lm}) \tag{A.5}$$

where G is the gravitational constant; ρ_w is the mean density of the sea water; R_e is the mean equatorial radius; k_l' is the load deformation coefficient for degree l; m is the order of the coefficient; μ is the ocean tide constituent index; $\bar{\beta}(t) = [\tau\ s\ h\ p\ N'\ p_1]$ are the Doodson arguments which define lunar and solar ephemeris and t is the time; $\bar{n} = [n_1\ n_2\ \cdots\ n_6]$ are integer multipliers of Doodson arguments; $\bar{C}^{\pm}_{\mu lm}$ are amplitudes of ocean tide constituents; and $\bar{\varepsilon}^{\pm}_{\mu lm}$ are the phase angles.

Eq. (A.5) can be expressed conveniently in the form of geopotential representation, given by Eq. (A.1) as a correction to the spherical harmonic coefficients, $\bar{C}_{lm}, \bar{S}_{lm}$. Another reason to adopt this form is that the adjustable ocean tide parameters would then be in a linear form instead of a nonlinear form expressed in terms of amplitude and phase. Let

$$\begin{bmatrix} C^{\pm} \\ S^{\pm} \end{bmatrix}_{\mu lm} = \bar{C}^{\pm}_{\mu lm} \begin{bmatrix} \sin \\ \cos \end{bmatrix} (\bar{\eta}_{\mu} \cdot \bar{\beta}(t) \pm m\lambda) \tag{A.6}$$

$$\begin{bmatrix} A \\ B \end{bmatrix}_{\mu lm} = \begin{bmatrix} C^{+} + C^{-} \\ S^{+} - S^{-} \end{bmatrix}_{\mu lm} \cos(\bar{\eta}_{\mu} \cdot \bar{\beta})$$

$$+ \begin{bmatrix} S^{+} + S^{-} \\ C^{-} - C^{+} \end{bmatrix}_{\mu lm} \sin(\bar{\eta}_{\mu} \cdot \bar{\beta}) \tag{A.7}$$

and

$$F_{lm} = \frac{4\pi R_e^2 \rho_w}{M} \left[\frac{(l+m)!}{(n-m)!(2l+1)(2-\delta_{0m})} \right]^{1/2}$$

$$\left[\frac{1+k_l'}{2l+1} \right] \tag{A.8}$$

where M is the mass of the earth; $C^{\pm}, S^{\pm}$ are ocean tide coefficients; and δ_{0m} is the Kronecker delta function, $\delta = 1$ for $m = 0$; $\delta = 0$, otherwise.

The total potential, U, that includes gravitational and tidal potential can thus be expressed through Eqs. (A.1), (A.5), (A.6), (A.7) and (A.8) as the following

$$U = \frac{GM}{r} \sum_{l=0}^{\infty} \sum_{m=0}^{l} \left[\frac{R_e}{r} \right]^l \bar{P}_{lm}(\sin\phi) \left[\left[\bar{C}_{lm} + F_{lm} \sum_{\mu} A_{\mu lm} \right] \cos m\lambda \right.$$

$$\left. + \left[\bar{S}_{lm} + F_{lm} \sum_{\mu} B_{\mu lm} \right] \sin m\lambda \right] \tag{A.9}$$

General Relativity

The dynamical effect of general relativistic satellite perturbation on an earth satellite is small but detectable. For example, BE-C with a semi-major axis of 7500 km and an eccentricity of 0.025 experiences a periapse advance rate due to relativistic effect of 0°.031/day with earth considered to be the single central body. The dynamical relativistic equations of motion, which require the earth, moon, sun and the planets as perturbing bodies, are quite complex [*Moyer*, 1968] and will not be discussed here. A simplified example involving a massless particle moving under the influence of a point mass in a circular orbit is given by *Moyer* [1968]. The acceleration of the particle including relativistic effects is given as follows

$$\ddot{\bar{r}} = \frac{-GM}{r^3} \left[1 - \frac{3GM}{c^2 r} \right] \bar{r} \tag{A.10}$$

where c is the speed of light in vacuum, and M is the mass of the central body.

If the relativistic effect is to be neglected, Eq. (A.10) reduces to the Newtonian acceleration. Eq. (A.10) shows that the relativistic perturbation is on the order of $1/c^2$ and is dominantly in the radial direction. In fact, the relativistic perturbation increases slightly the effective GM of the earth and can thus be absorbed by estimating GM in the orbit determination process.

Nongravitational Perturbations

In addition to the gravitational forces due to the earth, moon and planets, other forces must be modeled. These are briefly discussed in the following sections.

Atmospheric Drag

The dominant feature of atmospheric resistance for most satellites is a drag force in the direction opposite to the relative wind. Drag is usually modeled as

$$\bar{F}_D = -\frac{1}{2} \rho \left[\frac{C_D A}{m} \right] V_r^2 \bar{u} \tag{A.11}$$

where ρ is the atmospheric density, C_D is the drag coefficient, A is the cross-sectional area perpendicular to $\bar{V}_r$, m is the satellite mass, V_r is the speed relative to the atmosphere, and $\bar{u}$ is a unit vector in the $\bar{V}_r$ direction. The drag force is considerably larger than lift forces which, if they exist, would be perpendicular to $\bar{V}_r$. The parameter $(C_D A/m)$ is sometimes referred to as the ballistic coefficient, B, and A/m is the area-to-mass ratio. The drag coefficient, C_D, is a function of the geometry of the satellite and the Mach number, the ratio of the vehicle speed to the speed of sound. The atmosphere is quite rarefied at satellite altitudes, however, and the air flow may not be well represented by hypersonic flow theory. The C_D and A will change if the satellite spins or tumbles, producing difficult to model time variations of B. In addition, the density ρ is a complicated function, probably best represented by empirical tables. Drag has an important effect on the orbit producing a secular change in the semimajor axis and the eccentricity which causes the orbit to decay, i.e., the orbit tends to spiral inward with perigee becoming smaller.

From the data collected on motions of satellites, especially balloon satellites, and in situ measurements, some basic characteristics of the upper atmosphere density have emerged. These characteristics are:

1. A nearly diurnal variation produced by the sun. The sub-solar region can be modeled with an atmospheric bulge, the axis of which approximately coincides with the earth-sun line.

2. Solar activity has an important influence through production of disturbances in the upper atmosphere. The primary sources of these disturbances are solar flares and solar plasma events. The frequencies of these phenomena follow the eleven-year solar cycle and the 27-day solar rotation period.

3. The density is influenced by geomagnetic activity and interaction with the charged particles in the upper atmosphere. This phenomena can produce significant changes in the density at time scales of a few hours to a day.

4. Seasonal changes in the density, including both annual and semi-annual variations, have been observed, although individual models may not account for all these effects.

Most of these phenomena cannot be predicted accurately in advance of their occurrence. The consequence of this difficulty is that considerable uncertainty exists with long-term prediction of satellite lifetimes and that precise orbit computations can only be performed after the fact, i.e., after data have been collected on solar and developed geomagnetic activity.

Solar Radiation Pressure

Radiation from the sun produces a small force by transferring momentum from particles streaming out from the sun to a satellite. This force for a spherical symmetric satellite is approximately

$$\bar{F} = -P(1+\eta)\,\frac{A}{M}\,\nu\,\bar{u} \tag{A.12}$$

where $\bar{F}$ is the direct solar radiation force per unit mass, P is the momentum flux due to the sun, A is the cross-sectional area of the satellite normal to the sun, $\bar{u}$ is the unit vector pointing from the satellite to the sun, η is the reflectivity coefficient with values between 0 and 1, and ν is an eclipse factor such that $\nu = 0$ if the satellite is in shadow of the sun, $\nu = 1$ if the satellite is in sunlight, $0 < \nu < 1$ if the satellite is in partial shadow or penumbra. The passage of the satellite from sunlight to shadow is not abrupt, but the interval of time spent in partial shadow will be very brief for near-earth satellites.

A simple cylindrical shadow model can be used to determine the eclipse factor. Consider the sun forms a cylindrical shadow region behind the earth; the position of the satellite can thus be computed and determined whether it is in the sunlight or in complete shadow. A more sophisticated model, such as the model described by *Kogut et al.* [1977], is required to model the penumbra region.

For composite satellites that are not spherically symmetric and carry rotating solar panels or other antenna arrays, Eq. (A.12) needs to be modified to include area variation effects as in the case of the drag force. The parameter, η, can also be expressed as a linear time-varying quantity,

$$\eta = \eta_o + \dot{\eta}\, t \tag{A.13}$$

where η_o is the reflectivity coefficient at epoch, and $\dot{\eta}$ is the time rate change in η.

It is seen from both Equations (A.2) and (A.3) that the (A/m) of a satellite is an important factor in the magnitude of the drag and radiation pressure force experienced by a satellite. If A/m can be kept small, the associated orbit determination errors also can be kept small. For example, A/m for GEOS-3 was 0.004 m^2/kg, while for Seasat, it was approximately 0.012 m^2/kg. Hence, other things being equal, the orbit of GEOS-3 could be much more accurately estimated than that of Seasat. For Topex, A/m for the satellites being considered averages 0.01 m^2/kg.

Earth Radiation Pressure

For a close-earth satellite, the solar radiation pressure perturbation due to sunlight reflected from the earth is sometimes not negligible. The radiation pressure of a spherically symmetric satellite due to the earth albedo can again be expressed in the form of Eq. (A.12) as follows

$$\bar{F} = -P'(1+\eta)\,\frac{A}{M}\,\nu\,\bar{r} \tag{A.14}$$

where $\bar{F}$ is the earth albedo radiation pressure per unit mass, ν is the eclipse factor, P' is the solar momentum flux due to the earth albedo, and $\bar{r}$ is the radial position vector normal to the effective reflecting "disc" on the earth to the satellite. Eq. (A.12) is highly simplified, since it treats the effective reflecting surface as a reflecting spherical disc and assumes reflective property of the surface to be the same. P' is a function of the earth albedo, γ, and the angle of incidence of the reflected sunlight. γ can be approximated within five-percent precision by a five-parameter model given by

$$\gamma(\phi) = a_o + a_2 \sin^2(\phi + c_2) + a_4 \sin^4(\phi + c_4) \tag{A.15}$$

where ϕ is the geodetic latitude; a_o, a_2, a_4 are arbitrary constants; and c_2 and c_4 are phase angles.

Eq. (A.14) can be modified to take into account the two factors for more accurate representation of the earth albedo model. First, the radiation model can be better formulated by intensity flux due to the optical radiation and the infrared radiation. Second, the reflective surface considered in Eq. (A.14) can be subdivided into smaller surfaces in a finite element approach, and the momentum flux can be properly integrated over each surface area to ascertain a better representation of the earth albedo on the surface.

Again, the assumption made in arriving at Eq. (A.14) is that the satellite is spherically symmetric. Note that the surface area, A, and the reflectivity coefficient, η, can be different from that of the direct solar radiation pressure in Eq. (A.12) for a geodetic satellite bus with rotating solar panels.

REFERENCES

Bierman, G. J., *Factorization Methods for Discrete Sequential Estimation*, Academic Press, 1977.

Gelb, A. (Ed.), *Applied Optimal Estimation*, MIT Press, 1974.

Jazwinski, A. H., *Stochastic Process and Filtering Theory*, Academic Press, New York, 1969.

Lawson, C. L., and R. J. Hanson, *Solving Least Squares Problems*, Prentice Hall, 1963.

Liebelt, P. B., *An Introduction to Optimal Estimation*, Addison-Wesley, 1967.

Schlee, S. F., C. J. Standish, and N. F. Toda, *AIAA J.*, *5*, 1114–1120, 1967.

Tapley, B. D., Statistical Orbit Determination Theory, *Recent Advances in Dynamical Astronomy*, D. Reidel Publishing Co., 396–425, 1973.

Combination of Satellite, Altimetric and Terrestrial Gravity Data

Richard H. Rapp

Department of Geodetic Science and Surveying
The Ohio State University
Columbus, Ohio 43210

1.0 INTRODUCTION

The first gravity measurements on the earth were made with pendulums. In the 1930's gravimeters were developed that led to measurements of gravity differences over large areas. These differences led to gravity values and then gravity anomalies over broad land and ocean areas. The gravity anomaly values are stored as point or mean values with coordinates (ϕ, λ, H) given in some specified datum (both horizontal and vertical). The mean anomaly value are computed for a variety of compartment sizes ranging from $6' \times 10'$ to $5° \times 5°$. Although this data set is far from uniform in terms of areal coverage and accuracy it provides substantial information on the earth's gravity field.

In the late 1950's analysis of the perturbations of artificial satellite also led to information on the earth's gravitational potential as represented, for the most part, by a finite series of spherical harmonic coefficients. The maximum degree of the expansion has continually increased as data accuracy has increased and as the availability of satellites to analyze has increased. In the early 1960's models to degree 4 were being computed. In the early 1980's the models were extended to degree 20 (with additional terms) and today's solutions are being developed to degree 50. From these models we can determine gravity (or gravity anomalies) at the surface of the earth. These anomalies are band limited in the sense they have only the frequency content of the harmonics being solved for from the satellite data.

Other data types are available for the determination of the gravity field. Satellite altimetry is a key data type because of it's ocean wide coverage. Satellite to satellite tracking was carried out on a limited basis in the 1970's. And topography (elevations) is also a source of information once some hypothesis are made on the relationship between gravity anomalies and topography.

The problem addressed in this paper is related to the combination of these data types to arrive at an accurate representation of the earth's gravitational potential. The problem is complicated by the different spectral contents of the data. In addition the data may have bias' or correlations that need to be considered for the optimal estimation of the parameters.

For some discussions in this paper we will assume that potential coefficient analysis from satellite data has been carried out. Such solutions are called satellite alone fields. Such solutions, in the past have been, GEM9, GEML2, and recently GEMT1 (Marsh et al, 1988a). We also assume that the error covariances matrix of the parameters (potential coefficients) is also available.

With the above information we will first examine the role of terrestrial gravity data in the combination models. We will then consider the altimeter data and then other data types. Our results will include the methods to obtain models to the same degree as the satellite field or to estimate high (360) degree expansions.

2.0 Representation of the Gravitational Potential

We need to adopt a representation for the potential. The spherical harmonic representation has been most useful in satellite geodesy. It can also be used with surface gravity data but increasing care must be taken as accuracy requirements increase. Recent suggestions focus on the use of ellipsoidal harmonics which reduces some of the problems that arise with spherical harmonics. We therefore consider both representations.

2.1 Spherical Harmonics

We define the polar coordinates of a point as r, θ, and λ where r is the geocentric distance; θ is the geocentric co-latitude (90°-θ, where θ is the geocentric latitude); and λ is longitude. The θ and λ values must be given with respect to a precisely defined rectangular coordinate system. Such a system may be the Conventional Terrestrial System, but most likely is a system defined by a specific group for their satellite analysis. In this system the Newtonian potential is:

$$V(r,\theta,\lambda) = \frac{kM}{r}\left[1 + \sum_{n=2}^{\infty} \left(\frac{a}{r}\right)^n \sum_{m=0}^{n}(C_{nm} \cos m\lambda + S_{nm} \sin m\lambda)\, P_{nm} (\cos \theta)\right] \tag{1}$$

where:

- kM geocentric gravitational constant
- a scaling parameter associated with the potential coefficients
- C,S fully normalized potential coefficients
- P_{nm} fully normalized associated Legendre functions

A more compact form for (1) may be obtained with the following substitution:

$$C_{nm}^{\alpha} = \begin{Bmatrix} C_{nm}, & \alpha=0 \\ S_{nm}, & \alpha=1 \end{Bmatrix} \tag{2}$$

$$Y_{nm}^{\alpha} (\theta,\lambda) = \begin{Bmatrix} P_{nm} (\cos \theta) \cos m\lambda, & \alpha=0 \\ P_{nm} (\cos \theta) \sin m\lambda, & \alpha=1 \end{Bmatrix} \tag{3}$$

Other forms are possible. Define Y_{nm} as follows:

$$Y_{nm} (\theta,\lambda) = P_{n|m|}(\cos \theta) \begin{Bmatrix} \cos m\lambda, & m\geq 0 \\ \sin |m|\lambda, & m<0 \end{Bmatrix} \tag{3A}$$

Let $C_{n|m|} = \begin{Bmatrix} C_{nm}, & m \geq o \\ S_{n|m|}, & m < o \end{Bmatrix}$

then

$$V(r,\theta,\lambda) = \frac{kM}{r}\left[1 + \sum_{n=2}^{\infty} \left(\frac{a}{r}\right)^n \sum_{m=-n}^{n} C_{nm} Y_{nm}(\theta,\lambda)\right] \tag{4A}$$

We now can write (1) as:

$$V(r,\theta,\lambda) = \frac{kM}{r}\left[1 + \sum_{n=2}^{\infty} \left(\frac{a}{r}\right)^n \sum_{m=0}^{n} \sum_{\alpha=0}^{1} C^{\alpha}_{nm} Y^{\alpha}_{nm}(\theta,\lambda)\right] \tag{4}$$

We will use both forms (i.e. 4 and 4A) in this paper.

We now can define the disturbing potential T at the point r,θ,λ:

$$T(r,\theta,\lambda) = V(r,\theta,\lambda) - U(r,\theta,\lambda) \tag{5}$$

where U is a reference potential, that is usually implied by a rotationally symmetric, equipotential ellipsoid. Assuming the mass of the reference ellipsoid and the earth are the same we have:

$$T(r,\theta,\lambda) = \frac{kM}{r} \sum_{n=2}^{\infty} \left(\frac{a}{r}\right)^n \sum_{m=0}^{n} \sum_{\alpha=0}^{1} C^{\alpha}_{nm} Y^{\alpha}_{nm}(\theta,\lambda) \tag{6}$$

where the C^{*}_{21} coefficients have the reference field coefficients removed.

2.1.1 Spherical Potential Coefficients and Gravity Anomalies

Numerous discussions have taken place in the literature, in the past few years, on the fundamental boundary conditions expressing anomalies as a function of the disturbing potential coefficients. In this discussion we follow Pavlis (1988).

We define the "surface free-air anomaly" as:

$$\Delta g = |\vec{g}_p| - |\vec{\gamma}_q| \tag{7}$$

where $|\vec{g}_p|$ is the magnitude of observed gravity and $|\vec{\gamma}_q|$ is the magnitude of theoretical gravity at the point on the telluroid. In terms of the disturbing potential we can write:

$$\Delta g = -\left[\frac{\partial T}{\partial h}\right]_Q + \frac{1}{\gamma_Q}\left[\frac{\partial \gamma}{\partial h}\right]_Q T_Q - \frac{1}{\gamma_Q}\left[\frac{\partial \gamma}{\partial h}\right] \Delta W + \varepsilon_p \tag{8}$$

where :

$$\Delta W = W_p - U_q = W_o - U_o \tag{9}$$

W_o = geoid potential

U_o = surface potential of the ellipsoid

and:

$$\varepsilon_p = \left[(1 - \cos\theta)\frac{\partial W}{\partial h} - \frac{\xi}{M}\frac{\partial W}{\partial \phi} - \frac{\eta}{N\cos\phi}\frac{\partial W}{\partial \lambda}\right]_P \tag{10}$$

where θ is the total deflection of the vertical and ξ and η are the usual meridean and prime vertical components. The height h, is in the direction of the ellipsoidal normal. The geometry associated with the boundary condition is shown in Figure 1 (Pavlis, ibid, Figure 3).

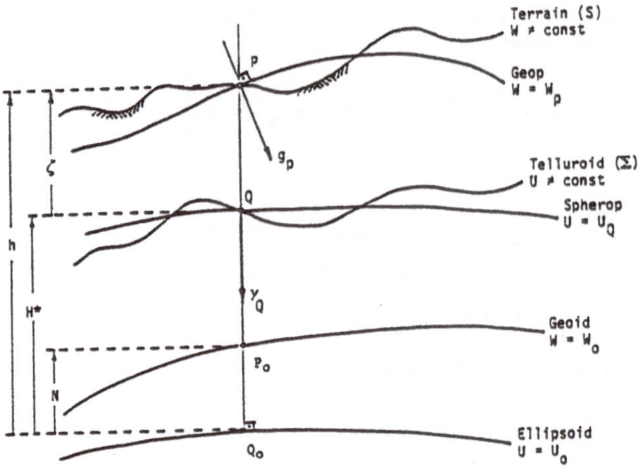

Figure 1. The Geometry Associated with Boundary Condition (10) (Pavlis, 1988)

Taking into account the definition of the normal potential of the ellipsoid one can show that (8) can be written in the following form:

$$\Delta g = -\left[\frac{\partial T}{\partial r}\right]_Q - \frac{2}{r_Q} T_Q + (\varepsilon_h)_Q + (\varepsilon_\gamma)_Q + \frac{2}{r_Q} \Delta W$$

$$- \left[6J_2 \frac{a^2}{r^3} P_2(\cos\theta_Q) - \frac{3\omega^2 r_Q^2}{kM} \sin^2\theta_Q\right] \Delta W + \varepsilon_p \tag{11}$$

where:

$$\varepsilon_h = e^2 \sin\theta \cos\theta \left[\frac{1}{r}\frac{\partial T}{\partial\theta}\right] \tag{12}$$

$$\varepsilon_\gamma = \left[6J_2 \frac{a^2}{r^3} P_2(\cos\theta) - \frac{3\omega^2 r_Q^2}{kM} \sin^2\theta\right] T \tag{13}$$

If $\Delta W(W_0-U_0)$ were zero and we were to neglect ε_h, ε_γ, ε_p the resultant equation would be the usual spherical approximation of the boundary condition. If we substitute (6) into (11) we can write:

$$\Delta g = -\frac{k\delta M}{r_Q^2} + \frac{kM}{r_Q^2} \sum_{n=2}^{\infty} (n-1) \left(\frac{a}{r_Q}\right)^n \sum_{m=-n}^{n} C_{nm} Y_{nm}(\theta,\lambda)$$

$$+ (\varepsilon_h)_Q + (\varepsilon_\gamma)_Q + \frac{2}{r}\Delta W - \left[6J_2 \frac{a^2}{r^3} P_2(\cos\theta) - \frac{3\omega^2 r^2}{kM} \sin^2\theta\right]_Q \Delta W + \varepsilon_p$$

where $\delta kM = k(M-M')$ with M = mass of the actual earth and M′ is the mass of the reference ellipsoid.

It is beyond this discussion to examine the small terms given in (14) (i.e. ε_h, ε_γ, ε_p). Pavlis (ibid) and Cruz (1986) have shown how these terms can be represented in a spherical harmonic series based on approximate values of the potential coefficients. Since these quantities are small and long wave length in nature they will be regarded as known for further discussion.

For many applications the anomaly given as the observable is an area mean anomaly usually defined in the following sense:

$$\overline{\Delta g}_{ij} = \frac{1}{\Delta\sigma_{ij}} \iint_{\sigma ij} \Delta g \sin\theta \, d\theta d\lambda \tag{15}$$

Applying (15) to (14) we have (Pavlis, 2.104):

$$\overline{\Delta g}_{ij} = \frac{1}{\Delta\sigma_{ij}} \frac{GM}{\overline{r}_{ij}^{2}} \sum_{n=2}^{\infty} (n-1) \left(\frac{a}{\overline{r}_{ij}}\right)^{n} \sum_{m=-n}^{n} C_{nm} \, IY_{nm}$$

$$+ \, I\hat{E}_{\ell}^{ij} + I\hat{E}_{\gamma}^{ij} + a_1 \, G\delta M + a_2 \, \Delta W + IE_{p}^{ij} \tag{16}$$

where:

$$\Delta\sigma_{ij} = \int_{\theta_i}^{\theta_{i+1}} \sin\theta d\theta \int_{\lambda_j}^{\lambda_{j+1}} d\lambda$$

$$IY_{nm} = \int_{\lambda_j}^{\lambda_{j+1}} \begin{Bmatrix} \cos m\lambda, & m\geq 0 \\ \sin m\lambda, & m<0 \end{Bmatrix} d\lambda \int_{\theta_i}^{\theta_{i+1}} P_{nm}(\cos\theta)\sin\theta d\theta$$

$$a_1 = -\frac{1}{\overline{r}_{ij}^{\,2}} \tag{17}$$

$$a_2 = \frac{2}{\overline{r}_{ij}} - \left[3J_2 \frac{a^2}{\overline{r}_{ij}^{\,3}} + \frac{\omega^2 \overline{r}_{ij}^{\,2}}{kM}\right](\cos^2\theta_i + \cos\theta_i \cos\theta_{i+1} + \cos^2\theta_{i+1}) + 3\left(J_2 \frac{a^2}{\overline{r}_{ij}^{\,3}} + \frac{\omega^2 \overline{r}_{ij}^{\,2}}{kM}\right)$$

The values for the other terms represent the integrated expressions for the correction terms. In writing (16), the variation of r_{ij} within the $\Delta\sigma$ block is neglected.

Two points need to be made before the final expression is written. First, the $a_1 \, k\delta M + a_2\Delta W$ term acts essentially as a constant which can be written in the following form:

$$a_1 k\delta M + a_2\Delta W = -\frac{k\delta M}{\overline{r}_{ij}^{\,2}} + \frac{2}{\overline{r}_{ij}} \Delta W \tag{18}$$

Second, the anomalies computed from (16) refer to an earth with no atmosphere. Since the atmosphere is implicitly considered in the computation of the terrestrial anomaly a suitable correction must be applied to (10) or to (7). If we wish (16) to yield the anomaly comparable to the terrestrial anomaly we must <u>add</u> the atmospheric correction, δg_A, to the terrestrial anomaly. Using (17) we now can write (16) as (Pavlis, 2.142):

$$\overline{\Delta g}_{ij} + \overline{\delta g}^{\S}_{ij} = \frac{1}{\Delta\sigma_{ij}} \frac{kM}{\overline{r}_{ij}^{2}} \sum_{\substack{n=0 \\ n\neq 1}}^{\infty} (n-1) \left(\frac{a}{\overline{r}_{ij}}\right)^{n} \sum_{m=-n}^{n} C_{nm} \, IY_{n|m|} + a_3 \, \Delta W$$

where:

$$\overline{\delta g}^{\S}_{ij} = \delta g_A^{ij} - (I\overset{n}{E}_A + I\overset{n}{E}_{\gamma} + I\overset{n}{E}_p)^{ij} \tag{20}$$

We note that the left hand side of (20) is regarded as known and that (20) represents a precise formulation of the boundary condition in terms of the observed anomaly and potential coefficients. Also note that it is possible to have different ΔW values for different vertical datums.

2.1.2 Spherical Harmonics and Orthogonality Relationships

Consider a form of (14) where all the terms on the right-hand side, except the summation term, are brought to the left-hand side to define a modified anomaly:

$$\Delta g + \delta g_{ij} = \Delta g^* = \frac{kM}{r_Q^2} \sum_{n=2}^{\infty} (n-1) \left[\frac{a}{r_Q}\right]^n \sum_{m=-n}^{n} C_{nm} Y_{n|m|} (\theta,\lambda) \tag{21}$$

Now let $r_Q = a$ so that we regard the anomalies Δg^* given on a sphere of radius a. Then (21) becomes:

$$\Delta g^* = \frac{kM}{a^2} \sum_{n=2}^{\infty} (n-1) \sum_{m=-n}^{n} C_{nm} Y_{n|m|} (\theta,\lambda) \tag{22}$$

We can apply the usual spherical harmonic orthogonality relationship (Heiskanen and Moritz, 1967, page 31) to calculate the potential coefficients given Δg^*. We have:

$$C_{nm} = \frac{1}{4\pi \frac{kM}{a^2} (n-1)} \iint_\sigma \Delta g^* Y_{nm}(\theta,\lambda)d\sigma \tag{23}$$

where σ is now the sphere of radius a. The evaluation of (23) requires Δg^* to be given in a continuous fashion. In practice mean Δg^* values are used which requires (23) to be approximated by a discrete form such as:

$$C_{nm} = \frac{1}{4\pi \frac{kM}{a^2} (n-1)q_n} \sum_{i=0}^{N-1} \sum_{j=0}^{2N-1} \overline{\Delta g}^*_{ij} \iint_{\sigma_{ij}} Y_{n|m|} (\theta,\lambda)d\sigma \tag{24}$$

where q_n is a quantity dependent on n and the block size (σ_{ij}), that is designed to reduce the approximation in going from (23) to (24). N is the number of latitude belts in which the equiangular anomalies are given. Various techniques have been designed to obtain optimum values of q_n. Colombo (1981, p.76) suggested that the quadrature weights q_n be set as follows:

$$q_n = \beta_n^2; \quad 0 \leq n \leq N/3$$

$$q_n = \beta_n; \quad N/3 < n < N$$

$$q_n = 1; \quad n > N \tag{25}$$

where $N = 180^\circ/\theta^\circ$, θ is the block size, and β_n is the Pellinen smoothing operator given by:

$$\beta_n = \frac{1}{1-\cos\psi_0} \frac{1}{\sqrt{2n+1}} \left[P_{n-1}(\cos\psi_0) - P_{n+1}(\cos\psi_0)\right] \tag{26}$$

where ψ_0 is the radius of a spherical cap whose area is the same as that of a block whose sides are θ°. Specifically:

$$\sin\left[\frac{\psi_0}{2}\right] = \left[\frac{\theta \sin\theta}{4\pi}\right]^{1/2} \tag{27}$$

Pavlis (ibid) described some assumptions that lead to an alternate determination of q_n. Specifically he (eq.4.46) suggested q_n is degree and order dependent:

$$q_{nm} = \left[\sum_{i=0}^{N-1} \sum_{j=0}^{2N-1} \frac{(Y_{nm}^{ij})^2}{\Delta\sigma_{ij}} \right]^{-1} \tag{28}$$

The calculation of q_{nm} is more complex than the q_n computation but need be done only once after the block size and maximum degree of expansion is given. Numerical tests by Pavlis suggest that (28) gives slightly better results than (25). The best results, in terms of consistency between given potential coefficients and anomalies is to regard the $\overline{\Delta g}^*$ values as referring to the center point of the cell and expressing (24) in the following form:

$$C_{nm} = \frac{1}{4\pi \frac{kM}{a^2}(n-1)\ \beta_n} \sum_{i=0}^{N-1} \sum_{j=0}^{2N-1} Y_{nm}(\theta,\lambda)\ \Delta g^* \sigma_{ij} \tag{29}$$

Additional studies are needed to understand proper ways to evaluate quadrature type formulas using block data and that contains contributions from all parts of the frequency spectra and that has noise (errors) associated with it.

2.2 Ellipsoidal Harmonics

The earth is closer in shape to an ellipsoid than a sphere. Consequently one must ask if ellipsoidal harmonics are more advantageous to use than spherical harmonics. For potential representations for satellite geodesy (orbit calculations), spherical harmonics are still the obvious choice. However, when data is given on an ellipsoid (such as some interpretations of terrestrial anomalies), ellipsoidal harmonics may play a more appropriate role. Discussions on the use of ellipsoidal harmonics may be found in Jekeli (1981, 1988) and Gleason (1988).

We start with the ellipsoidal coordinates u, δ, and λ where:

u is the semi-minor axis of a confocal ellipsoid;

δ is the reduced colatitude;

λ is the longitude.

These coordinates are related to the rectangular coordinates by:

$$\begin{aligned} x &= \sqrt{u^2+E^2}\ \sin\delta\ \cos\lambda \\ y &= \sqrt{u^2+E^2}\ \sin\delta\ \sin\lambda \\ z &= u\ \cos\delta \end{aligned} \tag{30}$$

where E = ae. A function, F, satisfying Laplace's equation, in this ellipsoidal coordinate system can be written as (Gleason, ibid, eq(1.12)):

$$F(u,\ \delta,\ \lambda) = \sum_{n=0}^{\infty} \sum_{m=-n}^{n} \bar{\bar{f}}^{e}_{nm} \frac{\bar{Q}_{n|m|}\left(\frac{iu}{E}\right)}{\bar{Q}_{n|m|}\left(\frac{ib}{E}\right)} \bar{Y}_{n,m}\ (\delta,\lambda) \tag{31}$$

where:

$$\bar{Y}_{n,m}(\delta,\lambda) = \bar{P}_{n|m|}(\cos\delta)\begin{cases}\cos m\lambda \text{ if } m\geq 0\\ \sin m\lambda \text{ if } m<0\end{cases} \tag{32}$$

The $\bar{Q}_{n,m}$ are the fully normalized (in the usual sense as $\bar{P}_{nm}$) Legendre function of the second kind and $\bar{f}^e_{n,m}$ are real fully normalized ellipsoidal harmonic coefficients. Given the function F on the surface of the ellipsoid, and using the standard orthogonality relationships for $\bar{Y}_{nm}$ we can write:

$$\bar{f}^e_{n,m} = \frac{1}{4\pi}\iint_\sigma F(u=b,\ \delta,\ \lambda)\ \bar{Y}_{n,m}(\delta,\lambda)d\sigma \tag{33}$$

where $d\sigma = \sin\delta\ d\delta d\lambda$.

Using spherical harmonic coefficients the solution to Laplace's equation is:

$$F(r_p,\ \theta_p,\ \lambda_p) = \sum_{n=o}^{\infty}\sum_{m=-n}^{n}\left(\frac{a}{r_p}\right)^{n+1}\bar{f}^s_{n,m}\ \bar{Y}_{n,m}(\theta_p,\ \lambda_p) \tag{34}$$

If the potential function is given on the sphere of radius a the coefficients can be found using the usual orthogonality relationships:

$$\bar{f}^s_{n,m} = \frac{1}{4\pi}\iint_\sigma F(r=a,\ \theta,\ \lambda)\ \bar{Y}_{n,m}(\theta,\lambda)d\sigma \tag{35}$$

where $d\sigma = \sin\theta d\theta d\lambda$. The problem examined by Jekeli, Gleason and others was to relate the spherical and ellipsoidal coefficients recognizing the uniqueness of the potential outside some boundary surface. An operational conversion between $\bar{f}^s_{n,m}$ and $\bar{f}^e_{n,m}$ is derived by Gleason. He shows that (ibid, 1.22):

$$\bar{f}^s_{n,m} = \sum_{k=o}^{s'}\frac{1}{\bar{S}_{n-2k,|m|}\left(\frac{b}{E}\right)}L_{n,m,k}\ \bar{f}^e_{n-2k,m} \tag{36}$$

All terms are defined in Gleason with the L term computed by a recurssive relationship. Equation (36) is important to us because it will enable us to convert $\bar{f}^e$ values found from (33) to $\bar{f}^s$ values which are compatible with the satellite derived potential coefficients.

2.2.1 Ellipsoidal Harmonics and Gravity Anomalies

Let the gravity anomaly of interest be the radial component. From (14), considering (16) we have:

$$\Delta g^*_E = \Delta g_s + \delta g_A - \varepsilon_h - \varepsilon_\gamma - \varepsilon_p - \frac{d\Delta g}{dh}h \tag{37}$$

where the last term reduces the surface anomaly to the ellipsoid. If Δg^*_E were upward continued to the sphere of radius a we would have the Δg^* values given by (22) with the spherical potential coefficients given by (23).

Now write (21) in the following form:

$$\Delta g^*(r_p,\ \theta_p,\ \lambda_p) = \sum_{n=2}^{\infty}\left(\frac{a}{r_p}\right)^{n+2}\sum_{m=-n}^{n}\bar{g}^s_{n,m}\ \bar{Y}_{nm}(\theta,\lambda) \tag{38}$$

where

$$\bar{g}^s_{n,m} = \frac{kM}{a^2}(n-1)\ \bar{C}^s_{n,m} \tag{39}$$

Multiplying (38) by r_p we have:

$$r_p \; \Delta g^*(r_p,\theta_p,\lambda_p) = a \sum_{n=0}^{\infty} \left(\frac{a}{r_p}\right)^{n+1} \sum_{m=-n}^{n} \bar{g}^s_{n,m} \; \bar{Y}_{n,m} \; (\theta,\lambda) \tag{40}$$

The left hand side is a harmonic function which can be expressed in a spherical harmonic series (as it is now) or in an ellipsoidal harmonic series. The latter can be written as (Gleason, ibid, 2.7):

$$r_p \; \Delta g^*(u_p,\delta_p,\lambda_p) = a \sum_{n=0}^{\infty} \sum_{m=-n}^{n} \frac{\bar{S}_{n|m|}\left(\frac{u_p}{E}\right)}{\bar{S}_{n|m|}\left(\frac{b}{E}\right)} \; \bar{g}^e_{n,m} \; \bar{Y}_{n,m} \; (\delta_p,\lambda_p) \tag{41}$$

The coefficients "a $\bar{g}^e_{n,m}$" represent ellipsoidal harmonic coefficients related to the bounding ellipsoid. To calculate these coefficients we can use (33) where F is the left-hand side of (41) evaluated on the surface of the ellipsoid divided by a. We then can write:

$$\bar{g}^e_{n,m} = \frac{1}{4\pi a} \iint_{\sigma} r\Delta g^*(u{=}b,\delta,\lambda) \; \bar{Y}_{n,m} \; (\delta,\lambda) d\sigma \tag{42}$$

These ellipsoidal coefficients can be converted to the spherical coefficients using (36). The spherical potential coefficients can then be obtained from (39).

The technique discussed in this section represents an improvement over the spherical analysis where the anomaly on the ellipsoid was analytically upward continued to the bounding sphere of radius a. Although this procedure was done in terms of potential coefficient correction terms, the studies by Gleason indicate the procedure using ellipsoidal harmonics is more correct.

3.0 Data Definition

3.1 Satellite Data

As pointed out in the introduction we primarily view the satellite data in terms of the resultant normal equations formed from such data. The typical data types that one can find in a satellite alone solution include: optical tracking data; laser tracking data; Doppler observations, radar data, satellite to satellite tracking etc. In addition solutions may be carried out with satellite altimeter data but these solutions require methodology not needed in the solutions using the data previously described. The accuracy of these satellite solutions depends on many items including the distribution of satellite inclinations; data density, distribution and accuracy; observation models; and the general parametrization of the problem.

3.2 Terrestrial Gravity Data

The primary terrestrial data type we will use will be free-air gravity anomalies. To this point the anomalies have been defined by (7). We now consider a more precise formulation. Let H* be the normal height of the point P and H the orthometric height. Since H*-H is small we will approximate H* by H. Then normal gravity at the telluroid is:

$$\gamma_Q = \gamma_{Q0} + \left(\frac{d\gamma}{dh}\right)_{Q_0} H + \frac{1}{2!}\left(\frac{d^2\gamma}{dh^2}\right)_{Q_0} H^2 + - - - \tag{43}$$

where γ_{Q_0} is normal gravity on the ellipsoid. The surface anomaly is then:

$$\Delta g = g_P - \left[\gamma_{Q_0} + \left(\frac{d\gamma}{dh}\right)_{Q_0} H\right] - \frac{1}{2!}\left[\frac{d^2\gamma}{dh^2}\right]_{Q_0} H^2 \tag{44}$$

In practice almost all organizations neglect the H^2 term. Pavlis (ibid) has shown that the neglect of this term can cause systematic errors in the potential coefficient solutions which imply geoid errors on the order of 2 cm with a maximum error of 1.8 m in the Himalaya region. Fortunately it is possible to correct the given anomaly by the last term in (44).

The terrestrial data is usually given in 1°x1° mean values or 30′x30′ values. A variety of techniques, some complex, some simple, are used to evaluate the mean anomalies and their accuracy. The data are usually regarded as independent although it has been shown (Weber and Wenzel, 1982) that in fact there is an error correlation between the anomalies. This correlation is neglected in all combination solutions so far. Systematic errors due to vertical datum inconsistencies also cause errors, some of which can be modeled, but usually are not.

The location of 48,955 1°x1° anomalies based on terrestrial measurements only is shown in Figure 2 (Despotakis, 1986). The accuracy estimates range from ± 1 mgal to ± 62 mgal. There are an additional set of 5,684 anomalies that have been estimated in the early 1960's using geophysical correlation techniques. The location of these anomalies is shown in Figure 3. There is some evidence, to be discussed later, that there are long wavelength errors present in this geophysically predicted anomalies. This is why combination solutions are usually made with and without such anomalies so evaluations can be done to see which solutions provide the best overall results.

Computations have also started for 30′ data files (Despotakis, 1986). Land data is available for North America, Europe, Australia, India, New Zealand, portions of South America and Africa, Japan, and other regions. The coverage of such data will not be as complete as that of 1°x1° data. Coverage in the ocean areas will be fairly good due to the use of altimeter data to derive the anomalies.

Another data type not used so far in combination solutions is topographic heights. Improved elevations in 1°x1° and 30′x30′ cells have recently become available. Rummel et al (1988) has shown there is a good (greater than 0.6) correlation between the observed potential and that implied by a topographic/isostatic potential estimated with an Airy isostatic hypothesis with D = 30Km. Such information may be used to calculate gravity anomalies that could be used to replace the geophysically predicted anomalies and to infer anomalies in areas where no other information is available. Tests are now underway to evaluate the value of such data recognizing that the longwave length information will be missing from the topographic data but the short wavelenth may prove valuable.

4.0 Data Combination

4.1 General Principles

We derive a model where all parameters are estimated taking into account all data with correct accuracies and error correlations considered. This ideal statement is not followed in practice but various attempts are made to achieve it. The basic principle followed in combination solutions to date has been the least squares principle with some prior information taken into account on selected parameters such as potential coefficients. Other techniques are possible and a number are now under development and testing.

4.2 Least Squares Principles

We define in this section our basic mathematical models that will be applied in a subsequent section in several forms. We start with some definitions:

- F a set of functions relating observations and parameters;
- L_ℓ a set of observations,
- L_{x^o} a given set (approximate or observed) of parameters;
- V_ℓ a set of residuals to be added to L_ℓ to obtain the adjusted observations, $L_{\ell a}$;
- V_x a set of residuals to be added to L_{x^o} to obtain the adjusted parameters L_{xa};
- W the misclosure vector, $W=F(L_\ell,L_{x^o})$.

The mathematical model is written as:

$$F = F(L_{\ell a},\ L_{x^o}) = 0 \tag{45}$$

which is linearized to yield the observation equation:

$$B_\ell V_\ell + B_x V_x + W = 0 \tag{46}$$

where

$$B_\ell = \frac{dF}{dL_\ell};\ B_x = \frac{dF}{dL_x} \tag{47}$$

If we designate P_ℓ and P_x as the weight matrices for the observations and parameters respectively, the weighted least squares condition for solutions is:

$$V_\ell^T\ P_\ell\ V_\ell + V_x^T\ P_x\ V_x = \text{a minimum} \tag{48}$$

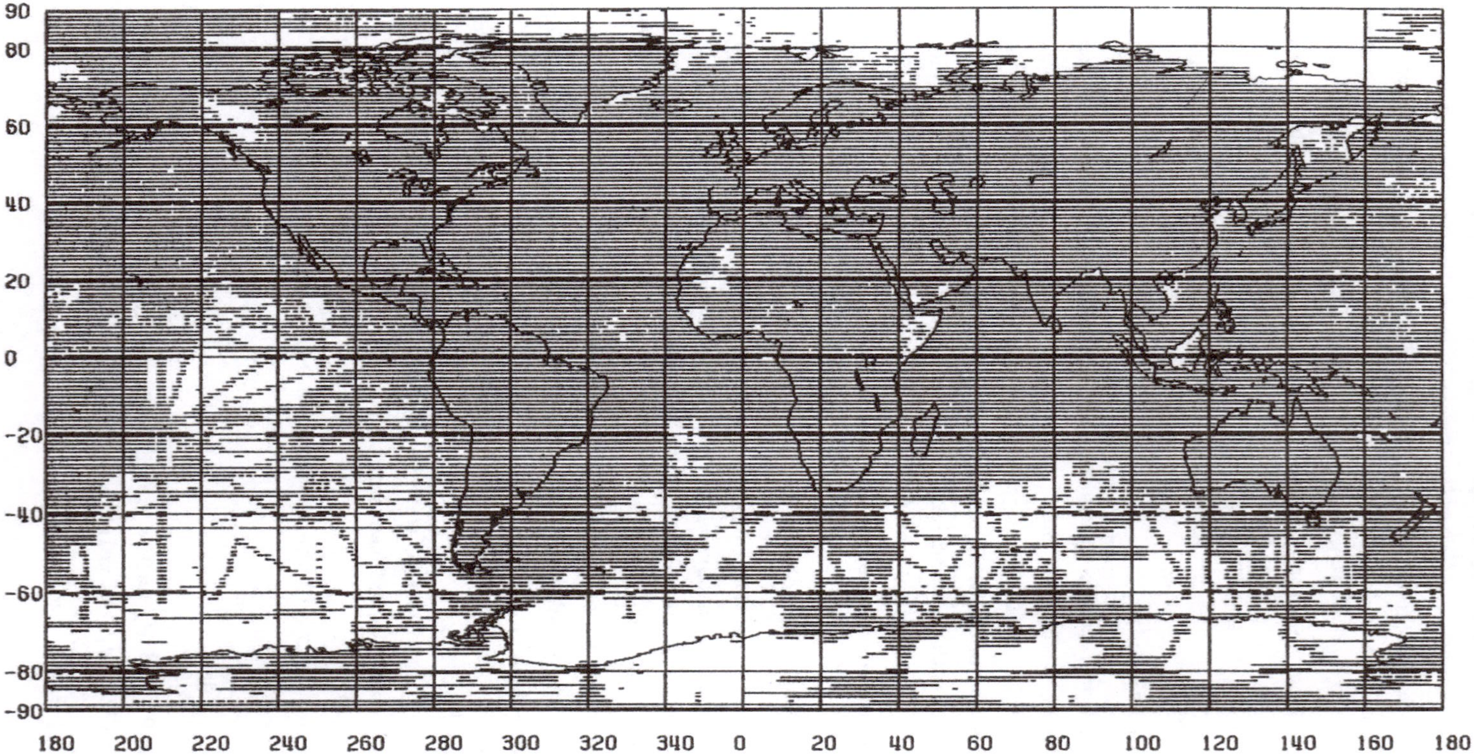

Figure 2. Location of 48955 1° x 1° Anomalies in the June 1986 Data Set.

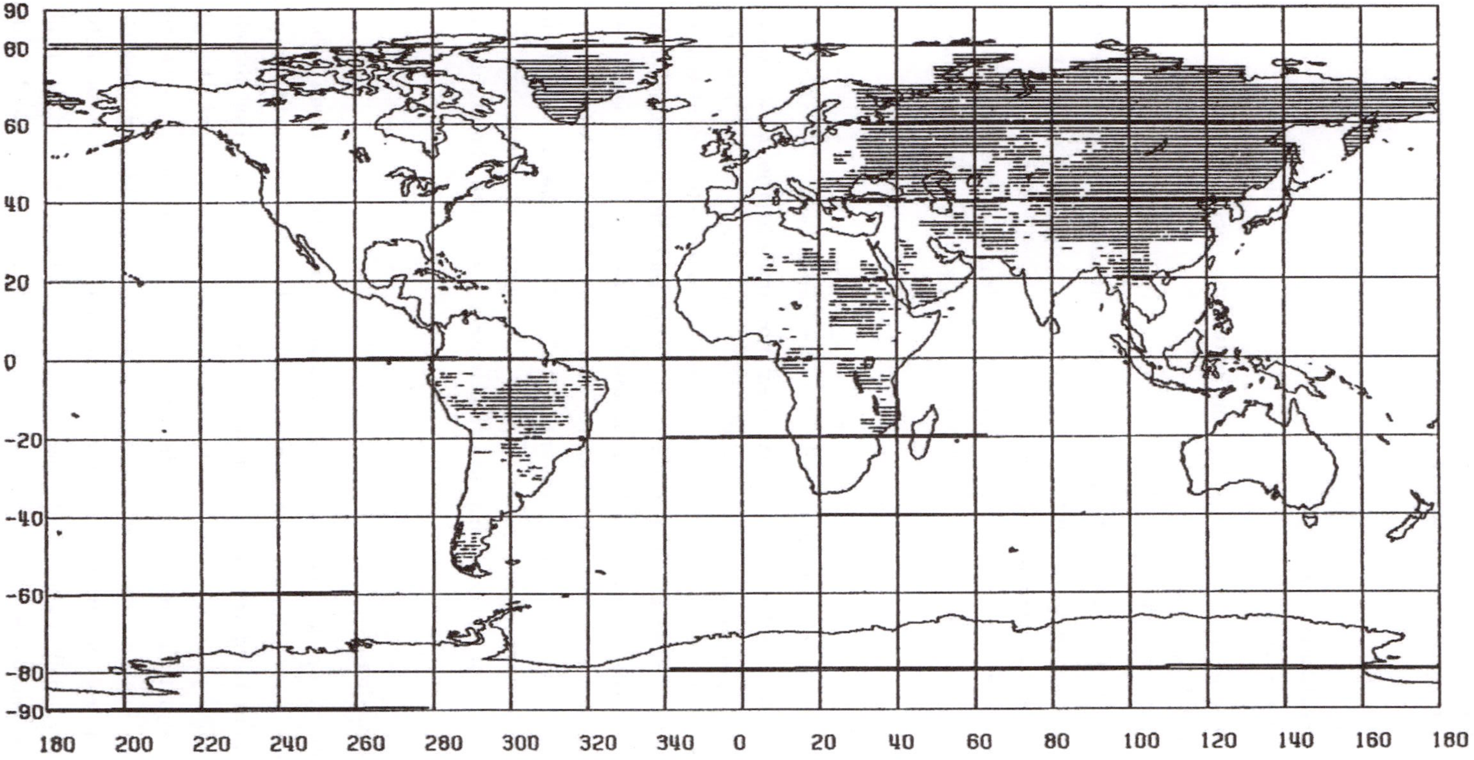

Figure 3. Location of 5684 1°x 1° Geophysically Predicted Anomalies in the June 1986 Data Set.

The solution for V_x is:

$$V_x = -(B_x^T M^{-1} B_x + P_x)^{-1} B_x^T M^{-1} W \tag{49}$$

where

$$M = B_\ell P_\ell^{-1} B_\ell^T \tag{50}$$

The observation residuals are:

$$V_\ell = -P_\ell^{-1} B_\ell^T M^{-1} (B_x V_x + W) \tag{51}$$

The error variance-covariance matrix for the solutions vector would be:

$$\Sigma_{xx} = m_0^2 (B_x^T M^{-1} B_x + P_x)^{-1} \tag{52}$$

where m_0^2 is the variance of unit weight.

These equations will be applied to two specific combination procedures shortly. However it is instructive to consider an alternate combination solution where the normal equations from various data sets are combined. Let N_i be the i th normal equations from a specified data type. Assuming that the variance of unit weight is the same for all data we can write the combined normal equations in the following form:

$$X = (\sum_i N_i + P_{xi})^{-1} (\sum_i U_i) \tag{53}$$

where P_{xi} is the a priori weight matrix for the parameters. In this case the normal equations for certain data types may not include the same parameters found in another data type. For example N_i terms from orbital analysis may include as parameters potential coefficients; station coordinates; tidal parameters; etc., while N_i from terrestrial gravity data may include potential coefficient parameters only. The merger of the individual normal equations must be done carefully and must be done with factors that would allow different weighting for different data types. The advantage of (53) is the inclusion in one simultaneous adjustment of all possible data. Alternate solutions are possible where all but common parameters are eliminated from the individual normal equations.

4.3 Optimal Estimation

Our concept of optimal estimation is analogous to least squares collocation estimation where we recognize pre-existing information on the primary gravity field parameters to be determined, and the covariances between the signals that constitute the observations. The usual collocation solutions require the solution of systems of equations whose size is equal to the number of observations. Colombo (1981) has shown that proper ordering of the parameters and observations can lead to structured covariances matrices that can be solved in an efficient way. Still substantial computational effort is needed to implement these optimal estimation solutions. In some cases it has been found that such solutions smooth the resultant gravity field parameters too much. Additional study is needed to define more precisely the best way to implement optimal estimation procedures in combination solution.

5.0 Observation Equation Formation

In this section we will examine two models that can be used for the combination of satellite and gravity data. Each model will lead to a set of observation equations that will lead to a set of normal equations.

5.1 Combination Procedure- Method A

The mathematical structure of this method is established by forming the difference between a set of spherical potential coefficients, $L_x{}^o$, and an estimated, $L_{x_c}^{*}$: computed through a global set of gravity anomalies. In our discussions in section 2.2.1 we defined a procedure of anomaly correction (eq. (37)); calculation of ellipsoidal coefficients (eq. (42)); the calculation of the spherical coefficients (eq. (36)) and finally the spherical potential coefficients (eq. (39)). We write for the adjusted case:

$$F = L_x{}^a - L_{x_c}^{a} \tag{54}$$

We let $L_x{}^o$ be the spherical potential coefficients estimated from satellite data. We then calculate $L_x{}^e$ using the procedure described above. The misclosure vector would then be:

$$W = L_x{}^o - L_x{}^c \tag{55}$$

We also have:

$$B_x = \frac{dF}{dL_x{}^o} = I; \quad B_\ell = -\frac{dL_x{}^c}{dL_\ell} \tag{56}$$

The L_ℓ values are the gravity anomalies Δg^* and the $L_x{}^c$ are the spherical potential coefficients. The evaluation of B_ℓ is complicated by the conversion from ellipsoidal harmonics to spherical harmonics. An approximate form (sufficient for observation equation coefficients) can be formed by recognizing $L_{n,m,k}$ is $0(e^2)$ so we will consider only the care of k=0. From (36) we would have:

$$\bar{f}^s_{n,m} \approx \frac{\bar{f}^e_{n,m}}{\bar{S}_{n,|m|}\left(\frac{b}{E}\right)} \tag{57}$$

We now evaluate (56) using (42), (57), and (39). We have as an element of B_ℓ:

$$[B_e] = \frac{-1}{4\pi\frac{kM}{a^2}(n-1)} \; \frac{1}{\bar{S}_{nm}\left(\frac{b}{E}\right)} \left(\frac{r}{a}\right) \bar{Y}_{nm}(\delta,\lambda)\,d\sigma \tag{58}$$

One may evaluate the coefficients in (58) by integrating over the block σ_{ij} or using the center point evaluation discussed earlier.

The solution vector is now found from (49):

$$V_x = -((B_\ell P_\ell^{-1} B_\ell^T)^{-1} + P_x)^{-1} (B_\ell P_\ell^{-1} B_\ell^T)W \tag{59}$$

These V_x are added to the initial satellite implied, spherical potential coefficients to obtain the adjusted potential coefficients. The anomaly residuals may be computed from (51) or the equivalent expression:

$$V_\ell = P_\ell^{-1} B_\ell^T P_x V_x \tag{60}$$

These values are added to the initial Δg^* values to obtain the adjusted Δg^* values on the ellipsoid.

In this procedure we assume that the surface anomalies are reduced to the ellipsoid using the following approximation:

$$\Delta g_E = \Delta g_S - \frac{d\Delta g}{dh} h \tag{61}$$

where h is the ellipsoidal height of the anomaly, or mean compartment if, as would be expected, mean anomalies are used. Other correction terms as shown in (37) must also be applied. Wang (1988) has carried out a global computation of the correction the term working with the Molodensky g_1 term writing:

$$\Delta g_E = \Delta g_S + g_1 \tag{62}$$

Wang used 5′ elevation data and Fourier techniques to calculate g_1 terms for both 30′ and 1° anomalies on a global basis. The root mean square g_1 value is ± 1.5 mgals for 30′ cells and ± 1.2 mgals for 1° cells.

5.2 Combination Procedure– Method B

We first express (18) in the following form:

$$\overline{\Delta g}^*_{ij} = \frac{1}{\Delta\sigma_{ij}} \frac{kM}{\bar{r}^2_{ij}} \sum_{\substack{n=0 \\ n\neq 1}}^{\infty} (n-1) \left(\frac{a}{\bar{r}_{ij}}\right)^n \sum_{m=-n}^{n} C_{nm} \, IY_{nm} \tag{63}$$

The $\overline{\Delta g}^*$ anomalies represent the mean anomalies after correction for the atmospheric correction and other terms related to a precise interpretation of the boundary condition. (See eq. 20). Note that (63) involves a summation to ∞ which clearly is not possible. It is necessary to truncate the series to some N_{max}. One choice of N_{max} would be to have it compatible with the frequency content of the $\overline{\Delta g}$ values. If such anomalies are given in $\theta°$ cells, N_{max} could roughly be given as $180°/\theta°$. For anomaly blocks given in 1° cells, for example, the number of parameters to be solved for would be too large for most computers. (For example, if N_{max} = 180, there would be 32761 potential coefficients to be estimated). However such solutions are possible if the parameters are ordered in an optimum way and only near diagonal terms are considered in the solution of the system of equations.

An alternative selection of N_{max} is one that makes the number of potential coefficients to be solved for consistent with that found in the analysis of the satellite data. To do this one needs to remove the signal from $N_{max}+1$ to ∞ with the latter being replaced by some existing high degree potential coefficient model. Let dg_{hf} be defined as follows:

$$(\overline{dg}_{hf})_{ij} = \frac{1}{\Delta\sigma_{ij}} \frac{kM}{\bar{r}_{ij}^2} \sum_{n=N_{max}+1}^{\bar{N}} (n-1) \left(\frac{a}{\bar{r}_{ij}}\right)^n \sum_{m=-n}^{n} C_{nm}\ IY_{nm} \tag{64}$$

We then use (63) with ∞ replaced by N_{max} and $\overline{\Delta g}*$ replaced by:

$$dg_{ij} = \overline{\Delta g}^*_{ij} - (\overline{\delta g}_{hf})_{ij} \tag{65}$$

The advantage of this procedure is the reduction of the effect of the high frequency values in $\Delta g*$ on the estimated parameters.

Our general mathematical model (see Section 4.2) now becomes:

$$F = L_\ell - f(L_x) \tag{66}$$

where L_ℓ is the observed gravity anomaly (with suitable corrections (see 65)) and L_x are the potential coefficients derived from a satellite analysis. An element of B_x (for degree n and order m would be):

$$[B_r] = -\frac{1}{\Delta\sigma_{ij}} \frac{kM}{\bar{r}_{ij}^2} (n-1) \left(\frac{a}{\bar{r}_{ij}}\right)^n IY_{nm} \tag{67}$$

Since the observable is the anomaly we have B_ℓ=I. The solution vector follows from (49):

$$V_x = -(B_x^T P_\ell B_x + P_x)^{-1} B_x P_\ell W \tag{68}$$

where $W = [dg_{ij} - dg^c_{ij}]$ where the C designates the anomaly computed from the a priori potential coefficients. Note that the residuals on the anomalies can be computed from (51). However they will reflect only the changes in degrees 0 to N_{max} in the potential coefficients. To obtain the adjusted $\overline{\Delta g}*$ value the high frequency effects must be added back in. Unfortunately such effects may be incorrect and the $\Delta g*$ values will not reflect an adjusted value in all frequency bands.

A modification of the above procedure takes place when (53) is used for the normal equations. In this case a simultaneous solution is made for all parameters of interest for satellite data and for terrestrial data. In this case the normal equation contribution from terrestrial gravity data would be:

$$(B_x^T P_\ell B_x) X_g = - B_x P_\ell W_a \tag{69}$$

where X_g represents the potential coefficient parameters.

Pavlis (ibid) has studied the form of the normal equations generated from the terrestrial data (with and without the geophysically predicted anomalies) considering different parameter ordering schemes. Some schemes minimize the band width and should be used if a rigorous inversion is not being done. As the terrestrial data increases the normal equations become increasingly diagonal dominant. However the current real world data coverage yields a situation where the coefficients recovered from a terrestrial solutioon only may be highly correlated. Additional discussion on the structure of the normal equations for specified data coverage may be found in Bosch (1987).

5.3 Comment

Other combination solutions have been described in the literature. Sjoberg (1981) has discussed the spectral combination of the coefficients from satellite and terrestrial data. Schaffrin and Middel (1987) have discussed robust estimation procedures. All the combination methods suffer common problems of approximate weighting. All solutions made to date regard the anomalies to be independently estimated. As noted earlier. there are error correlatioons that cause problems with the combination solution. One ad hoc method to reduce the problem was used by Rapp and Cruz (1986) where they multiplied the given anomaly standard deviations by 2.5 and scaled the resultant standard deviation to fall within the interval 8 to 38 mgal. Such scaling seems to have worked in practice.

Another point relates to the computer time needed for the formation of the normal equations and the subsequent inversion. If we include the geophysically predicted anomalies there are 48955 values in the June 86 data set. The calculation of the normals can take a substantial time on scalar processors and only vector processors will yield reasonable times for N_{max} values of interest for current solutions. Pavlis has provided the computer times on a CRAY XMP 2/4 computer (Table 1) when forming the normal equations using 1°x1° anomalies in the formation of the anomaly only normal equations of the Method B combination solution.

Table 1. Cray Computer Times Related to Method B For Various Degrees of Expansion Using 1°x1° Anomalies.

N_{max}	Number of Unknowns	CPU Time (seconds) Normals	Inversion
24	622	432	1.86
36	1366	2119	16.40
50	2598	8644	168.70

The times in Table 1 need to be multiplied by about 60 to obtain the equivalent times on a scalar, IBM 3081D, machine. It is clear that such processing can only be done on supercomputers.

6.0 The Development of High Degree Potential Coefficient Models

To this point we have discussed methods that can be used to combine satellite information and terrestrial gravity data. But the rigorous solutions have been limited to the highest degree present in the satellite derived fields. We now examine the computation of high degree fields whose frequency content is compatible with that of the original data.

Consider combination Method A described in Section 5.1 where the correction to the coefficients are given by (59) and the corrections to the "observed" anomalies are given by (60). We have

$$L_{x^a} = L_{x^o} + V_x \tag{70}$$

where L_{x^a} are the adjusted <u>spherical</u> potential coefficients and L_{x^o} are the approximate (a priori) values. The adjusted anomalies, on the ellipsoid, are found by adding the residuals from (60) to the corrected anomalies on the ellipsoid. Using (20) and (62) we have for the adjusted anomalies:

$$\overline{\Delta g}_a^* = \overline{\Delta g}_{ij} + \delta g_A^{ij} - (\hat{IE}_A + \hat{IE}_\gamma + \hat{IE}_p)^{ij} + g_1 + [V_\ell] \tag{71}$$

or

$$\overline{\Delta g}_a^* = \overline{\Delta g}_{ij} + \overline{\delta g}_{ij}^e + g_1 + [V_\ell] \tag{72}$$

We now use these values in (42) to find $\bar{g}_{n,m}^e$ which are then used in (36) to find the corresponding spherical coefficients. These values can then be used in (39) to find the corresponding spherical harmonic potential coefficients. These coefficients should be the same as the adjusted coefficients from (70). This procedure removes the need for the ellipsoidal correction terms proposed by Cruz (1986) and used in Rapp and Cruz (1986, ibid).

The key equation in this evaluation is (42). The actual evaluation is done in a manner analogous to that used in (24) or (29). Decisions must be made on the q_n factors and the integration of the Y_{nm} functions. The HARMIN type of calculation suggested by Colombo (1981) can be used provided the integration in (42) is done in the (δ,λ) coordinate system. Using 1°x1° anomalies the expansion can be taken to degree 180 while 30′ data can yield expansions to degree 360. In both cases one finds that lack of data in some areas will mean that the expansions will not, in reality, have the same spectral content in all regions.

These methods have the unfortunate attribute of having no accuracy estimates except for those coefficients that were part of the original adjustment. Studies need to be made to show efficient, and realistic ways to obtain the accuracies of the potential coefficients at high degree.

7.0 The Role of Satellite Altimeter Data

Satellite altimeter data represents a measurement from a satellite to the instantaneous ocean surface. This data can be used in orbit determination procedures or by itself as a means of estimating the gravity anomaly field in the ocean areas. A dominant limitation on the use of these altimeter measurements is in the accuracy of the original satellite orbit, or in the ability of the orbit determination process to recover accurate orbits. We now consider, in general terms, the use of altimeter data for gravity field improvement.

We first consider the use of altimeter data in the orbit determination process following Marsh et al. (1988b). Let A be the observed range from the satellite to the ocean surface. The altimeter observation equation is then represented in the following form:

$$R - (R_e + N + \zeta + T) + B \cong A + v \tag{73}$$

where:

- R is the radial distance of the satellite to the center of mass of the Earth. This distance is a function of the usual orbit parametization including station coordinates and gravity field parameters which are adjusted in the solution;
- R_e is the distance from the center of mass to the sub-satellite point;
- N is the geoid undulation computed from the potential coefficients which are adjusted in the solution;
- ζ is the sea surface topography as parameterized by (e.q.) a low degree spherical harmonic expansion whose coefficients are estimated;
- T is the tidal correction;
- B is the altimeter bias, including an equatorial radius correction, which is adjusted;
- A is the altimeter measurement corrected for various environmental effects with high degree geoid information removed so that the frequency content of A is the same as in the potential coefficient parametization.

The observation equations from (73) yield a set of normal equations that are combined with normal equations from other data (both satellite and terrestrial gravity) that then leads to a general parameter estimation using (53).

An alternate method for using the altimeter data is to first remove as much remaining orbital error as possible. Typically this has been done using cross-over methods solving for bias' and tilts in altimeter arcs. This procedure is not completely satisfactory because it does not completely remove geographically correlated errors (Engelis, 1987). After the adjustment the sea surface heights can be corrected, in only an approximate way, to geoid undulations. This conversion is approximate only because sea surface topography is inadequately known. The resultant heights can then be used to derive gravity anomalies with a spacing consistent (in some reasonable way) with the altimeter track spacing. Rapp (1986) describes such conversions that yielded anomalies on a 0.125 grid. These anomalies can be used to fill in empty areas in the oceans or to completely replace existing terrestrial data in the oceans by the more reliable altimeter data. In this context more reliable refers to mean anomalies (such as 1°x1° or 30′x30′) and not necessarily point values acquired from modern ship data. However the accuracy of point anomalies from altimeter data increases with data density, but the spectral content of the altimeter data will be more limited than that of ship data.

The use of altimeter derived anomalies is especially useful in developing an anomaly data base for the Method A Combination Solution (Section 5.1). For the Method B (Section 5.2) solution the geoid heights could be directly used provided the spectral content is made compatible with the potential parametization. This procedure is similar to the method used by Wenzel (1985) in developing the GPM2 model.

In the use of altimeter data one must be careful not to include duplicate information. This would happen if one used altimeter data in orbit estimation in combination with anomalies calculated from the altimeter data.

8.0 Comparison of Satellite and Terrestrial Gravity Anomaly Fields

In Section 1 we described the very general ideas of gravity determinations from satellite and terrestrial measurements. It is instructive to compare such measurements before carring out any combination solution to detect data or model errors. Since satellite gravity data is more band limited than terrestrial data the most informative comparisons can be made in the spectral domain. The most recent such comparisons have been described by Pavlis (1988) where the satellite data was GEMT1 (complete to degree 36) (Marsh et al, 1988a) and the terrestrial data was the June 1986 1°x1° data set.

Pavlis took the 1°x1° data and estimated a set of potential coefficients using the method described in Section 5.2. The anomalies implied by the coefficients were then compared to the anomalies implied by GEMT1. Figure 4 shows the location where differences exceeded 10 mgals when geophysical anomalies are included in the solutions, and Figure 5 shows the comparison when the geophysical anomalies are included. Comparison of Figure 4 with the location of the geophysically predicted anomalies (Figure 3) shows a moderately good correlation indicating some inconsistency between the geophysically predicted anomalies and the satellite implied anomalies.

These anomaly discrepencies are also reflected in undulation discrepencies. These discrepencies reach 20 m in east central Africa and 10 m in the Himalaya Mountain regions. Tests described by Pavlis for Africa shed strong doubt on the credibility of the terrestrial gravity data in portions of Africa and Asia.

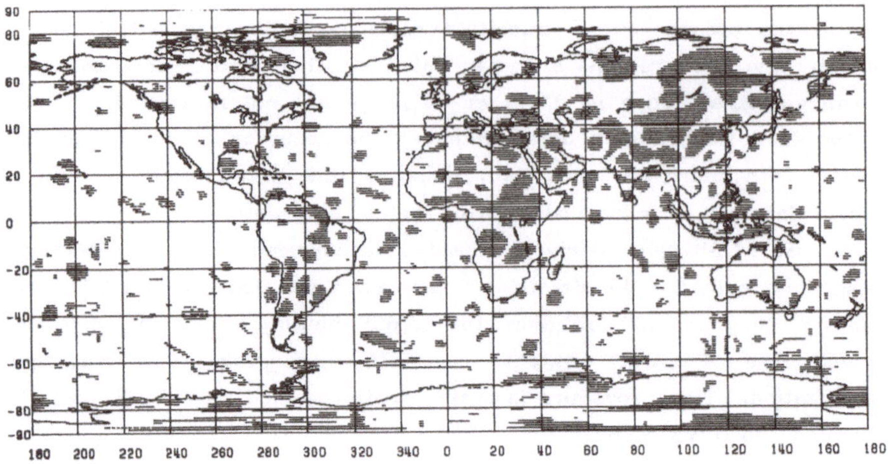

Figure 4. Location of 9745 1°x1° Blocks Where $|\Delta g(GEMT1) - \Delta g(A20)|$ Exceeds 10 mgals. (Locations Corresponding to Geophysically Predicted Anomalies are included in the Comparison).

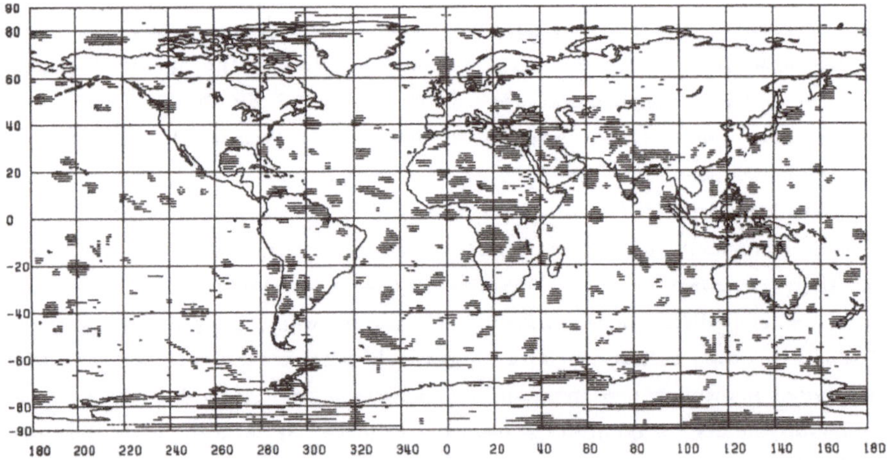

Figure 5. Location of 7545 1°x1° Blocks Where $|\Delta g(GEMT1) - \Delta g(A40)|$ Exceed 10 mgals. (Locations Corresponding to Geophysically Predicted Anomalies are Excluded from the Comparison).

9.0 Conclusion

The purpose of this paper has been to describe the method that might be used in the combination of satellite and terrestrial data. The emphasis has been on theory. Several high degree fields are in use today. These include OSU81 (180); OSU86 e/D (250); OSU86 E/F (360) GEM10C (180); GEM2 (200) where the numbers in parenthesis give the highest degree in the expansion. A comparison between some of these models may be found in Wenzel (1985), Rapp (1986) and Rapp and Cruz (1986a, 1986b).

Hopefully it is clear from the discussions in this paper that there are many choices to be made in carrying out combination solutions. Some of these choices lead to approximate solutions but no other alternatives may exist. The success of existing high degree fields implies that more fields will be developed using new satellite models and improved terrestrial data. Careful evaluation of these models is needed to assess their accuracy and the assumptions made in developing the model. Although computer resources continually increase it is doubtful if expansions above degree 360 will prove of value but this type of prediction has proved unreliable in the past.

REFERENCES

Bosch, W., High Degree Spherical Harmonic Analyses by Least Squares, in Proc, of IAG Symposia, Vol. I, IAG/IUGG Meeting, August, 1987, Paris

Colombo, O., Numerical Methods for Harmonic Analysis on the Sphere, Report No. 310, Dept. of Geodetic Science and Surveying, The Ohio State University, Columbus, 1981

Cruz, J.Y., Ellipsoidal Corrections to Potential Coefficients Obtained from Gravity Anomaly Data on the Ellipsoid, Report No. 371, Dept. of Geodetic Science and Surveying, The Ohio State University, Columbus, 1986

Despotakis, V., The Development of the June 1986 1°x1° and the August 30′x30′ Terrestrial Mean Free-Air Anomaly Data, Internal Report, Dept. of Geodetic Science and Surveying, The Ohio State University, Columbus, 1986

Engelis, T., Radial Orbit Error Reduction and Sea Surface Topography Determination Using Satellite Altimetry, Report No. 377, Dept. of Geodetic Science and Surveying, The Ohio State University, Columbus, 1987

Gleason, D., Comparing ellipsoidal corrections to the transformations between the geopotential's spherical ellipsoidal spectrums, manuscripta geodaetica, 13, 2, 11-412, 1988

Heiskanen, W., and H. Moritz, Physical Geodesy, W.H. Freeman, San Francisco, 1967

Jekeli, C., The Downward Continuation to the Earth's Surface of Truncated Spherical and Ellipsoidal Harmonic Series of the Gravity and Height Anomalies, Report 323, Dept. of Geodetic Science and Surveying, The Ohio State University, Columbus, 1981

Jekeli, C., The exact transformation between ellipsoidal and spherical harmonic expansions, manuscripta geodaetica, 13, 2, 106-113, 1988

Marsh, J., et al., A New Gravitational Model for the Earth From Satellite Tracking Data, GEM-T1, J. Geophys. Res., 93, B6, 6169-6215, 1988a

Marsh, J., et al., Dynamic Topography, Gravity and Improved Orbital Accuracies from the Direct Evaluation of SEASAT Altimetry, paper presented at the AGU meeting, Baltimore, May, 1988b

Middel, B., and B. Schaffrin, Robust Determination of the Earth's Gravity Potential Coefficients, in Proc. of IAG Symposia, Vol I, IAG/IUGG Meeting, August 1987, Paris

Pavlis, N., Modeling and Estimation of a Low Degree Geopotential Model from Terrestrial Gravity Data, Report No. 386, Dept. of Geodetic Science and Surveying, The Ohio State University, Columbus, 1988

Rapp, R., Gravity Anomalies and Sea Surface Heights Derived from a Combined Geos 3/Seasat Altimetric Data Set, J. Geophys. Res, 91, B5, 4867-4876, 1986

Rapp, R.H., and J.Y. Cruz, Spherical Harmonic Expansions of the Earth's Gravitational Potential to Degree 360 Using 30′ Mean Anomalies, Report No. 376, Dept. of Geodetic Science and Surveying, The Ohio State University, Columbus, 1986

Rapp, R.H., and J.Y. Cruz, The Representation of the Earth's Gravitational Potential in a Spherical Harmonic Expansion to Degree 250, Report No. 372, Dept. of Geodetic Science and Surveying, The Ohio State University, Columbus, 1986a

Rapp, R.H. and J.Y. Cruz Spherical Harmonic Expansions of the Earth's Gravitational Potential to Degree 360 Using 30′ Mean Anomalies, Report No. 376, Dept. of Geodetic Science and Surveying, The Ohio State University, Columbus, 1986b

Rummel et al., Comparisons of Global Topographic/Isostatic Models to the Earth's Observed Gravity Field, Report No. 388, Dept. of Geodetic Science and Surveying, The Ohio State University, Columbus, 1988

Sjoberg, L., Least Squares Combination of Satellite and Terrestrial Data in Physical Geodesy, Annales de Geophysique, Vol. 37, 25-30, 1981

Weber, G., and H.G. Wenzel, Estimation of Error Properties, in Validation of Seasat-1 Altimetry Using Ground Truth in the North Sea Region, Detusche Geodatsche Kommission, Reihe B: No. 263, Frankfurt Au Main, 1982

Wenzel, H.G., Hochauflöesende Kugelkunktionsmoddelle für das Gravitationspotential der Erde, Wiss. Arb. der Fachrichtung Vermessungswesen der Universitat Hannover, Nr. 137, Hannover, 1985

SUMMER SCHOOL LECTURES ON SATELLITE ALTIMETRY

Carl A. Wagner
National Geodetic Survey
Charting and Geodetic Services
National Ocean Service, NOAA
Rockville, Md. 20852 USA

Lecture 1. Purposes and Motivation, The Altimetric Equation, Radial Perturbations.

As you know satellite altimetry has two principal uses, the metric one of measuring the distance between the spacecraft and the earth's surface, and the characteristic one of learning something of the state or condition of that surface. The first use, concerned only with the mean time of return of the radar pulse is what interests us here. It is the form of the return pulse of course that gives important information on the surface state. But we will ignore this and trust that the engineers have given us a correct reading of the distance as essentially half the round trip time of the pulse times the speed of light.

What can we do with these metric observations? Intuitively we realize that if we could compute the position of the spacecraft at the same time, these observations would immediately yield the position of all the surface elements illuminated by the radar beam. Of course there is a large gap to be filled before that intuition can be realized to a useful accuracy. Nevertheless we all know that beginning with Skylab in 1974 and continuing with Geos 3 (1975), Seasat (1978) and Geosat starting in 1985, most of the earth's surface has indeed been so mapped to precisions of the order of centimeters over oceans and meters over land.

Recall the definition of the measures involved in satellite altimetry as shown in Figure 1. I write the fundamental "altimetric equation" there for the surface

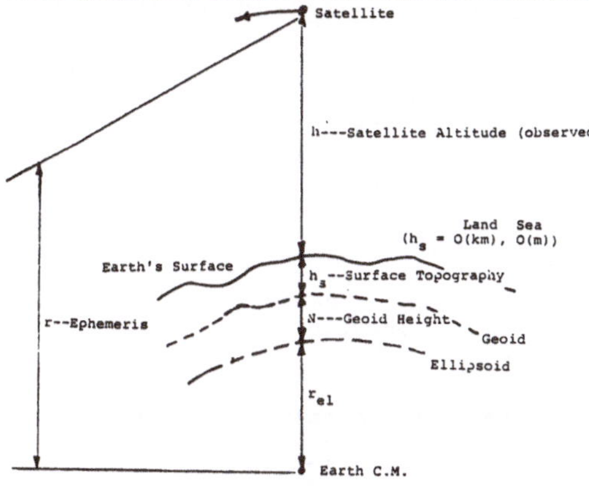

$$h_s = r - r_{el} - N - h : \text{Surface Topography From Altimetry}$$

Figure 1. Metrics of Satellite Altimetry. The altimetric equation is solved for surface topography [assuming small geoid (N) and radial (r) errors]. Applications are in land and sea mapping, sea topography yielding information on surface currents, eddies, tides and general circulation. In sea areas where the anomalous water column height is known (from bottom gages and/or good hydrographic control) the equation can also be solved for N to improve the geoid.

topography, perhaps the most interesting byproduct of the actual satellite altitude measurement. Over land this is the usual topographic height above a reference level surface, conventionally called the geoid which is within about a meter of mean sea level. In fact the distance between mean sea level itself and the geoid is just the "permanent sea surface topography" (PST) whose determination is what I call the most important unsolved problem in geodesy today.

Why is it important? Because this permanent water height above a level reference can only be maintained by permanent external forces acting on the ocean's surface layers, either prevailing wind stresses or dynamic forces associated with ocean currents such as the coriolis forces which accompany the great ocean gyres that result from equator to pole heat transport in the open oceans. Knowing the constant part of this topography would essentially determine the circulation of the surface waters in the deep ocean with great benefits to marine science, navigation and climatology.

What are the gaps in knowledge that must be filled to solve this equation? Look at the terms. Only one is perfectly known by definition, the radius to a conventional mean earth ellipsoid. The measurement itself (h) is subject to a multitude of uncertainties starting right at the satellite where the distance from the center of the radar antenna to the center of mass of the spacecraft which carries the calculated ephemeris information (r) may have significant errors.

Table 1, adapted from a Seasat paper by Tapley, Born and Parke (1982), shows a large number of these error sources, (perhaps exhaustive of the dominant ones) and their estimated remedies as of 1982. Notice for example that the item clouds/rain

Table 1. Altimeter Altitude Error Budget

Type of Error	Source of Error	Amplitude of Full Efect cm	Residual Error cm	Wavelength km
Geoid	Mass Dist. of Earth	10,000	200	500-20,000
Orbital	Gravity	10,000,000	140	40,000
Orbital	Drag	300	30	20,000
Orbital	Sol. Radiation	300	30	20,000
Orbital	Station Location	---	10	40,000
Ocean Tide	Luni-Solar Gravity	100	10	500-20,000
Sea State Bias	Waveheight and Trackers	<70	<20	500-1000
Timing	Data time tag	---	5	20,000
Altimeter	Noise	---	5	50
Troposphere	Water Vapor	10-40	3	50-500
Ionosphere	Free electrons	2-20	3	50-10,000
Depression of Sea	Atmospheric Pressure	50	3	200-1000
Solid Earth Tide	Lunisolar gravity	20	2	20,000
Altimeter	Bias	10	2	-----
Troposphere	Mass of air	240	1	1000
Liquid Water	Clouds, rain drops	50-100	?	10-100

suggests only a draconian cure: elimination of the patient. However current opinion is that only actual rain causes problems of the order of 1 meter (from backscattering off raindrops) and this occurs rather rarely and only over a few tens of kilometers.

The radius of the satellite (r) of course undergoes its largest changes from gravity, the geopotential predominantly, but also including lunisolar gravity both direct and from tidal influences. Lesser changes are due to solar radiation pressure and atmospheric drag. Note that after modeling, residual errors due to these sources are still dominated by the earth's geopotential. Later I will demonstrate that using the latest geopotential models should result in radial errors for Geosat or Seasat of about 60 cm with greatest power at long wavelengths (near 1 cycle per revolution). Without even considering errors in the geoid we will see that currently radial orbit correction (e.g., from the altimetry itself) will be necessary to obtain a meaningful value of PST which over the oceans has a long wave amplitude also of the order of about 50 cm (away from the strongest boundary currents).

Geoidal errors in this equation at the time of Seasat (as shown in Table 1) were estimated to be of the order of a few meters. Soon after though, perhaps 1 meter would have been a fairer assessment with use of the detailed Geos 3-Seasat marine geoids such as those developed by Dr. Rapp (e.g., Rapp, 1986). In most areas of the oceans, away from the trenches and rough bottom topography, these geoids should yield errors of 50 cm or even better. Again though, it appears necessary to solve for geoid corrections as well as radial orbit corrections to drive overall errors sufficiently below 50 cm to reveal the sea topography signal clearly. But we will see in later lectures that using only a limited part, namely the low frequency domain of the sea topography for which geoid errors are especially small, will greatly aid in the solution of our problem.

Some people have remarked that the dilemma we face (finding the permanent part of from altimetry) is insoluble since sea surface topography and geoid errors have the same surface expression and thus the geometric component of altimetry can make no distinction between them. But we will also see that the dynamic component of the orbit error that is uniquely tied to the geoid error provides the distinguishing information to solve the problem in theory. I don't have to tell you though that there are still many practical problems in the way of such a simultaneous solution or we would have seen it already in the 13 years since Geos 3 altimetry has made this approach feasible.

What we have seen in those years are approaches which solve the altimetric equation by subtracting a fixed geoid from sea surface heights obtained after resolving the orbit error as the first step of an iterative process. These attempts have proven remarkably suggestive even with the use of geoids that are not current. For example, Figure 2 shows two such "subtraction surfaces" derived a few years ago from Seasat altimetry and a fixed Gem L2 geoid (Lerch et al, 1985). The

a

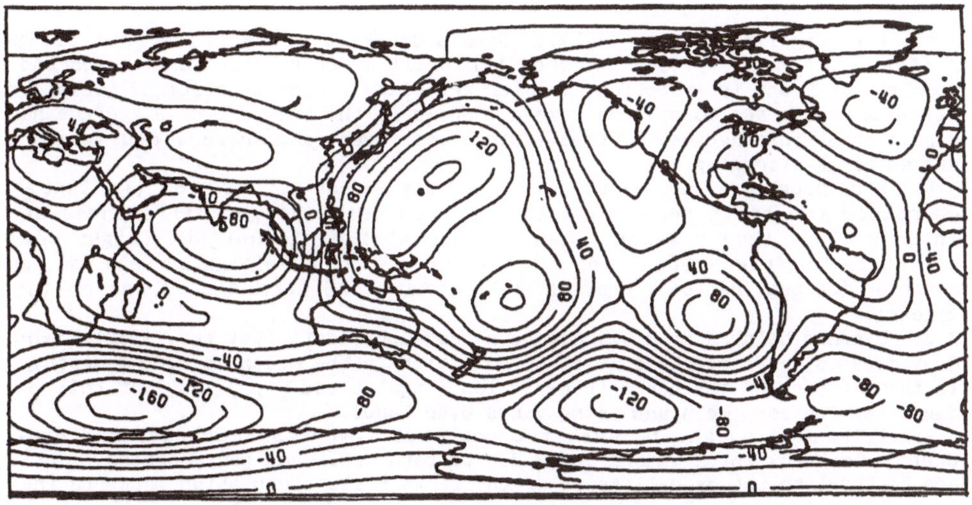

(Figure 2: Caption and continuation on the next page)

b

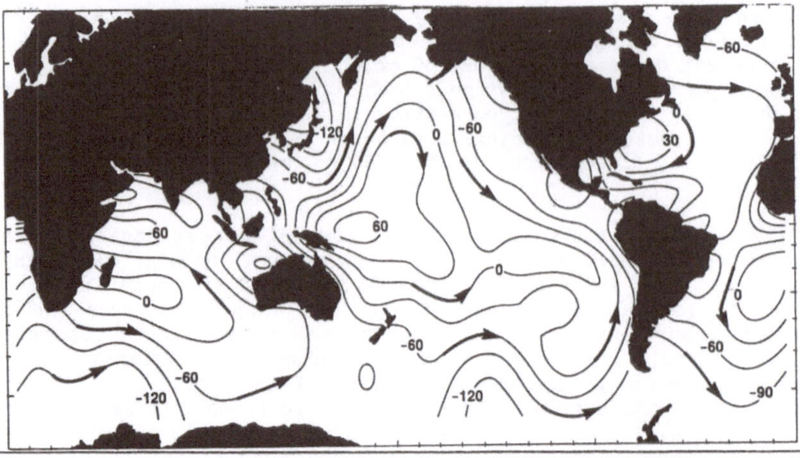

c

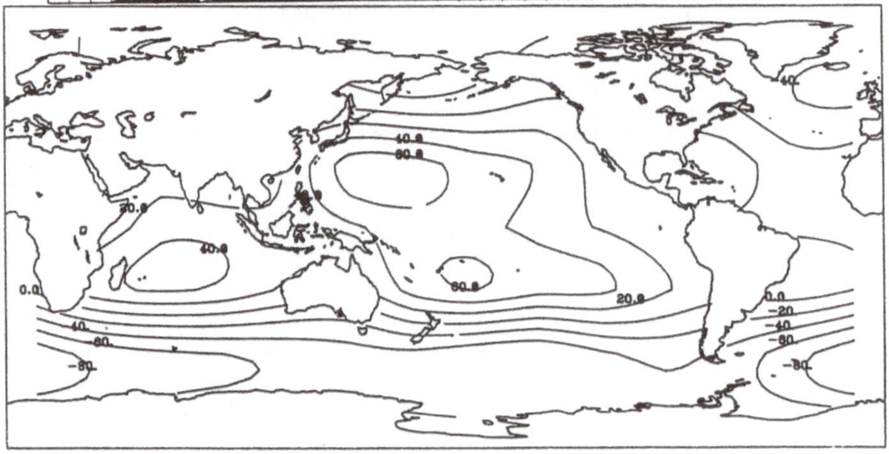

Figure 2. Representations of "Permanent" Sea Topography From Seasat Altimetry using the Gem L2 Geoid, in cm.

a. From Engelis and Rapp (1984): Uses a detailed sea surface from all Seasat Altimetry (summer 1978) corrected for orbit error by short arc biases and tilts resolved at track intersections (master arc fixed); subtracts Gem L2 geoid; adjusts difference surface to have zero mean. Shown is the 6x6 part of a surface harmonic expansion of this reduced surface (using zeros over land areas), equivalent to 30 degree block averaging.

b. From Douglas et al. (1984): Uses a 3-day Seasat sea surface (in Sept. 1978) corrected for orbit error by a time series using height differences at track intersections (heavy weights) and direct altimetric heights (light weights) elsewhere. After subtracting the Gem L2 geoid, resulting surface was Gaussian smoothed (20 deg. radius), equivalent to about a 5x5 harmonic expansion (35 deg. blocks).

c. Dynamic heights from 1x1 degree hydrographic data (Levitus, 1982) taken over many decades, referred to a level of no motion in the ocean at 2200 m depth. Shown is the 6x6 part of the harmonic expansion (Engelis and Rapp, 1984) of this surface (adjusted for zero mean and using zeros over land).

Satellite maps a. and b. seem too powerful (even as smoothed) and show some unreal features which appear to be due to both orbit and geoid errors. Still they are faithful to the overall sense of the large scale circulation patterns indicated in map c. Only the Engelis-Rapp map, however, begins to catch the circumpolar nature of the strong Antarctic current.

Gem L2 field is a satellite-only model (from over 30 orbits including heavy Lageos tracking) complete to 20x20 with higher degree and order zonal and resonance terms. The Engelis and Rapp, 1984 surface (Figure 2a) used an orbit correction scheme consisting of one revolution of orbitally uncorrected altimetry tied to a set of short crossing arcs of data whose radial errors were modelled by independent biases and tilts. These parameters were determined by matching (in a least squares sense) the computed differences of altitude residuals at the crossover points of the trajectory. These differences should not be affected by errors in the permanent features of the geoid or sea topography, only by the orbit errors themselves or other time varying effects such as ocean eddies and tides.

Unfortunately it was not recognized until after this work was published that the crossover differences were also not affected by certain permanent features of the orbit error itself, a problem that was first solved analytically by Rosborough (1986). Thus the Engelis-Rapp map already must contain a defect. The map of Douglas et al. (1984) on the other hand (Fig. 2b) was derived from an empirically determined global Fourier series for the orbit error and used altimetry not only at the crossovers but direct height data (h) as well, which contains both geoid and sea topography error at all frequencies. The Fourier series has the advantage of including (as we shall see) the correct orbit frequencies due to the geopotential. Unfortunately the unavoidable use of direct altimetry with a signal significantly biased by sea surface topography and geoid errors undoubtedly caused distortions in the orbit correction process. On top of these errors of course, the Gem L2 geoid was known to be so poor beyond about degree 6 that no sea topography derived without geoid correction would have meaningful information at shorter wavelengths.

Also in Figure 2c is a map of the Levitus (1982) dynamic sea topography (to the same global wave number) determined from oceanographic data. Both satellite altimetry maps of PST are probably too powerful, contain spurious high frequency detail and have curious long wavelength north-south distortions that may well relate to inadequate determination of primary orbit error at 1 cy/rev. In the Engelis-Rapp map this error is fundamentally ill determined. In the map of Douglas et al. (1984) it may be poorly determined due to aliasing from higher frequency geoid error, or even from the PST itself.

How do we make these sweeping judgements? First of all we know that the motion of the satellite is governed mainly by the earth's gravitational field whose high frequency effects at altitude are severely diminished compared to low frequency effects. Time series-frequency analysis of the motion then offers not only a greater understanding of the altimeter signal but the possibility of replacing a complex numerical calculation of all geopotential effects with a selective one of only the few dominant, generally low frequency terms.

The approach I have taken to the frequency analysis of the radial signal is the classical one starting with the familiar representation of the radius to an object in an elliptic orbit:

$$r = a(1-e^2)/(1+e\cos f) \qquad (1)$$

(See Fig. 3). So, the harmonic analysis of the radial variation in a two-body orbit is given simply in terms of the fundamental orbit period and all its overtones dying away in powers of the eccentricity. This theme is also followed with the more complex perturbations of the radius due to the actual field of the earth.

As we know the spinning earth is noticably oblate with asphericity of order 1/1000. For an inclined orbit the oblateness field will have two highs (at the equator crossings) and so induce a significant effect at twice the orbital rate. What can we say about the higher harmonics of gravity due to the density inhomogeneities in and on the earth? The lowest degrees of these are of order 10^{-6} and they decline in power from there rather slowly, roughly inversely as the

Mean Anomaly M proceeds uniformly in time

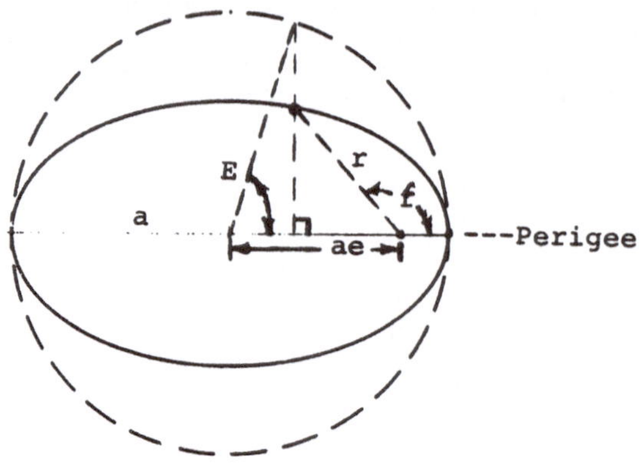

$$r = \frac{a(1-e^2)}{1+e\cos f} = a(1-e\cos E)$$

$$= a(1+e^2/2 + \cos M(-e+\ldots)+\cos 2M(-e^2/2 + \ldots)+\ldots$$

Figure 3. Radial Variations in an Elliptic Orbit. The two-body solution time series contains all integer multiples of the fundamental period.

degree square (the famous Kaula's rule). This slow decline represents the fact that the planet inside and out is nearly as rough as it can stand with gravitational stresses generally close to failure for its constituent matter. In fact an ultimate decline with a power of 2 implies infinite point gravity anomalies somewhere.

It is interesting to consider the general form the radial variation takes with respect to the geopotential and the eccentricity of the orbit. The detailed theory shows that:

$$r=a[1+O(J_2)+O(J_{\ell m})+O(J_2.e)+O(e^2)+O(J_2^2)+O(J_2J_{\ell m})+O(J_2.e^2)+O(J_{\ell m}.e)+\ldots] \qquad (2)$$

Thus for $a=7000$ km and a near circular orbit with $e=.001$ we see that the simplest theory (including only the J_2 term) will yield results good to about $7000\times10^{-6}=7$ meters. While this may be adequate for many purposes we may need to consider higher order effects in e and interactions with oblateness to obtain results necessary to interpret centimeter altimetry properly.

Even before the advent of altimetry the simplest development of just the radial variation due to oblateness was available from Kozai (1959) where it was used in the first analytical satellite tracking programs. The derivation of it was based on combining results from the perturbations of the classical Kepler elements, an approach I followed also for the effect of the general $J_{\ell m}$ term on the near circular orbit and also followed by Rosborough (1986) for the effect of $J_{\ell m}$ on an orbit of any eccentricity (<0.6627...). Since Kaula's development of the first order theory to $J_{\ell m}$ was available in the early 1960's, it was astonishing that the radial variations weren't also worked out by the mid 1970's when Skylab's altimetry was first analyzed. We suspected of course that because of the attenuation with altitude the dynamic orbit effects would be insignificant compared to the geometric (from the geoid) at frequencies higher than about 2 cy/rev. But as far as I am aware these spectra weren't actually derived till the middle 80's in preparation for TOPEX.

Figure 4 shows an example I gave (Wagner, 1985) of the radial effect on Seasat from Gem 9 errors (as well as the M2 Tide) geometric as well as dynamic using this formalism. Gem 9 (Lerch et al., 1979) is a satellite-only model (using over 30 distinct orbits but without Lageos for geopotential information) complete to 20x20 with a few higher degree resonance and low order terms. It illustrates the concentration of dynamic orbit error at low frequency, as contrasted with the essentially "white noise" aspect of the geoidal signal in altimetry. Even the full tidal signal is generally insignificant (both dynamic and geometric aspects) compared to that due to Gem 9 errors.

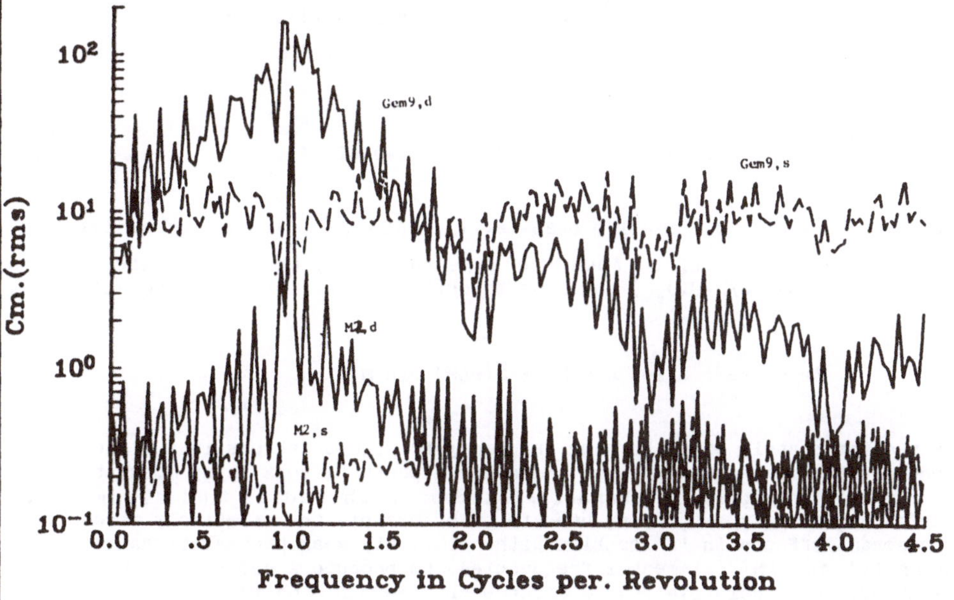

Figure 4. Radial Errors on Seasat from Gem 9 Variances and The M2 Ocean Tide. Shown are the spectra accumulated in bins of 0.02 cy rev. Solid lines are the dynamic part of the radial orbit spectrum due to Gem 9 (Gem 9,d) and the M2 tide (M2,d). Terms at and close to 1 cy/rv are excluded from Gem 9,d. Dashed lines are spectra for geoid height measurements due to Gem 9 (Gem 9,s) and the M2 tide (M2,s). Oribt parameters: I = 108 degrees, 785 km altitude, e = .001, 36x36 fields were used but tidal effects were computed from degree variance data only. Note the concentration of power in dynamic spectra near and below 1 cy/rev and the flatness of the surface spectra (from Wagner, 1986.)

I first approached this problem in its simplest form, as the radial perturbations of a near circular orbit, as this was the form of all past, present and projected altimetric satellite orbits. This theory was readily worked out; the frequencies themselves were easy to discuss and classify.

The convenient place to start is with the familiar 2-body spectra:

$$r = a[1 + e^2/2 + (-e + \ldots)\cos M + (-e^2/2 + \ldots)\cos 2M + \ldots] \qquad (3)$$

Limiting this even further to terms of order e:

$$r \doteq a(1-e\cdot\cos M) \tag{4}$$

Imagine that a,e and M are on a reference orbit and that the Kepler perturbations Δa, Δe, and ΔM are available, say from Kaula's theory. Then, to first order they cause perturbations Δr given by:

$$\Delta r = \frac{\partial r}{\partial a}\Delta a + \frac{\partial r}{\partial e}\Delta e + \frac{\partial r}{\partial M}\Delta M \tag{5}$$

where

$$\frac{\partial r}{\partial e} = -a.\cos M, \quad \frac{\partial r}{\partial M} = a.e\sin M, \quad \text{and } \frac{\partial r}{\partial a} = (1-e.\cos M) \tag{6}$$

Thus:

$$\Delta r = \Delta a(1-e\cdot\cos M) + \Delta e(-a\cos M) + \Delta M(a\cdot e\cdot\sin M). \tag{7}$$

But in keeping with our small e case ($e<<1$), $e\Delta a=O(eJ_{\ell m})$ and can be ignored. On the other hand ΔM will be found to have an 'e' in the denominator of its perturbations so that $e\Delta M=O(J_{\ell m})$ and must be retained. We then have for the near circular orbit:

$$\Delta r = \Delta a - a(\Delta e)\cos M + a(e\Delta M)\sin M \tag{8}$$

We see immediately that the same frequency for perturbations in a,e and M does not produce the same frequency in r, the e and M perturbations being modulated at 1 cycle. Partial shifting is characteristic of both radial and along-track frequencies and takes some getting used to. For example the relatively strong long period resonant effects in 'a' combine with relatively weak perturbations of short period in 'e' and 'M' (elemental frequencies) to produce a radial perturbation of low frequency, while relatively weak short-period variations of 'a' combine with relatively strong changes of long-period in 'e' and 'M' (elementals) to yield a strong radial effect near 1 cycle per revolution (realized frequency).

What is the relation between the fundamental or elemental frequencies and the realized radial ones? Recall Kaula's specification of the fundamental gravitational frequencies:

$$\dot{\alpha}_{\ell mpq} = -q\dot{w} + (\ell-2p+q)(\dot{M}+\dot{w}) + m(\dot{\Omega}-\dot{\theta}_e) \tag{9}$$

where $\dot{w}$ and $\dot{\Omega}$ are the precession rates of perigee and node due to the earth's dominant oblateness (what might be called the zeroth order solution to the Lagrange equations). The other terms relate to the general geopotential: $\dot{\theta}_e$ is the earth's rotation rate, p is an index related to the rotation of the potential from the equatorial system to one oriented along the plane of the orbit, and q is an index accounting for the final transformation of the potential from true to mean anomaly (effectively creating a true time series for the potential along the satellite's path). Another way of writing this specification is:

$$\dot{\alpha} = q\dot{M} + (\ell-2p)(\dot{M}+\dot{w}) + m(\dot{\Omega}-\dot{\theta}_e) \tag{10}$$

It turns out that to first order in $J_{\ell m}$ only q=0 for Δa need be included while for Δe and $e\Delta M$ only q= ± 1 are present. Geopotential orbit frequencies such as radial for low eccentricity orbits are thus most simply classified by (m,k) where k = 1 - 2p:

$$\dot{\alpha}(\text{orbit frequency}) = k(\dot{M}+\dot{w}) + m(\dot{\Omega}-\dot{\theta}_e) \tag{11}$$

Furthermore because it also turns out that the $\pm\dot{M}$ parts of the elemental frequencies for Δe and $e\Delta M$ are exactly cancelled by the one cycle/rev modulations of the radial perturbation, this form also applies to them.

Rather unexpected was the fact that while the 1 cy/rev term dominates the 2-body variation, there is evidently none to first order in the anomalous geopotential for low eccentricity. Of course we expect to find the 1 cycle variation from the more complete theory but exactly how this term enters for the near circular orbit is not clear since the orbit may no longer be elliptical (J_2 being the same order as the eccentricity). We hope to clarify this issue in what follows.

Also of concern with the classical derivation is the use of ordinary Kepler elements which are singular for a purely circular orbit. In fact though, the radius itself should be an exceptionally well behaved variable, not differing by more than about a thousandth (relatively) from the semimajor axis in each revolution. Particular longitude arguments though, such as the argument of perigee and the mean anomaly, might undergo great changes in an orbit. But the radius as well as the other position and velocity coordinates should not suffer such indignities. Still it pays us to treat the problem from a completely different viewpoint (using nonsingular elements) to verify the classical description and gain a stronger insight into the real behaviour of the near circular orbit.

The nonsingular variables to choose will be apparent from the approximate 2-body description; $r=a(1-e \cdot \cos M)$, because defining $\lambda=w+M$ as the mean argument of latitude, we have:

$$r = a[1-e\cdot\cos(\lambda-w)] = a[1-(e\cdot\cos w)\cos\lambda - (e.\sin w)\sin\lambda] \tag{12}$$

and instead of three singular variables e,w,M we will follow the perturbations of the non-singular ones λ, e.sin w, e.cos w as well as 'a'. Historically, Kozai (1959, 1961) was the first to give this particular treatment of the near circular orbit (J_2, J_3 only), although G. E. Cook (1966) deserves the credit for bringing out all the interesting features of the evolution [J_2, all J(odd)] except those near critical inclination ($I \doteq 63.4$ and 116.6 degrees). What these earlier studies showed were that the odd zonal harmonics as distinct from the even zonals had no secular effects on the longitude arguments but significant long period ones with period of the argument of perigee. Such effects were of order J(odd)/J_2 or 10^{-3} and aside from resonance with the earth's rotation were the only effects that need to be included in a first order theory good to 10^{-6} besides the secular and short period J_2 terms. In these circumstances since nonzonal resonance will be brought in later, we are concerned now with only J_2 and the J (odd) zonals giving long period terms.

For simplicity consider only the disturbing function for J_2 and J_3:

$$R = (3/2)(\mu/r)J_2(r_e/r)^2[(1/3)-\sin^2 I.\sin^2(f+w)] + (5/2)(\mu/r)J_3(r_e/r)^3.$$

$$.[\sin(f+w).\sin I\{(3/5)-\sin^2 I.\sin^2(f+w)\}]. \tag{13}$$

[In equation (13), μ is the earth's Gaussian gravity constant and r_e is its mean equatorial radius.]

We can conveniently separate R into three parts, the first constant over one revolution and the second and third of short and long period (with respect to one revolution), the long period part being a sinusoid in just the argument of perigee. Kozai (1959) showed that after solving the appropriate Lagrange planetary equations in non-singular variables with this disturbing function, the long period evolution of e.cos w, e.sin w was on a circle in the plane of those coordinates with the center of the circle on the e.sin w axis. Kozai (1961) included short period J_2 terms in the solution and Cook (1966) extended the long period result to all odd zonals. Cook gave the location of the center of the motion in terms of Legendre polynomials, but in fact we can also give it in terms of Kaula's inclination and eccentricity functions.

To do this we note that with error of order e^2 the low-e solution of Cook (1966) is identical to the familiar linear solution of the Lagrange equations, given by Kaula (1966, p.40) for e sufficiently large that circulation of 'w' occurs with $\dot{w}$ essentially constant. The important point is that Cook's center of motion is just the amplitude of the large-e motion in Kaula's first order theory. Kaula (1966, p.41) gives this amplitude specifically as:

$$\Delta e_\ell = -[\mu(1-e^2)^{1/2}/(n_0 a^3 e\dot{w})] \sum_{p=(\ell-1)/2}^{p=(\ell+1)/2} (r_e/a)^\ell F_{\ell,0,p}(I) G_{\ell,p,2p-\ell}(e) . S_{\ell,0,p,2p-\ell} \qquad (14)$$

where:

$$q = 2p - \ell = \pm 1 \text{ and } S(q = \pm 1) = \mp C_{\ell,0} \sin w.$$

But the 'e' in the denominator of Eq. (14) is cancelled by the leading term in the G function which is of order e for these q=±1 effects. Further, we note that the two inclination functions are the same with opposite signs. In addition, the G functions (to first order in e) are identical. Finally, using the result in Allan (1967):

$$G_{\ell,(\ell-1)/2,-1} = G_{\ell,(\ell+1)/2,+1} = (e/2)(\ell-1) + O(e^3) , \qquad (15)$$

we find that the amplitude of the perturbation over all odd zonals is:

$$\Delta e = (n_0/\dot{w}) \sum_{\ell \text{ odd}} C_{\ell,0}(r_e/a_0)^\ell (\ell-1) F_{\ell,0,(\ell-1)/2}(I) = C/\dot{w} , \qquad (16)$$

the 'circulation' center of Cook (1966).

The conclusion reached by Kozai (1961) and others bears repeating; the treatment of orbit or coordinate perturbations leads to the same result for the near circular orbit whether the development is done in terms of singular or nonsingular elements. Writing it down for the first order evolution of the radius to errors of order e^2, J_2^2, eJ_2 and $J_{\ell m}$:

$$R = R_0 + C_n \cos n_0 t + S_n \sin n_0 t + C_\lambda \cos \dot{\lambda} t + S_\lambda \sin \dot{\lambda} t + C_{2\lambda} \cos 2\dot{\lambda} t + S_{2\lambda} \sin 2\dot{\lambda} t, \qquad (17)$$

where:

$$R_0 = a_0[1-(3/2)C_{20}(r_e/a_0)^2\{(3/2)\sin^2 I - 1\}]$$

$$C_n = -Aa_0 \cos(\lambda_0 - \gamma)$$

$$S_n = Aa_0 \sin(\lambda_0 - \gamma)$$

$$C_\lambda = -(C/\dot{w})a_0 \sin\lambda_0$$

$$S_\lambda = -(C/\dot{w})a_0 \cos\lambda_0$$

$$C_{2\lambda} = (a_0 C_{20}/4)(r_e/a_0)^2 \sin^2 I \cos 2\lambda_0$$

$$S_{2\lambda} = (-a_0 C_{20}/4)(r_e/a_0)^2 \sin^2 I \sin 2\lambda_0 \qquad (18)$$

The mean elements a_0, λ_0 are determined from osculating values at epoch ($t=0$) as:

$$a_0 = a + (3/2)aC_{20}(r_e/a)^2 \sin^2 I \cos 2\lambda$$

$$\lambda_0 = \lambda + (3/4)C_{20}(r_e/a_0)^2[(5/2)\sin^2 I - 1]\sin 2\lambda$$

$$\dot{w} = (-3/4)n_0 C_{20}(r_e/a_0)^2(5\cos^2 I - 1)$$

$$\dot{\Omega} = (3/2)n_0 C_{20}(r_e/a_0)^2 \cos I$$

$$n_0 = \mu^{1/2}(a_0)^{-3/2}[1-(3/2)C_{20}(r_e/a_0)^2\{1-(3/2)\sin^2 I\}]$$

$$\dot{\lambda} = n_0 + \dot{w}$$

$$A = [(e_0 \cos w_0)^2 + \{e_0 \sin w_0 - (C/\dot{w})\}^2]^{1/2}$$

$$\gamma = \mathrm{Tan}^{-1}[\{e_0 \sin w_0 - (C/\dot{w})\}/(e_0 \cos w_0)]$$

$$e_0 \cos w_0 = e \cdot \cos w + (3/2)C_{20}(r_e/a_0)^2[1 - (5/4)\sin^2 I]\cos\lambda$$

$$+ (7/8)C_{20}(r_e/a_0)^2 \sin^2 I \cos 3\lambda$$

$$e_0 \sin w_0 = e \cdot \sin w + (3/2)C_{20}(r_e/a_0)^2[1-(7/4)\sin^2 I]\sin\lambda.$$

$$+ (7/8)C_{20}(r_e/a_0)^2 \sin^2 I \sin 3\lambda \qquad (19)$$

and

$$C = n_0 \sum_{\ell\ \mathrm{odd}} C_{\ell,0}(r_e/a_0)^\ell(\ell-1)F_{\ell,0,(\ell-1)/2}(I). \qquad (20)$$

The mean elements $e_0 \sin w_0, e_0 \cos w_0$ are strictly speaking only short term (1 revolution) mean values. They have long period variation:

$$(e_0 \cos w_0)_{t>0} = A \cos(\dot{w}t + \gamma)$$

$$(e_0 \sin w_0)_{t>0} = (C/\dot{w}) + A \sin(\dot{w}t + \gamma) \qquad (21)$$

which is on a circle about $(0, C/\dot{w})$ in the $e.\cos w, e.\sin w$ plane.

The one interesting aspect of the development in nonsingular elements is that it shows explicitly how the zonals, both odd and even determine the 1 cycle/revolution variation of the radius for a near circular orbit. Note particularly that in the exact so called "frozen" orbit configuration where the long term oscillation of e.sin w and e.cos w ceases, A=0 and there is no 1 cycle motion of the radius. What this means is that the frozen commensurate (or geostationary) orbit [A=0, $\dot{\lambda}$= R revolutions in D synodic (or nodal) days each of length $2\pi/(\dot{\theta}_e-\dot{\Omega})$] allows a true repetition of radial variations over R revolutions. Unfortunately, since the gravity field is not known perfectly, the true geostationary orbit can only be achieved on iteration after launch by correcting the C (or lumped odd zonals) of the reference orbit after a number of days of tracking and making an appropriate small maneuver to put the mean perigee at the correct frozen point. These issues are further discussed by Colombo (1984a.)

How accurate is the first order estimation of the radial variation? Recall our estimate (e<<1) of about 7 meters for a typical altimetric satellite. To check this I have tested the theory against a numerically computed 20 day ($2\pi/\dot{w}$) trajectory in a low orbit at 173 km altitude (16 revs/day) of 5 degree inclination. The J_2,J_3 field used was from Gem 10b (Lerch et al., 1981) and I targeted the initial e,w for a frozen orbit (w_0=90°, $e_0=C/\dot{w}$) at this altitude and inclination. (Gem 10b is a 36x36 combined satellite-surface gravimetry/altimetry model that uses the same satellite information as Gem 9). Figure 5 shows the evolution of the mean values e_0,w_0 for the 20 days as computed from the first order J_2 variation above (Eq. 19) using the osculating elements from the numerical integration. Clearly the

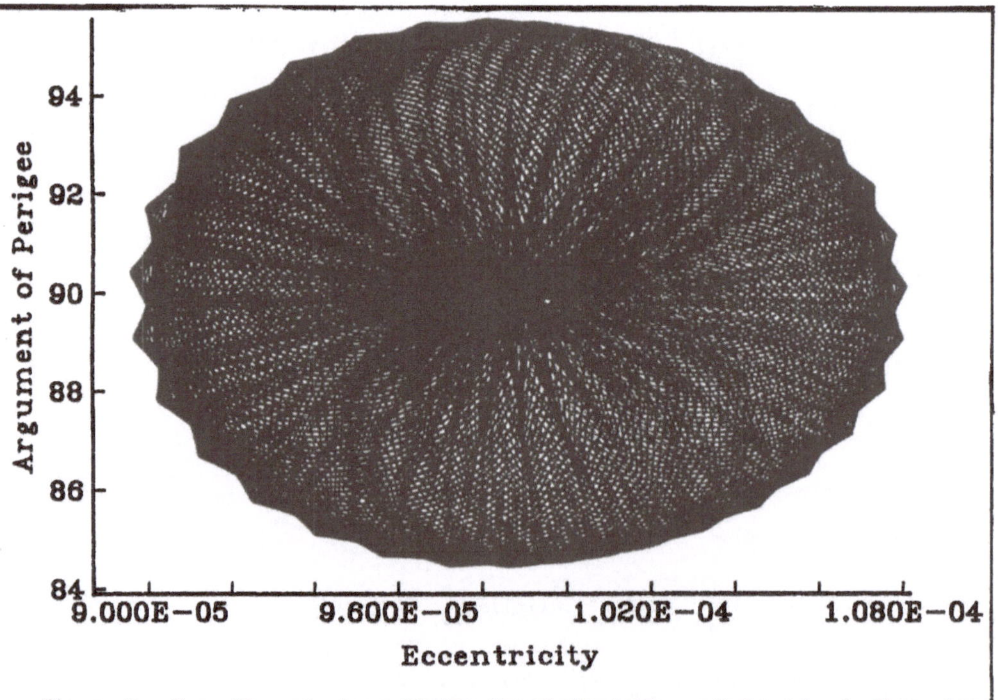

Figure 5. Mean Elements in a 20-day Zonal Trajectory Using J_2,J_3 From Gem 10b. The period of rotation of perigee is also 20 days (orbit altitude = 173 km, Inclination = 5 degrees, other in-plane elements targeted for a "frozen" orbit, A≐0, see Eqs. (19). Shown are 'one revolution' or short term mean elements calculated from the first order theory (Eqs. 19). Theoretically they should be stationary at w_0 = 90 degrees. That they are not indicates that an important second order term at or near 1 cy/rev is missing in the theory. This is the same trajectory that produced the radial results shown in Table 2.

formula lacks an additional component near 1 cycle; the mean elements do not describe simple circulation about $(0, C/\dot{w})$. Figure 5 shows that the true long period libration constants A, γ in this trajectory were $A=4.8\times10^{-6}$, $\gamma=2.6°$. The theoretical harmonic analysis with the target and actual values of A, γ is shown in Table 2.

Table 2. Numerical Test of First Order Theory for Radial Variation (J_{20}, J_{odd})

Results of harmonic analyses of the radial variation in a numerically generated 20-day trajectory (h=173km, I=5(degrees)), targeted for a "frozen" orbit (J_2, J_3 from Gem 10b, A=0). Rms variation of r = 10,010 meters. Frequencies calculated from first order theory using slightly different values of initial semimajor axis (a_i). Theoretical harmonics calculated using these frequencies and slightly different values of initial "circulation constants" (A, γ)

Cases:	Measured	Theory
	a. Freqs. with true a_i	Frozen Orbit (A=0)
	b. Freqs. with tuned a_i	Frozen Orbit (A=0)
	c. Freqs. with true a(init.) (J_2^2 secular rates)	Actual A, γ of orbit (Fig. 5) (J_2^2 secular rates)

Case	Frequency (Cy/Rev)	C (m) Measured	C (m) Theory	S (m) Measured	S (m) Theory
a	r_0	-9999.	-9972.	-------	------
b	(constant)	-9999.	-9972.	-------	------
c		-9999.	-9972.	-------	------
a	n_0	30.9	0.0	-4.2	0.0
b	(1.000)	31.0	0.0	-0.2	0.0
c		31.0	31.4	-0.4	-1.4
a	$\dot{\lambda}$	-11.3	0.0	-649.6	-651.3
b	(1.003)	0.0	0.0	-649.7	-651.3
c		0.6	0.0	-649.7	-651.3
a	$2\dot{\lambda}$	12.7	12.8	0.4	0.0
b	(2.006)	12.7	12.8	0.0	0.0
c		12.7	12.8	0.0	0.0
	Residuals of Fit (m)				
a	3.4				
b	0.2				
c	0.3				

In addition a number of numerical harmonic analyses were performed with frequencies $n_0, \dot{w}, \dot{\lambda}$ determined using a: the osculating elements at epoch and equations (18), b: an initial semimajor axis tuned to give the smallest radial residuals [trajectory values minus those computed on application of equations (18)] and c: the original osculating elements but with $n_0, \dot{w}, \dot{\lambda}$ determined using the $J_2{}^2$ secular terms from Brouwer (1959). As seen in Table 2 these secular effects in the mean motions go a long way to resolve the errors of the secondary effects in the n_0

and $\dot{\lambda}$ terms. Additional higher order effects in the constant and short period terms should improve the theory further. Engelis (1987) discusses this development and provides some second order formulas for the elements but his expression for the radius still lacks an important term. Kozai (1962) gives a complete second order expression for the radius of an orbit of any eccentricity (low degree zonals only) but not explicity as a time series. Similar effects of second order in e.sin w, e.cos w are needed to improve the targeting for the frozen orbit.

The additional secular rates to $\dot{w}$ and n_0 due to $J_2{}^2$ terms are (from Brouwer, 1959, for e<<1):

$$\Delta n_0 = (3/128)(C_{20})^2(r_e/a_0)^4[26 - 156 \cos^2 I + 274 \cos^4 I] \tag{22}$$

$$\Delta\dot{w} = (3/128)(C_{20})^2(r_e/a_0)^4[14 - 228 \cos^2 I + 790 \cos^4 I]. \tag{23}$$

We see in Table 2 that the addition of these rates dramatically improves the n_0, and $\dot{\lambda}$ frequencies but discrepancies still remain at the meter level. Fortunately the largest discrepancy is in the constant term which is not of oceanographic consequence.

The formulae given here (Eqs. 17-22) may be augmented by the linear perturbations in radius for $J_{\ell m}$ worked out by Wagner (1985) for the near circular orbit or to any eccentricity (<.663) by Rosborough (1986) (including the Kozai (1962) solution for $J_2{}^2$ effects in the radius). For example the time series for the low-e perturbations in $J_{\ell m}$ is given as a sum of frequencies $\alpha = (m,k)$:

$$\Delta r = \sum_{\alpha} S_\alpha \sin \dot{\alpha}t + C_\alpha \cos \dot{\alpha}t \tag{24}$$

where,

$$(S_\alpha, C_\alpha) = \sum_{\ell=\ell \min}^{\ell \max, k \text{ par.}} H_{\ell mk} \left\{ \begin{vmatrix} -C_{\ell m}, S_{\ell m} \\ S_{\ell m}, C_{\ell m} \end{vmatrix} \sin \alpha_0 + \begin{vmatrix} S_{\ell m}, C_{\ell m} \\ C_{\ell m}, -S_{\ell m} \end{vmatrix} \cos \alpha_0 \right\} \begin{matrix} \ell-m \text{ even} \\ \ell-m \text{ odd} \end{matrix} \tag{25}$$

and,

$$\dot{\alpha} = k\dot{\lambda} + m(\dot{\Omega}-\dot{\theta}_e) \tag{26}$$

and the initial orbit phases are given as:

$$\alpha_0 = k\lambda_0 + m(\Omega_0-\theta_{e,o}) \tag{27}$$

The amplitudes of these oscillations are given as $J_{\ell m}H_{\ell m}$ where

$$H_{\ell mk}=a_0(r_e/a_0)^\ell F_{\ell,m,(\ell-k)/2}(I)[\{\beta_{mk}(\ell+1)-2k\}/\{\beta_{mk}(\beta_{mk}^2-1)\}] \tag{28}$$

where the β_{mk}'s are frequencies normalized to 1 cycle/rev:

$$\beta_{mk}=\dot{\alpha}_{mk}/n_0. \tag{29}$$

When I first derived these perturbations I was concerned about two resonance (long period) cases; when $\beta \doteq 0$ (nonzonal resonance) and $\beta \doteq \pm 1$ (odd zonal 'resonance'). We have just succeeded in banishing the ogre of the odd zonal case (at least for inclinations outside the narrow bands at critical inclination). An evaluation of $H_{\ell mk}$ for $(m,k)=(0,\pm 1)$ shows that the large term when $\beta \doteq \pm 1$ is exactly the same as that from the small-e theory just given. However, the linear theory here does break down for those cases close enough to zero frequency (and exactly 1 cycle) that small forcing terms can disrupt the steady rates (in $\dot{w}$ and $\dot{M}$) on which the theory relies. For nonzonal resonance the disruption is caused by those harmonics directly, for zonal resonance the cause is the wavering of the mean perigee rate due to its interaction with the long period $J(\text{odd})/J_2$ term in the inclination (of argument 'w'). But this term has 'e' as a multiplying factor so for the near circular orbit the non-linear resonance band around critical inclination is extremely narrow. Gedeon (1969) has postulated that for most of the nonzonal resonances of the earth, if the resonant beat period $[2\pi/\dot{\alpha}(\text{res})]$ is less than about 100 days the linear theory is adequate.

In any case though, when there are appreciable long period terms in the radius not accounted for by the reference orbit (with its gravity field) the computed reference trajectory can (by altering its initial elements) adjust itself to minimize these discrepancies over a short arc. With regard to terms in the radius with period much longer than such a tracking period (T), we can compute a worst case scenario and assume that the radial discrepancy is changing at its maximum rate (and remains at that rate over T) when we adjust the initial elements. In this case the maximum velocity for the (m,k) frequency due to the harmonic (ℓ,m) is:

$$|\Delta\dot{r}|_{max} = 2J_{\ell m} k a_0 n_0 (r_e/a_0)^{\ell} F_{\ell,m,(\ell-k)/2}(I) \tag{30}$$

where, at (nonzonal) resonance

$$k \doteq m\dot{\theta}_e/n_0, \tag{31}$$

since, then $\dot{\alpha}\doteq 0 = k(\dot{w}+\dot{M}) + m(\dot{\Omega}-\dot{\theta}_e)$, and $\dot{w},\dot{\Omega} \ll \dot{M}$. Faced with a steady increase of 'r' over a tracking arc length T, it is easily shown that minimizing just the radial residuals with an adjustment of the initial semimajor axis Δa will result in the following residuals over T:

$$\Delta r_{\ell,m,k(\dot{\alpha}\doteq 0)}(\text{rms}) = 2J_{\ell,m} k a_0 n_0 (r_e/a_0)^{\ell} F_{\ell,m,(\ell-k)/2}(I)T/(12)^{1/2} \tag{32}$$

How large are these nonzonal resonant perturbations likely to be? A rough estimate can be made for a given resonance using an order of magnitude estimate for the resonant inclination function F.

The inclination functions are closely related to the Fourier coefficients of a unit potential $U_{\ell m}$ (e.g.: $\mu, C_{\ell m}, S_{\ell m} = 1$) sampled uniformly on the satellite track over a unit "frozen" sphere (e.g., Wagner, 1983; Goad, 1987).

It is easy to show for example that the power of the set $F_{\ell mp}$; $0 < p < l$, is just $\langle U^2_{\ell m}\rangle$ along the same track. Thus if the resonant inclination function is a typical example of the set, and the track is a typical sample of the global field:

$$F_{\ell,m,(\ell-k)/2}(I) \doteq (\ell+1)^{-1/2} \langle U^2_{\ell m}(\phi',\lambda')_{track}\rangle = (\ell+1)^{-1/2}, \tag{33}$$

if $U_{\ell m}$ is a fully normalized potential harmonic (and F a similarly normalized function) since then,

$$\langle U^2_{\ell m} \rangle_{global} = 1. \tag{34}$$

Using $k_{res}=m_{res}\dot{\theta}_e/n_o$, Kaula's rule for fully normalized harmonic coefficients $J_{\ell m}$, and the specification for n_0 as the commensurate ratio R/D where the track repetition is achieved over R revolutions in D days, I find that (over a tracking period of T days):

$$\Delta r_{\ell,m,R/D}(\text{meters,rms})=(2170)mT\ell^{-2}(\ell+1)^{-1/2}(R/D)^{2(\ell-1)/3}(290.5)^{-1/3} \tag{35}$$

For example, for GPS orbits, R/D = 2 and ℓ,m = 2,2 is the leading resonance term so that for T = 10 days: Δr(rms) = 230 meters. But for Topex, where R/D = 127/10, the leading exact resonance is for (ℓ,m) = (127,127) and for this term $\Delta r_{rms} = 4.2\times10^{-11}$ meters. But the Topex repeat period (10 days) was set by oceanographic considerations and the first exact resonance is of extraordinarily high order.

The second case of "resonance" would arise for terms with frequencies near 1 cycle/revolution. These are fundamental long period effects in Δe and $e\Delta M$ modulated to near 1 cycle by the orbit frequency. They fall into two categories. Those from odd zonal harmonics (m,k=0,±1) are generated by the slow secular motion of perigee which can turn libratory near the two critical inclinations. The second case involves the much more prevalent non-zonal resonances for the radius, here involving the q=±1 terms that are not seen well in ordinary tracking which generally focuses on resolving along-track variations. In either case these near 1 cycle effects can be ameliorated (and are in the orbit determination process) by a small adjustment of initial in-plane elements such as e,M and/or w, to produce residuals which resemble a "bow-tie" over a tracking arc, being a 1 cycle/rev. variation (from the orbit correction) modulated at long period (see, e.g., Colombo, 1984a or Engelis, 1987).

To estimate the necessary adjustment and the resulting residual after orbit adjustment, it is convenient to consider the near 1 cycle term with frequency $\dot{\alpha}$ (for odd zonals, $\dot{\alpha}=\dot{\lambda}$) to be:

$$\dot{\alpha}= n_0+\dot{\Delta} \tag{36}$$

Then we can expand the radial perturbation for α,($C_\alpha \cos \dot{\alpha}t + S_\alpha \sin \dot{\alpha}t$) as:

$$\begin{aligned}\Delta r_\alpha = {} & \cos n_0 t[C_\alpha \cos \dot{\Delta}t+ S_\alpha \sin \dot{\Delta}t] \\ & + \sin n_0 t(-C_\alpha \sin \dot{\Delta}t+ S_\alpha \cos \dot{\Delta}t]\end{aligned} \tag{37}$$

which is an amplitude modulation of n_0 at slow frequency $\dot{\Delta}$. However, at this point this modulation is wholly mathematical since the result (prior to the orbit correction) is an unaltered $\dot{\alpha}$ oscillation. But if we impose a 1 cycle correction term (C_n,S_n) on this perturbation arising from an initial orbit adjustment, then the residual (perturbation minus correction) will be a true slow modulation which, when $\dot{\Delta} \doteq 0$ becomes (to first order in $\dot{\Delta}t$):

$$\delta r = \cos n_o t[(C_\alpha-C_n)+S_\alpha\dot{\Delta}t] + \sin n_o t[(S_\alpha-S_n)-C_\alpha\dot{\Delta}t], \tag{38}$$

the familiar "bow tie" effect.

An interesting aspect of this result is that it appears to say that as $\dot{\Delta}$ goes to zero (the fundamental period of the perturbation becomes infinite) it should be possible to "absorb" all of the perturbation in a 1 cycle adjustment, no matter how long the tracking arc. But this appearance is illusory since at the same time that $\dot{\Delta}$ goes to zero, the amplitude of the perturbation becomes unbounded as can be seen from the sensitivity function $H_{\ell mk}$(Eq. 28). In fact the orbit accommodation process always produces a limited residual signal no matter how close to resonance the pertubation is. To first order in $\dot{\Delta}T$ over a tracking arc of length T I find the residual signal with least power (determining C_n, S_n) to be:

$$\delta r = \cos n_0 t[S_\alpha \dot{\Delta}\{(T/2)+t\}] + \sin n_0 t[C_\alpha \dot{\Delta}\{(T/2)-t\}] \tag{39}$$

where (to first order in $\dot{\Delta}T$):

$$C_n = C_\alpha + S_\alpha(\dot{\Delta}T/2)$$

$$S_n = S_\alpha - C_\alpha(\dot{\Delta}T/2) \tag{40}$$

gives the necessary 1 cycle orbit adjustment to the perturbation. Notice in the linear theory C_α, S_α can still be unbounded but this occurs when $\dot{\Delta} \doteq 0$ and there, near critical inclination, the motion is not simply periodic (but is bounded) and the linear theory no longer describes the evolution.

Further, the residual after adjustment is:

$$\delta r(\mathrm{rms}) = [\dot{\Delta}T/(24)^{1/2}](C_\alpha^{\ 2}+S_\alpha^{\ 2})^{1/2} \tag{41}$$

But as stated above this result does not really depend on $\dot{\Delta}$ as $\dot{\Delta}$ goes to zero since for $\dot{\alpha} = n_0 + \dot{\Delta}$, $\beta = \dot{\alpha}/n_0 = 1 + \dot{\Delta}/n_0$, and using the sensitivity H yields:

$$\delta r(\mathrm{rms})_{\ell,m,k\ (\beta \doteq 1)} = [(6)^{1/2}/24]J_{\ell m} a_0 n_0 T(\ell+1-2k).(r_e/a_0)^\ell F_{\ell,m,(\ell-k)/2}\ (I) \tag{42}$$

In the case of odd zonal 'resonance', k= −1, while for the strongest (lowest m) nonzonal resonances (near 1 cycle), k=0. So these near 1 cycle terms tend to be stronger than the corresponding resonant radial effects at low frequency ($\beta \doteq 0$).

To complete the evaluation of the resonant radial errors after orbit adjustment, we can estimate the adjustment (for in plane elements) necessary to account for C_n, S_n. For long period (maximum velocity) behaviour the necessary adjustment of 'a' is:

$$\Delta a = J_{\ell m} a_0 k n_0 T(r_e/a_0)^\ell, \tag{43}$$

the average residual perturbation over T. To evaluate the initial orbit correction required to minimize δr in the 1 cycle cases it is necessary to use nonsingular elements to avoid the nonlinearities in M at low eccentricities. Working with e.sin w, e.cos w and w+M and keeping w+M constant in the adjustment I find the following corrections to initial osculating elements:

$$a_0(e\Delta w)_{cor} = C_\alpha[-\sin M_o + (\dot{\Delta}T/2)\cos M_o] - S_\alpha[\cos M_o + (\dot{\Delta}T/2)\sin M_o]$$

$$a_0(\Delta e)_{cor} = -C_\alpha[\cos M_o + (\dot{\Delta}T/2)\sin M_o] + S_\alpha[\sin M_o - (\dot{\Delta}T/2)\cos M_o]$$

$$(\Delta M)_{cor} = (-\Delta w)_{cor} \tag{44}$$

Lecture 2. Frequency Classification and Observability of Radial Variations

In the applications of the first order radial theory to altimetry it is important to understand how the frequencies (α:m,k) are organized and realized. Their structure can give us insight into both the observability of radial effects on the satellite and the recovery of geoidal and PST information from the altimetric equation.

Essential to the argument is the relationship of the gravitational frequencies $\dot{\alpha}$ to Fourier harmonics periodic in some time interval. We can anticipate that the condition of a commensurate orbit is what is required to make this conversion.

Let the satellite make exactly R revolutions in time DT_e, where D is a whole number and $T_e = 2\pi/(\dot{\theta}_e - \dot{\Omega})$. T_e, the time it takes the earth to make a complete rotation with respect to the satellite's orbit plane, is called the nodal day. The period (T_g) of a gravitational frequency is given by:

$$T_g = 2\pi/[k\dot{\lambda} + m(\dot{\Omega} - \dot{\theta}_e)] \tag{45}$$

But since $\dot{\lambda} = 2\pi R/DT_e$,

$$T_g = DT_e/(kR - mD), \text{ or } T_g(kR - mD) = DT_e \tag{46}$$

which says that every D nodal days the gravitational effect represented by the frequency (m,k) goes through kR - mD = N, an integer number of cycles. In other words, to first order in the eccentricity, the radial frequencies are Fourier harmonics on the commensurate orbit R,D. Only the one cycle "orbit frequency" n_0 will not repeat in D days unless D = the period of the rotation of apsides or the inclination is critical.

There is another requirement on the commensurate orbit which applies to an intuitive "observability" argument and at the same time to the most efficient realization of a field determined from continuous tracking data (the approach first discussed by Colombo, 1984b). For a "band limited" gravitational field (L X L, say) we would like to be able to specify just those commensurabilities R,D which will yield a unique wave number (N) for each frequency (m,k). Considering two different (m,k)'s m_1,k_1 and m_2,k_2 and a commensurability R,D, nonuniqueness would occur if:

$$m_1D - k_1R = m_2D - k_2R \, . \tag{47}$$

Excluding the case $m_1 = m_2 = 0$, $k_1 = -k_2$ which yields the same Fourier harmonics (in one revolution) for all orbits (commensurate or not) I find that uniqueness is always assured if R > 2m(max) or R > 2L.

Let us assume we have such a commensurable orbit, then the unique frequencies of the field can be illustrated by the diagram in Figure 6. Examining this diagram we see that for each surface harmonic m there are always the same number of Fourier harmonics (total of odd+even degree parity, sine+cosine terms) 2(2L+1) except for the zonals for which there are only L+1 harmonics. But the surface harmonics producing these Fourier harmonics decline from 2L (at m=1) to 2 (at m=L). Again, the zonals are unique with only exactly as many surface harmonics (L+1) as Fourier harmonics. Overall there are about four times as many Fourier as surface harmonics in a band-limited field.

Imagine also that we have a time series of the first order radial perturbations on the R,D orbit, exactly repeating every D nodal days. If we could exactly match these unique frequencies with an earth fixed surface function $S(\phi',\lambda')$, where $\phi'$is

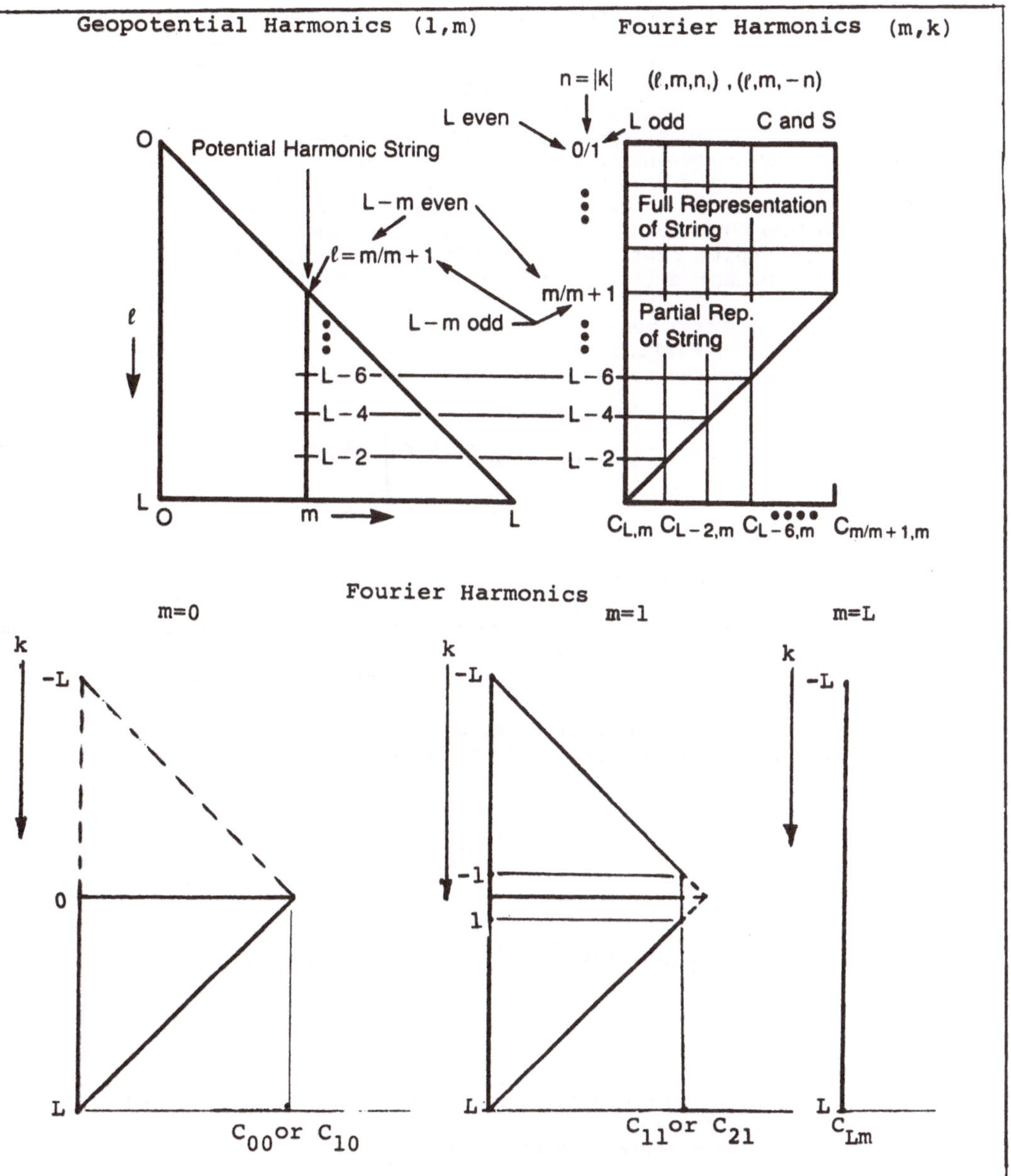

Figure 6. Gravitational Frequency Classification. Shown are the coefficients influencing each frequency α =(m,k) in the linear first order theory for perturbations of a near circular orbit. Note that the ratio of the spherical harmonics to Fourier harmonics is greatest for low order (m) as is the ratio of the strongest (|k| small) to weakest (|k| large) frequencies. For m=1 this discordance is further exaggerated because there is conventionally no (1,1) geopotential spherical harmonic. As a result the string of frequencies [1,k(odd)] can have a particulary strong representation by a mean geographic (surface) function.

geocentric latitude and λ' is geographic longitude, these radial perturbations would be unobservable from crossover altimetry. But such a function, as seen by the satellite, has the same frequencies as produced by the (earth-fixed) geopotential field. If we were to match an arbitrary surface function to the specific radial time series from the (band limited) geopotential we would see that (referring to Figure 6), 1; the matching surface function would always have the same band limitation as the geopotential, and 2; except for its zonal harmonics, the surface function would always leave residuals at every frequency. Furthermore, the spherical harmonic components of the surface function only appear in the same frequency strings as the geopotential components so that the closest match of the two functions (aside from m=0 which can be perfect) will be at m=1 while the worst will be at m=m(max) providing the power in the frequencies is about the same. In fact we might anticipate that the m=1 match will be even stronger because the surface function can have a (1,1) term which the geopotential lacks.

Let us make these ideas more definite by working out the case for first order zonal information. Recalling Equations (24)-(29) a typical $(\ell,0)$ term in the geopotential is represented at frequency (m,k) by the Fourier coefficients for radial effects:

$$(C_\alpha, S_\alpha) = C_{\ell,0} H_{\ell,0,k} \begin{vmatrix} \cos k\lambda_0, & -\sin k\lambda_0 \\ \sin k\lambda_0, & \cos k\lambda_0 \end{vmatrix} \begin{matrix} \ell \text{ even} \\ \ell \text{ odd} \end{matrix} \tag{48}$$

Now, for each ℓ parity there are $\ell(\max) + 1$ frequencies in the geopotential string $m, L[=\ell(\max)]$ running from $k(\min)=-L$ to $k(\max)=L$. Seemingly there are two independent Fourier coefficients for each frequency, or more information in the gravity perturbations than can be absorbed by a set of matching surface zonal harmonics. But we see immediately that the negative k frequencies are not independent of the positive ones (for m=0). In fact they can be lumped very simply together since every k (positive) has a negative counterpart. The result is that a given term $(\ell,0)$ is represented by the radial series:

$$\Delta r_{\ell,0} = \sum_k^{\ell \text{ par.}} C_{\ell,0} \sin k\dot{\lambda}t \begin{vmatrix} -\sin k\lambda_0 \\ \cos k\lambda_0 \end{vmatrix} [H_{\ell,0,k} {+ \atop -} H_{\ell,0,-k}] \begin{vmatrix} \ell \text{ even} \\ \ell \text{ odd} \end{vmatrix}$$

$$+ C_{\ell,0} \cos k\dot{\lambda}t \begin{vmatrix} \cos k\lambda_0 \\ \sin k\lambda_0 \end{vmatrix} [H_{\ell,0,k} {+ \atop -} H_{\ell,0,-k}] \begin{vmatrix} \ell \text{ even} \\ \ell \text{ odd} \end{vmatrix} \tag{49}$$

We have reduced the frequencies by half (not counting the constant term) but there still appears to be twice as many zonal Fourier harmonics as surface ones. But note the following symmetries for the zonal inclination functions:

$$\begin{aligned} F_{\ell,0,(\ell-k)/2} &= -F_{\ell,0,(\ell+k)/2} \; : \ell, \text{ odd} \\ F_{\ell,0,(\ell-k)/2} &= F_{\ell,0,(\ell+k)/2} \; : \ell, \text{ even} \end{aligned} \tag{50}$$

The other factor of H depending on k and β remains unchanged for negative k's so that we have finally:

$$\Delta r_{\ell,0} = \sum_k^{\ell \text{ par.}} 2C_{\ell,0} H_{\ell,0,k} \begin{vmatrix} \cos k(\dot{\lambda}t+\lambda_0) \\ \sin k(\dot{\lambda}t+\lambda_0) \end{vmatrix} \begin{matrix} \ell \text{ even} \\ \ell \text{ odd} \end{matrix} \tag{51}$$

Thus for ℓ even, only the cosine series is generated and for ℓ odd, only the sine. In sum, there are exactly the same number of Fourier as surface harmonics, and the surface function can completely absorb the radial zonal information.

To be more explicit for the general case of the (m,L) string of L+1 frequencies, the Fourier harmonics for them (generated by the geopotential) can be written as the matrix equation:

$$r_{2(L+1)x1} = H^r_{2(L+1)xE}\; C_{\ell mi,Ex1} \tag{52}$$

where the geopotential vector $C_{\ell mi}(C_{\ell m1}=C_{\ell m}, C_{\ell m2}=S_{\ell m})$ has dimension E=L-m+1 if L,m are of the same parity and L-m if L,m are of different parity. For all nonzonal cases E < L+1 and the equations cover both sine and cosine coefficients. Thus if the Fourier coefficients 'r' were given, except for the zonals, the $C_{\ell mi}$ would be overdetermined.

H^r is given by the product of the sensitivity factors $H_{\ell mk}$ (in Equation 28) and an appropriate orbit phase, sin α_0 or cos α_0. In matching 'r' to a spherical surface function $S(\phi',\lambda')$, with sensitivities $H^s_{\ell mk}= F_{\ell,m,(\ell-k)/2}$ times the same orbit phases) to these same frequencies, an unweighted least squares solution of the overdetermined equations:

$$r_{2(L+1)x1} = H^s_{2(L+1)xE}\; S_{\ell mi,Ex1} \tag{53}$$

would define the 'best fitting' (dimensioned) surface harmonics (of order m, parity ℓ) as:

$$S_{\ell mi}=[(H^s)^T(H^s)]^{-1}(H^s)^T H^r C_{\ell mi} \tag{54}$$

Then finally, at the surface point (ϕ',λ') the 'best fit' radial perturbation would be:

$$\tilde{r}(\phi'\lambda') = \sum_{\ell}\sum_{m}\sum_{i} S_{\ell mi} Y_{\ell mi}(\phi',\lambda') \tag{55}$$

or at each geographic point the geographic or 'best fit' component of the perturbation (so defined from the best fit to the time series) can be expressed as a linear sum of geopotential harmonics.

This approach, while interesting and intuitive has significant drawbacks. The most serious is that it fails to characterize the variable component of the radial perturbation at a specific geographic point in terms of geopotential harmonics. A complete solution to the problem of the geographic nature of orbit perturbations was first given by Rosborough (1986) and further elaborated by Engelis (1987).

In this new development, which also yields the variable component as a geographic function, the starting point is also the spectral representation (Equations 24-29). The aim is to replace the longitude arguments of that representation by their equivalent geographic posiitions. While this is readily accomplished for geographic longitude which always proceeds steadily (as a single valued function of time or orbit argument) it is not so for the latitude which is a multiple valued function of time. But Rosborough removes the ambiguities in the conversion to latitude by developing the ascending tracks (which do proceed uniquely, with repetition at multiples of the orbit period) separately from the descending. The average of the two solutions, the geographic mean, is often called the geographically correlated component. The functional form of his solution is:

$$r_{\ell m} = \sum_k H_{\ell mk}\{\Phi^M_{\ell mk}(\phi')[C_{\ell m}\cos m\lambda' + S_{\ell m}\sin m\lambda'] +$$

$$\pm\Phi^V_{\ell mk}(\phi')[C_{\ell m}\sin m\lambda' - S_{\ell m}\cos m\lambda']\} \tag{56}$$

where Φ^M and Φ^V are the mean and variable components and with the positive sign for ascending and negative for descending passes.

As Rosborough points out, this form suggests, and it is confirmed in simulations that where the mean perturbations are maximum, the variability is minimum and visa-versa. Rosborough's studies confirm that besides the m=0 harmonics which are perfectly correlated geographically, the m=1 harmonics have a particularly strong mean component while the higher orders give rise to about equal contributions of variable and mean parts. We have already given an intuitive explanation of the strength of the m=1 mean component but in fact the largest radial errors in any actual orbit determined from tracking is likely to be those from initial orbit error which we have seen will bias itself to absorb long period (and generally large amplitude) information which is in error in the reference trajectory. These induced initial orbit errors will of course produce a 1 cy/rev frequency which is not strictly 'gravitational'.

Consider any such 1 cycle/rev perturbation due to initial orbit error. It would have the form:

$$P_n = C_n\cos n_0 t + S_n\sin n_0 t \tag{57}$$

(In the following argument, for the next three paragraphs, we ignore the distinction between n_0 and $\dot\lambda$). Notice we proved (see Eq.51) that we could completely absorb odd or even zonal radial information in the sinusoids $\cos k(\dot\lambda t+\lambda_0)$ for k or ℓ even and $\sin k(\dot\lambda t+\lambda_0)$ for k or ℓ odd. The easiest way to see what these statements mean is to consider $\lambda_0 = \pi/2$ at $t = 0$ or the perturbation referenced to maximum latitude. Then both the even and odd zonal variations proceed from cos kt. But with respect to maximum latitude, a crossover will occur for a purely circular orbit at some time ±t(cross) where t(i)=t(cross) and t(j)= $(2\pi N/\dot\lambda)$-t(cross) (see Figure 7), where the j'th orbit is N revolutions past the i'th. But we see then that any zonal information is both perfectly symmetric proceeding backwards and forwards from maximum latitude and perfectly periodic in the nodal period $(2\pi/\dot\lambda)$, exactly as is the geographic latitude of the satellite referred to maximum latitude. Thus at any crossover the zonal information will be the same in either the ascending or descending pass. In other words, the zonal information is purely a function of latitude, or entirely geographic, which we surmised previously from the intuitive surface function approach to characterizing the radial perturbations of the geopotential.

In applications it is more convenient to reference the 1 cycle terms to the equator crossing and from the above argument we see then that $\sin\dot\lambda t$ is not observable at crossovers while $\cos\dot\lambda t$ is wholly resolvable by such observations with their maximum power on the equator and no power at max/min latitude. Since the unobserved sine component mimics the (1,0) zonal harmonic it can be called the north-south componenet of 1 cycle variation. The cosine term changes sign at every equator crossing and can be called the east-west component.

These simple facts should be encouraging to those who use a net of crossover altimetry to solve for orbit error referenced to a fixed orbital arc of presumably low error (e.g., Rapp, 1986). This is because the reference arc has been fit to direct tracking data predominantly away from the equator where the power of the component that will be unobserved in crossing arc analysis is greatest. It is likely then that this error component will be relatively small in the reference arc and consequently in the resulting adjustment. The other component should have no difficulty in being resolved by the crossing arc method as long as equatorial crossovers are used liberally (as they usually are).

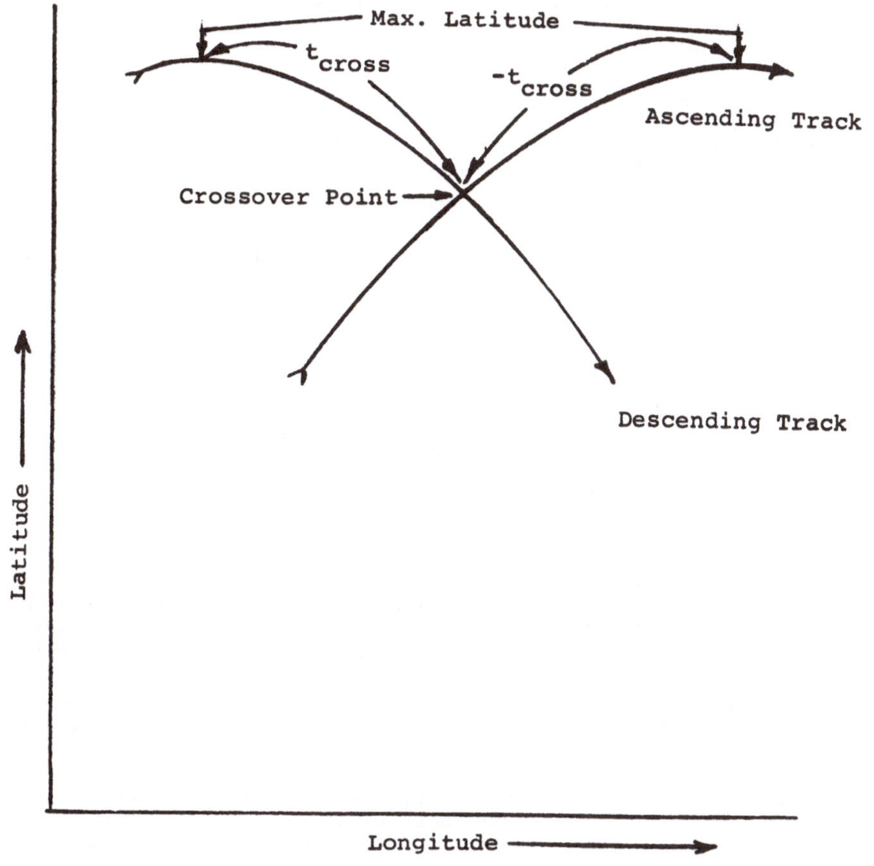

Figure 7. Observability of 1 cy/rev Perturbations at Track Intersections. Note that the intersections are always at equal time increments [±t(cross)] from maximum latitude points. Therefore the cosine term of 1 cy/rev variations (referred to max. lat.) yields the same value there on ascending or descending tracks. This term is therefore a purely geographic function and not observable by differences of it at these intersections.

Of course the actual 1 cycle orbit error is not quite of nodal frequency $\dot{\lambda}$ (as long as the inclination is not critical). In fact all of the 1 cycle error would be easily observable in crossovers if a net spanning the apsidal period were used in the adjustment. While this is generally too long for practical purposes (212 days for Seasat and Geosat), as long an orbital arc as possible should be used for the net. We will return to the question of "observability" of radial perturbations later in connection with the recovery of the PST.

A last consideration of the radial spectrum concerns its use with the full covariance matrix of a geopotential solution (we have already seen in Figure 4 an example of its use with just the variances). These error propagations can be made as a time series along the orbit but the use of the full matrix (necessary to achieve the full benefit or detriments of correlations) is an expensive calculation. A considerable economy is realized if we are only interested in a mean square error over a certain time. If that time extends sufficiently to discriminate the significant frequencies of the error spectrum, then it is easy to show that this mean error is just the power spectrum of radial errors:

$$< \Delta r^2 > = (1/2)\sum_{\alpha} [\Delta C_{\alpha}^2 + \Delta S_{\alpha}^2] \tag{58}$$

where ΔC_{α} and ΔS_{α} are sums of geopotential harmonic errors ($\Delta C_{\ell mi}$) pertaining to each resolved frequency (α=m,k). Taking the expected value and evaluating this sum in terms of the variance-covariance matrix ($\sigma C_{\ell mi} C_{rsj}$) of the gravitational field yields the following convenient formula (Wagner, 1985):

$$E < \Delta r^2 > = (1/2) \sum_{m,k} \lfloor H_{\ell mk}^2 \sigma^2 C_{\ell mi} \rfloor + \sum_{m,k} [\sum_{\ell,i,r>\ell} H_{\ell mk} \sigma C_{\ell mi} C_{rmi}] \tag{59}$$

But this drastic simplification (even over the repetition period D of a geostationary orbit) is only valid if each (m,k) frequency is unique. And it ignores correlation with a 1 (or even 2) cycle/revolution orbit error effect which generally would not be orthogonal with any gravitational frequency over a repetition period.

In any case, orbit determination over any finite tracking arc T, as we have seen, will try to accommodate errors at very low frequencies and those close to n_0 of which the long period odd zonal effects at $\dot{\lambda}$ is usually (but not always) the closest. We have indicated how to calculate these for both long period and close-to-1 cycle terms under ideal (perfect tracking) circumstances (Equations 32 and 42), but the induced effect of this orbit accommodation may no longer be periodic (the "bow tie" phenomenon) which would then introduce additional coupling with all other periodic terms.

If the averaging period is too short, even though the frequencies are resolvable (say D=1,2, or 3 days) not only will there be a certain number of sensitive frequencies which share the same information (i.e., have the same wave number) but it is likely that the deep nonzonal resonance with zero wave number over D (where kR-mD=0, for m≠ 0) will have low enough order to have strong effects not well gaged by the linear theory. This latter occurred for the R/D = 43/3 day geostationary orbit for Seasat in Sept-Oct 1978 [m(res)=43] but analysis has yet to elucidate this resonance. As to the former problem, the formula can be expanded to include all the (m,k)'s within the same wave numbers once these are identified. In fact we have already done this for the zonals (see equation 51) restricting k, in this case to just the positive integers.

In spite of these caveats the fact that actual satellite geopotential covariance matrices are block-diagonal in the sense discussed for the ideal high order geostationary orbit implies that over the ensemble of tracking arcs on many orbits in a large satellite solution, the decoupling of the frequencies is quite good. The formula has also been checked with numerical integration on a proposed Topex orbit (127 revolutions/10 days) using the Gem L2 and Gem T1 covariances in the presence of an orbit adjusted to a global set of simulated Doppler observations (Marsh et al, 1987). Gem T1 is a 36x36 geopotential model (and includes the solution for 66 principal ocean tidal components) from the laser, electronic and optical tracking of 17 satellites with primary emphasis on Lageos and Starlette. For a 3-day arc of such data, the radial errors along the orbit (rms) were within a few centimeters of the value from Equation (59) evaluated without long period effects. In this case, however the Topex orbit is at critical inclination where the long-period odd zonal term might be perfectly accommodated by the orbit adjustment. At other inclinations the situation is not so sanguine and it turns out that the long-period and/or resonance terms may contribute significantly to the radial error, even in an arc short in comparison to the periods of these effects.

At any rate the formula provides an ideal answer which may be acceptable in the case of most of the past, current and future altimetric satellites that have long repetition periods (D > 3 days). To illustrate the covariance propagation of radial errors (compared to just the variances), I have projected (In Figs. 8 and 9) errors of the Goddard Earth Model Topex 1 (GEM T1) (Marsh et al., 1987) into the radial variations for the Geosat Orbit (244 revs/17 days), which is close to Seasat's in July-Aug 1978. Gem T1 has a well calibrated covariance matrix which is

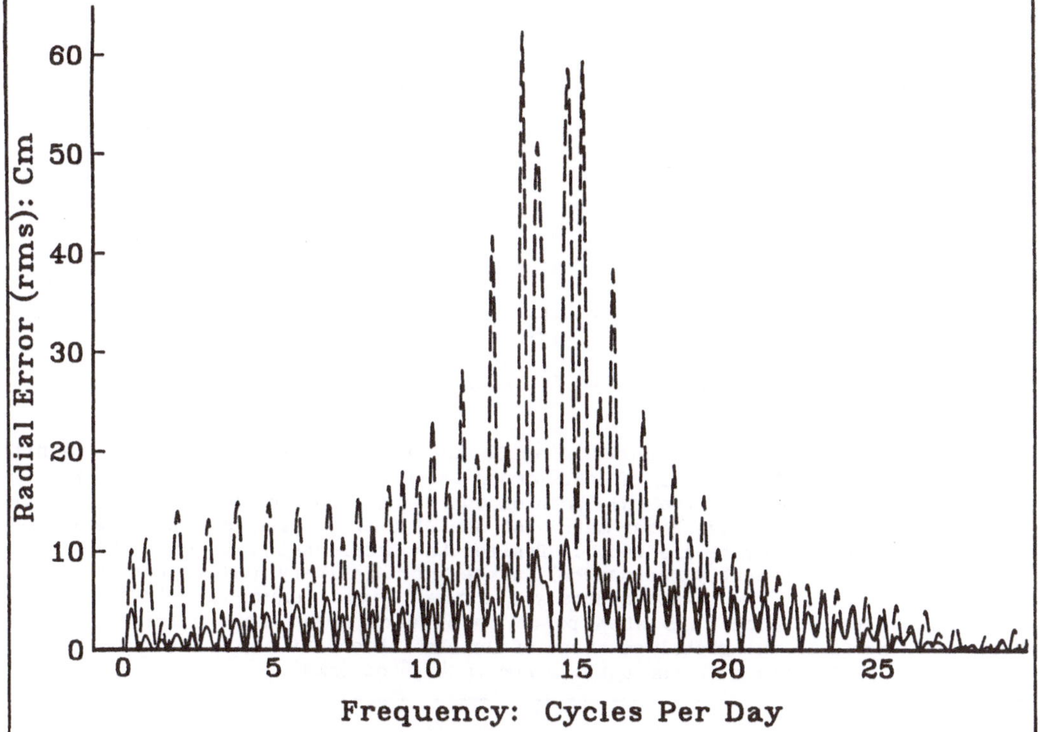

Figure 8. Radial Error Spectrum on Geosat from Gem T1. Accumulated in bins of 0.2 cy/day. Dashed Spectra from variances only. Solid from the full covariance matrix of Gem T1, total error = 45.5 cm (rms) but no long period (or close to 1 cy/rev) terms considered. Orbit altitude = 785 km, Inclination=108 degrees, mean motion = 244 revs/17 days (14.35 rpd). The great reduction in error considering the covariances of Gem T1 is due to the heavy influence of Seasat tracking in this model. Virtually no error is seen higher than 2 cy/rev. Equivalent spectra from the covariance matrix of Gem L2 accumulated 1.5 m (rms).

used here. The resulting small covariant errors compared to those from variances-only is due to the heavy Doppler tracking of Seasat in Gem T1. The result is that almost every radial frequency in a Gem T1 Geosat ephemeris will be about equally well determined, as shown in both the line and order spectra. This calculation also was done without long period terms (or those close to 1 cycle/rev.)

Figure 10 shows these results on a Geosat-type orbit (of 244 revs/ 17 days) over a wide range of inclinations for just the overall radial error. The

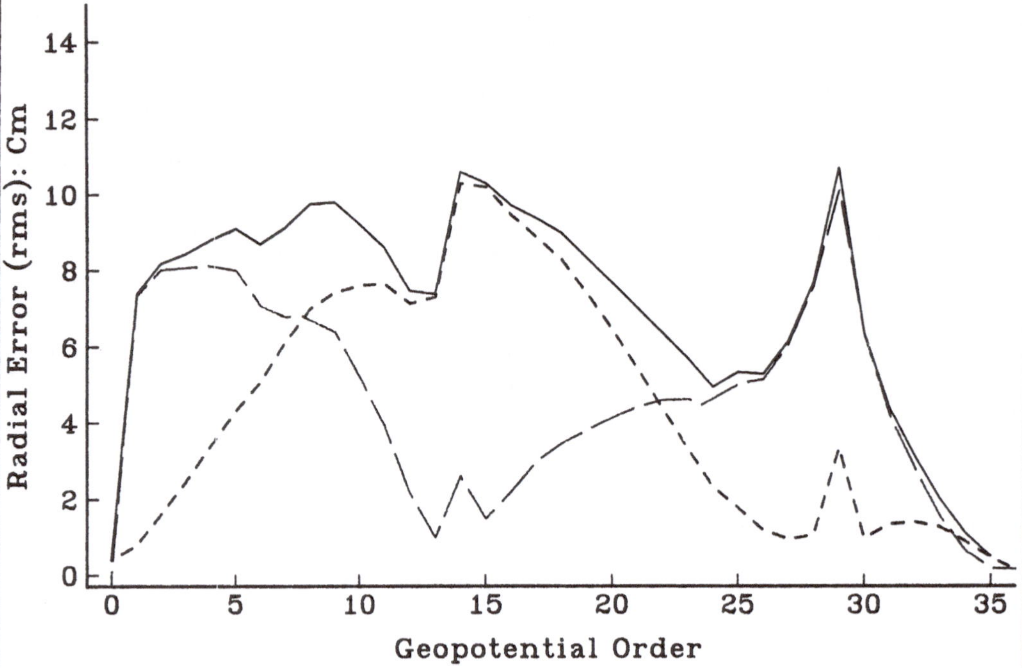

Figure 9. Radial Error Spectrum on Geosat from Gem T1 (By Order). Orbit specifications (see Figure 8). At each order is shown the accumulation of error in all its unique frequencies (solid), both of odd degree (long dash) and even degree (short dash) from the covariance matrix of Gem T1 (36x36) without consideration of long term (or near 1 cy/rev) effects. Besides the usual resonance peaks, here at order 14 and 29 (especially pronounced), there are relatively high errors at orders above m=1. Normally the lowest orders yield the largest errors, aside from the resonant orders. From this standpoint this spectrum is more nearly "flat", a reflection again of the global coverage of Doppler observations on the Seasat orbit in Gem T1. As in Figure 8, the total error is 45.5 cm (rms).

pronounced minimums in the solid curve are due to the heavy Seasat tracking at 108 degrees (and its polar mirror at - 64 degrees) as mentioned above, but also from heavy Starlette tracking (in Gem T1) at 50 degrees (with its mirror at 140 degrees). But note the large effects from long period terms at most inclinations not well observed by GemT1 orbits. The tracking period assumed (17 days, in the Exact Repeat Mission) may be rather long but will be attempted with this model (the drag and radiation forces on Geosat are only moderate, the solar cycle currently is near minimum). The dashed curve predicts no significant increase in radial errors for tracking over the ERM cycle from the long period odd zonal term (shifted to near 1 cycle). This is because Seasat data at the same inclination as Geosat is a strong constituent of Gem T1.

Incidently, Gem T1 orbits computed for Seasat (6-day arcs) show crossover errors of about 50 cm compared to about 2 m for orbits computed with only limited laser and S-Band data using a field (PGS-S2) determined from this and other extensive tracking on other satellites (Lerch et al., 1982). The 50 cm includes the observable part of the 1-cycle error which in any case is not assessed in Figures 8 and 9. Inasmuch as 50 cm is already at the level of global PST power we

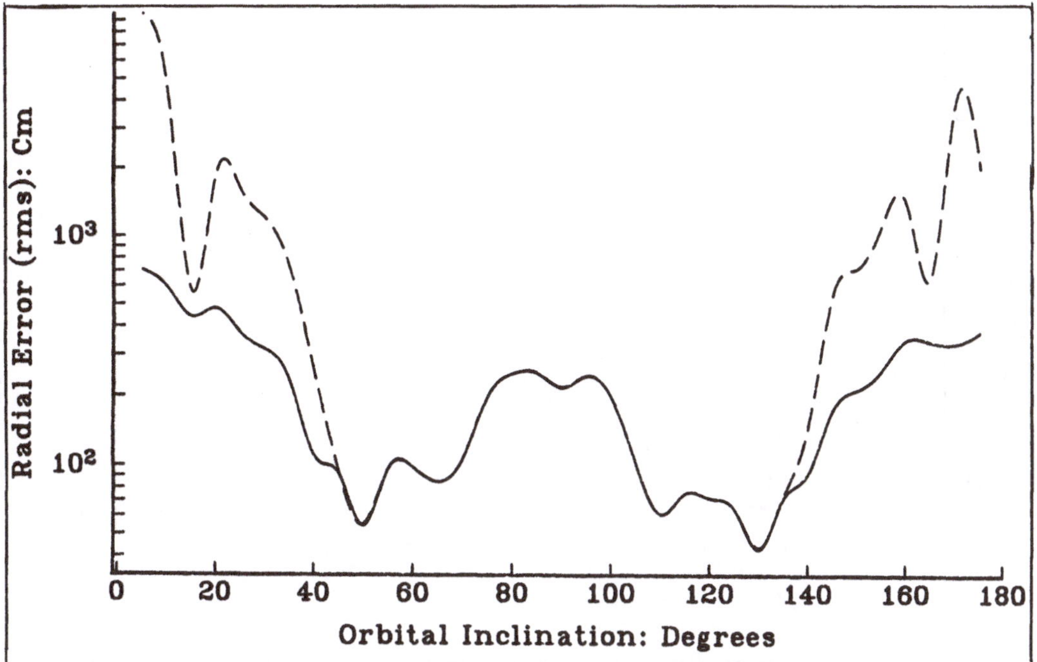

Figure 10. Radial Error on a "Geosat Orbit" Across Inclinations From Gem T1. Shown is the total commission error (rms) on orbits of 244 revs/17 days; solid line, without considering effects greater than a 17-day tracking period; dashed line, considering all effects. Note the large errors estimated at inclinations not strongly represented in Gem T1 (below 40 deg., near 90 degrees and their polar reflections). The polar symmetry of perturbations is due to the close similarity of geopotential sampling in direct or retrograde orbits between the same latitude bands. Similarly, note the sharp minimums at 50 and 108 degrees and their polar reflections of the strong tracking of Seasat Geos 3 and Starlette in Gem T1. Of particular interest, the errors from the odd-zonal long period term on Geosat actually dominates at inclinations not well covered by Gem T1 tracking.

are encouraged to believe that with only a small amount of improvement with the use of altimetry, the PST should emerge without undue ambiguity from the altimeter equation when Gem T1 orbits are used for Seasat (and Geosat).

How much further can we drive the orbit error down using altimetry? Rosborough's studies suggest that using only crossover altimetry, perhaps half of the orbit error will remain unobserved to infect a PST solution projected on the ocean's surface upon subtraction of the altimeter and geoid heights. This brings us squarely to the crucial questions concerning the errors inherent in our solution of the altimeter equation. We have assessed the level of the orbit term (r). What is the level of the geoid errors (N) in this equation?

Figure 11 gives a spectral view of these errors worldwide for Gem T1 compared to a PST spectrum derived from oceanographic data (using zeros for land values) and the geoid errors of an earlier satellite model, Gem L2 (Lerch et al., 1985). This figure suggests that the geoid part of the equation now allows an altimetric solution of PST (by subtraction of N) limited on the surface to about wave number 7 or 8 compared to about 4 in the recent past. This information also alerts us to the need to filter the output of any simple subtraction process or the higher frequency geoid errors will drown out the PST signal. Armed with these results let us look at a number of possible schemes for deriving PST from altimetry. .

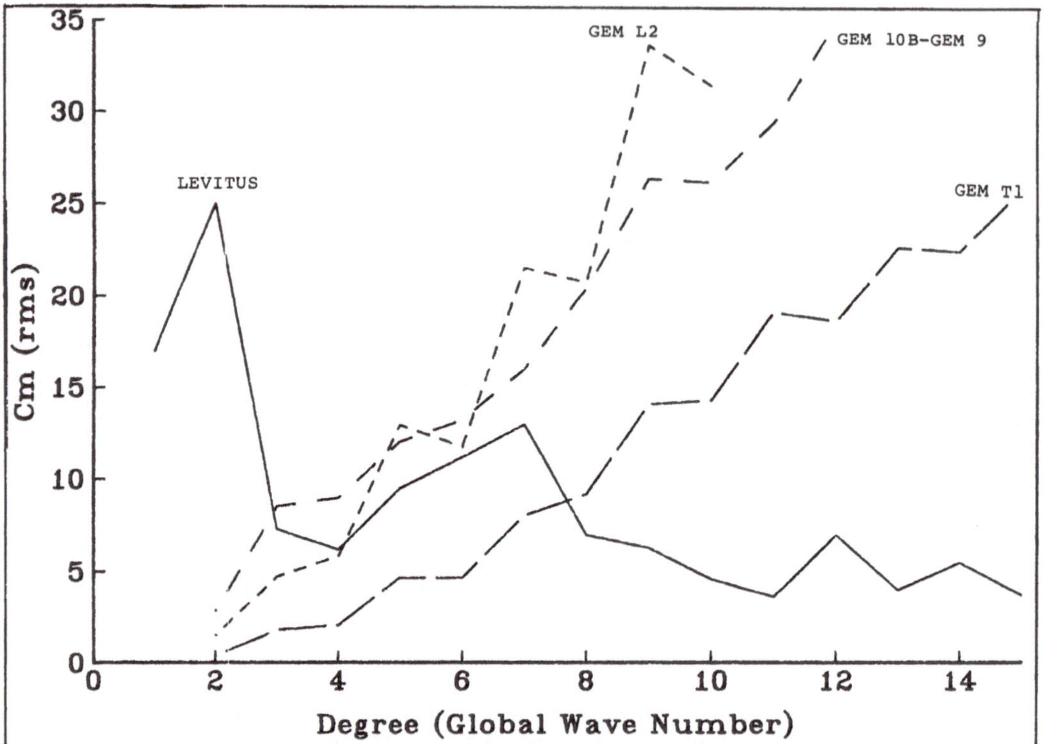

Figure 11. Global Geoid Error and PST Signal Spectra. The geoid spectra are degree variances (rss, worldwide) from the covariant models for Gem L2 (short dash), Gem T1 (long dash) and the simple difference field Gem 10b- Gem 9 (medium dash). The solid line is from a global expansion of the Levitus 1982 "Permanent" Sea Topography, reduced to zero mean and with zeros over land, due to Engelis and Rapp (1984). In general the geoid error of satellite-only models increases strongly with degree due to upward attenuation of the geopotential. Conversely the PST signal declines with degree because the currents that maintain it are mainly circulations of the broadest basins. Note that Gem T1 apparently has almost twice the resolving power for sub-basin oceanographic features as Gem L2. Note also that if Gem10b-Gem9 represented the true errors of current knowledge, all such features would be obscured in the simple method of PST determination by subtraction of the geoid from the orbit corrected altimetric sea surface.

Lecture 3. Determination of Permanent Sea Topography From Altimetry 1: Removal of Orbit Error

Methods for doing this consist of two general types involving solutions of the altimeter equation for PST: 1. by subtraction of an unadjusted geoid, and 2. with simultaneous resolution of the geoid. The first (simpler) scheme (which is the only one that has actually been tried (e.g., Mather et al., 1978; Tai and Wunsch, 1984; Engelis and Rapp, 1984; Douglas et al., 1984; Cheney et al., 1984) actually can be iterated to accomplish the second. Heretofore, subtraction schemes have used more or less self-contained methods of removing orbit error; they have relied as little as possible on both normal orbit determination from ordinary tracking data (ranges and range rates to fixed stations) and direct use of altimetry. The latter caveat was taken so that, at least in the first iteration the data used to correct the orbit were uncorrupted by an assumed PST. In the subtraction schemes, as we have said, the final surface is heavily smoothed to remove high frequency geoid error (see Figure 11). In the simultaneous schemes different parameterizations of the PST can be tried (block anomalies, surface harmonics, basin functions) and solutions with and without crossover data can be attempted (Wagner, 1986; Engelis, 1987).

Let's first discuss orbit correction with the subtraction schemes. We have already talked about the use of arc biases and tilts to remove orbit error. An even more primitive scheme is to do nothing but average the results of many surface profiles of sea height referred to a large number of separate arcs of data (e.g., Cheney et al, 1984). Since the strictly orbit error components at 1 cycle will tend to be random, the hope is a large number of orbits computed at different times and under different circumstances will reduce this source of error when averaged to yield an accurate mean surface. Perhaps the 'best' picture of the PST (Certainly the only one that spans more than a few months) was produced from such a crude grand average of Geos 3 and Seasat altimetry in 1976-78 by Cheney et al.(1984) (Figure 12). The interesting feature of this PST is its clear delineation of the

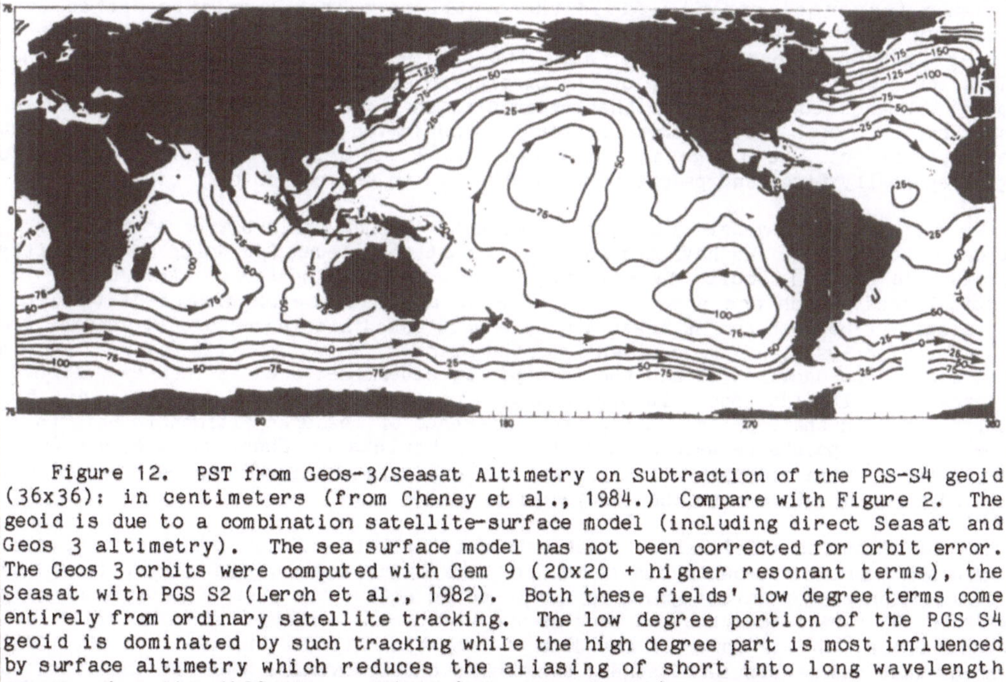

Figure 12. PST from Geos-3/Seasat Altimetry on Subtraction of the PGS-S4 geoid (36x36): in centimeters (from Cheney et al., 1984.) Compare with Figure 2. The geoid is due to a combination satellite-surface model (including direct Seasat and Geos 3 altimetry). The sea surface model has not been corrected for orbit error. The Geos 3 orbits were computed with Gem 9 (20x20 + higher resonant terms), the Seasat with PGS S2 (Lerch et al., 1982). Both these fields' low degree terms come entirely from ordinary satellite tracking. The low degree portion of the PGS S4 geoid is dominated by such tracking while the high degree part is most influenced by surface altimetry which reduces the aliasing of short into long wavelength errors when the difference surface (sea-geoid height) is formed. The subtraction surface shown here has been smoothed to the equivalent of 45x45 degree block means or 4x4 in spherical harmonics. The waviness of the contours is an artifact of the plotting routine. This surface clearly shows the global nature of the circumpolar Antarctic current probably because the averaging of orbit results has reduced the net error of the (1,0),(1,1) and similar long wavelength harmonics of the mean geographic radial orbit variation.

strong circumpolar Antarctic current, not well seen in the Douglas et al.(1984) and Engelis and Rapp 1984 PST's from just Seasat altimetry (Figures 2a and 2b). This current results in a general elevation of the whole southern ocean relative to the northern, reflected in a strong (1,0) surface harmonic which has no geoidal counterpart (if the reference earth ellipsoid is earth centered).

One of the intrinsic difficulties of extracting PST from (just) marine altimetry is that this surface component will be indistinguishable from the "unobservable" (from crossovers) north-south part of the 1-cycle/rev. orbit error. The only hope of seeing the constant signal of the PST at (1,0) without additional direct tracking data is to use a number of different orbital arcs with random north-south error and average the results. The success of Cheney and his colleagues in reproducing this important current is undoubtedly due in large part to such averaging.

The PST has other powerful components in even zonal harmonics [(2,0) especially] which are due to the great gyres in both the northern and southern oceans. However, for a near circular orbit the 2-cycle/rev. orbit error should be small and so these strong PST components should be readily distinguishable from their even zonal geoidal counterparts because 1: the PST terms have no dynamic effects on the orbit, and 2: the geoid error (and its consequences on the orbit) from even zonals should be especially small. Such harmonics are among the best determined from satellite tracking (e.g. Marsh, et al, 1987).

Probably the most interesting conclusion that we can draw from the observability analysis of radial error is that from a single satellite, using a single (errored) gravity field there will be a characteristic rather long wavelength pattern of geographically fixed orbit error which will be difficult or impossible to determine using satellite altimetry in the crossover mode alone. This error will be independent of the particular satellite arc. It will persist over all arcs computed from that orbit with that gravity field. For example, we saw that all zonal gravity error is totally unobserved in crossover altimetry. But also, much of the large m=1 gravitational errors on the orbit are also geographically fixed and pass through unobserved in the crossover process.

The attempt to use crossovers alone to define the orbit error function is doomed to failure. We have mentioned that most techniques for doing so implicitly recognize the difficulty by relying on one or a small number of fixed unadjusted orbital arcs which are presumed to be accurate because of excellent tracking coverage. (The need for at least a single unadjusted sloping reference surface for crossover analysis, was recognized from the start). A net of crossovers between the fixed reference arc(s) and the more poorly determined arcs is then established. Biases and tilts (and possible higher degree curves) are then used to model the orbit error in the poor arcs using the difference of measured altimetric heights at the crossover points between such arcs as constraints on these parameters (e.g. Rapp, 1986; Marsh and Martin,1982). All ambiguity in the system is resolved by the crossover constraints with the unadjusted arc(s).

But Cloutier (1981) showed that mathematically it was not necessary to assume any continuous or discontinuous model for the orbit error. He showed that the discrete crossover discrepancies alone could define the error function uniquely (within an overall constant) as long as an additional minimizing constraint was imposed, say on the power of the discrete solution itself. Sandwell et al.(1986) showed that the particular discrete constraint on the overall power of the solution proposed by Cloutier could be interpreted simply as requiring that the error function be as smooth as possible (have minimum slope power). Unfortunately, the "observability" results show that any such purely internal-looking constraint used with only crossover information can never recapture more than about half of the orbit error. Sandwell et al.(1986) actually demonstrated these pitfalls of crossover altimetry with simulated data but went on to show that only a very few absolute satellite heights (with respect to the earth's center) are necessary to remove most of the ambiguity in the crossover problem introduced by the geographic component of orbit error. The method of biases and tilts pinned to one or more "accurate" (but unadjusted) arcs may not remove all of these "physical" ambiguities of the problem. It only certainly removes the "mathematical ones".

It is undoubtedly true that the reason only a few absolute heights can make a dramatic difference in crossover analysis is because the fundamental nature of the mean geographic error is at long wavelength, eg., at near 1 cycle or fewer per revolution. Yet our "intuitive" analysis of the unobservable part in the frequency domain also shows that some unobservable signal remains at all wavelengths. How much? Referring to Equations (53) and (54), let:

$$H' = H^s[(H^s)^T H^s]^{-1} (H^s)^T. \tag{60}$$

Then referring to the radial harmonics $\delta r_{2(L+1)x1}$ for each m,L string generated by the surface function $S_{\ell mi}$ that has been fitted to the full radial perturbation data, we can show that the relation between the covariances of these "geographic frequencies" and the geopotential covariances is given by:

$$E\ (\delta r \delta r^T) = H' H^r \sigma C_{\ell mi} C_{rsj} (H^r)^T (H')^T. \tag{61}$$

Based on these results we could lay out a network of transponders on the ground track of the geostationary orbit so as to capture these (presumably) few significant frequencies (which would otherwise be unobservable from crossovers) in an optimal way.

Sandwell et al. (1986), following Goad et al. (1980) and Douglas et al. (1984) have used a strict Fourier series to model the orbit error with crossovers, a global rather than a local discontiuous function (as with biases and tilts). If the data were continuous this would obviously be the most efficient method but since it is not, I have found it more convenient to work with the gravitational frequencies themselves. We have seen that these are always a subset of the Fourier harmonics on a geostationary orbit. To them it is only necessary to add a few strictly orbit frequencies which (unfortunately) are not orthogonal to the gravitational ones in an arbitrary repetition period (D).

The actual difficulties with the time series (or Fourier) methods of resolving the orbit error have been due more to the desire by the groups doing this research to get their results without doing any traditional orbit determination with external tracking data. Yet without some direct height data, we have seen that a good part of the orbit error is unrecoverable from a crossover scheme. The only absolute external heights available to date (of sufficient precision) have been the direct marine altimeter heights themselves. But the modeling of these heights by the reference sea surface (geoid+tides+variable and constant sea surface topography) at the time the Seasat data were first analyzed, was perhaps at the 2 meter level. The dominant non-orbit error source was the short wavelength part of the geoid (at the time, available only from inadequate Geos 3 altimetry). Since then a new sea surface ("geoid") model has been developed by Rapp (1986) incorporating Seasat altimetry as well. A continuing and disquieting thought about these surface models though is that they all have small but unknown amounts of orbital error in them since they were constructed by the straight averaging or the bias and tilt master arc method just mentioned.

Thus, the try for complete independence in the time series method is also illusory. However, what the time series researchers found fortunately was they needed only a small amount of this direct altimetry but well distributed to minimize aliasing (into the orbital and gravitational frequencies) from the surface model errors. But there has been no simulation analysis to account for aliasing in this method just as there has been none to account for unobservability using biases and tilts. I will now try to remedy this situation for the promising time-series method.

Lecture 4. Determination of PST from Altimetry 2: Simulation of a Subtraction Method.

The outline of the simulation is as follows. A perturbed "Geosat" trajectory (to a repeat track after 72 revolutions in 5 days) was generated numerically [using a Runge-Kutta 8th order scheme (Fehlberg, 1966) in rectangular coordinates] using the Gem 10b field (36x36) (Lerch et al., 1981). In this trajectory, every 50 seconds, satellite heights were taken above a "sea surface" consisting of the Gem 10b geoid plus a 36x36 version of the Levitus, 1982 PST (Engelis and Rapp, 1984). To these 8666 observations, pseudo-random noise at a 10 cm level was added. A similar reference trajectory was then computed from nearly the same starting elements using the Gem 9 field (Lerch et al., 1979), with the energy of the two trajectories being the same. Heights in the reference trajectory were taken with respect to just the Gem 9 geoid and then differenced from those in the perturbed trajectory to form residual "sea height" errors. At the same time I computed two other files of pure radial errors and pure surface height errors, the first consisting of just the orbital part and the second of the surface part of the total height error (Gem 10b geoid + Levitus - Gem 9 geoid). The pure radial orbit error in the Gem 9 (reference) trajectory was 256 cm (rms). The pure surface part of this error was 208 cm (rms global). As mentioned these values were typical of such errors a few years ago.

The first question I wanted to answer was: what were the significant frequencies in these two distinct time series? In Figure 13 I show the result of an harmonic analysis of them. From a total of over 2600 gravitational frequencies

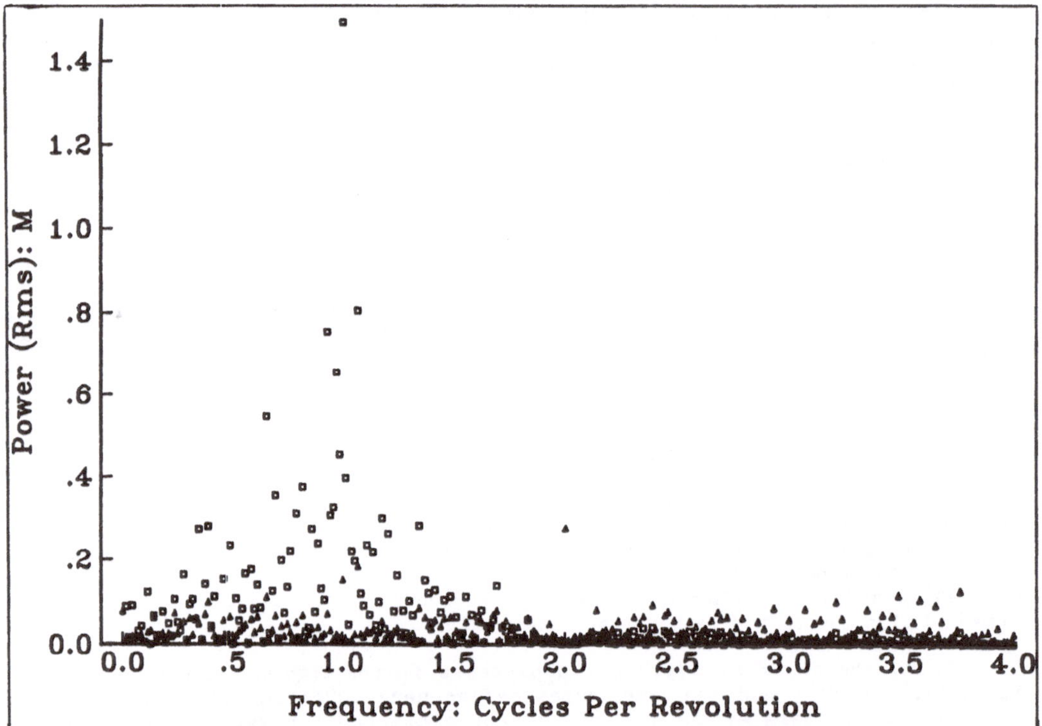

Figure 13. Spectra of Discrepancies in Two Geosat Trajectories. Shown is the power (rms) in frequencies determined from two 5-day trajectories: Boxes are from pure radial orbit differences in the two ephemerides, Gem10b(36x36)-Gem 9(20x20), total difference = 256 cm (rms). Triangles are for only the pure residual sea height along the common ground track, Gem10b-Gem 9 geoid +Levitus, 1982 sea topography (36x36). The dynamic radial discrepancy spectrum is sharply peaked to 1 cy/rev with little power over 2 cy/rev while the residual sea height spectrum is essentially flat (out to 36 cy/rev, not shown above 4 cy/rev) with noticeable strength only at 2 cy/rev from the raised equatorial gyres of the oceans. The marked distinction between these two spectra gives hope of separating the geoid part of the surface function (which is tied to the orbital effects) from the PST. Orbit: I=108 degrees, h=785km, mean motion=72 revs/5 days.

in this data, only 122 had power over 2 cm (rms) in the pure radial error spectrum, all of them of less than 2 cycles/ revolution and strongly peaked near 1 cycle. The surface height error spectrum had a completely different shape, showing basically a white noise structure except for the rather prominent line at 2 cycles/revolution (1 revolution = 6050 seconds) arising from the strong northern and southern ocean gyres. These distinct spectra were what encouraged me three years ago to believe that the direct altimeter height measurements which combine these error sources could be made to separate the two fields of information PST and geoid which seem hopelessly bound together (on the surface). The idea was that in a joint solution the mild (and flat) geoid part of the surface component must be made repsonsible for the dramatic peaking, while the PST component, with declining power at higher frequencies, would have no consequences on the radial orbit error spectrum (see also Figure 4).

The next question concerned the inherent power of crossover altimetry to recover the pure radial orbit error. I first formed 367 well distributed differences of the radial error at marine crossovers in the 5 day arc. These differences amounted to 248 cm (rms). It is interesting that these differences are actually smaller than the 256 cm (rms) value for the pure radial error. Even including the land crossovers, the differences of the full set are also slightly less than the absolute radial error. If there were no geographic component of orbit error the expected differences of error at crossovers would be $(2)^{1/2}$ greater than the absolute error. Again this illustrates the rule that about 1/2 of the orbit error is unobserved by crossover altimetry.

I now attempted to recover the 122 leading gravitational frequencies (including 1 cycle orbit terms) by a constrained least squares solution from just the crossover data with the help of the mildest a priori information on the Fourier harmonic errors (300 cm). These latter were necessary to obtain a solution in the face of a near singular normal matrix. My interest was simply to find how many external (direct height) constraints were needed to resolve the 'singularity' occasioned by the 'unobservable' geographic error in the radial error residuals. Since the parameter covariance matrix obtained was highly correlated I first looked at the 244 independent eigenvalues of this matrix (Fig. 14).

If V is the variance-covariance matrix for a vector of parameters 'x' whose errors are correlated (i.e., V is non diagonal), then a vector of parameters 'y' whose errors are uncorrelated can be formed as a linear transformation of 'x': y = Bx, such that:

$$E\, \Delta y\, \Delta y^T = BVB^T = \lambda \tag{62}$$

where λ (here) is the diagonal matrix of the eigenvalues or variances of 'y' corresponding to the eigenmatrix B of V. The so called eigenvectors of V (to which the λ correspond) are the columns of the B^T matrix.

Figure 14 shows that (essentially) all but 8 of the 244 independent frequency parameters are well determined by the crossover information, confirming the results of Sandwell et al. (1986). But specifically, what frequencies were causing these (relatively) few difficulties? Figure 15 shows the variances (rms) of the recovered (dependent) Fourier harmonic solution. The eight clearly offending frequencies evidently, taken as a set, coalesce to yield nearly the same radial residual at the crossover times. The fact that the critical sine and cosine correction terms always have about the same (large) error merely means that it is a linear combination of these that (essentially) define the well determined independent parameters 'y'. Considering the remarks earlier it is not surprising that most of these poorly determined frequencies belong to low-m strings. It should also not be a surprise, but it still was something of a shock nevertheless to see how small the recovered radial error differences were at the crossovers (17

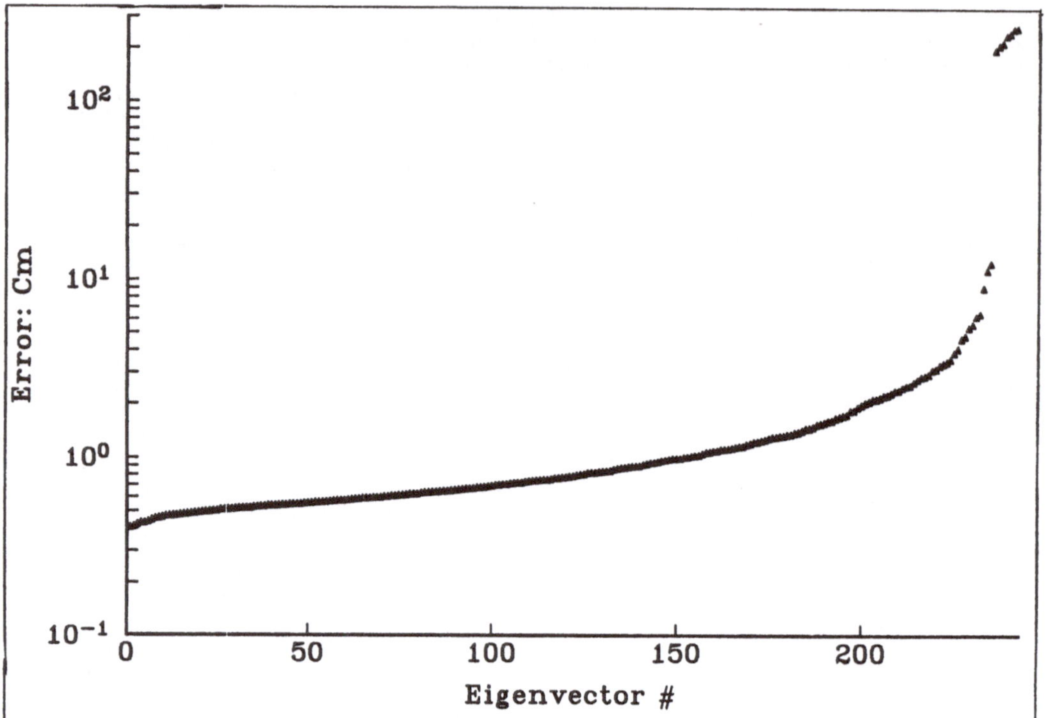

Figure 14. Eigenvalues of the Recovered Significant Radial Difference Spectrum from two Geosat Trajectories, Gem 10b-Gem 9 in cm. (See Figure 13 for specifications of the simulation.) Only the 122 frequencies which had power over 2 cm were determined here from differences of discrepancies at 367 marine crossovers in the 5 day arc. The resulting Fourier harmonics (determined by least squares fitting) were transformed to 244 independent parameters using the covariance matrix of the solution because a fair number of them were highly correlated. Of the eigenvalues (errors) of these independent parameters, only the last eight are seen to be ill determined (over 100 cm errors), the others are determined to about 10 cm or better. The a priori error of the (dependent) Fourier harmonics used in the Bayesian least squares recovery was 300 cm. After the radial difference recovery, crossover residuals were 17 cm. But the resulting radial residuals (original discrepancies-recovered) were 366 cm (rms), worse than the original differences!

cm, rms) and yet how poor was the actual error recovery (366 cm globally, rms). We started with the errors from about 2500 frequencies with combined power of 12 cm, rms that were not adjusted in the solution. Yet the solution error was actually worse than the original trajectory error which was only 256 cm, rms.

It should be noted that none of these poorly determined frequencies were truly 'singular' since they all started with a priori errors of 300 cm in each component. Even the worst of the recoveries (of m,k = 1,-1) still had sufficient non-geographic or variable information to reduce its formal error from 300 to 225 cm.

Further experiments with this crossover data augmented by direct residual height data globally distributed (as from a laser tracking network or from ground transponders or lake reflections or even ocean returns from areas with good absolute tide control) show that only a small set of these "absolute" heights (8-16, also weighted at 10 cm) can reduce the overall level of orbit error from the 2-3 meter level (in this rather poorly determined reference trajectory) to about 30-40 cm. But no optimization of this procedure (with respect to the geographic location of these external heights) was attempted, to take advantage of the

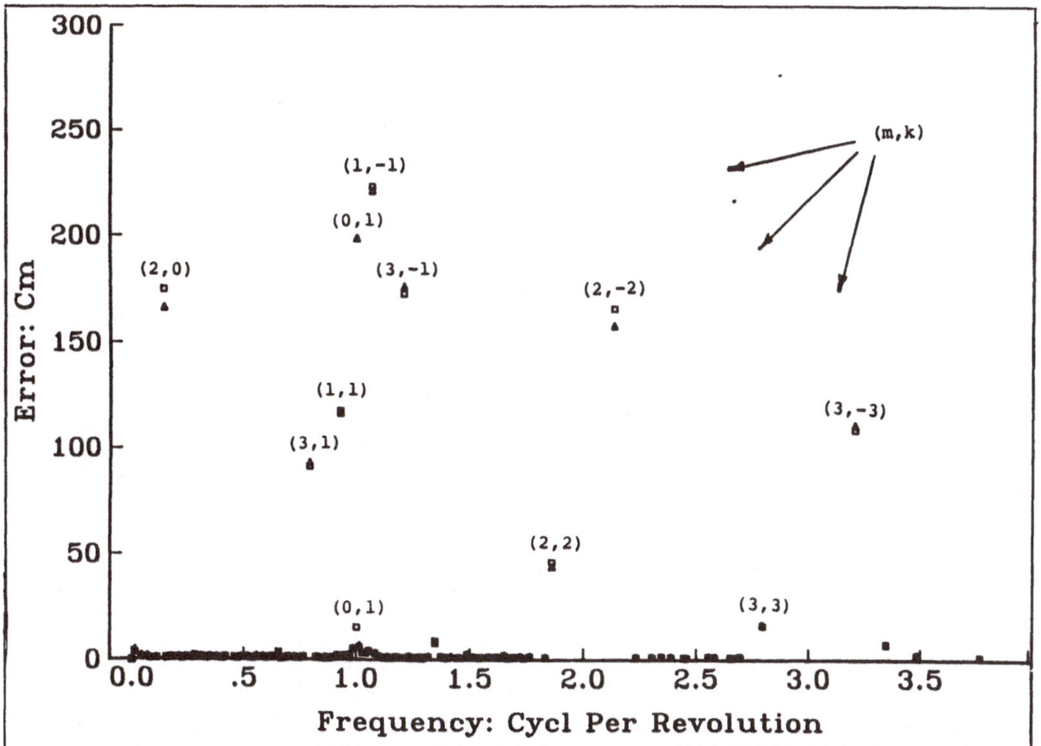

Figure 15. Error Spectrum (Fourier Harmonics) of Radial Orbit Recovery from Simulated Geosat Trajectories. (See Figures 13 and 14 for orbit and frequency specifications.) Errors in the Fourier coefficients (boxes for cosines, triangles for sines) of the radial difference recovery for 244 components from a least squares fit to residual differences at 367 marine crossovers. The most poorly determined (m,k) frequencies are noted. They appear to carry the bulk of the "constant" geographic information in the radial discrepancies (Gem 10b-Gem9). But only (m,k)=(3,1),(1,1) and (1,-1) also have a strong orbital signal in this simulation (see Fig. 13.) Note that all the critical frequencies are of low order (m<4). But even these have some observability (at crossovers) since the a priori errors used for all terms (in the fit) was 300 cm.

specific geographic correlation found in the results from the crossover-only solution. In fact we do not have access to such accurate (10 cm) data and must use the marine altimeter heights themselves which as mentioned before could only be modeled a few years ago to the 2 meter level (worse in the ocean trenches, better in the zones with smooth bottom topography). The fact that these direct height data also contain PST information is not so important because we hope to iterate the solution with our own values of PST later. More serious will be the PST information in the geoid part of the reference sea surface which will be left out of the final solution unless we iterate a new geoid as well as a PST. In any case, to resolve the few frequencies poorly determined from the crossovers, the direct heights need only be weighted lightly, but should cover wide areas of the ocean to reduce the aliasing caused by non uniform sampling.

Another aid to the use of direct marine altimetry that seems reasonable and would not necessitate a mask to avoid all rough ocean bottom areas (trenches, seamounts, and fracture zones) would be to low pass filter (or smooth) the heights and reduce the conditioning on the orbital frequencies in the presence of such data. The simplest filter to employ would be a straight running average one (called a box filter) which for a time series yields the following amplitude reductions:

$$(C_n', S_n') = (C_n, S_n)[\sin(n\Delta_t/2)/(n\Delta_t/2)] \quad : n \neq 0$$
$$= 1 \quad : n=0 \tag{63}$$

where Δ_t is the averaging interval and the reference time 't' is at the center of the interval.

How much direct altimetry, with how much smoothing and how much weight (compared to the crossover data) is necessary to achieve a respectable reduction of the orbit errors over the oceans? Smoothing aims to suppress unwanted short wavelength geoid error while remaining sensitive to the low frequency information in the radial dynamics. In Figure 13 the geoid plus PST error spectrum actually is almost "white" out to the 36x36 limit of the fields (36 cycles/rev). Ideally we would like to suppress all of it but the low frequency part must be retained to resolve the orbit dynamics. A target averaging interval to span at least 2 cycles/rev (>3000 seconds) seems a reasonable compromise. However, we might anticipate difficulties with such severe smoothing because there are observability problems with a few orbital frequencies of second and third order (m) of shorter period than this (see Figure 14 again). Of course knowing what those critical frequencies are in this case (Gem 10b-Gem 9) we might also just suppress (or not solve for) them when we condition the crossover data. But as Figure 13 shows, their power is negligible in any case. So in conditioning the crossover data but not knowing that this particular frequency is suspect, the only fair thing we might do is to enforce small a priori errors on all high orbital frequencies.

In fact, being guided by such spectra as shown in Figures 4 and 13 I have used in subsequent data reductions a priori errors for all such frequencies consisting of a composite power law which is surpressive at both high and low ends of the spectrum (20 cm at the low, 10 cm at the high) and peaking at 1 cycle (100 cm). There is a price to be paid, however for this rather indiscriminate use of a priori information. Since its use is a prejudged zero constraint on those parameters, the strong crossover data are not permitted to exert their full power to resolve the many more (relatively) 'observable' frequencies. The result is, of many frequency recoveries with both crossover and direct marine altimetry with varying a priori information, amount and weight and degree of smoothing, a reduction of orbit error (over the oceans) was achieved from the 2-3 meter level to 30-40 cm. At the same time in almost all cases it was possible to keep the crossover discrepancies to under 20 cm, close to the pseudo-random noise level. But unfortunately, because of the aliasing from the significant low frequency portion of the surface errors (see Figure 13 again) still present in the smooothed direct altimetry, that level could not be pushed down further.

In the best frequency solution, using 661 crossovers (weighted at 14 cm each) and 1854 smoothed altimetric heights (to a sea surface consisting only of the Gem 9 geoid) weighted at 500 cm each, the orbit error solution still had discrepancies of 40 cm (rms over ocean areas) from the true radial errors in the trajectory. The solution had crossover discrepancies of 23 cm (rms) and the smoothed marine height residuals were driven down from 235 cm (as simulated) to only 47 cm (rms). Clearly it would be more desirable to have a solution for the frequencies where the crossovers were fit better and the heights worse since the latter are meant only to resolve the frequency ambiguities. Still, even if a perfect solution were obtained, the geoid errors in Gem 9 would preclude a decent PST recovery (by this subtraction method) beyond about wave number 4.

To illustrate this geoidal limitation on PST recovery, Figure 16 shows the 36x36 model of oceanographically determined PST that we used (Engelis and Rapp, 1984) to simulate the actual sea surface addition to the Gem 9 geoid. Obviously, if we left the Gem 9 geoid unadjusted and unfiltered, its high degree errors

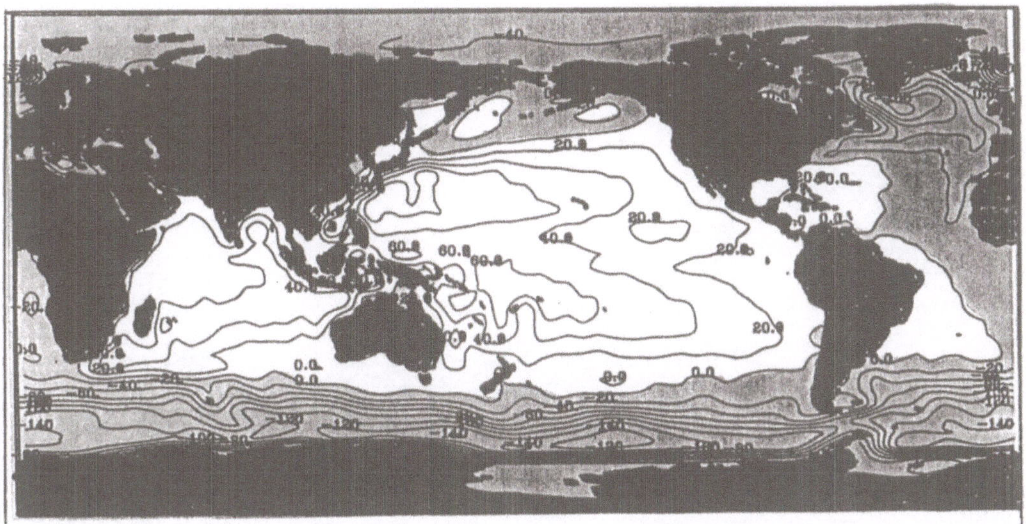

Figure 16. "Permanent" Sea Topography from Hydrographic Data (36x36). In cm (from Levitus, 1982); harmonic expansion from Engelis and Rapp (1984) (data adjusted for zero mean in oceans, uses zero's on land). The data,1x1 degree stearic anomalies referred to a surface of no motion at 2200 m) represents averages over decades. Note the strong circumpolar rise representing the Antartic current and the broad ciculation patterns in the northern and southern subtropical oceans (geostrophic flows circulate clockwise around highs in the north and counterclockwise in the south).

(almost the same as those of Gem L2 in Fig. 11) would almost entirely erase the PST here and this scheme would yield almost no reliable information on it. As already mentioned, even with current knowledge, the geoid is perhaps known in greatest detail only at the 1 meter level and at that contains some amount of corrupting PST information which will be lost in an uniterated scheme. But we might hope to recover at least the long wave ocean basin information (the dominant power in the PST spectrum) by smoothing the resulting surface. If so, and Figure 11 encourages us to believe that we can, how do we measure how well we have done by such smoothing?

In the 'subtraction' method I assume here that the best we can do with a reduction scheme (including smoothing) is to compare it to the same reduction on perfectly known data. For example, Figure 17 shows the best residual sea surface that would result from smooothing the sum of the Levitus surface and the Gem 10b-Gem 9 geoid sampled only at marine points (and every 50 seconds) on the 5-day Geosat trajectory. If the reference trajectory error were driven to zero by our solution, the resulting subtraction surface smoothed by fitting surface harmonics (6x6) to that particular "true sample" would result in the surface of Figure 17. Contrast this "rough" picture (distorted by the geoid errors of Gem 9, albeit smoothed but not sufficiently) to Figure 18 which gives a view of the Levitus, 1982 surface (36x36 version) alone sampled over all ocean areas and smoothed by a 6x6 (least squares fitted) set of surface harmonics. Finally contrast both of these to Figure 19 which gives the residual topography smoothed to 6x6 after using the 'best' orbit solution for the radial frequencies in the simulation (from the direct and crossover altimetry). Certainly the equatorial highs do show grossly in this picture, but the circumpolar aspect of the Antarctic current is lost and many anomalous features are spurious. Comparing Figures 18 with 17 and 19 however, we see that most of this distortion is due to geoid error, not poorly modeled orbit error.

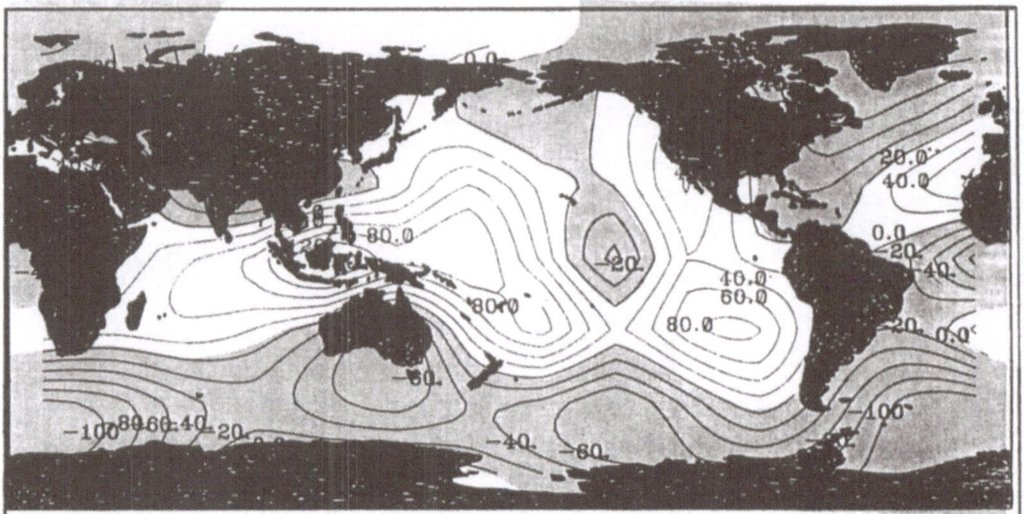

Figure 17. Best Achievable Sea Topography (6x6) in Subtraction Simulation. The contour interval in this and subsequent figures (except 25 and 27) is 20 cm. This surface, Gem10b-Gem9 geoid + Levitus is the sea topography that would result from subtracting the Gem 9 geoid from a perfectly determined sea surface if Gem 10b was the correct gravity field and Gem 9 the reference (known) model, the conditions of the subtraction simulation. The surface was determined from a least squares fit (of 6x6 surface harmonics) to marine sampling of the full difference geoid and Levitus models along track in the simulated Geosat 5 day trajectory. Refer to Figure 11 and 16 and note particularly the great distortion of the Antarctic current, the introduction of spurious circulation patterns in the Southwest Pacific and the generally greater power of all the patterns in the Indian and Pacific Oceans.

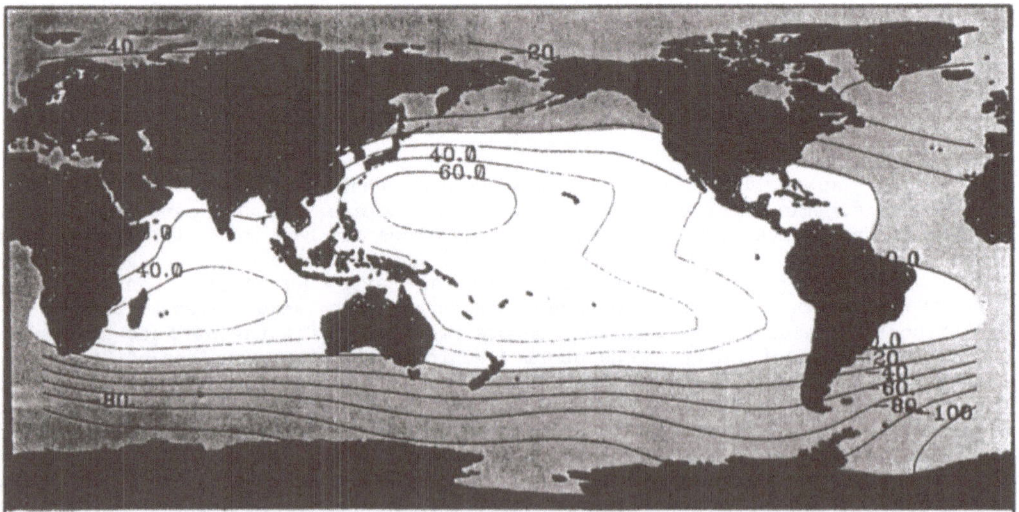

Figure 18. "Permanent" Sea Topography from Hydrographic Data, in centimeters from a least squares fit (6x6) to the 36x36 Levitus, (1982) surface in marine areas only. Compare with Figure 17. The (36X36) model (Fig. 16) was used in the simulation of sea heights (with Gem 10b) from the perturbed Geosat trajectory.

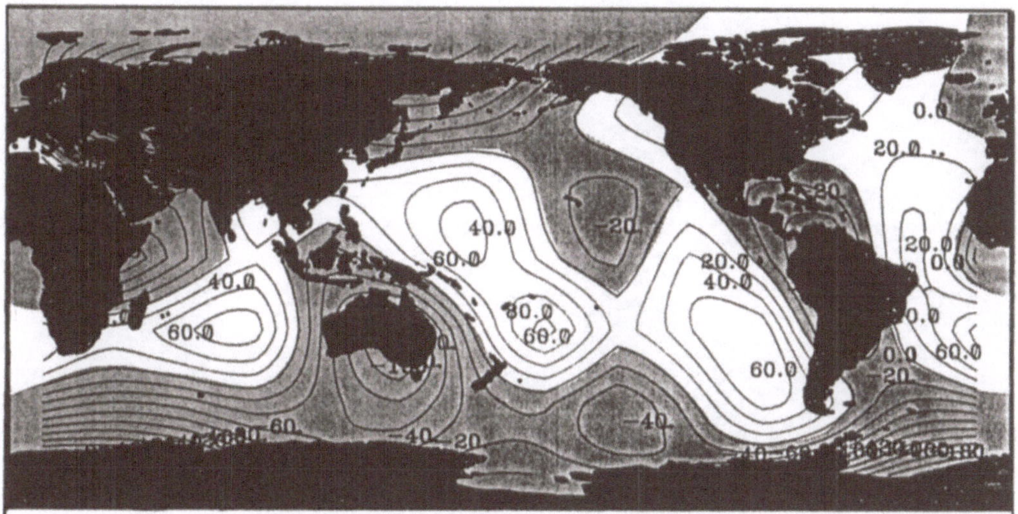

Figure 19. Residual Sea Topography (6x6) in centimeters. A least squares fit to orbit corrected sea heights minus the Gem 9 geoid from the 5 day simulated reference trajectory (along track sampling). The "error" of the corrected orbit (using crossover and direct altimetry) in this simulation was 40 cm (rms over marine areas). The "best" topography that can be achieved in this subtraction method is given in Figure 17 (assumes zero orbit error). Figure 18 gives the "true" 6x6 representation of the PST used in the simulation. The unresolved "orbit error" has added features in the Atlantic Ocean that are not present in either the "best" achievable or the "true" PST, but the Indian Ocean pattern is actually closer to the truth (fortuitously) than in the best achievable map.

At a scale of 4x4 in smoothing, the comparisons are even better, as you might expect, but at a sacrifice of some real detail. The PST surface for the oceans at 4x4 is shown in Figure 20 . Note how the structure of the South Pacific gyre, in particular, is lost compared to Figure 18 . The best residual topography possible (without geoid adjustment) at this scale is shown in Figure 21 (using along-track sampling). Here finally we see the smoothed geoid errors no longer break up the Antarcti current though geoid error distortions still remain (note the relative high in the North Atlantic and the false extensions of the gyres in the Southeast Pacific and Indian Oceans). Now also the comparisons of the surface found after the orbit solution, with both these "best" surfaces are also better (Figure 22). The distortions seen in it are clearly due to geoid errors even though the (unsmoothed) orbit solution itself is in error at the 40 cm level. Evidently the geographic component of this error is small. But it is hard to predict under what circumstances (of direct height data sampling and weight) this fortunate occurrence can be made permanent.

Lecture 5. Determination of PST from Altimetry 3: Simulation of a Simultaneous solution for the Geoid.

We noticed in the last 'subtraction' experiment that we were severely limited in the PST we could extract from altimetry because 1: the direct sea heights we were forced to use to resolve 'singularities' in a few of the orbital frequencies introduced errors in all of them because of large sea surface modeling errors, and 2: part of the same modelling errors, from the unadjusted geoid, caused

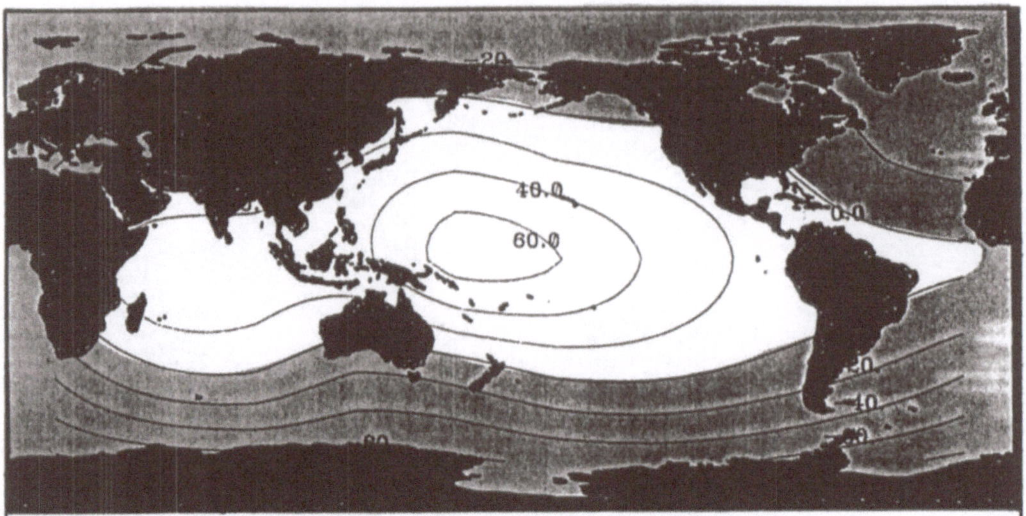

Figure 20. PST from Hydrographic Data(4x4) in centimeters. From a least squares fit to the full Levitus model (36x36) in marine areas only. Compare with Fig. 18. Important details of the Indian, Pacific and Atlantic ocean circulation patterns have been lost with this additonal smoothing.

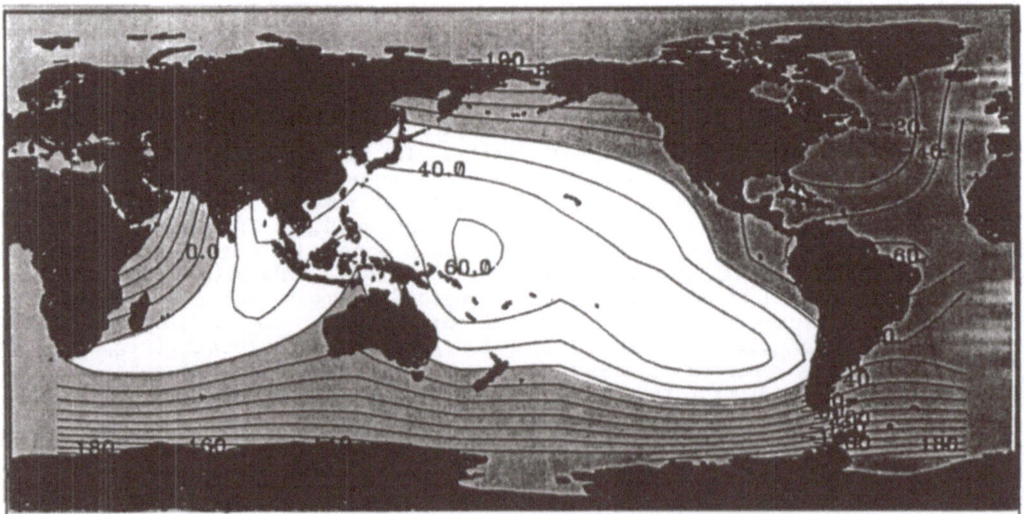

Figure 21. Best Achievable Sea Topography (4x4) in Subtraction Simulation, in centimeters. Derived from fitting along track sampling of Gem10b- Gem9+Levitus full height data on Geosat 5 day trajectory. Compare with Figure 17. At this additional smoothing the strong Antarctic current stands out clearly and overall (unresolved) geoid errors cause only minor distortions in Pacific and Atlantic circulation patterns. The exaggerated strength of this surface in polar regions is due to a lack of along-track sampling there.

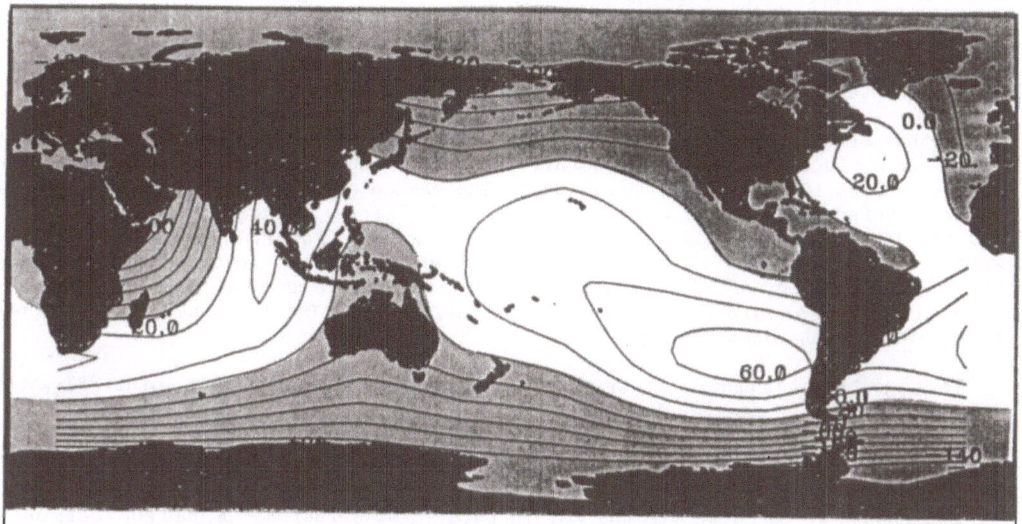

Figure 22. Residual Sea Topography (4x4), in centimeters. Derived from the orbital solution in the subtraction simulation. Compare with Figure 19 and Figure 21. This realization is now much closer to both the best achievable and the smoothed PST (Fig. 20) than the 6x6 version.

unacceptably large errors in the resulting PST surface upon subtraction of the orbital solution.

What is necessary for a better PST solution from this process is a geoid adjustment either iteratively after the first PST solution or simultaneously with it. The principal is this: As Figure 13 shows, even though the frequencies of PST, geoid and orbit error fields in the radial measurements are the same, the spectra should be sufficiently distinct and the two surface fields sufficiently overdetermined (in their spectral information) that a joint conditioning should separate them adequately. Wagner (1986) showed through simulations with direct altimetry alone that this is possible though there are problems with (1,0) of the PST in particular. But in his study the effects of sea surface modeling errors from unadjusted geoid terms were not directly simulated. It was concluded though that smoothing to reduce these anticipated high frequency surface errors would probably be necessary in the conditioning of the data to the solution parameters.

Engelis (1987) also showed through simulations that an improved solution could be obtained by using crossover data to further separate the geoidal from the PST fields by sharpening the necessary connection of the geoidal part of the surface variation with orbital errors and also by using a well based geopotential covariance matrix to precondition the gravity solution. In both of these studies the separation of the surface solution (geoid and PST) was also aided by a priori information on the errors of the PST solution based on oceanographic data. Again though, the effects of necessary truncation of the adjusted sea surface model was not considered.

In considering truncation through an error analysis of a simplified global soution using data over an ideal world-ocean, however, Wagner (1986) found that these high degree effects propagated into the lower degree fields to such an extent the recovered PST would be unacceptable unless 1: the geoid adjustment were carried to degrees higher than 50, or 2: the direct altimetry data was smoothed to the resolution of the recovered geoid.

But since the geoidal truncation error (following Kaula's rule) is roughly $64/\ell(max)$ meters in a band limited field (Chovitz, 1973), even in a 180x180 model the truncation error is an appreciable fraction of the PST. Therefore smoothing of the direct altimetry will probably always be necessary, at least in the first step of an iterated solution.

Other strategies proposed to overcome truncation error such as masking out rough sea surface areas or using detailed geoids to model the direct altimetry seem promising also but have their drawbacks. The masked areas are often in regions of wanted surface currents and the detailed geoids, since they have all been realized (to date) with altimetry that have not been corrected for PST, are contaminated with a certain amount of this information which will be lost in the subsequent first stage recovery of PST. Again, however, the whole process of PST recovery from altimetry should be viewed as an iterative one. The lost information should be recoverable at a later stage.

To gage the seriousness of truncation error in the context of a joint PST-geoid solution I used the same radial trajectory data as in the "subtraction" experiment, but now solved directly, not for frequencies of orbit error, but for a limited set of PST parameters and geopotential coefficients from the direct and crossover altimetry.

How should we characterize the PST in this joint solution? A surface harmonic function $S_{\ell mi}$ is an obvious choice since the power of the PST is mainly at low degree (Fig. 11). But because these functions would compete with the gravity coefficients for the same signal frequencies (at least in the direct altimetry), perhaps other parameters (e.g., basin functions or block anomalies) should be considered since these could be more easily constrained by external hydrographic information. It is even possible to make the problem fully constrained by the requirements of physical oceanography (following Wunsch and Gaposchkin, 1980) and by oceanographic data (such as bottom gages and sounders). But exactly how this should be done is a topic for the future.

My own experimentation with block anomalies (Wagner, 1986) led me to the conclusion that by themselves, that is without particular interblock constraints, they offer no special advantage over surface harmonics in representing the PST. At least the harmonics can be preconditioned according to a reasonable declining power law (similar to Kaula's rule).

An objection to the time series used (e.g., in Wagner, 1986) for surface harmonic representation is that the sensitivity coefficients of the surface function to the (m,k) frequency are not simply $r_e F_{\ell,m,(\ell-k)/2}$ (Engelis, 1987, p.91). This approximation assumes the satellite is travelling uniformly in a circular orbit over a spherical earth. More realistically the satellite moves with a slight 2 cy/rev wavy motion (from earth oblateness). The variation (along-track) is of order ±40km (see Eqs. 19). The radial variation over the earth ellipsoid is ±10 km, producing a change in geoid sensitivity of 10% for 30th degree terms. Similarly the mislocation of the smallest blocks from the wavy along track motion can be as much as 12% of the size of the smallest resolvable features for 30th degree terms. Unfortunately the geoid errors which need most correction from altimetry are those of high degree, and in total these may accumulate to over 1 m in some marine areas. Thus if analytic solutions from the time series representation of the geoid are seriously pursued, these second order effects will need to be addressed if a 10 cm (or better) geoid is the goal. Here we will only discuss the ellipsoidal correction and leave the more complicated problem of the effect of the unsteady satellite motion to the future.

Rapp (1986, Equation 112) has given the geoid height on an earth reference surface in terms of geopotential harmonics (using Bruns' formula) as:

$$N = (\mu/r\gamma)\sum_{\ell}\sum_{m}\sum_{i}(r_e/r)^{\ell}C^{*}_{\ell mi}Y_{\ell mi}(\phi',\lambda') \tag{64}$$

where γ is here the gravity at the surface point and the $C^*_{\ell mi}$ are the usual geopotential harmonics except for the low degree zonals which are those harmonics minus their reference values on the surface (generally taken to be an ellipsoid). Linearizing Equation (64) by retaining only terms linear in the flattening constants f,m (Heiskanen and Moritz, 1967), I find:

$$N \doteq r_e[1-f_1-f_2\sin^2\phi'] \sum_\ell \sum_m \sum_i (r_e/r)^{\ell-1} C^*_{\ell mi} Y_{\ell m} \tag{65}$$

where $r=r_e(1-f.\sin^2\phi')$ is the radius to the mean earth ellipsoid of flattening f and mean equatorial radius r_e at geocentric latitude ϕ', $f_1=f-(3/2)m$, $f_2=(5/2)m-3f$, $f \doteq 1/298.257$ and $m=(\dot{\theta}_e^2 r_e)/\gamma_{eq.} \doteq 0.00345$. But $(r_e/r)^{\ell-1}=(1-f.\sin^2\phi')^{1-\ell} \doteq [1-f(\ell-1)\sin^2\phi']$ which is adequate till $\ell >> 100$. If 'u' is the argument of latitude (argument of perigee plus true anomaly) in an orbit of inclination 'I', we have $\sin^2\phi'=\sin^2 u.\sin^2 I=(1/2)(1-\cos 2u)\sin^2 I$. Then the geoid height is given as:

$$N \doteq r_e \sum_\ell \sum_m \sum_i \{1-f_1+[f(\ell-1)-f_2](1/2)\sin^2 I(1-\cos 2u)\} C^*_{\ell mi} Y_{\ell mi} \tag{66}$$

The leading sum of Eq. (66) is just the familiar spherical approximation to the geoid height. The correction term in the flattening is given as:

$$N_c = r_e\{-f_1+[f(\ell-1)-f_2](1/2)\sin^2 I[1-\cos 2(\lambda_0+\dot{\lambda}t)]\} \sum_\ell \sum_m \sum_i C^*_{\ell mi} Y_{\ell mi} \tag{67}$$

The correction to the frequencies of the spherical approximation is now clear. It will consist of two parts, each of the order of the flattening and each linear in ℓ. The first merely corrects the sensitivity of the ℓ,m,k terms. But the second introduces at each frequency (m,k) two additional frequencies $k'=k \pm 2$ due to $2\dot{\lambda}$, the frequency of the ellipsoidal sampling. In the simulations here however, for simplicity I have used only the spherical formulation of the geoid height.

Since the simulated data were generated from 36x36 fields (PST and gravity) a solution for a truncated version will illustrate the effects of omission error on the joint recovery. The PST with declining signal power was solved to degrees less than 10 in these solutions with little difference since their harmonics were constrained with a declining power law in any case. The major problem was in the gravity truncation and unfortunately the error field Gem10b-Gem9 had large increasing errors to well beyond degree 15 (e.g., Fig. 11 for $\ell<14$). Even with gravity solutions to 14x14 and direct altimetry smoothing to reduce the aliasing of higher geoid harmonics, the joint solution while superior to the subtraction results, still left significant distortions in the recovered PST.

Figure 23 shows this result as a byproduct of the most satisfactory joint solution. In this experiment 1862 direct marine altimetric heights were used (weighted at 75 cm for values being the average of data over 550 seconds along track) with differences of heights at 367 marine crossovers (weighted at 50 cm for each height).

Compared with the equivalent "best" 6x6 PST with the subtraction method (orbit correction only), this joint solution is a vast improvement as seen in its fidelity to the equivalent filtering of the Levitus surface (see Figures 18 and 19 again). In Figure 23 the Antarctic current has the proper strength and is clearly circumpolar. The flow in the major ocean basins have the correct sense too (with some small distortions and an overall diminished power). In retrospect we might have relaxed the a priori PST errors to allow more actual surface information to

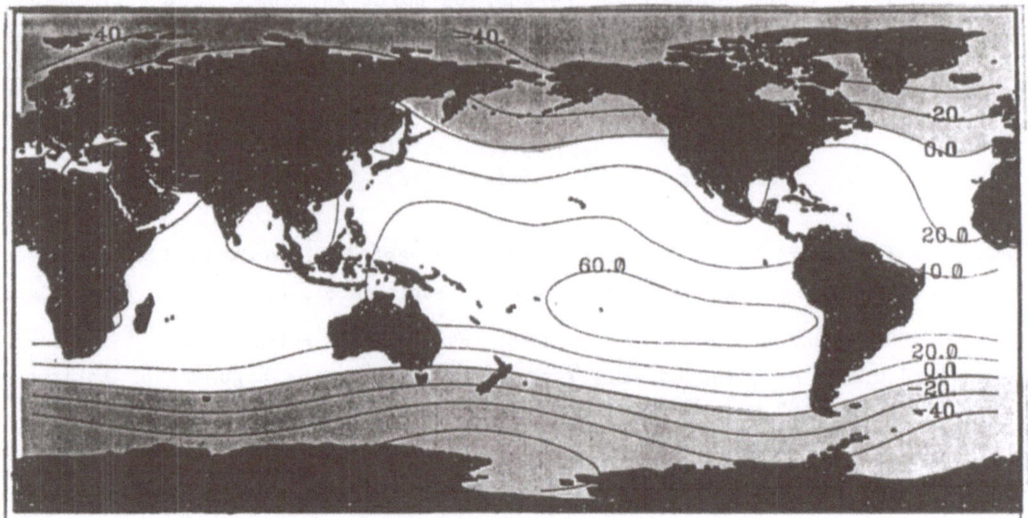

Figure 23. PST (6x6) from Joint Solution (Geoid 14x14), in centimeters. A much closer approximation of the simulated topography at this resolution than with the subtraction method (compare this map and Figure 19 to Figure 18). In addition to the displayed fidelity, the solution inverse (scaled by the residuals after the fit) gives a conservative estimate of the actual differences between this surface and that of Figure 18).

influence the solution. The direct data smoothing was equivalent to shutting off all surface harmonic information of wave number shorter than 11 which left geopotential harmonics from degree 11 to 14 to be determined essentially by the crossover altimetry that was not weighted nearly enough to do this job properly (the dynamic sensitivity to high degree terms being small to begin with).

There are many more options available to us considering the truncation problem in a joint solution than when we merely have to correct the orbit error. The trickiest involves the conditioning of the important (1,0) PST surface harmonic which can only enter the solution through the single frequency m,k=0,1 determinable soley from direct altimetry. This is the harmonic which sets off the South Pacific Ocean at a higher level than the North due to the Antarctic current. First recall from the discussion of observability that there are only just enough frequencies to determine all the zonals uniquely (which makes the radial variations "geographic" over a geostationary orbit). Now when we add PST harmonics to the solution we see that the combined zonals are about twice underdetermined from the available frequencies. A priori information therefore is essential for these since no information is gained for them from the crossover data. But the (1,0) PST harmonic has an additional difficulty. Not only must this order 15 cm surface harmonic compete with the odd zonal geopotential for the direct information but it also must distinguish its frequency from the nearby much stronger fundamental orbit error. And to make matters still worse, the competition is from that part of the orbit error which is unobservable from crossover data! We should also add (as if the difficulty were not bad enough yet) that the odd zonal geopotential error is especially sensitive to the m,k=0,1 frequency since it carries its long period information (shifted to $\dot{\lambda}$ by the orbit frequency). A small dodge which has proven useful to this almost hopeless task (with just marine altimetry in these two modes) is to remove this frequency from the geopotential conditioning leaving the generally small errors in the odd zonals to be corrected from the higher odd integer cycle/rev effects in the direct data.

A Footnote on New Results From the Subtraction Method

While I believe the future of the determination of PST and its consequence, the general circulation of the oceans, lies with the joint solution method (ultimately including oceanographic data and physical constraints), the recent development of better detailed geoids and more accurate low degree satellite-only geopotential models has afforded a better look at the PST from mean sea surfaces of the past. Teo Engelis (private communication, 1986) has looked at some of this developement in some detail and I want to show briefly the kind of improvement which he has highlighted.

I refer here specifically to the surfaces computed from Seasat altimetry during the summer of 1978 implicit in the PST maps of Engelis and Rapp (1984) (Fig. 2a), Douglas et al. (1984) (Fig. 2b) and Tai and Wunsch (1984). Tai and Wunsch used Rapp's mean sea surface for Seasat and the less accurate Gem 9 satellite geoid model, yet produced a highly credible PST whose features are similar to those we have seen and will now discuss with respect to the newer geoid models. One of the disadvantages of the subtraction method is that the two principal sources of error, orbit and geoid must be discussed separately. This has proven difficult in the past because the parameterization of both surfaces has been so disjoint. The orbit error implicit in the sea surfaces is a function not only of the original tracking model but of the peculiarities of the ad hoc methods used to correct for it. No rigorous method of projecting these errors has yet been attempted. Even the geoid error in these maps is difficult to assess because most maps use a combination of a low degree satellite model with a higher degree geoid which has already been contaminated with direct altimetric observations that contain PST information. The situation is not that critical when only smoothed versions of such a composite geoid are actually used since most of the error then will be in the low-degree satellite portion. This is fortunate since the errors of the high degree detailed portion are not as well calibrated. All this difficulty in error analysis is contrasted with the simplicity of at least the "noise only" result in the joint solution which solves for the sea surface (effectively) and the geoid simultaneously from a common reference geopotential. That result is just the familiar inverse of the normals of the least squares fitting process, the first (simulated) example of which was given in Wagner (1986).

But in the subtraction methods the global error variances (spectra) of these geoid models do give a quick and useful summary of likely PST resolving power (Fig. 11). And from the well calibrated satellite-only models we can also show examples of the spatial variation of geoid error possible with these models. This kind of analysis in the context of the maps that have been produced has always been possible.

Referring to Figure 11 we see that GemL2 truncated even at 6x6 should project significant errors into a PST surface derived by subtraction, while the new model Gem T1 should not. In Figure 24 I show the result of a subtraction of the 6x6 portion of Gem L2 [together with a background model, the OSU86E (Rapp, 1986) from degree 7 to 360] from the 1x1 degree mean Seasat sea surface used by Engelis and Rapp (1984). This map closely resembles the Engelis and Rapp PST (Figure 2a) as indeed it should, the only fundamental difference between the two is that here a better, more detailed background geoid has been used that is more compatible with GemL2. Engelis and Rapp (1984) used an earlier 180x180 model based on Gem9 for the low degree portion. The major differences between Figures 24 and 2a however seem to be due to the way the land areas were treated. Engelis and Rapp's map is based on a strict spherical harmonic expansion of the difference surface (sea heights-geoid) with zero values over land. Figure 24 is merely a least squares fit of 6x6 surface harmonics to the marine surface only. It attempts to be as faithful as possible to just the data available. The largest differences in the two maps appear to be in areas closest to land, for example in the southernmost parts of the oceans where the Engelis-Rapp map shows the Antartic current broken by large circulation patterns which may be artifacts (Gibb's phenomenon) since they do not

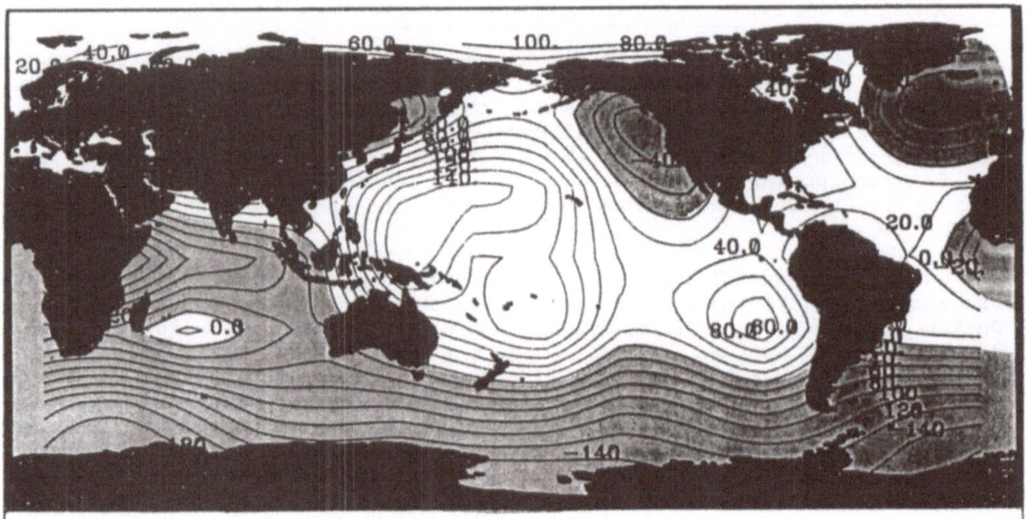

Figure 24. PST (6x6 Least Squares): Seasat Hts.-GemL2(6x6)/OSU(7-360) in centimeters. The Seasat sea heights were 1x1 degree gridded values (Engelis and Rapp, 1984). The geoid heights were computed from OSU86E,360x360 (Rapp, 1986). Differences at the sea height grid points were taken and averaged into 5x5 degree blocks means (weighted according to the number of 1 degree points in each block). Sea heights were rejected if formal errors for them were over 0.8 m. The 5 degree means were adjusted (by a constant) to have zero average (geographically), then fit by least squares to 6x6 surface harmonics. The difference geoid (6x6): GemL2-OSU86E was then computed and subtracted from the 6x6 fitted difference surface just described. The result is shown and agrees quite well with previously derived maps using GemL2 and satellite derived sea surfaces.

appear in the original Levitus data (reproduced in Figure 25). The problem of the proper representation of the PST at various scales is not a trivial one since the "borders" of the oceans are literally everywhere at the broadest scales which is

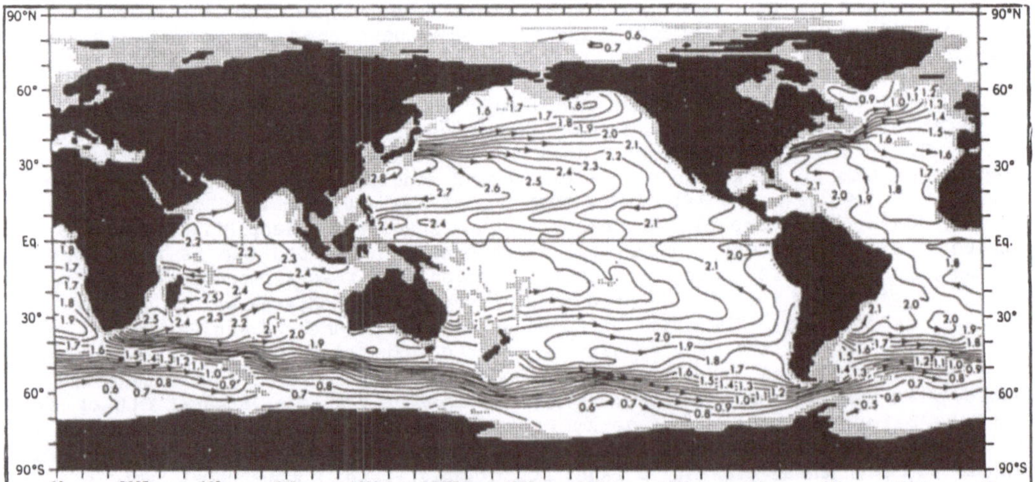

Figure 25. Mean Oceanographic Topography Over Decades (C.I. = 0.1 m). (From Levitus, 1982). Referred only to ocean depths over 2000 m.

all we can fairly see at present. Engelis (1987) presents more information on this matter, but quite an interesting discussion was already given in Mather et al. (1978).

At any rate my principal reason for showing the GemL2 result was to compare it with an identically derived map using Gem T1. This is presented in Figure 26 and while the result is similar to Gem L2 the gains in "reality" appear significant.

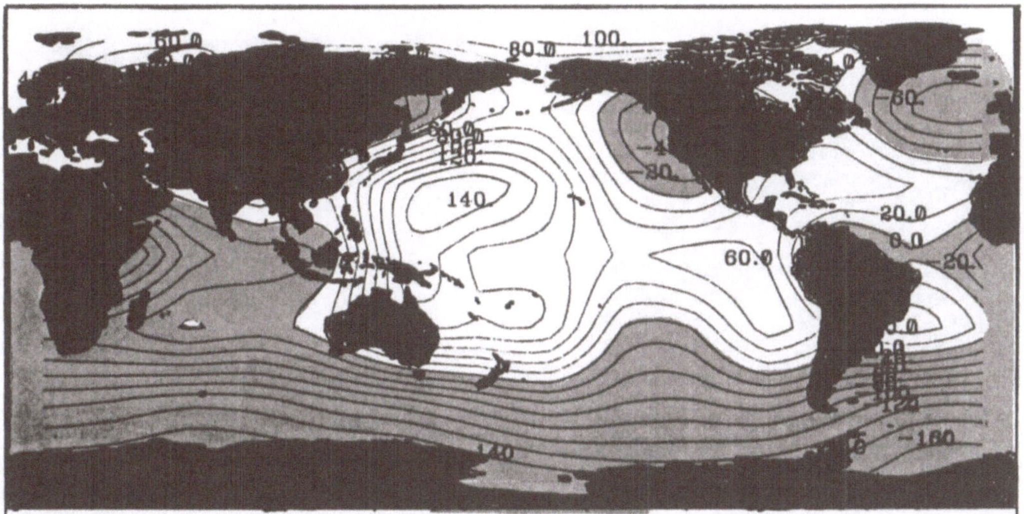

Figure 26. PST (6x6, Least Squares): Seasat Heights-GemT1(6x6)/OSU(7- 360), in centimeters. The origin of this map is the same as in Figure 24 except GemT1 has replaced GemL2 in the low degree portion of the geoid. The principal change (from the GemL2+ geoid) is the reduction of the strong circulation feature in the Southwest Pacific.

Most striking is the large reduction of the circulation pattern in the Southwest Pacific which was prominent on all previous maps including the Cheney et al (1984) one referenced to the mean surface of 1976-78 (Figure 12). Smaller but still important gains appear to be the reduction of strong flow patterns in the Indian Ocean and the introduction of "new" ones in the South Atlantic which appear to be realistic. Again, the overall power of this surface exceeds that of the Levitus at the same scale (Fig. 18). Engelis and Rapp (1984) noted this fact and attributed it to possibly stronger currents in the summer of 1978 but the same excessive overall power is also present in the Cheney surface (1976-78) so this may be a true reading of broad scale ocean set-up. However, in a comparison of mean sea surfaces produced for Seasat and for Geos 3/Seasat, Sailor (1985) found just such significant long wavelength differences which could account for the extra power. So the possibility of spurious orbit or surface bias in such internally corrected surfaces remains an unsolved problem. The only true resolution of where the mean sea surface stands above the earth, of course, lies in the joint solution which mixes tracking data from well determined land stations with marine altimetry.

Finally it is instructive to see what the possible broad scale changes in the Gem T1 map are, due to its well estimated covariance matrix. It is well known that the covariant errors of satellite geoids tend to be zonal in character with little longitude structure (e.g., Wagner and Lerch, 1978). A simple way to display the scale of allowable changes is to use a clone of Gem T1 as a possible error model. The clone was derived from a random number generator applied to the eigenvalues of the Gem T1 covariance matrix (transforming the "normally" perturbed independent parameters back to conventional geopotential parameters by way of the matrix of eigenvectors). The result of differencing this clone with the parent (Gem T1) at 6x6 is shown in Figure 27. At a level of about 10 cm most of the features of Figure 26 are changeable from just Gem T1 geoid error. A more thorough analysis of the complete geoid error could also be made but in view of the uncertainties mentioned above in the sea surfaces, this does not appear to be warranted at this time.

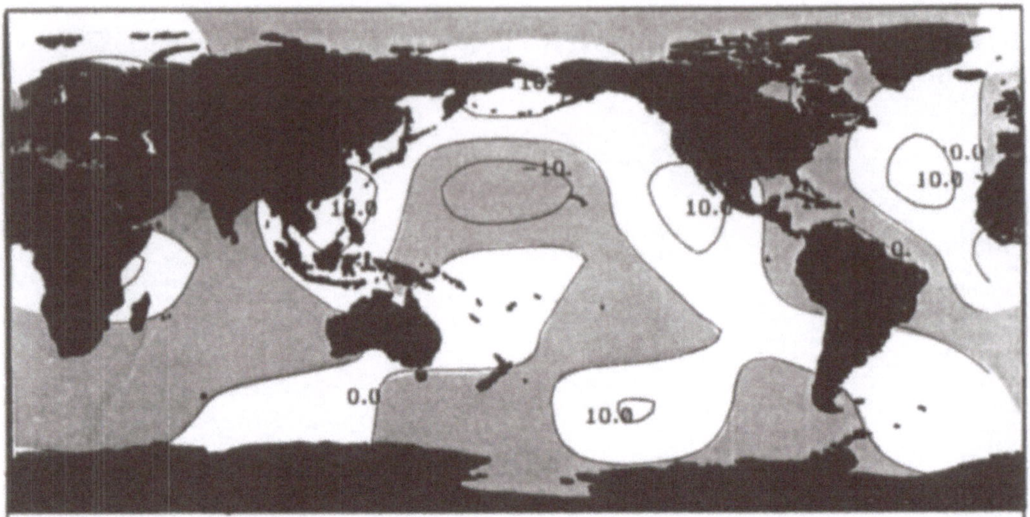

Figure 27. Possible errors in PST from GemT1 errors (6x6), C.I.= 10 cm, as assessed by the difference of GemT1 and a (normal deviate) clone of it. In the clone, no eigenvalue deviation greater than 3 sigma was permitted. Note the level of possible errors at this scale is no more than 10 cm over regions no larger than about 40 degrees.

References.

Allan, R.R. (1967), "Satellite Resonance with Longitude Dependent Gravity II: Effects Involving the Eccentricity", Planet. and Sp. Sci. 15, 1843-1844.

Brouwer, D. (1959), "Solution of the Problem of Artificial Satellite Theory Without Drag", Astron. Journ. 64, 378-397.

Cheney, R.E., B.C. Douglas, D.T. Sandwell, J.G. Marsh, T.V. Martin, and J.J. McCarthy (1984), "Applications of Satellite Altimetry to Oceanography and Geophysics", Marine Geophys. Researches 7, 17-32.

Chovitz, B.H. (1973), "Downward Continuation of the Potential from Satellite Altitudes", Bollettino Di Geodesia Scienze Affini Vol. 32(#2), 81-88.

Cloutier, D.T. (1981), "A New Technique for Correcting Satllite Ephermeris Errors Indirectly Observed from Radar Altimetry", Tech. Report TR 246, Naval Oceanographic Office, Bay St Louis, Mississippi.

Colombo, O. (1984a), "Altimetry, Orbits and Tides", NASA Tech. Memo. 86180, Goddard Space Flight Center, Greenbelt, Maryland.

Colombo, O. (1984b), "The Global Mapping of Gravity with Two Satellites", Monograph of the Netherlands Geodetic Commission, New Series Vol. 7(#3).

Cook, G.E. (1966), "Perturbations of Near Circular Orbits by the Earth's Gravitational Potential", Planet. and Space Sci. 14, 433-444.

Douglas, B.C., R.W. Agreen and D.T. Sandwell (1984), "Observing Global Ocean Circulation with Seasat Altimeter Data", Marine Geodesy 8, 67-83.

Engelis, T. (1987), "Radial Orbit Error Reduction and Sea Surface Topography Determination Using Satellite Altimetry," Report No. 377, Dept. of Geodetic Sciences and Surveying, Ohio State Univ., Columbus, Ohio.

Engelis, T. and R.H. Rapp (1984), "Global Ocean Circulation Patterns Based on Seasat Altimeter Data and the Gem L2 Gravity field", Marine Geophys. Res. 7, 55-67.

Fehlberg, E. (1966), "New One-Step Integration Methods of High Order Accuracy Applied to some Problems in Celestial Mechanics", NASA Tech. Report TR-R-248, Marshall Space Flight Center, Huntsville, Alabama.

Gedeon, G.S. (1969), "Tesseral Resonance Effects on Satellte Orbits", Celestial Mechanics 1, 167-189

Goad, C. (1987), "An Efficient Algorithm for the Evaluation of Inclination and Eccentricity Functions", Manuscripta Geodaetica 12, 11-15.

Goad, C.C., B.C. Douglas and R.W. Agreen (1980), "On the Use of Satellite Altimetry for Radial Ephemeris Improvement", J. Astronaut. Sci. 27(4), 419-428.

Heiskanen, W.A. and H. Moritz, (1967), Physical Geodesy, Freeman & Co., San Francisco, California, 74-77.

Kaula, W.M. (1966), Theory of Satellite Geodesy, Blaisdell Co., Waltham, Mass.

Kozai, Y. (1959), "The Motion of a Close Earth Satellite", Astron. Journ. 64, 367-377.

Kozai, Y. (1961), "Note on the Motion of a Close Earth Satellite with a Small Eccentricity", Astron. Journ. 66, 132-134.

Kozai, Y. (1962), "Second Order Solution of Artificial Satellite Theory Without Air Drag", Astron. Journ. 67, 446-461.

Lerch, F.J., S.M. Klosko, G.B. Patel and C.A. Wagner (1985), "A Gravity Model for Crustal Dynamics", Journ. of Geophys. Res. 90(B11), 9301-9311.

Lerch, F.J., J.G. Marsh, S.M. Klosko and R.G. Williamson (1982), "Gravity Model Improvement for Seasat", Journ. of Geophys. Res. 72(C5), 3281-3296.

Lerch, F.J., B.H. Putney, C.A. Wagner and S.M. Klosko (1981), "Goddard Earth Models for Oceanographic Applications (Gem 10b and Gem 10c)", Marine Geodesy 5(#2), 145-187.

Lerch, F.J., S.M. Klosko, R.E. Laubscher and C.A. Wagner (1979), "Gravity Model improvement using Geos 3 (Gem 9 and 10)", Journ. of Geophys. Res. 84(B8), 3897-3916

Levitus, S. (1982), "Climatologic Atlas of the World Ocean", NOAA Geophysical Fluid Dynamics Laboratory Professional Paper No. 13, Rockville, Maryland.

Marsh, J.G., F.J. Lerch, B.H. Putney, D.C. Christodoulidis, B.V. Sanchez, T.L. Felsentreger, D.E. Smith, S.M. Klosko, T.V. Martin, E.C. Pavlis, J.W. Robbins, R.G. Williamson, O.L. Colombo, N.L. Chandler, K.E. Rachlin, G.B. Patel, S. Bhati and D.S. Chinn (1987), "An Improved Model of the Earth's Gravitational Field: Gem T1", Journal of Geophys. Res. 93(B6), 6169- 6215.

Marsh, J.G. and T.V. Martin (1982), "The Seasat Altimeter Mean Sea Surface Model", Journ. of Geophys. Res. 87, 3269-3280.

Mather, R.S., F.J. Lerch, C. Rizos, E.G. Masters and B. Hirsh (1978), "Determination of some Dominant Parameters of the Global Dynamic Sea Surface Topography from Geos 3 Altimetry", NASA Tech. Memo. 79558, Goddard Space Flight Center, Greenbelt, Maryland.

Rapp, R.H.(1986), "Global Geopotential Solutions", Lecture Notes on Earth Sciences, Vol.7, Springer Verlag, Berlin.

Rosborough, G. (1986), "Satellite Orbit Perturbations due to the Geopotential", Center for Space Research Report CSR-86-1, 57-76, University of Texas at Austin, Texas

Sandwell, D.T., D.G. Milbert and B.C. Douglas (1986), "Global Nondynamic Orbit Improvement for Altimetric Satellites", J. of Geophys. Res. 91(B9), 9447-9451

Sailor, R.V. (1985) "Comparison of Altimetric Global Mean Sea Surface Maps", EOS (Trans. Am. Geophys. Un.) 66(#18), 246-247.

Tai, C.K. and C. Wunsch (1984), "An Estimate of Global Absolute Dynamic Topography", Journal of Phys. Oceanography 14(2), 457-463.

Tapley, B.D., G.H. Born and M.E. Parke (1982), "The Seasat Altimeter and its Accuracy Assessment", J. of Geophys. Res. 87(C5), 3179-3188

Wagner, C.A. (1986), "Accuracy Estimate of Geoid and Ocean Topography Recovered Jointly from Satellite Altimetry", Journ. of Geophys. Res. 91 (B1), 453-461.

Wagner, C.A. (1985), "Radial Variations of a Satellite Orbit due to Geopotential Errors: Implications for Satellite Altimetry", Journ. of Geophys. Res. 87 (C5), 3179-3188.

Wagner, C.A. (1983), "Direct Determination of Gravitational Harmonics From Low-Low GRAVSAT Data", Journ. of Geophys. Res. 88, 10309-10321.

Wagner, C.A. and F.J. Lerch (1978), The accuracy of geopotential models, Planet. and Sp. Sci. 26, 1081-1140.

Wunsch, C. and E.M. Gaposchkin (1980), "On using Satellite Altimetry to Determine the General Circulation of the Oceans with Application to Geoid Improvement", Reviews of Geophys. 18, 725-745

ADVANCED TECHNIQUES FOR HIGH-RESOLUTION MAPPING OF THE GRAVITATIONAL FIELD(*)

Oscar L. Colombo

EG&G, WASC, Inc., 5000 Philadelphia Way, Suite J
Lanham, Maryland 20706, U.S.A.

1. BASIC TECHNIQUES FOR GATHERING DATA ON A GLOBAL BASIS

To obtain full coverage of the earth, two main ideas have been considered:

a) Satellite-to-satellite tracking:

- One more spacecrafts in high orbits (say, GPS), tracking another in a low orbit (sensitivity increases with height); or
- two spacecrafts tracking each other in the same low orbit.

b) Satellite gravity gradiometry:

- A collection of accelerometers rigidly connected which provide information on the second gradients of the gravitational potential V.

Some mention will be made of technique (A) in later lectures. Most of the course will be dedicated to the gradiometers given their greater present chances of implementation, and the richness of the topic&^/Da8I@Bp.6)d^|# (H c: the USA (GRM mission, Ho Paik's superconducting device) and in Europe (CNES/ESA GRADIO instrument (A. Barnard, France)), which operates at "room temperature" with somewhat less accuracy than Paik's (but may offer greater potential as an instrument to be carried in interplanetary probes to study the solar system), for the ARISTOTELES mission. Both projects are currently in their preliminary stages, with the effort concentrating on the design and construction of the instruments themselves.

Why use a gradiometer?

In the free-fall conditions inside a spacecraft, one can only measure the difference in the acceleration of gravity between the points where an accelerometer is located, and the center of mass of the satellite (this is the

(*) [Editor's note] These lecture notes are essentially in the same form as the draft distributed by Oscar Colombo during the Summer School, since, due to new engagements, the Author could not write down the final version.

tidal field in a frame fixed to the spacecraft). But to make sense of this measurement, one needs to know the precise location of both the accelerometer and the center of mass. The latter is difficult to do accurately. Also certain forces, acting on the surface of the satellite (drag, solar radiation pressure) will be measured and corrupt the data severely.

So instead one uses two accelerometers precisely located with respect to each other. If their separation is a distance d, their sensitive axes (the direction in which they measure acceleration) are aligned parallel to the x direction, and the line between them has the same orientation as the sensitive axes, then

$$V_{xx} = (\text{accel. } 1 - \text{accel. } 2)/d \qquad (\text{to first order in } d^{-1})$$

Similarly, if the separation is along the direction y, the same calculation gives the cross gradient V_{xy} (which is the same in value as V_{yx}).

A full-tensor gradiometer will measure all nine gradients, or second derivatives of V, even when only three of the cross gradients are different, and one of the other three (sometime called "in-line") is linearly dependent on the other two, because of Laplace's equation. A full-tensor device provides the maximum redundancy and information. All other configurations are partial realizations of this case.

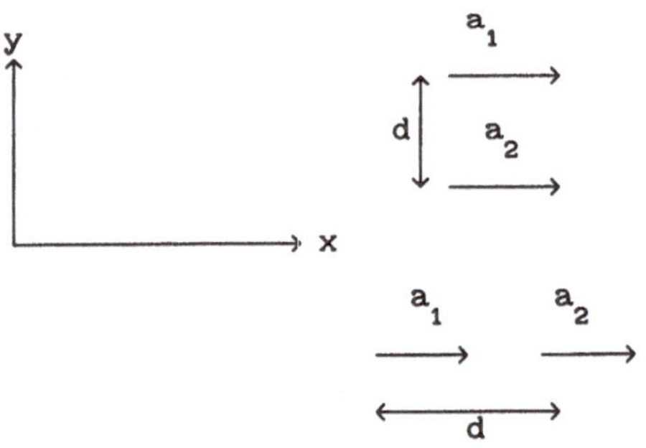

$$\frac{1}{d}(a_1 - a_2) \simeq V_{xy} = V_{yx} \equiv \frac{\partial^2 V}{\partial x\, \partial y} \tag{1.1}$$

$$\frac{1}{d}(a_1 - a_2) = V_{xx} \equiv \frac{\partial^2 V}{\partial x^2} \tag{1.2}$$

Fig. 1 - Gradiometry based on pairs of accelerometers.

Fundamental problems

a) The satellite, and the instrument in it, rotates. The result is that what is actually measured is gravity, not gravitation (this situation is similar to that encountered in physical geodesy, when dealing with "earth-fixed" data, but the problem is more difficult to solve, as will be axplained).

b) One needs to know where the measurements are taken (the orbit).

c) The orientation of the instrument and spacecraft with respect to an external frame is needed if all the information in the measurements is to be used.

Accuracy required

On earth, and in near-earth orbit (160-220 Km) the second gradients of gravity are of the order of

3×10^3 Eötvos Units (E.U.)

where

$$1 \text{ E.U.} = 10^{-9} \text{ m/sec}^2\text{/m} = 10^{-9} \text{ sec}^{-2} \tag{1.3}$$

Most of the 3×10^3 E.U. is due to the "central force" term GM/r in V.
The fine details on the earth crust and upper mantle contribute 10^{-7} of the whole, or less. Taking into account that many measurements will be taken, and their errors averaged, the required accuracy for an instrument that can give meaningful data is of not less than a few hundredths of E.U. Current specification for the GRADIO instrument (France, ESA) is of 10^{-2} E.U., and for the NASA supercooled device, of 10^{-4} E.U. If the separation between accelerometers is of 10 cm the required precision of the accelerometers themselves is of the order of 10^{-12} m/sec^2 , or better than one part in ten trillion (million million) of the normal acceleration opf gravity, in the case of GRADIO, and one part per quadrillion (thousand million million) for the NASA instrument.

Problems limiting the accuracy

The following list is illustrative of the complexity of the task facing those trying to build these super-accurate instruments:

- thermal noise in spring and electronics;
- vibrations;
- alignment of the accelerometers (cancellation of drag, etc.);
- drift (due to mechanical deformation, change in electrical components, etc.);
- gravitation excerted by satellite parts and by the instrument on itself, specially complex changes with time of the mass distribution (emptying fuel tanks);
- relative scale errors in each pair of accelerometers;
- errors in determining attitude in space, and change in attitude (rotation).

These are very difficult problems, and have to be reduced through engineering design and data processing (corrections made by estimating nouisance parameters).

The null-point principle

To achieve the extreme instrumental accuracy and stability required, one principle of design most helpful is the "null-point" idea illustrated below:

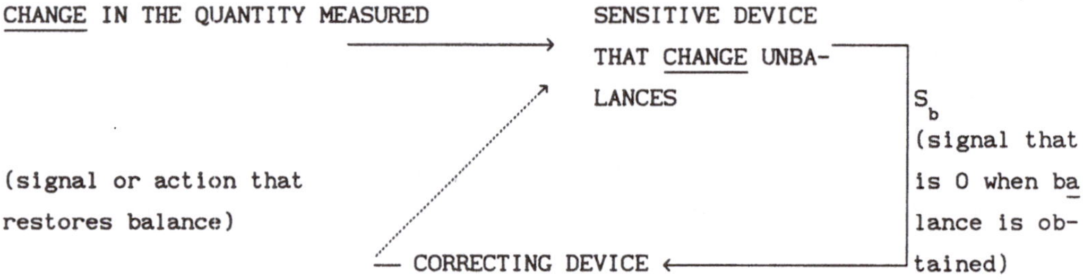

The desired properties obtained with this configuration are due in part to the use of a negative feed-back loop to maintain the balance in the sensitive device under changes in the quantity measured; the actual measurement is that of the restoring or corrective signal on the left. Negative feed-back allows, in general, for greater rejection of noise and higher stability in the face of unstable parts of the loop . The null-point (where signal at right is zero) can be obtained very exactly by increasing the sensitivity with which S_b is measured as equilibrium is approached and this signal becomes smaller and smaller.

Every-day examples

The precision balance used in laboratories

Signal of lack of "balance" is deflection of needle from central mark in scale.
Restorative action is the addition of successively smaller counterweights.
Measurement of restorative action when equilibrium is reached once more: the needle is back on the center, the total restorative action is the sum of the compensating weights. If weight changes, then counterweights are added or removed, to maintain the needle centered, and a more or less continuous

estimate of the changing weight is obtained in this way. (Notice that the deflection of the needle, while itself a measure of weight, and a direct one, is too much inaccurate to be used; this ignoring of a direct indicator is typical of this technique).

Likewise one could mention:

- the use of the principle of spirit levelling to measure tilt;
- the phase-locked loop discriminators used in many high-precision electronic measurements (including GPS receivers);
- electric bridge-type devices; bridges can be resistive, inductive or capacitive; Paik's instrument is based on an inductive bridge, Barnard's on a capacitive bridge.

To be able to measure very exactly with this principle, the usual design requires that the proof-masses of the accelerometers be restrained with counteracting springs (these spring/proof-mass examples act as spring balances, giving direct measurements; once more, direct measurements are too coarse, and only the signal that restores the balance to the "null-point" is used).

The use of the "null-point" principle, by itself, does not make the instrument reach the desired accuracy. It is described here as one example of how good design principles are integral parts of the development of these exquisitely sensitive devices.

Measuring Gravity

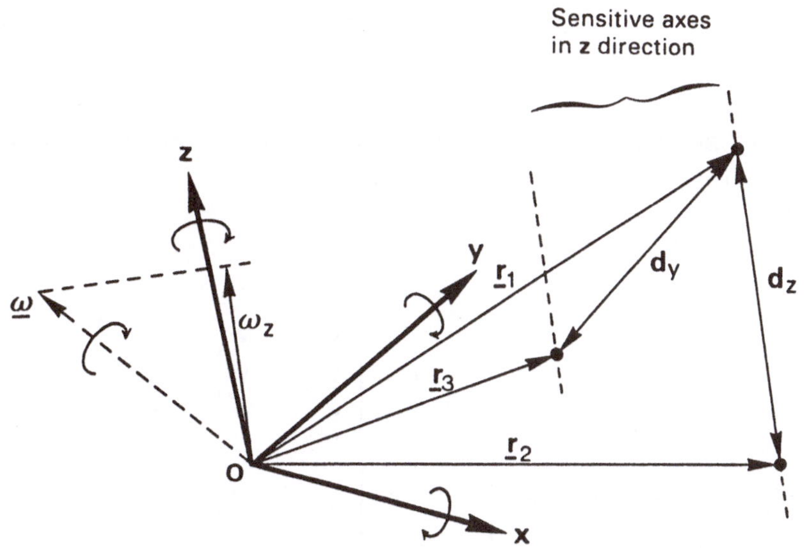

Fig. 2 - Measuring with an accelerometer pair in a rotating frame.

Notice: no $\dot{\underline{r}}_i$ because of restraining springs, so no Coriolis term.

$$\underset{\substack{\uparrow\\ \text{total}}}{\ddot{\underline{r}}_i} = \overset{\substack{\text{gravitational}\\ \downarrow}}{V\,V} - \underline{\omega} \times \underset{\substack{\uparrow\\ \text{centrifugal}}}{(\underline{\omega} \times \underline{r}_i)} + \dot{\underline{\omega}} \times \underset{\substack{\uparrow\\ \text{linear (accelerations)}}}{\underline{r}_i} , \qquad i = 1,2,3 \tag{1.4}$$

$$\underline{\omega} = \underline{e}_x \omega_x + \underline{e}_y \omega_y + \underline{e}_z \omega_z = \text{angular velocity vector;} \tag{1.5}$$

$$\dot{\underline{\omega}} = \underline{e}_x \dot{\omega}_x + \underline{e}_y \dot{\omega}_y + \underline{e}_z \dot{\omega}_z = \text{angular acceleration vector;} \tag{1.6}$$

$\underline{e}_x, \underline{e}_y, \underline{e}_z$ unit vectors aligned with axes.

If $\underline{e}_z$ is in the direction of the sensitive axis $z(||\underline{e}_z|| = 1)$:

$$\ddot{r}_{zi} = \ddot{\underline{r}}_i \, \underline{e}_z \ . \tag{1.7}$$

Differencing the accelerometer measurements and dividing by d_z:

$$\frac{1}{d_z} (\ddot{r}_{z1} \ \ddot{r}_{z2}) \simeq V_{zz} - (\omega_x^2 + \omega_y^2) \ . \tag{1.8}$$

Similarly for the cross-gradient (line between accelerometers in the y direction):

$$\frac{1}{d_y} (\ddot{r}_{z3} \ \ddot{r}_{z2}) \simeq V_{zy} + \omega_z \omega_y - \dot{\omega}_x \simeq V_{yz} + \omega_z \omega_y - \dot{\omega}_x \ . \tag{1.9}$$

These are gradients of <u>gravity</u> (gravitation + rotation). If all the second derivatives of gravity are measured in this way, one gets, in matrix form, the expression of the full-tensor gradiometer:

$$\Lambda = \Gamma + \Omega^2 + \dot{\Omega} \tag{1.10}$$

where

$$\Lambda = \begin{bmatrix} \Lambda_{xx} & \Lambda_{xy} & \Lambda_{xz} \\ \Lambda_{yx} & \Lambda_{yy} & \Lambda_{yz} \\ \Lambda_{zx} & \Lambda_{zy} & \Lambda_{zz} \end{bmatrix} ; \quad \Gamma = \begin{bmatrix} V_{xx} & V_{xy} & V_{xz} \\ V_{yx} & V_{yy} & V_{yz} \\ V_{zx} & V_{zy} & V_{zz} \end{bmatrix}$$

$$\Omega^2 = \begin{bmatrix} -(\omega_y^2+\omega_z^2) & \omega_x\omega_y & \omega_x\omega_z \\ \omega_y\omega_x & -(\omega_x^2+\omega_z^2) & \omega_y\omega_z \\ \omega_z\omega_x & \omega_z\omega_y & -(\omega_x^2+\omega_y^2) \end{bmatrix} ; \quad \Omega = \begin{bmatrix} 0 & -\dot{\omega}_z & -\dot{\omega}_y \\ \omega_2 & 0 & -\dot{\omega}_x \\ \omega_y & \omega_x & 0 \end{bmatrix} . \tag{1.11}$$

Some basic relationships

$$\frac{1}{2}(\Lambda + \Lambda^T) = \Gamma + \Omega^2 \quad , \tag{1.12}$$

$$\frac{1}{2}(\Lambda - \Lambda^T) = \dot{\Omega} \quad , \tag{1.13}$$

$$\Lambda_{xx} + \Lambda_{yy} + \Lambda_{zz} = -2\,(\omega_x^2 + \omega_y^2 + \omega_z^2) = -2\,||\underline{\omega}||^2 \quad . \tag{1.14}$$

Also

$$V_{xx} + V_{yy} + V_{zz} = 0 \tag{1.15}$$

by Laplace's equation (valid in vacuum). Moreover $(\Gamma + \Omega^2)$ trasforms under an orthogonal coordinate trasformation (such as rotation) as a second order tensor:

$$(\Gamma + \Omega^2)_2 = R_{12}\,(\Gamma + \Omega^2)_1\,R_{12}^T \quad , \tag{1.16}$$

and this is also true for Γ :

$$\Gamma_2 = R_{12}\,\Gamma_1\,R_{12}^T \quad . \tag{1.17}$$

where R_{12} is the orthogonal matrix of the trsformation.

Which way is up in free fall?

Let us assume that the effects of attitude and change of attitude have been sorted out, and we are left with a measurement of the gradient in some direction we can change at will. As we turn this direction about (let assume that the two accelerometers have their axes on the same line between them, so one mesures "in-line" gradients, such as V_{ss} for the direction s), if the earth is regarded as a sphere (which, to a close a approximation it is), then the reading from this one-axis device will be maximum (about 3000 E.U.) when it is pointed towards the center of the earth (geocenter). That is the "down" direction, the opposite being "up"; thus a gradiometer allows to define a connection (partial) between the frames of the instrument and an earth-fixed frame. The horizontal (any horizontal) is the direction where the reading is a minimum (about half that of the maximum). This is an old idea, going back to the early days of space activities, in the late fifties. The principle is related to that of inertial navigation, and the similitudes do not stop here (after all, both inertial navigation systems and gradiometers are arrangements of precise accelerometers and other devices, such as gyroscopes). If one had a full-tensor gradiometer, then the eigenvector of the measured matrix Γ associated with the largest eigenvalue (again, equal to 3000 E.U.) would point to the center of the earth, and no search for this direction would be needed. The value measured in that direction being $2GM/r^3$, if one also measured r with radar, determining one component of position, then the gravity field, described fully in this spherical case by GM, could be determined directly. Conversely, knowing the field (GM), one could determine r from the measurement of the gradiometer. As the direction to the geocenter is also known in the frame of the instrument (that of the principal eigenvector) one has the relative position of instrument and earth in the frame of the instrument. Through external attitude sensors (star-trackers) and knowledge of precession, nutation, polar motion and earth rotation, one could find the position of the instrument in earth-fixed coordinates, in inertial coordinates, and so on. Therefore, with a gradiometer (full-tensor):

- if one knows the gravitational field, one knows also the position (at least, relative, in the instrument frame);
- if one knows the position (orbit) one can determine the gravity field;
- if one knows both field and position only approximately, it may be possible to improve the knowledge of both using the gradiometer measurements, as explained later.

The statements above are important, because they hold true in more complex fields than that of a sphere.

While this suggests immediately the idea of using the gradiometer as a self-tracking device, the idea of using it to model gravity and position has a more realistic application in the possible simultaneous estimation of orbit error and field parameters, where the orbit would be determined, not as a substitute for that obtained with ordinary tracking, but as a way of removing its unwanted effect from the data. (The orbit obtained in this way could be quite bad, it is enough that the effect of orbit errors be removed to the extent that the gradiometer is sensitive to them, and no further.)

Prospecting in the Asteroids

The ideas presented in the previous paragraph are little more than "thought experiments". But similar ideas have been proposed in recent years as applications for gravity gradiometers in space probes sent to explore the solar system. Once a small body, such as an asteroid, comes to be in the proximity of the probe, a combination of radar and gradiometry would provide an estimate of GM, and thus of M (rough, of course). At the same time, photographs taken of the asteroid as it tumbles about in space, and as the probe moves pass it in its orbit, would allow, once received on earth, to calculate the volume (by finding the shape from the profiles of the various images). From mass M and volume V it would be possible to determine the mean density, at least (more detail than just GM could be found if the asteroid tumbled fast enough while near enough to the gradiometer for finer features of its field to be sampled and resolved). This estimate of density would allow to type the asteroid, guess at its composition, build a more complete picture of the make of the solar system and its history, and even identify possible mineral resources that a more advandced stage of technical civilization might use, one day, as it expands its reach outwards from our planet.

Dealing with orbit error and attitude / Rotation to estimate gravitation

a) The orbit error

To make sense of the measurements of the gradients, it is necessary to know where they have been taken along the orbit. In general, before the final processing, one would have an approximate knowledge of this orbit from tracking of the low, gradiometer-bearing satellite by means of (primarily) much higher satellites (e.g. the geostationary TDRSS, and the 12 hours GPS spacecrafts). Ideally, the orbit would be drag-free, through the action of special sensors (such as individual accelerometers) used to determine when to fire small rocket engines to compensate for the momentum lost to drag.
Two main approaches have been proposed so far for dealing with the errors in the computed orbit:

A) To eliminate them from the problem altogether by making cetain combinations of gradients where the error substracts out.
B) To include a mathematical representation of the orbit error in the observation equations of the adjustment, side by side with the terms representing the gravity field, and do one joint adjustment of orbit, field, and other parameters that might have to be considered (such as those describing instrument drift, scale errors, or rotation, as explained later).

Consider the case (for semplicity's sake) where the spacecraft (or at least that part of it containing the gradiometer) is kept in an earth-pointing orientation.

Further, assume that either the instrument axes have the orientation shown in the figure, known as local orbital frame (x across, normal to the orbit plane that conatins the geocentric position and the velocity vectors; y along, perpendicular to both the vertical and the x axes, and aligned with the velocity vector if the orbit is exactly circular; and z vertical, or radial, pointing away from the geocenter, along the position vector)

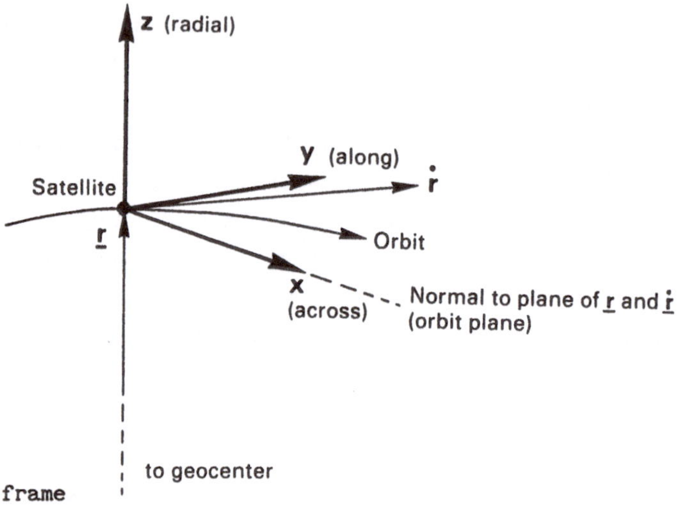

Fig. 3 - The local orbital frame

In this frame, ignoring orientation/rotation effects, the gradiometer matrix Λ is:

$$\Lambda = \begin{bmatrix} V_{xx} & V_{xy} & V_{xz} \\ V_{yx} & V_{yy} & V_{yz} \\ V_{zx} & V_{zy} & V_{zz} \end{bmatrix} - \begin{bmatrix} \Delta z & 0 & \Delta x \\ 0 & \Delta z & \Delta y \\ \Delta x & \Delta & -2\Delta z \end{bmatrix} 3 \frac{GM}{r^4} , \qquad (1.18)$$

where Δx, Δy and Δz are the position orbit errors in the same system of local coordinates. Clearly, one could substract Λ_{xx} from Λ_{yy} , or any other two diagonal elements of Λ , and get a quantity where the orbit error (Δz in this case) cancels out and disappears. This is one example of approach (A). One problem with this is that it is, in principle, too wasteful of information: eliminating the unknown orbit error in this way is mathematically equivalent to solving for one unknown per error component (Δz here) per observation time. While one does not notice it, the adjustment is actually burdened by all those unseen unknowns, with the consequent loss of degrees of freedom, and of accuracy in the estimates of the gravitational parameters, which are adjusted explicitly. (This is the same situation encountered when eliminating clock errors in GPS by differencing them out, forming single or double differences). A way of strengthen the adjustment is to reduce the number of unknowns assigned to the orbit error; but this requires bringing these unknowns explicitly into the adjustment. (In GPS, this is done by replacing the clock errors with correlated process variables, or polynomials, etc.).

To do this one needs some kind of mathematical model for the orbit error.

If the spacecraft is not drag-free (different from drag-compensated; the second merely means that the orbit is maintained by periodically "pushing" with rocket engines until the former orbit is restored; the drag-free case is one where this compensation is so finely tuned that the drag force inside the spacecraft becomes virtually undetectable), then no satisfactory model may exist at the very low altitude of this type of mission, where the irregular force of air-drag is very strong, and the best one can do is to eliminate all trace of orbit error using approach (A). In the drag-free case, one can model the orbit as a function of errors in the initial conditions and on the gravity field model. This type of dynamic model greatly reduces the number of additional orbit variables (to, in fact, the six unknown errors in the initial position and velocity vectors: Δx_o, Δy_o, Δz_o; $\Delta\dot{x}_o$, $\Delta\dot{y}_o$, $\Delta\dot{z}_o$). This is approach (B). While this is best done, from the point of view of the results, by integrating numerically the perturbation equations (linearized equations of motion, about the known orbit), an analytical approach (which could be used for data reduction in some cases as well) will be used here to explain the concept.

For a nearly circular orbit (such as it is most likely to be chosen, as it will be explained in a future lecture), the equations of motion in the rotating local orbit coordinates have the following linearized form, as an approximation, known as Hill's equations (related to the famous equations of celestial mechanics)

$$\begin{aligned} \Delta\ddot{x} &= -\,n_o^2\,\Delta x + f_x \\ \Delta\ddot{y} &= -2n_o\,\Delta\dot{z} + f_y \\ \Delta\ddot{z} &= 3n_o^2\,\Delta z + 2n_o\,\Delta\dot{y} + f_z\;; \end{aligned} \tag{1.19}$$

f_x, f_y, f_z = components of disturbing forces.

Notice that they are similar to those of undamped oscillators. Not surprisingly, their free response, representing the effect of the initial state errors by themselves, has the oscillatory form:

$$\Delta x = \Delta x(o) \cos n_o t + \frac{\Delta \dot{x}(o)}{n_o} \sin n_o t$$

$$\Delta y = \frac{2}{n_o} \Delta \dot{z}(o) \cos n_o t + \left[\frac{4}{n_o} \Delta \dot{y}(o) + 6 \Delta z(o) \right] \sin n_o t +$$

$$+ \left[\Delta y(o) - \frac{2}{n_o} \Delta \dot{z}(o) \right] - \left[3\Delta \dot{y}(o) + 6 n_o \Delta z(o) \right] t$$

$$\Delta z = \frac{\Delta \dot{z}(o)}{n_o} \sin n_o t - \left[\frac{2}{n_o} \Delta \dot{y}(o) + 3\Delta z(o) \right] \cos n_o t \quad .$$

Here, $n_o = (GM/r^3)^{1/2}$, according to Kepler's law for a spherical earth. So the forced response must be resonant at the natural frequencies of zero and once per revolution. As a result, most of the error, regardless of cause, will have its spectrum peaking at near zero and once per revolution, and decaying quickly at increasing frequencies. This confinement of most of the undesired signal caused by the orbit error to what are, in the present context, low frequencies, is shared by other effects, such as drift, scale errors and, with proper design. attitude and rotation-casued errors. This suggests that, with more or less continuous measurements being taken over a long period of time (as, in fact, is now planned), one could more or less do a Fourier analysis of the signal, and eliminate terms below some suitably low frequency (as shown later, this is likely to be 3 cycles per revolution). This approach is intermediate between modelling only six parameters for the orbit, and using the combinations of method (A), with one unknown per observation. There are many more frequencies eliminated between zero and n_o than the six orbit parameters, and each one can be regarded as an extra unknown (they also convey information on the gravity field, so the loss of that information when they are left out is obvious).

b) Dealing with attitude and rotation errors

To explain the main concepts, let us a take a particulary simple example, which may very well correspond to the situation in an actual gradiometry mission. Assume that the x,y,z axes are the local orbital Cartesian axes of Fig. 3, and further assume that the principal axes of inertia coincide with both the instrument axes and with the local orbital axes, except for small departures from the latter. Let ψ_x, ψ_y and ψ_z be three small rotations about the instrument/inertia axes, by which they depart from those of the local coordinates. To obtain a dynamical model, in terms of differential equations, that can be used to represent these angles and their rates of change and accelerations (in the same way as the orbit errors Δx, Δy and Δz have been treated in the preceding paragraph when explaining approach (B)), one starts with the rotational equivalent of Newton's equation of motion. These rotational equations of motion, in the frame of the principal axes of inertial that has been chosen here, are known as Euler's equations:

$$
\begin{aligned}
I_x\dot{\omega}_x + (I_z - I_y)\,\omega_y\omega_z &= N_x \\
I_y\dot{\omega}_y + (I_x - I_z)\,\omega_x\omega_z &= N_y \qquad (1.20)\\
I_z\dot{\omega}_z + (I_y - I_x)\,\omega_y\omega_x &= N_z
\end{aligned}
$$

where N_x, N_y and N_z are the components of the external torque applied by various forces (primarily gravity) on the satellite. When the momenta of inertia I_x, I_y and I_z are in the right proportions (which is a question of how the spacecraft is built) amongst each other, the satellite becomes gradient stabilized: the z axis of the satellite/instrument is kept aligned with that of the local coordinates (except for the small rotations ψ_x, ψ_y and ψ_z) by the torque excerted by the gravitational field itself.
Then, the angular (pointing) errors obey, to first order, the linearized form of Euler's equations

$$
\begin{aligned}
I_x(\ddot{\psi}_x + n_o\dot{\psi}_z)+(n_o\dot{\psi}_z - n_o^2\psi_x)(I_z - I_y) &= 3n_o^2\psi_x\,(I_z - I_y) \\
I_y(\ddot{\psi}_y + 2n_o e\,\sin n_o t) &= -3n_o^2\psi_y\,(I_x - I_z) \qquad (1.21)\\
I_z(\ddot{\psi}_z - n_o\dot{\psi}_x)+(n_o\dot{\psi}_x + n_o^2\psi_z)(I_y - I_x) &= 0 \ .
\end{aligned}
$$

where e is the mean eccentricity of the orbit (while small for nearcircular orbits, its effect has to be considered here). With proper choice of the values of the principal momenta, the solutions of these equations are oscillations with frequencies of the order of n_o (once per revolution).
As in the case of the orbit errors, these equations depend on six initial conditions: ψ_{xo}, ψ_{yo}, ψ_{zo}, and $\dot{\psi}_{xo}$, $\dot{\psi}_{yo}$, $\dot{\psi}_{zo}$.
In this way, one has a model for the rotation that can be used to incorporate ψ_{xo}, ψ_{yo}, ψ_{zo}, and $\dot{\psi}_{xo}$, $\dot{\psi}_{yo}$, $\dot{\psi}_{zo}$ as unknowns to be adjusted together with the gravitational parameters, to account for the effect of rotation in the

measurements. This is what has been called previously approach (B): to introduce the various sources of non-gravitational signal in the data <u>explicitly</u> in the observation equations by adding relatively few additional nuisance parameters. Approach (A), where lots of extra unknowns are introduced <u>implicity</u> (they are not seen in the observation equations, but they are there all the same) has an equivalent in this case of a gravity stabilized satellite. With gravity stabilization, the angular velocity relative to inertial space is primarily $\omega_x = n_o$. This is much larger than ω_y and ω_z, and the square of ω_x^2 is even larger than those of the other two. For this reason, if one substracts Λ_{yy} from Λ_{zz} the term ω_x^2 cancels out, and one is left only with the squares of the other two small quantities. Using $\Lambda_{yy} - \Lambda_{zz}$ as data, one introduces one unknown per observation equation to account for the vanishing ω_x^2 , but only implicitly. One only notices that these unknowns are present because the adjustment becomes weaker (in other words, information is lost in differencing). How small have the rotations $\omega_y = \dot{\psi}_y$ and $\omega_z = \dot{\psi}_z$ to be so that their effects are at the noise level of the instrument? The units for centrifugal force and gravitational gradients being the same (1/sec^2), except for the scale factor 10^{-9} , one would have to control the rotations about the z axis (yaw) and the y axis (roll) to having rates of less than $3\text{x}10^{-5}$ rad/sec for a 10^{-2} E.U. instrument, and less than $3\text{x}10^{-6}$ rad/sec for a 10^{-4} E.U. device.

If the attitude control mechanism cannot achieve these very low rates, then one might try incorporating the linearized model for these two rotations explicitly into the main adjustment, as in approach (B), ending up with a hybrid of both methods. (Notice that $\Lambda_{yy} - \Lambda_{zz}$ is <u>not</u> free of orbit error while $3\Lambda_{xx} - (\Lambda_{yy} - \Lambda_{zz})$ is "free" of that, too). [See equations (1.11) and (1.18)].

When using approach (B) to include explicitly rotation, orbit and other non-gravitational effects in the observation equations, one finds that the terms for these effects are mostly low-frequency (order of n_o, and even lower for instrumental drift). So it is also possible to eliminate them by filtering out all frequencies low enough to be affected by them, and use the higher frequencies, which are almost purely gravitational, to map the gravity field, as explained in the next lecture. Other data that has information on rotation and orbit errors, from attitude sensors and from tracking data, can be incorporated as observations in approach (B) to advantage. In particular, tracking data (most likely from higher satellites) will have information on the low-frequency gravitational effects not affected by the rotation of the instrument, its drift etc., and can strengthen the detrmination of longwavelengths of the field, accordingly.

2. GLOBAL DATA ANALYSIS

The gravitational potential in equatorial spherical coordinates with origin at the center of mass of a planet has a spherical harmonic expansion of the form:

$$V(\theta,\lambda,r) = \frac{GM}{a} \sum_{n=0}^{\infty} \sum_{m=0}^{n} \sum_{\alpha=0}^{1} \left[\frac{a}{r}\right]^{n+1} \bar{Y}_{nm\alpha}(\theta,\lambda)\, \bar{C}_{nm\alpha} \tag{2.1}$$

with

$$\bar{Y}_{nm\alpha}(\theta,\lambda) = \bar{P}_{nm}(\cos\theta)\cos\left[m\lambda - \alpha\frac{\pi}{2}\right] . \tag{2.2}$$

where R is the mean planetary radius, M the planetary mass, G the gravitational constant, $\bar{P}_{nm}$ the normalized Legendre function of degree n and order m, and r, λ, θ are the planetocentric distance, the longitude and the spherical co-latitude, respectively, The $\bar{C}_{nm\alpha}$ are the normalized spherical harmonic coefficients.

Along a spacecraft orbit with a ground-track that repeats almost exactly at regular intervals, the departures of the potential and its gradients from their central mass values, due to the non-spherical irregularities in the field expressed by the $\bar{C}_{nm\alpha}$, can be represented by Fourier series of the form

$$V = \sum_{n=0}^{\infty} \sum_{m=0}^{n} \sum_{\alpha=0}^{1} \bar{C}_{nm\alpha} \sum_{p=0}^{n} \sum_{q=-\infty}^{\infty} \Gamma^{nmpq}_{x_i x_j} \cos\left[[(n-2p+q)N_R + m\, N_D]\omega_r t + \phi'_{nmpq\alpha}\right] , \tag{2.3}$$

where ω_r is the repeat frequency of the orbit, whose period $T_r = 2\pi/\omega_r$, can be as long as the mission itself. N_r is the number of orbit revolutions that take place over the N_D days of an orbit repeat. Both N_R and N_D are relative primes integers, to ensure that there are no repetitions of the ground-track over periods of less than N_D days. This condition makes sure that the finest possible coverage of the earth is achieved during the length of one N_D day repeat. $\phi_{nmpq\alpha}$ is the value of the argument of the cosine at time t=0. Notice that the amplitude V_{nmpq} does not depend on α. This representation assumes that the orbit's mean excentricity and argument of perigee have been chosen so that they remain constant: what is known as a "frozen" orbit (Cook, 1966, Colombo, 1986). Such orbit are near-circular for planets with moderate departures from a perfect sphere.

If the observed quantities are second gradients of the potential, they will also have Fourier expansions of the form of equation (2.3),

$$\Gamma_{x_i x_j} = \sum_{n=0}^{\infty} \sum_{m=0}^{n} \sum_{\alpha=0}^{1} \bar{C}_{nm\alpha} \sum_{p=0}^{n} \sum_{q=-\infty}^{\infty} \Gamma_{x_i x_j}^{nmpq} \cos\left[[(n-2p+q)N_R+m\ N_D]\omega_r t+\phi'_{nmpq\alpha}\right] , \tag{2.4}$$

with $\Gamma_{x_i x_j}$ indicating the gradient with respect to the directions of the axes x_i, x_j in a cartesian system. In wath follows, assume that the axes are: x across the orbit plane, z away from the mass center of the planet, y in the along direction, perpendicular to the other two (for near circular orbits, y is almost parallel to the velocity vector). The phase angle ϕ' may differ from ϕ in (2.3) by some integer multiple of $\pi/2$. The Fourier coefficients in (2.4) can be calculated with formulae like those given in the next section for the simple case of a circular, polar orbit.

It is usually the case that the instrument measures the values of the gravity gradients averaged over an interval Δa of a few seconds, which here is assumed to be no larger than the sampling interval between measurements. To account for this time-averaging, the Fourier coefficients in (3.3) must be replaced with smoothed coefficients $\hat{\Gamma}_{x_i y_j}^{nmpq}$, by multiplying the original coefficients by the smoothing factors derived in the next section, equation (3.7).

If the $\bar{C}_{nm\alpha}$ are unknown, they can be estimated from the measurements by solving simultaneously a system of redundant equations like (2.4) above by a method such as least squares. The expansions in (2.4) have to be truncated at suitably high values of q and of n. The degree of truncation nmax is determined by the strength of the gravitational signal, that becomes progressively more attenuated with degree at altitude because of the factor $(r/R)^{n+1}$ in (2.1) and the decrease of the size of the $\bar{C}_{nm\alpha}$, which follows approximately the exponential decay law $|\bar{C}_{nm\alpha}| \simeq 10^{-5}/n^2$. As shown in the next section, if nmax is high enough, an error analysis will show that most coefficients with a degree higher than nmax will be estimated, for a given altitude and instrumental accuracy, with more than 100% error. Ideally, nmax will be the lowest degree for which the error is 100%. In practice, for the accuracies and altitudes considered here (above 160 Km, 10^{-2} Eötvos units (E.U.)), a degree nmax = 360 was found to be satisfactory. As for the subscript q, for low eccentricity orbits the required upper limit for q is small, and increases gradually with nmax. If namx = 360, $-3 \le q \le 3$ seems adequate.

To estimate the coefficients according to (2.4), one has to form the system of normal equations

$$\underline{v} = H\,\underline{c} \quad , \tag{2.5}$$

where H is the normal matrix, $\underline{v}$ is the vector of observed values (data), and $\underline{c}$ is the vector of unknown coefficients (and other unknowns corresponding to non-gravitational effects such as attitude errors, drift, etc.). Because of (2.4), the part of the matrix corresponding to the $\bar{C}_{nm}$ will have each element $h_{n_1 m_1 \alpha_1 n_2 m_2 \alpha_2}$ associated with a pair of unknown potential coefficients $\bar{C}_{n_1 m_1 \alpha_1}$ and $\bar{C}_{n_2 m_2 \alpha_2}$ of the general form:

$$h_{n_1 m_1 \alpha_1 n_2 m_2 \alpha_2} = \begin{cases} k \sum\limits_{ij} \sum\limits_{pq} \Gamma^{n_1 m_1 pq}_{x_i x_j} \; \Gamma^{n_2 m_2 pq}_{x_i x_j} & \text{if } m_1 = m_2 \,,\; \alpha_1 = \alpha_2 \\ \\ 0 & \text{otherwise,} \end{cases} \tag{2.6}$$

where K is a constant that depends on the number and accuracy of the measurements, the altitude, etc., and where the sum is over all combinations of p and q that give different values for the frequency in (2.4), $([n-2p+q]\, N_R + m\, N_D)\, \omega_r$. Expression (2.6) shows that many elements of the matrix will be zero, because of the orthogonality of trigonometric functions either of different frequncies, or else in quadrature with each other, provided that no frequency in (2.4) is higher than half the sampling rate of the data. If the unknown $\bar{C}_{nm\alpha}$ are grouped within $\underline{c}$ by oder m, and by subindex α, the submatrix corresponding to these coefficients will consist of diagonal blocks that, for each order m, are of dimension no largher than (nmax+1)-m. Moreover the blocks for α=0 and α=1 are identical for m larger than 0 (only the block for α=0 is present if m=0). Consequently, the normal matrix is both very sparse and redundant. This makes it possible both to set up H (using expressions such as (2.6) to calculate the elements directly), and to invert it, in spite of the very large dimension of H (about $nmax^2$, or in the order of 10^5 for high resolution field modelling).

If one assumes that the non-gravitational effects are confined to a low-frequency band, by appropriate design of the instrument and of the spacecraft attitude control system, one can filter out such band with a high-pass filter, and use the higher frequency residuals for the analysis. Those residuals can be represented by the terms in the Fourier series (2.4) with frequencies above the cutoff of the filter. The structure of the normal matrix is now purely block diagonal (it might not have been with the non-gravitational effects unfiltered, as these may not have the same periodicity as the orbit). This is the idea used in the error analyisis discussed in the next section. It is a rather wasteful approach, as it means throwing away information on the field itself, which is present at frequencies as low as that of the orbit repeat, ω_r . An alternative to filtering is adding possibly non-periodical terms in (2.4) to represent the non-gravitational part of the signal, with extra parameters to be adjusted (drift rates, initial attitude errors, orbit state errors, etc.), assuming that a satisfactory formulation for such terms can be found. In that case, the normal matrix will consist of the block-diagonal part of expression (2.6) for the $\bar{C}_{nm\alpha}$ plus two "edges" that give the non-zero part the shape of an arrow (Fig. 5). There are very efficient techniques for handling matrices of this type (see, for example, Colombo (1984), Chapter 3). If the spacecraft is drag-free, the orbital errors due to the unknown potential coefficients will have Fourier expansions with the same frequency terms as (2.4). The effect of these errors on the measurements (scaled by factors of the order of GM/r^4 E.U./m) are likely to be insignificant compared to the terms in (2.4) for frequencies above 3 cycles/rev. For lower frequencies, those effects should be included. In keeping with the analytical approach followed here, one may use analytical expressions for the Fourier coefficients of those orbit errors. Such expressions can be found in Colombo (1984) and (1986). The introduction of these gravitational orbital effects does not change the frequencies in (2.4), so the sparse structure implied by (2.6) is maintained.

3. MISSION ERROR ANALYSIS FOR A 10^{-2} E.U., FULL-TENSOR INSTRUMENT

An error analysis based on the theory presented in the previous section and simplified by assuming that the orbit is both polar and circular, was carried out to detrmine the accuracy with which potential coefficients up to degree and order 360 could be determined with a full-tensor gradiometer accurate to 10-2 E.U. in all nine components. What follows is the mathematical derivation of the main formulae used for this analysis.

3.1 Time series representation of the second gradients for a circular, polar, repeating orbit

If the orbit is circular, with inclination $I = \pi/2$, if $n_o = N_R\omega_R$ and $\dot{\theta} = N_D\omega_D$ are the orbit's and earth's angular frequencies, respectively, and if, for t=0, $\lambda_N=0$ for the ascending node, and the satellite is at the point of highest latitude $\phi_o=\phi_{max}$, then expression (2.2) can be written as

$$\bar{Y}_{nm\alpha}(t) = \sum_{j\geq 0}^{n} p_{nmj} \left\{\cos[(jn_o + m\dot{\theta})t + \phi_{mj\alpha}] + \cos[(jn_o - m\dot{\theta})t + \phi_{mj\alpha}]\right\}, \qquad (3.1)$$

where

j=0,2,4,...,n if n is even

j=1,3,5,...,n if n is odd

$$\phi_{mj\alpha} = \frac{\pi}{2}\left[\alpha + \frac{1-(-1)^m}{2}\right] . \qquad (3.2)$$

So now the observation equation (2.4) becomes

$$\Gamma_{x_i x_j} = \sum_{nm\alpha} \bar{C}_{nm\alpha} \sum_{j\geq 0}^{n} p_{nmj} \left\{\cos[(jn_o+m\dot{\theta})t+\phi_{mj\alpha}]+\cos[(jn_o-m\dot{\theta})t+\phi_{mj\alpha}]\right\}, \qquad (3.3)$$

which can be obtained by keeping only terms with q=0 (corresponding to zero eccentricity), and defining j as the absolute value of (n-2p).
If x is the across direction, then it is also true that on the unit circle

$$\frac{\partial \bar{Y}_{nm\alpha(t)}}{\partial x} = \sum_{j\geq 0}^{n} q_{nmj} \left\{\cos[(jn_o+m\dot{\theta})t+\phi_{mj\alpha}]+\cos[(jn_o-m\dot{\theta})t+\phi_{mj\alpha}]\right\}, \qquad (3.4)$$

The p_{nmj} are the inclination functions and the q_{nmj} are related to their derivatives; they can be obtained by appling the FFT to equispaced values of $\bar{Y}_{nm\alpha}$ and $\partial\bar{Y}_{nm\alpha}/\partial x$ on a unit circle of inclination I, with $\lambda_N=0$, $\phi_o=\phi_{max}$. (This calculation is greatly simplified if $I = \pi/2$.)
As already explained, measurements are usually time-averaged over an interval Δa of a few seconds. To account for this in the theory, one must consider first the Fourier expansion of the general spherical harmonic $\bar{Y}_{nm\alpha}$ over the time interval Δa:

$$\frac{1}{\Delta a}\int_{t-\frac{\Delta a}{2}}^{t+\frac{\Delta a}{2}} \bar{Y}_{nm\alpha}(t')\,dt' \;.$$

Using the formulae for the explicit integrals of sines and cosines, the Fourier coefficients in (3.1) and (3.4) become

$$\hat{p}_{nmj} = F(\pm m, j, \Delta a)\; p_{nmj} \tag{3.5}$$

$$\hat{q}_{nmj} = F(\pm m, j, \Delta a)\; q_{nmj} \tag{3.6}$$

with

$$F(\pm m, j, \Delta a) = \frac{2}{(jn_o \pm m\dot{\theta})\Delta a} \sin\left[(jn_o \pm m\dot{\theta})\;\Delta a/2\right] \;, \tag{3.7}$$

where $\pm m\dot{\theta}$ corresponds to the $(jn_o + m\dot{\theta})$ and $(jn_o - m\dot{\theta})$ terms in (3.1), (3.4) respectively.

Finally, from the preceding arguments and the fact that, if y is in the along direction and β is the angle counterclockwise from the point of highest latitude then, on the unit circle,

$$\frac{\partial}{\partial y} \cos(j\beta \pm m\lambda + \phi_{mj\alpha}) = -j \sin\,(j\beta \pm m\lambda + \phi_{mj\alpha}) \;,$$

it follows that the Fourier coefficients of the averaged time series of the second gradients $\Gamma^{nm\alpha}_{x_i x_j} = \frac{\partial \bar{Y}_{nm\alpha}}{\partial x_i \partial x_j}$ are

$$\Gamma^{nm\alpha}_{zz} \longrightarrow \hat{p}_{nmj}\;(n+1)\;(n+2) \tag{3.8}$$

$$\Gamma^{nm\alpha}_{yy} \longrightarrow -\hat{p}_{nmj}\;j^2 \tag{3.9}$$

$$\Gamma^{nm\alpha}_{xx} \longrightarrow -\hat{p}_{nmj}\left[(n+1)\;(n+2) - j^2\right] \text{ (from Laplace's equation)} \tag{3.10}$$

$$\Gamma^{nm\alpha}_{yx} = \Gamma^{nm\alpha}_{xy} \longrightarrow \hat{q}_{nmj}\;j(-1)^{m+1} \tag{3.11}$$

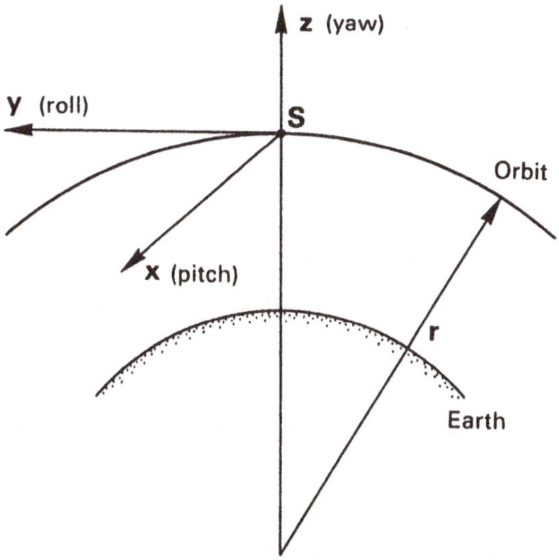

Fig. 4 - The rotating local orbital system of Cartesian coordinates.

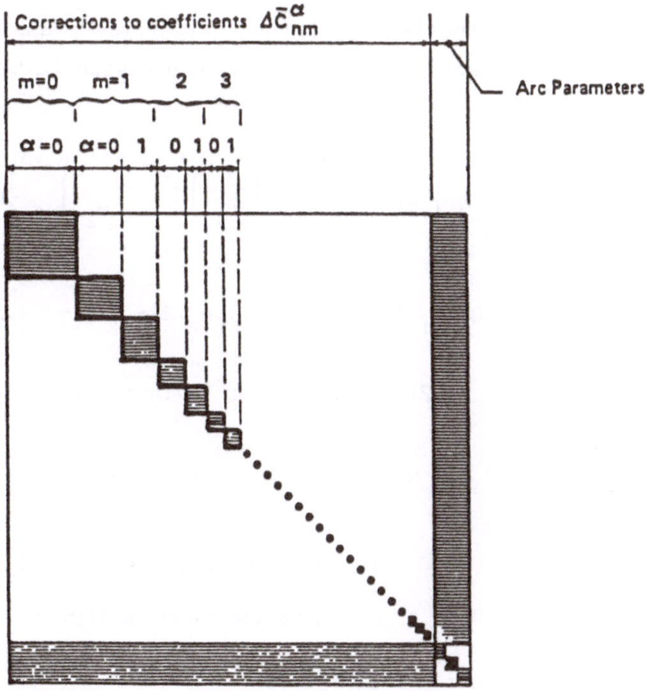

Fig. 5 - The arrow pattern of the normal matrix.

$$\Gamma^{nm\alpha}_{yz} = \Gamma^{nm\alpha}_{zy} \longrightarrow -\hat{p}_{nmj}\ (n+1)\ j(-1)^{m+1} \tag{3.12}$$

$$\Gamma^{nm\alpha}_{xz} = \Gamma^{nm\alpha}_{zx} \longrightarrow -\hat{q}_{nmj}\ (n+1) \tag{3.13}$$

Here " $\longrightarrow$ " indicates that the expression to the right is the Fourier coefficient of the gradient corresponding to one of the two frequencies ($jn_o \pm m\lambda$); x, y, z are the across, along, and radial local coordinates of the spacecraft.

3.2 The general element of the normal matrix H for the full tensor gradiometer

If $h_{n_1m_1\alpha_1n_2m_2\alpha_2}$ is the element of the normal matrix H corresponding to the unknowns $\bar{C}_{n_1m_1\alpha_1}$ and $\bar{C}_{n_2m_2\alpha_2}$, and assuming that the orbit repeats exactly after N_D days and N_R orbit revolutions (where N_D and N_R are relative primes), that $n_1 \le n_2$, that the sampling interval $\Delta t \ge \Delta a$ is sufficiently high, that the series (2.1) can be truncated at a degree n_{max} without compromising the results, that T_r is the lenght of the orbit repeat, and that all frequencies up to $j_{min}\, n_o$ have been eliminated. Then, expression (2.6) written out in full becomes :

$$h_{n_1m_1\alpha_1n_2m_2\alpha_2} = \begin{cases} \left(\frac{GM}{3}\right)^2 \frac{T_r}{2\Delta t} \frac{1}{\sigma^2} \sum\limits_{j \ge j_{min}}^{n_1} \left(\frac{a}{r_o}\right)^{n_1+3} \left(\frac{a}{r_o}\right)^{n_2+3} \Big\{ \hat{p}_{n_1mj}\, \hat{p}_{n_2mj} \\ \times \Big[2[(n_1+1)(n_1+2)(n_2+1)(n_2+2) + j^4] + \\ \quad + j^2[2(n_1+1)(n_2+1) - (n_1+1)(n_1+2) - (n_2+1)(n_2+2)] \Big] \\ 2\hat{q}_{n_1mj}\, \hat{q}_{n_2mj}\ [j^2 + (n_1+1)(n_2+1)] \Big\} \\ \qquad \text{if: } m_1 = m_2 = m\ ,\ \alpha_1 = \alpha_2 \text{ and } n_1 \text{ and } n_2 \\ \qquad \text{have the same parity;} \\ = 0 \qquad \text{otherwise} \end{cases} \tag{3.14}$$

Ordering the coefficients so they are grouped first by order m, then by α and, finally, by the parity (even, odd) of n, H becomes a very sparse block-diagonal matrix. There are $2n_{max}+1$ blocks that are different; the largest has dimension $(n_{max} - j_{min})$, the smallest is a scalar. The separation by m and by α holds if the orbit misclosure at the end of the repeat period is smaller then the smallest gravitational feature to be resolved. The separation by n-parity is true only in the circular-orbit approximation.

3.3 Rescaling for different altitude accuracies and mission lenghts

Instead of solving for the $\bar{C}_{nm\alpha}$, one could solve for the $\left[\frac{a}{r_o}\right]^{n+3} \bar{C}_{nm\alpha}$ (for a circular orbit), which are the potential coefficients at altitude. This means replacing the $\left[\frac{a}{r_o}\right]^{n+3}$ factor with ones in (20), resulting in a more numerically stable calculation. If one makes $T_r=1$, then the resulting "normalized" matrix, $H_{(norm.)}$, is related to H by

$$H_{(norm.)} = D^{-1} H D^{-1} , \tag{3.15}$$

where D is a diagonal matrix whose diagonal elemnts are of the form

$$d^{m\alpha}_{ii} = \left[\frac{a}{r_o}\right]^{n_i+3} \sqrt{T_r} . \tag{3.16}$$

Also the variance-covariance matrix of the estimated $\bar{C}_{nm\alpha}$ is

$$H^{-1} = D^{-1} H^{-1}_{(norm.)} D^{-1} . \tag{3.17}$$

The diagonal elements of D^{-1} are the inverses of those of D, and act as "scaling factors" to obtain the variances and covariances of the $\bar{C}_{nm\alpha}$ for different missions with various altitudes (r_o) and durations (T_r). The scaling is valid as long as $n_{max}(n_o+\Omega)\Delta a<<2\pi$ for all choices of r_o, which is the case in the present study $(n_o \simeq (GM/r_o^3)^{1/2})$.
Finally, for (3.14) follows that, if the accuracy of the instrument were σ' instead of σ, then

$$H^{-1} = \frac{\sigma'^2}{\sigma} D^{-1} H D^{-1} . \tag{3.18}$$

3.4 Calculating global RMS errors of area mean anomalies

Let δ^2_{nm} indicate the variance of the error in $\bar{C}_{nm\alpha}$ (same for α=0 or α=1), so δ^2_{nm} is a diagonal element of H^{-1}. Then, the degree variances of the errors and of the coefficients are, respectively,

$$\delta_n^2 = \delta_{no}^2 + 2 \sum_{m=1}^{n} \delta_{nm}^2 \quad , \tag{3.19}$$

and

$$\sigma_n^2 = \sum_{m=0}^{n} \sum_{\alpha=0}^{1} \bar{C}_{nm\alpha} \quad , \tag{3.20}$$

Also:

the global r.m.s. of mean anomaly on the unit sphere C is

$$\int_C (\text{mean anomaly})^2 \, dC \simeq \left[\sum_{n=n_{min}}^{n_T} \delta_n^2 \, \beta_n^2 \, K_n^2 + \sum_{n=n_T+1}^{\infty} \sigma_n^2 \, \beta_n^2 \, K_n^2 \right]^{1/2} . \tag{3.21}$$

Here

n_T = degree of truncation ($n_T < n_{max}$ if not all the terms in the solution are used),

$$K_n = \begin{cases} a^{\frac{GM}{2}(n-1)} & \text{for gravity anomalies} \\ a & \text{for geoid (undulation) anomalies} \end{cases}$$

and

$$\beta_n = \frac{1}{\cos\psi} \frac{1}{2n+1} \left[P_{n-1} (\cos\psi) - P_{n+1} (\cos\psi) \right] \tag{3.22}$$

is the n-degree smoothing factor, with $\psi = \Delta^\circ \sqrt{\pi}/180$ for area means over $\Delta^\circ \text{x} \Delta^\circ$ blocks ($\Delta^\circ \bar{=} \Delta$ in degrees). The first summation in (3.21) is the <u>commission</u> error, and the second one, the <u>omission</u> (or truncation) error.

Expression (3.21) further assumes a negligible error contribution from the harmonics with $n < n_{min}$, whose coefficients are kept fixed in the adjustment. The global r.m.s. is smalled when n_T, the degree at which the expansion is truncated, is $n_T = n_{100\%}$, the degree above which $\delta_n / \sigma_n > 1$ and the estimated coefficients are likely to be more than 100% in error. Expression (3.21) was calculated in this way to obtain the results shown in the figures that follow. For practical reasons, the infinite upper limit in the second summation was replaced with n=1000.

The values of the σ^2 were taken from field spectral analyses made by Prof. R.H. Rapp of Ohio University (see Rapp, (1979), and Colombo (1981), Ch. 3).

In addition, the mean error per coefficient per degrees:

$$\bar{\delta}_n = \left[\frac{\delta_n^2}{2n+1} \right]^{1/2} \quad , \tag{3.23}$$

and the mean percentage error per coefficient per degree is:

$$\bar{\delta}_n \% = \left[\frac{\delta_n}{\sigma_n} \right] \times 100 \quad , \tag{3.24}$$

Plots of (3.23) and (3.24) for the altitudes of 160 Km and 200 Km and for $\sigma=10^{-2}$ E.U., are shown in Fig. (2.6) and (2.7), as well as the corresponding plots of $\delta_n\%$ for GEM-T1 (Marsh et al., 1988), and of the previously studied satellite-to-satellite version of GRM (from Colombo, 1981). Up to degree n=100, the σ_n are actual values estimated from gravimetry and altimetry - this explains the bumpiness at low degrees of the percentage errors for the gradiometer.

4. IMPLICATIONS FOR THE STUDY OF THE EARTH OF THE RESULTS OF A GLOBAL ERROR ANALYSIS OF A FULL-TENSOR GRADIOMETER MISSION

The results of a global error analysis carried out as described in the preceeding lecture, taking full advantage of the symmetries of the spherical harmonics and of the repeating, circular orbit, are shown in Fig. 6,7,8 and 9. All these figures correspond to the expected accuracy of the French gradiometer, or 10^{-2} E.U. The first one shows the expected errors per coefficient, per degree up to degree and order n=360.

These uncertainties were calculated by first setting up and inverting the block-diagonal normal matrix of a full adjustment (equation (3.14) in previous lecture) assuming six months of data, sampled and averaged at 4 seconds' intervals, a repeating polar orbit at an altitude of 160 Km, and all nine second derivatives of the gravity potential, suitably corrected for orbit, attitude, rotation, drift, etc., by the simple expedient of filtering out all information below three cycles per revolution. This means that no potential coefficients below degree 4 could be estimated, so they were kept fixed in the solution. The graph in Fig. 1 can be scaled easily according to expression (3.18), previous lecture, replacing the whole of matrix H^{-1} with the mean errors derived from (3.14). So it is possible with very little additional calculation to use the results thus plotted to estimate what would happen if the accuracy of the instrument were different, or the length of the mission, or the height of the satellite (through a change in the elements of matrix D, which is diagonal). Since the radius of the earth a is another factor that can be changed quite readily, one can use this graph to estimate the accuracy of a similar mission carried out with a space probe orbiting another planet. One would have to change also GM in this case. From (3.14) it is clear that $(GM/a^3)^2$ is a factor common to all elements of the normal matrix H, so the scaling would require multiplying the values shown in Fig. 6 by the ratio of the above quantity's square root for the earth to that for the planet in question. An interesting consequence of this is that, for a low orbit near the planetary surface, same repeat period, sampling, etc., the same instrument will resolve the same harmonics to higher accuracy for a planet with smaller radius and density, and with less accuracy for a planet with a larger radius or less density (the period of the orbits near most planets are roughly the same, as their ratios depend on the ratios of the square roots of their mean densities by Kepler's third law).

Equally interesting is the fact that, in most of the other major components of the solar system, the atmosphere of planets and large moons are either less dense than ours, or altogether nonexistent. This means that a spacecraft can orbit much closer to the surface (the limit would be imposed by the inevitable lack of circularity of both orbit and planet). In many cases, such as that of our Moon, the gradiometer can orbit as close as 50 Km. Since the mass anomalies (mountains, for example) are measured from closer and, equally important, are not necessarily smaller because the world they happen to be is smaller than our own (in fact, the moon has mountains as high as on earth) the corresponding signals would be stronger. Moreover, for a given harmonic of degree n, the highest spatial frequency has a wavelength equal to the circumference of the planet divided by n. The result of this is that features of the same size will correspond to lower degrees in harmonics on the surface of the moon than on earth. To understand why the gradiometer will perform better on a smaller world, one should consider that the same number of equally sensitive measurements are taken from a closer distance over the altogether smaller area of the lesser world, so they are concentrated on a smaller space about features that are not necessarily smaller.

A characteristic of the plot of the coefficient accuracies is that they begin relatively large, then decrease as n increases, and reach a minimum at about n=70. Beyond this point, the errors increase more or less exponentially with n. The reason for the initial increase in sensitivity can be attributed to the greater signal content in the gradiometer signal with increasing frequency (double differentiation); as n increases, on the other hand, the harmonics become attenuated because of height, at the exponentially increasing rate $(a/r)^{n+3}$, which results in the errors increasing in proportion to the inverse

of this rate when calculated using (3.23) in the previous lecture.
As an example, consider the case when the gradiometer only measures the radial second derivative, V_{zz}. In this case, the harmonic expansion for this observable is the same as that for the potential, except for a scale factor, with each harmonic of degree n multiplied by (n+1)(n+2) (see expression (3.8), last lecture). Approximately, for large n (larger than 20), this is the same as n^2, so the initial tendency of the mean coefficient errors is to decrease as n^2 (for high degrees, the blocks in H are nearly diagonal matrices, from which fact follows the "error law" just given). But as n goes up, eventually the exponential factor takes over, which explains the shape of the curve at high degrees. A similar argument could be used when other gradients are measured. As horizontal derivatives are involved, the argument for them involves consideration of expressions like (3.9), (3.11), or (3.12), where the number of revolutions j, used as a measure of spatial frequency, enters to the square or to the first power (in which case, (n+1), for example, also appears, so there is always a "n^2" type factor). One needs also to use the asymptotic expressions for the Legendre functions when the degree tends to infinity, which say, in essence, that the function behave like sinewaves of a frequency approximately equal to jn_o. From all this discussion follows that the curve shown in Fig. 6 should be fitted rather well by an epression of the form:

$$\varepsilon_n \simeq \frac{F}{n^2}\left(\frac{r_s}{R}\right)^{n+3} \qquad r_s = \text{radius of satellite orbit} > R_E$$

This is indeed the case, and the best value for the scaling factor F is

$$F = 0.22 \times 10^{-8} \qquad \text{for } 10^{-2} \text{ E.U.}$$

The fit, in fact, is remarkably good, and has been used to "extend" the results of the error analysis far beyond the original degree n=360.
This gives some idea of the performance of the gradiometer at levels of resolution that would have been quite impossible to investigate directly by the global method of the last lecture. Remember that the inversion of a matrix (and, in the present case, also its setting up, in the first place) requires a time proportional to the 3rd power of the number of unknowns and that this number, with spherical harmonics, increases as the square of n.
The situation is alleviated somewhat by the sparseness of the matrix, so the time (number of computations) increases as the fourth power of n, rather than as the fifth, as would be the case for a full, unstructured normal matrix. The calculations up to n=360 needed about two hours; with the same program and

computer, it would require, therefore, some 30 hours to estimate directly the errors to n=600 (about 64 Km wavelength).

There are many geophysical features of interest (such as mountains, that have sizes between 100 Km and 50 Km, and it is important to study whether the accuracy of the instrument, height, length of mission, etc., are adequate to learn more about them. It would have been prohibitive ro use 32 hours of computer time for such a calculation, but the remarkable fit to the results by the simple expression above allows to make at least some educated guesses as to what one might be able to do with a gradiometer at spatial scales of high interest and very fine detail.

It should be remarked here that the results of the error analysis have been used only partially: only variances, and no covariances. Part of the reason is that, for studying global r.m.s. accuracies (see Fig. 7,8,9), averaging the error over the whole sphere eliminates cross-products between harmonics (orthogonality property), while for other studies (or interpretation of the averages), the justification is that the matrix has diagonal blocks that are reasonably close to being diagonal themselves, particularily for high orders.

Figure 7 shows, in bold line, the relative errors per degree, compared to the mean size expected for the spherical harmonic coefficients from what is known about the power spectrum of the gravity field (see expression (3.24) in previous lecture, and accompanying explanation). The plot tends to reflect the overall trends (first down, then up) in Fig. 6, although it is more irregular, as the quantities plotted in Fig. 6 have been divided by the power spectrum estimated empirically by Rapp from gravity and altimetry. The y axis is in percentage of the signal, so these are percentual "noise to signal" ratios.

There are two other curves also plotted: one showing the percentage errors in the mean GEM-T1 coefficients, indicative of the present situation with models derived from satellite data only, and the other showing the same type of plot for satellite-to-satellite tracking, assuming a mission of the same duration and height, and a tracking accuracy, for the Doppler system, of 1 micron/sec (Spacecraft separation is 300 Km). Clearly, the pair of mutually tracking spacecrafts have much greater sensitivity to low frequency information (the plot goes off the scale) but then lose this sensitivity quickly at about degree n=100; the "crossover" point is n=128, and from there on, the gradiometer is more sensitive. The weak side of gradiometry, clearly, is at very low degrees (below n=20), where the results show no substantial improvement over even contemporary models, like GEM-T1. But the satellite-to-satellite tracking curve shows that there is improvement where tracking is particulary strong (and at very low frequencies, resonant effects on the orbit cause the signal to noise ratio to improve even further). In an actual gradiometer mission, tracking will be conducted, most likely, from ground stations initially, but once the orbit is known very roughly, the tracking will be done by means of a GPS receiver in the spacecraft. In this way, many high satellites would be used to track the low one carrying the gradiometer. Because of the great accuracy that even now can be obtained with the carrier phase signal of GPS, and the number of satellites that can "track" simultaneously, all around the world, the information on the low degree and order harmonics of the field conveyed by this tracking should be enough to obtain much improved estimates of those low harmonics the gradiometer is less sensitive to. So a proper choice of tracking system (such as GPS, although other space systems might be available and, perhaps, work better, at the time of the mission) would provide a very important complement to the gradiometer.

Figures 8 and 9 show the accuracies implied by the error analysis results for (a) mean anomalies, and (b) mean geoid undulations. In calculating both according to equations (3.21) and (3.22), the degree of truncation n_{max} was chosen as that for which the percent error per coefficient in Fig. 7 was about 100%. The idea is that, as percentage errors raise monotonically (because n_{max} is always high enough to be in the exponentially raising part), the information about n_{max} is most likely drowned in noise. This is not necessarily true in all cases, because there may be gravitational features caused by, say, mountains, with coefficients much larger than the average errors at degrees higher than n_{max}. It is, nevertheless, a reasonable choice of the truncation degree. The assumption, then, is that the estimation is confined to terms where n is no higher than n_{max}. Because different mission lengths, heights, etc., shift the graph in Fig. 6 up or down as the scaling factor those things imply increases or decreases, the point where the bold curve in Fig. 7 crosses the 100% ordinate depends also on those factors. Figures 8 and 9 show the error in the estimation of the area means, as well as the degree of truncation (in brackets) for several heights and mission lengths, always assuming 10^{-2} E.U. accuracy for each measurement, and a full-tensor gradiometer. Since any change that moves the curve in Fig. 7 up will cause n_{max} to decrease (i.e., less accuracy in the instrument, shorter mission, higher altitude), the commission error, which is the sum of all errors in the computed coefficients (those errors to the left of n_{max}) will tend to decrease, while the omission error (due to the harmonics no included in the model), will increase. The result, as shown in the figures, is that the total error (commission+omission) they display, goes up very slowly with more unfavorable conditions (say, increasing height). The significance of the results shown is that they give a general idea of the accuracy of gravity maps that can be drawn, representing area means of anomalies quantities. Their importance is mostly for geodetic applications (say, levelling), for spotting features looking at the maps, and for oceanography (Fig. 9 shows that one can get less than 10 cm mean geoids), but are less clearly relevant to geophysicists interested in studying the mechanisms that build and support the existence of mountains and similar orographic features.

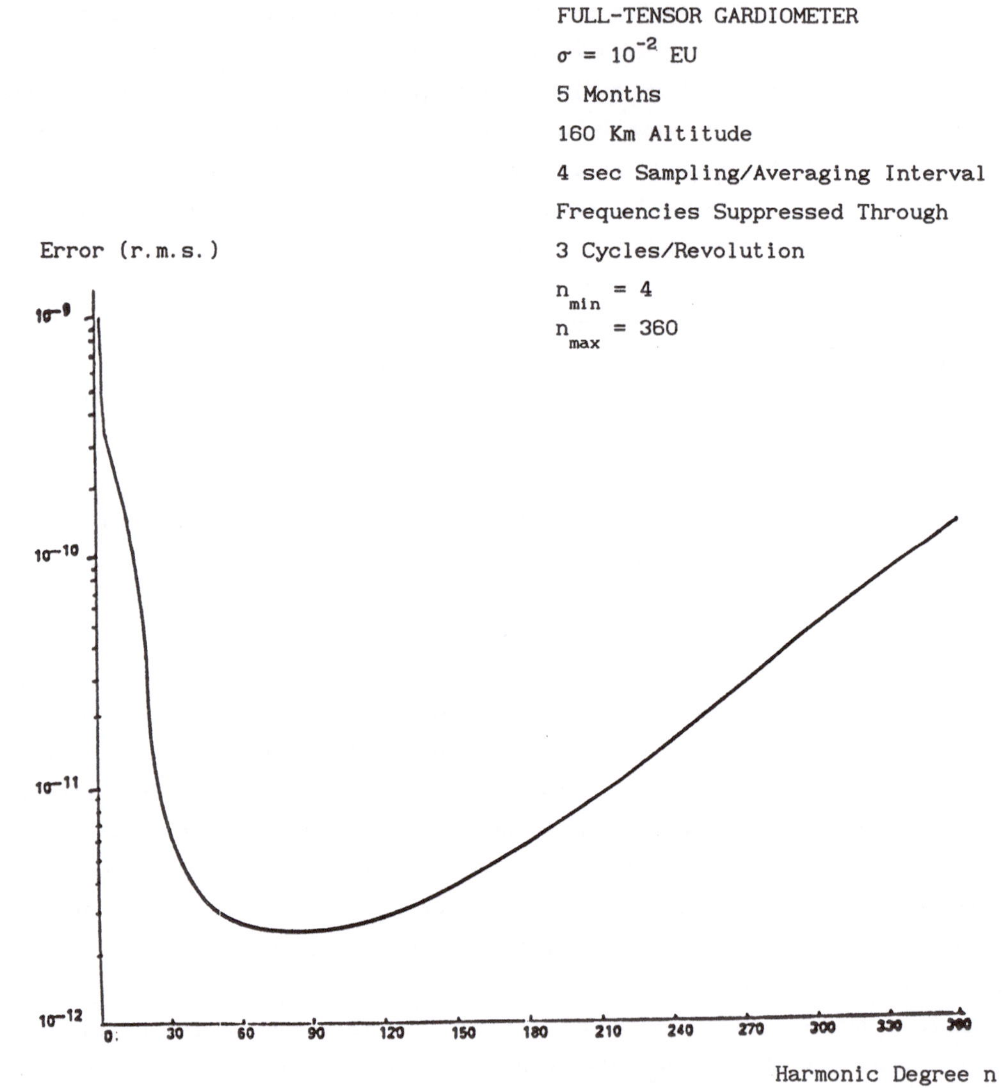

Fig. 6 - RMS Error per dimension less potential coefficient.

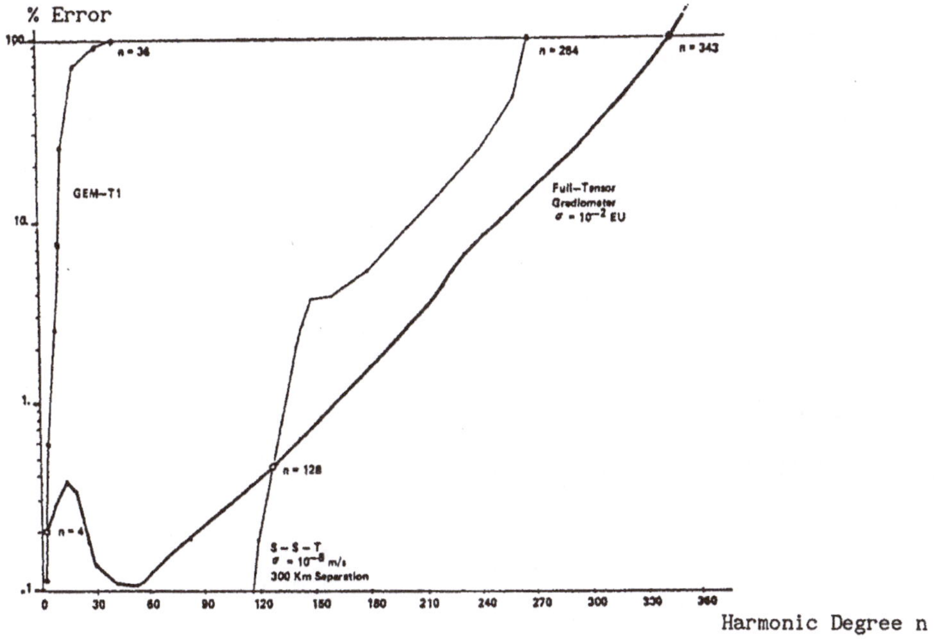

Fig. 7 - Comparison of percentage RMS error per coefficient vs. harmonic degree for GEM-T1, Gradiometer and SAT-SAT tracking

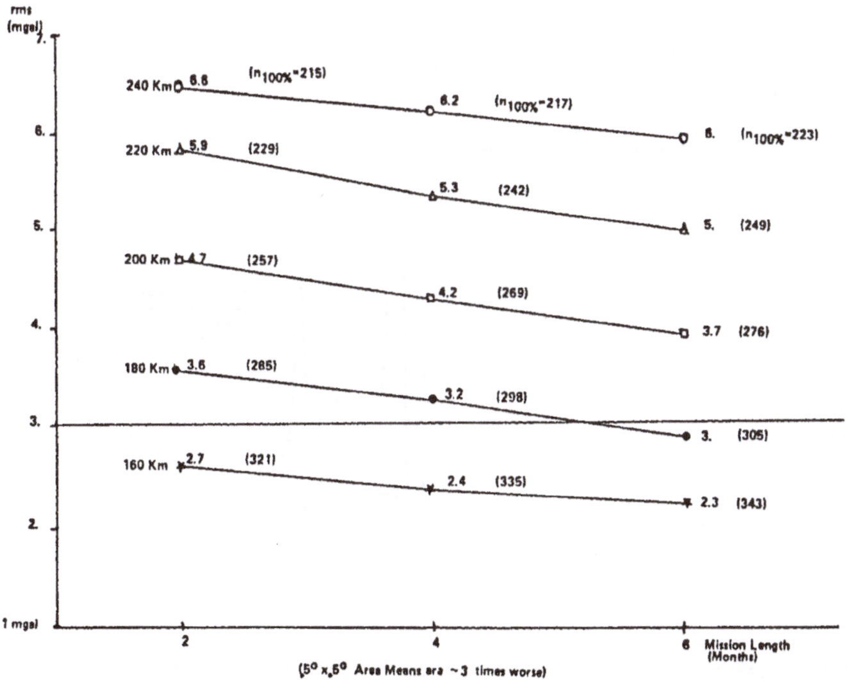

Fig. 8 - Accuracy of 1°x 1° area mean gravity anomalies vs. height and mission length (RMS omission + commission error)

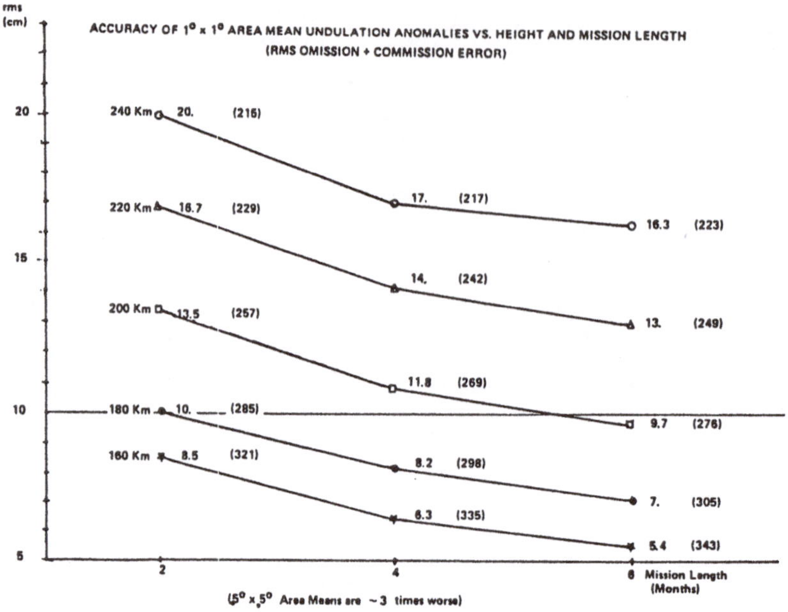

Fig. 9 - Accuracy of 1°x 1° area mean undulation anomalies vs. height and mission length (RMS omission + commission error)

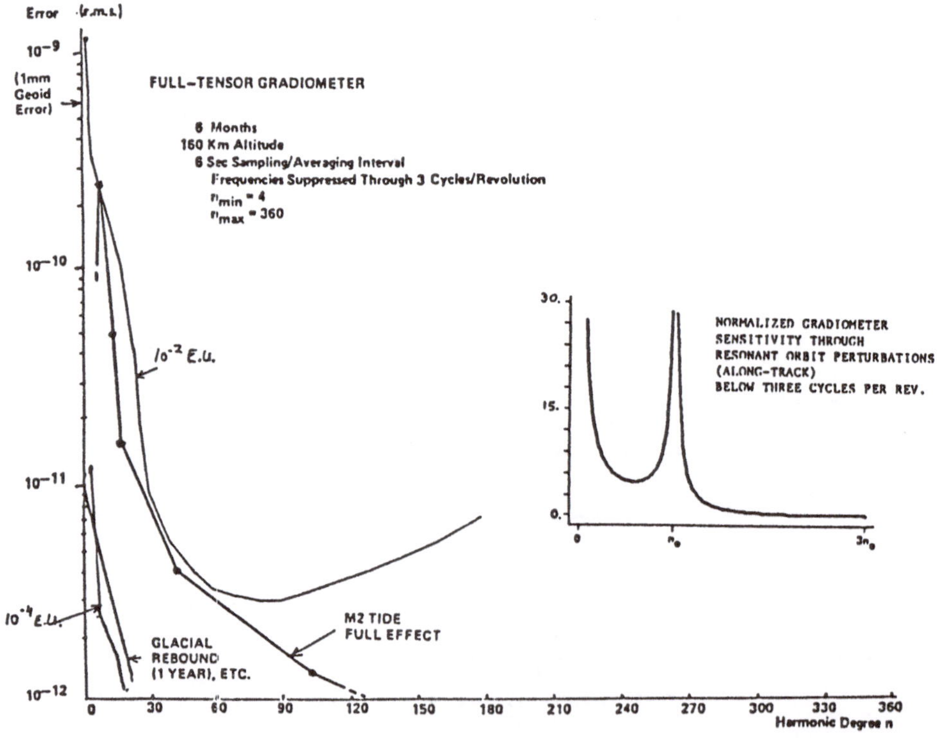

Fig. 10 - Tidal and other low degree and order signals of geophysical interest vs. RMS error per dimension less potential coefficient (no signal below three cycles per revolution)

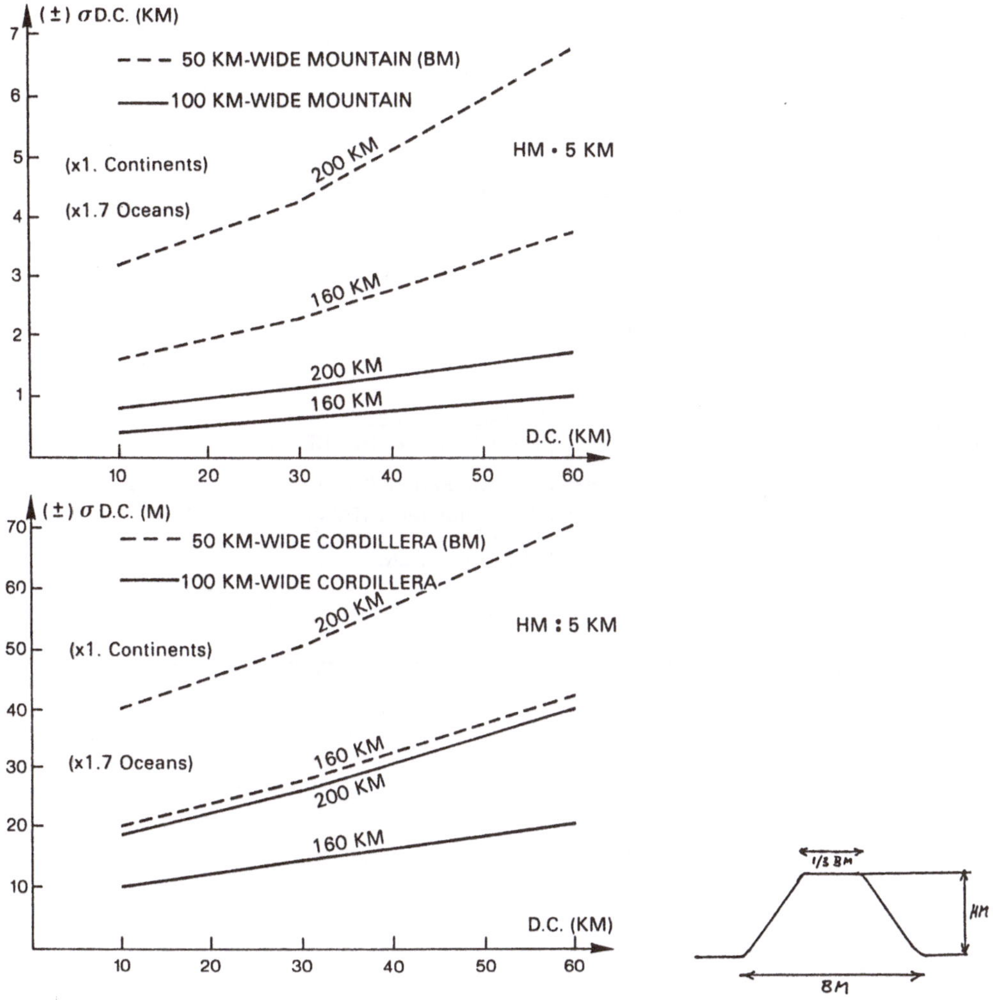

PROFILE OF MOUNTAIN (CORDILLERA)
BOTH ARE SURFACES OF REVOLUTION
MOUNTAIN PLACED AT NORTH POLE
CORDILLERA CIRCLES THE EQUATOR
COMPENSATION: AIRY-HEISKANEN
D.C. = DEPTH OF COMPENSATION
6 MONTH, FULL-TENSOR, 4 SEC.
SAMPLING / AVERAGING INTERVAL,
10^{-4} EU INSTRUMENT.

$$\Delta g_n = \frac{3GM}{R_\varepsilon^3} \frac{\rho_o}{\rho_E} \frac{H_o(n-1)}{(2n+1)} \left[1-\left(1-\frac{D.C.}{R_E}\right)^n\right]\left(\frac{R_E}{R_P}\right)^{n+2}$$

H_n = nth HARMONIC OF TOPOGRAPHY

$R_P = R_E$ (POINTS ARE ON EARTH'S SURFACE)

σ D.C. = ACCURACY OF D.C. WHEN ESTIMATED FROM RECOVERED POTENTIAL COEFFICIENTS TO $n_{max} = 600$

IF MOUNTAIN HEIGHT ≠ 5 Km,

σ D.C. = σ_{5Km} D.C. ≠ (5/Hm)

Fig. 11

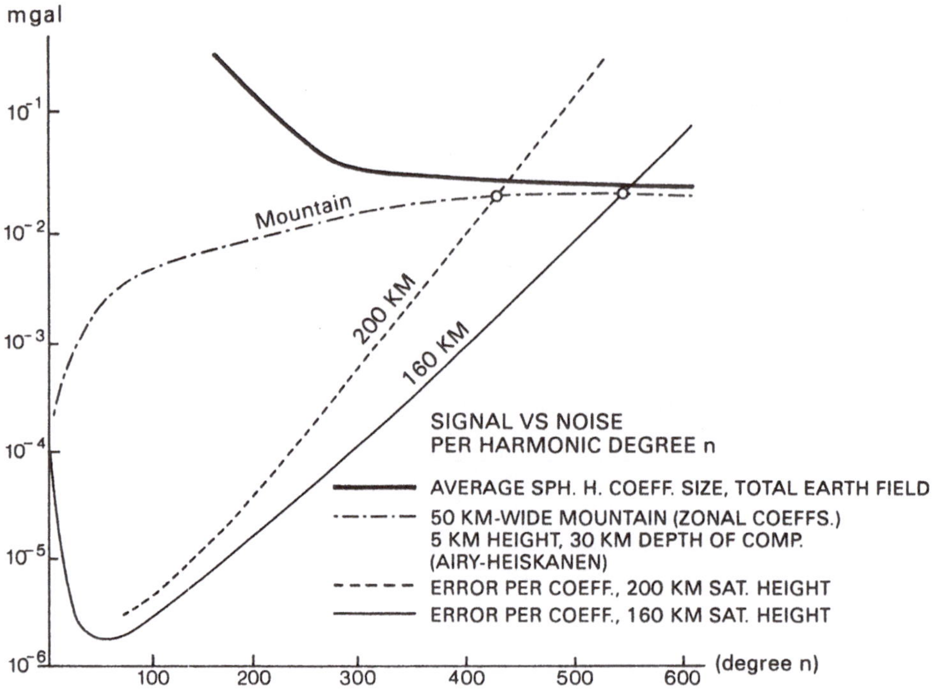

Fig. 12

REFERENCES

Chapter 1

Concerning the various developments in the design of gradiometers mentioned here:

- Paik H J (1981): Superconducting tensor gravity gradiometer for satellite geodesy and inertial navigation. Journal of Astronautical Science, 29, 1-17.
- Balmino G (1984): Le project GRADIO et la determination à haute resolution du geopotentiel. Bull. Géod., 58, 2, 151-179.

On the use of gradiometers to navigate in space and determine gravitational fields, here are a very old and a rather new reference:

- Carroll J J, Savet P H (1959): Gravity difference detection. Aereospace Engineering, 44-47.
- Rummel R, Colombo O L (1985): Gravity field determination from satellite gradiometry. Bull. Géod., 59, 233-246.

For mechanical principles and their application to spacecraft control, including a derivation of Hill's and Euler equations, gradient stabilization, etc.:

- Kaplan M (1976): Modern spacecraft dynamics and control. John Wiley & Sons, New York.

For a general coverage of the principles behind satellite gradiometry and a comprehensive list of references:

- Rummel R (1976): Satellite gradiometry. Lecture Notes in Earth Sciences, Vol. 7, 317-363, Ed.: H. Sünkel Springer-Verlag, Berlin/Heidelberg. (These were lecture notes of a course given at the Summer School on Geodesy in Admont, Austria, 1976).

Chapter 2,3,4

Cook E E (1966): Perturbations of Near Circular Orbits by the Earth's Gravitational Potential. Planetary and Space Science, 14, 433-444.

Colombo O L (1986): Notes on the Mapping of the Gravity Field using Satellite Data. Mathematical and Numerical Techniques in Physical Geodesy, Ed: H. Sünkel, 261-316, Springer, Heildelberg.

Colombo O L (1984): Altimetry, orbits and tides. NASA Technical Memorandum 86180, Goddard Space Flight Center, Greenbelt.

Colombo O L (1981): Numerical Methods for Harmonic Analysis on the Sphere. The Ohio State University, Dept. Geodetic Science, 310.

Rapp R H (1979): Potential Coefficient and Anomaly Degree Variance Revisited. The OhioState University, Dept. Geodetic Science, 293.

Marsh J G, Lerch F J, Putney B H, Christodoulidis D C, Felsentreger T L, Snachez B V, Smith D E, Klosko S M, Martin T V, Pavlis E C, Robbins J W, Williamson R G, Colombo O L, Chandler N L, Rachlin K E, Patel G B, Bathi S, Chinn D S (1988): An Improved Model of the Earth's Gravitational Field: GEM-T1. NASA Technical Memorandum 4019.

SEMINARS

THE INTEGRATED APPROACH TO SATELLITE GEODESY

Barbara Betti, Fernando Sansò
Istituto di Topografia, Fotogrammetria e Geofisica
Politecnico di Milano, P.za Leonardo da Vinci 32, I-20133 Milano

1. INTRODUCTION

The concept of Integrated Geodesy has been introduced in 1975 by J. Eeg and T. Krarup, and since that time it was recognized by several geodesists as one possible approach capable of giving a unified theoretical basis to Geodesy. One of the hot and controversal questions which were opened from the very beginning was the so-called "norm choice problem" or the fact that, in the authors' words, "...the norm problem has to be solved by a combination of statistical and mathematical and physical methods".

Some authors have strongly rejected any possible intervention of statistical concepts in the method, some others have underlined apparent paradoxes arising in the application of statistical methods (cfr. Tscherning, 1977): the formal problems have been solved in 1985 by F. Sansò (1986), where also the weak character of the stochastic interpretation has been illustrated.In any way concepts borrowed from statistics, like degree variances, powers, correlation lengths, etc., are widely used in geodesy and they are as well utilized in satellite geodesy from its fundation, as one can read in the milestone-book by W.M. Kaula (1966).

It is for this reason that, ever since six years ago the authors begun trying to understand why such a general scheme had not been applied to the reduction of satellite observations: the answer was soon found to be that the "non-locality" of satellite observations rendered the whole theory applicable only at the cost of a very heavy computational work.

It was for that reason that the convenient approach of parametrizing the unknown potential by means of harmonic coefficients in a truncated series has been almost exclusively followed, simply neglecting the influence of the non modelled terms. This approach in particular has got an even stronger push since the time that O. Colombo (1984) showed how to design a mission so that special simmetries could be taken into account and a large number of potential coefficients could be estimated.

Nevertheless the problem of computing a kind of average influence of the neglected (or erroneous) terms was always present in the theory of satellite geodesy for both purposes, to analyze actual residuals of adjusted data and to analyze the possible performance of a satellite mission. Moreover some attempts have been done to propose the idea of integrated geodesy to reduce satellite observations, also stimulated at the origin by nice discussions with the authors (Eissfeller and Hein 1985, Eissfeller 1985).

Meanwhile these problems are becoming more interesting, from a practical point of view too, since some projects have been proposed to fly very low satellites bearing drag-free gravity sensors, as it is in the analysis of missions of this kind (SST, Spaceborne gradiometry, etc.) that the neglected gravitational signal can have a larger importance.

This paper is an attempt to contribute to this difficult item, specially from the point of view of finding computable algorithms to calculate the covariances of satellite observations.

We will do that specially with the target in mind of:

- achieving the unification of the theory of satellite geodesy, under the main concept of integrated geodesy;
- establishing a suitable tool to analyze the propagation of gravity modelling errors to satellite orbits as well as to satellite observations;
- possibly improving the estimates of the parameters in the adjustment procedure and specially the estimate of their covariance matrix.

2. THE TYPICAL FORM OF SATELLITE OBSERVATION EQUATIONS

The first step in the integrated approach to Geodesy is to model properly the observation equations in their non linear form, taking into account the dependence from finite dimensional unknowns, i.e. the parameter vector p (e.g. point coordinates) and from the gravity field potential $u(\underline{x})$

$$Q_o = F[p,u] + \nu \quad ; \qquad (2.1)$$

ν is the observational noise which has to be modelled stochastically (typically ν is taken to be a white noise[1].

---

1 Obviously in the general form (2.1) ν is defined as "all what is not included in the model F[p,u] ". In this respect attention should be paid to many possible errors which can create very significant correlations in satellite observations (cfr. Betti et al., 1987).

Equation (2.1) has then to be linearized with respect to the "variation" of the unknowns of which we assume to know approximate values $\tilde{p},\tilde{u}$ (normal potential[2])

$$p = \tilde{p} + \delta p$$

$$u = \tilde{u} + \delta u$$

$$\delta Q_o = \tilde{F}_p\ \delta p + \langle \tilde{F}_u, \delta u \rangle + \nu \quad , \tag{2.2}$$

the scalar product being taken in a suitable space of harmonic functions [cfr. Sansò (1986) for a more precise meaning of (2.2)].

Satellite observations obviously fit such a general model, but in a specific form due to the fact that in part the dependence from the parameters and from the potential goes through a dynamical equation describing the time evolution of the state vector of the satellite; this is considered here as a point mass and therefore described by its position and velocity vector $\underline{x}(t)$, $\underline{x}(\dot{t})$ in a suitable reference system; that we shall assume to be inertial. Whence a "reasonable" model for satellite observations is

$$Q_o = F[p,\underline{x}(t),\dot{\underline{x}}(t);u] + \nu \tag{2.3}$$

$$\begin{aligned} \ddot{\underline{x}}(t) &= \nabla u[\underline{x}(t)] + \Pi[\underline{x}(t),\dot{\underline{x}}(t),q] \\ \underline{x}(o) &= \underline{x}_o \quad , \quad \dot{\underline{x}}(o) = \dot{\underline{x}}_o \end{aligned} \tag{2.4}$$

where p (parameter vector) can include

p -> • station coordinates in a terrestrial reference system

• several parameters to transform the terrestrial into an inertial reference system (e.g. precession, mutation, polar motion, LOD etc.)

• transmission coefficients (e.g. a parametric model for the refraction index) when the observation couples a "station" and the satellite by an e.m. wave

• various instrumental parameters (e.g. clocks' constants, time delay in electronic circuits, temperatures etc.).

Π, the model of perturbations of both non-gravitational and gravitational types, clearly depends on the dynamic variables as well as on other parameters collected in the vector q, and includes

---

2 More generally $\tilde{u}$ has to be considered as a model or reference potential which might be much closer to the actual potential than the usual normal potential.

II -> • athmospheric drag
- pressure of solar radiation
- pressure of earth's albedo
- e.m. forces
- relativistic corrections
- gravitational perturbations of external bodies (e.g. luni-solar perturbations)
- all kinds of tidal effects (athmospheric, oceanic, solid earth etc.)
- etc

Remark 2.1. When modelling a typical "geometrical" observation, like range or range rate from a ground station or from a high satellite, the operator F is finite dimensional and the dependence on the potential u, comes purely from the dynamical equation (2.4). When on the contrary the observation is done on board the satellite and includes sensors of the gravity field (e.g. gradiometer) we have a double dependence of Q_0 on u; one direct and one indirect by means of the dynamics.

Example 2.1. For instance the (non-linear) observation equations, in schematic form, for a range observation, for a complete gradiometric observation (i.e. observation of the whole tensor of second derivatives V_0, in an instrumental frame), for a radar altimetric observation are respectively

$$L_0 = |\underline{x}(t) - R\,\underline{x}_G| + \nu \tag{2.5}$$

$$V_0 = S\,U[\underline{x}(t)]\,S^+ + \nu \tag{2.6}$$

$$\begin{cases} H_0 = |\underline{x}(t) - R\,\underline{x}_G| + \nu \\ \dfrac{\underline{x}(t)-R\underline{x}_G}{|\underline{x}(t)-R\underline{x}_G|} = -\dfrac{R\nabla u(\underline{x}_G)}{|\nabla u(\underline{x}_G)|} \quad (3) \\ u(\underline{x}_G) = u_0 \quad , \end{cases} \tag{2.7}$$

where

3 Usually the direction of propagation is taken as the normal to the ellipsoid, rather than the normal to the sea surface, considered here as the geoid.

L_o = range observation
$\underline{x}_G$ = station position vector in the terrestrial system
R = rotation from the terrestrial to the inertial system
V_o = second derivatives observed in the instrumental system
U = second derivatives of the potential in a known reference system
S = rotation between the two previous systems
H_o = altimetric observation
$\underline{x}_G$ = footprint of the radar wave at the sea surface considered as the "geoid".

Obviously these equations are written for the pure sake of making an example and many important factors are neglected.

Remark 2.2. A typical difficulty in treating the dynamical equation (2.4), is in that the gravity potential u is more easily represented as a function of coordinates in an earth-fixed system, $u=u(\underline{x}_T)$: whence, to be more precise, if we define R(t) as the rotation matrix transforming a position vector from the terrestrial to the inertial reference system

$$\underline{x} = R(t)\ \underline{x}_T \quad , \tag{2.8}$$

in (2.4) we should more properly describe the gravitational term $\underline{g}= \nabla u(\underline{x})$ as

$$\underline{g}(t) = \underline{\nabla}_x\ u[R^+(t)\ \underline{x}]\Big|_{\underline{x}=\underline{x}(t)} \quad . \tag{2.9}$$

Remark 2.3. Here we have choosen for the sake of simplicity to express the dynamics in the form of Newton's equation; obviously any other description using properly six elements to describe the satellite state, is equivalent on condition that the dynamical equation is transformed accordingly (cfr. Kaula, 1966).

Now the linearization of equation (2.3) is the straightforward part of the problem: the general rule is formally

$$\delta Q_o = F_p\ \delta p + F_x\ \delta\underline{x}(t) + F_x\ \delta\underline{x}(t) + \langle F_u,\delta u\rangle + \nu \quad [4] \tag{2.10}$$

---

4 Particular attention must be paid when computing the variation of u, considering that when u refers to the orbit point $\underline{x}(t)$ we have more precisely
$\delta\{u[\underline{x}(t)]\} = \delta u[\tilde{\underline{x}}(t)] + \nabla\tilde{u}[\tilde{\underline{x}}(t)]\cdot\delta\underline{x}(t)$.

and we can illustrate this by applying it to the equations in Example 2.1.

Example 2.2. We linearize equations (2.5), (2.6), (2.7).

Linearized range equation

Setting:
$$\underline{x}(t) = \tilde{\underline{x}}(t) + \delta\underline{x}(t)$$
$$R = \tilde{R}(I+\delta R) \qquad (\delta R^+ = -\delta R)$$
$$\underline{x}_G = \tilde{\underline{x}}_G + \delta\underline{x}_G \quad ,$$

we find simply

$$\delta L_o = \frac{\tilde{\underline{x}}(t) - \tilde{R}\tilde{\underline{x}}_G}{|\tilde{\underline{x}}(t) - \tilde{R}\tilde{\underline{x}}_G|} \cdot \{\delta\underline{x}(t) - \tilde{R}\delta R\tilde{\underline{x}}_G - \tilde{R}\delta\underline{x}_G\} + \nu \quad . \tag{2.11}$$

The parameters δR, $\delta\underline{x}_G$ go into the parameter correction vector δp: the reference motion $\tilde{\underline{x}}(t)$ is in principle any approximate motion but it is usually computed from (neglecting smaller perturbations)

$$\begin{cases} \ddot{\tilde{\underline{x}}}(t) = \nabla\tilde{u}[\tilde{\underline{x}}(t)] \\ \tilde{\underline{x}}(o) = \tilde{\underline{x}}_o \quad , \quad \dot{\tilde{\underline{x}}}(o) = \dot{\tilde{\underline{x}}}_o \quad . \end{cases} \tag{2.12}$$

Linearized gradio observations

Setting:
$$S = \tilde{S}(I+\delta S) \qquad (\delta S^+ = -\delta S)$$
$$\tilde{U} = \tilde{U}[\tilde{\underline{x}}(t)]$$

and taking into account the note to formula (2.10), we get

$$\delta V_o = \tilde{S}\{\delta S\tilde{U} - \tilde{U}\delta S\}\,\tilde{S}^+ + \tilde{S}\delta U[\tilde{\underline{x}}(t)]\,\tilde{S}^+ + \tilde{S}\{\frac{\delta\tilde{U}}{\delta x}\,\delta\underline{x}(t)\}\,\tilde{S}^+ + \nu \quad . \tag{2.13}$$

Linearized altimetry observations

Linearizing the first of (2.7) we get

$$\begin{cases} \delta H_o = \tilde{\underline{e}}\cdot\{\delta\underline{x}(t) - \tilde{R}\delta R\,\tilde{\underline{x}}_G - \tilde{R}\,\delta\underline{x}_G\} + \nu \\ \tilde{\underline{e}} = \dfrac{\tilde{\underline{x}}(t) - \tilde{R}\tilde{\underline{x}}_G}{|\tilde{\underline{x}}(t) - \tilde{R}\tilde{\underline{x}}_G|} \; : \end{cases} \tag{2.14}$$

to make it simple, usually $\underline{\tilde{e}}$ (which by the way multiplies small quantities) is approximated with the ellipsoidal normal $\tilde{R}\underline{\nu}$ rotated into the inertial system and, furtheron, the variation

$$\underline{\tilde{e}} \cdot \tilde{R}\, \delta\underline{x}_G = \underline{\nu} \cdot \delta\underline{x}_G \quad ,$$

is expressed by the Brun's relation, which derives from the linearization of the third of (2.7)

$$\underline{\nu} \cdot \delta\underline{x}_G = \frac{\delta u(\underline{\tilde{x}}_G)}{\gamma(\underline{\tilde{x}}_G)} \qquad (\gamma(\underline{x}) = \nabla\tilde{u}(\underline{x}) = \text{normal gravity}) \quad .$$

Going back to (2.14) we eventually find

$$\begin{cases} \delta H_o = (\tilde{R}\underline{\nu}) \cdot \{\delta\underline{x}(t) - \tilde{R}\delta R\underline{\tilde{x}}_G\} - \dfrac{\delta u(\underline{\tilde{x}}_G)}{\gamma(\underline{\tilde{x}}_G)} + \nu \\ (\tilde{R}\underline{\tilde{x}}_G = \underline{\tilde{x}}(t) - H_o\tilde{R}\underline{\nu}) \quad . \end{cases} \tag{2.15}$$

The next point is the linearization of the dynamical equation (2.4).
To achieve this we first define the reference motion by (2.12) and we use the identity

$$\underline{\nabla}_{\underline{x}} u[R^+\underline{x}] = R(\nabla u)(\underline{x}_T) \qquad (\underline{x} = R\underline{x}_T) \quad : \tag{2.16}$$

accordingly, the gravity term is linearized as

$$\underline{\nabla}_{\underline{x}} u[R^+\underline{x}] = \underline{\nabla}_{\underline{x}}\, \tilde{u}[\tilde{R}^+\underline{\tilde{x}}] + \tilde{R}\, \delta R\, \nabla\tilde{u}[\tilde{x}_T] + \tilde{R}(\nabla\nabla\tilde{u})\, \delta\underline{x}_T + \tilde{R}(\nabla\delta u)[\underline{\tilde{x}}_T] =$$

$$= \underline{\tilde{\gamma}}(t) + \tilde{R}\delta R\tilde{R}^+\underline{\tilde{\gamma}}(t) + \tilde{V}\delta\underline{x}(t) + \underline{\nabla}_{\underline{x}}\delta u[\tilde{R}^+\underline{\tilde{x}}]$$

$$(\delta\underline{x}_T = \tilde{R}^+\delta\underline{x} \quad , \quad R = \tilde{R}(I+\delta R) \;) \quad ,$$

where $\underline{\tilde{\gamma}}(t)$ represents the normal gravity, $\tilde{V}=\tilde{R}\tilde{U}\tilde{R}^+$ the matrix of second derivatives of the normal potential, computed along the reference motion and expressed in the inertial reference system.
Going back to (2.4) and setting $\underline{\xi}(t) = \delta\underline{x}(t)$ for the sake of brevity, we find the (variational) linearized equation

$$\begin{cases} \ddot{\underline{\xi}} = \tilde{R}\delta R\,\tilde{R}^{+}\underline{\gamma}(t) + \tilde{V}\,\underline{\xi}(t) + \nabla\delta u[\tilde{\underline{x}}_T] + \Pi_{\underline{x}}\underline{\xi} + \Pi_{\dot{\underline{x}}}\dot{\underline{\xi}} + \Pi_q\delta q \\ \xi(o) = \delta\underline{x}_o = \xi_o \quad , \quad \dot{\xi}(o) = \delta\dot{\underline{x}}_o = \xi_o \quad . \end{cases} \tag{2.17}$$

Obviously $\delta u(\underline{x})$ can be possibly splitted into two parts, one of which is a finite series described in terms of harmonic coefficients, considered as unknown parameters,and another part which represents a residual signal

$$\delta u(\underline{x}) = \Sigma_{l,m}\,\delta u_{lm}\,Y_{lm}(\phi,\lambda) + \delta u_r(\underline{x}) \quad :$$

what we are really interested in, is the way in which δu_r influences the determination of the unknowns and ultimately the observable quantities. The idea is that δu_r can be treated as a "signal" in the scheme of the integrated geodesy and in order to take it into account we must find its covariance function and its propagation to $\underline{\xi},\dot{\underline{\xi}}$ and to the observables.

Since we are mainly interested in the direct gravitational effects, we shall simplify the dynamical model (2.17) to the point of writing

$$\begin{cases} \ddot{\underline{\xi}} = V\underline{\xi} + \nabla\delta u[\underline{x}_T] \\ \underline{\xi}_{(o)} = \underline{\xi}_o \quad , \quad \dot{\underline{\xi}}(o) = \dot{\underline{\xi}}_o \quad : \end{cases} \tag{2.18}$$

this is legitimate also taking into account that (2.17) is a linear problem so that the effects of the other unknowns can be separated in $\underline{\xi}$, leaving aside the treatment of the pure "signal" part which is in fact described by (2.18) (note that we have written back δu instead of δu_r).

3. THE "SPHERICAL FIELD-CIRCULAR MOTION" APPROXIMATION

In this paragraph we aim to make a further simplification of equation (2.18), which leads to a new equation explicitely solvable in various analytical forms.

This step consists in simplifying the coefficients of equation (2.18) referring them to a spherical reference potential $\tilde{u}_o$ and to a pure circular reference motion $\tilde{\underline{x}}_o$.

As it was pointed out, e.g. by R. Rummel (1987), this is strictly equivalent to the approximation procedure used in classical physical geodesy, where the (already linearized) fundamental boundary value problem is reduced first by taking the spherical potential as normal potential and then the boundary itself as a sphere, thus arriving at the analytically handable Stokes problem.

The point here is to establish to what extent these approximations are valid, possibly in computing the solution $\underline{\xi}(t)$, and particularly in computing its covariance function.

Then we start thinking that with the most precise available information on the gravity field (reference potential $\tilde{u}$), the reference motion $\underline{\tilde{x}}(t)$ has been computed, by by solving the problem

$$\begin{cases} \ddot{\underline{\tilde{x}}} = \nabla\tilde{u}[\tilde{x}(t)] \\ \underline{\tilde{x}}(o) = \underline{\tilde{x}}_o \quad , \quad \dot{\underline{\tilde{x}}}(o) = \dot{\underline{\tilde{x}}}_o \quad . \end{cases} \tag{3.1}$$

Furthermore, we observe that for short arcs (by that we mean arcs which are frac-fractions of a revolution), also taking into account that geodetic satellites have usually a very small eccentricity, we can approximate $\underline{\tilde{x}}(t)$ with a uniform circular motion $\underline{x}_c(t)$, i.e. the solution to the problem

$$\ddot{\underline{x}}_c = \nabla u_s[\underline{x}_c] = - \mu \frac{\underline{x}_c}{|\underline{x}_c|^3} \tag{3.2}$$

$$\left(u_s(\underline{x}) = \frac{\mu}{|\underline{x}|} \quad \text{purely spherical potential} \right) \quad ,$$

with suitable initial conditions (i.e. compatible with Kepler's third law).
The difference vector $\delta\underline{\tilde{x}} = \underline{\tilde{x}}(t) - \underline{\tilde{x}}_c(t)$ will accordingly satisfy, in linear approximation, an equation of the type (2.18), i.e.

$$\begin{cases} \delta\ddot{\underline{\tilde{x}}} = V_s[\underline{x}_c]\ \delta\underline{\tilde{x}} + \nabla\delta\tilde{u}[\underline{x}_c] \\ \delta\underline{\tilde{x}}(o) = \delta\underline{\tilde{x}}_o \quad , \quad \delta\dot{\underline{\tilde{x}}}(o) = \delta\dot{\underline{\tilde{x}}}_o \end{cases} \tag{3.3}$$

with

$$\delta\tilde{u}(\underline{x}) = \tilde{u}(\underline{x}) - u_s(\underline{x}) \quad : \tag{3.4}$$

no doubt the largest term in (3.4) is the J_2 term which can be seen (as we will show in the next paragraph) to produce as large corrections $\delta\underline{x}$ as some kilometers, even for short arcs.
In this respect the approximations in (3.3) are too rough to allow for an accurate computation of $\delta\underline{x}$ and we will use this equation only to perceive the order of magnitude of this vector.

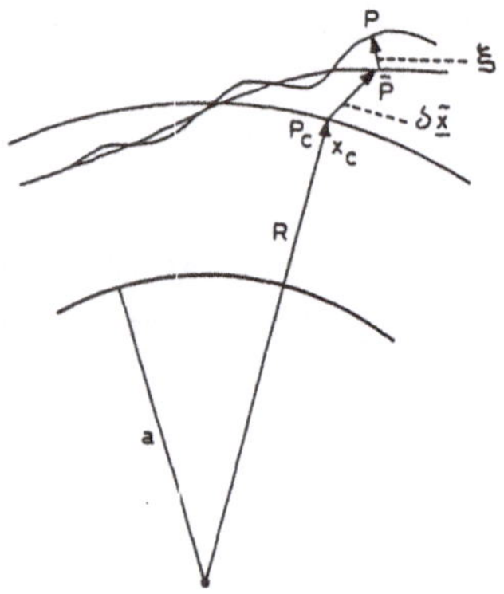

a = earth's radius

P = true position of the satellite

$\tilde{P}$ = moving point on the reference motion

P_c = moving point on the circular reference motion

$R = |\underline{x}_c|$

FIG. 3.1

We now go back to equation (2.18) which we rewrite as

$$\begin{cases} \ddot{\underline{\xi}} = \tilde{V}[\tilde{\underline{x}}]\underline{\xi} + \nabla\delta u[\tilde{\underline{x}}] \\ \underline{\xi}(0) = \underline{\xi}_0 \quad , \quad \dot{\underline{\xi}}(0) = \dot{\underline{\xi}}_0 \quad : \end{cases} \tag{3.5}$$

note that the subscript T has desappeared from the argument of $\nabla\delta u$ since it is clear without ambiguity that $\nabla\delta u$ has to be evaluated at the points of reference motion, whatever is the system in which we perform the computation. Without approximation we can write (3.5) in a form suitable to a perturbative solution

$$\begin{cases} \ddot{\underline{\xi}} = V_s[\underline{x}_c]\underline{\xi} + \nabla\delta u[\underline{x}_c] + \{\tilde{V}[\tilde{\underline{x}}] - V_s[\underline{x}_c]\}\underline{\xi} + \{\nabla\delta u[\tilde{\underline{x}}] - \nabla\delta u[\underline{x}_c]\} \\ \underline{\xi}(0) = \underline{\xi}_0 \quad , \quad \dot{\underline{\xi}}(0) = \dot{\underline{\xi}}_0 \quad . \end{cases} \tag{3.6}$$

One can think, first to solve the basic equation

$$\begin{cases} \ddot{\underline{\xi}} = V_s[\underline{x}_c]\underline{\xi} + \nabla\delta u[\underline{x}_c] \\ \underline{\xi}(0) = \underline{\xi}_0 \quad , \quad \dot{\underline{\xi}}(0) = \dot{\underline{\xi}}_0 \end{cases} \tag{3.7}$$

and then to use this solution in the neglected terms in (3.6) to compute a new solution and so on. This is perfectly equivalent to transforming (3.6) into an integral equation, by using the general solution of (3.7), and then solving it iteratively.

To study the convergence of this approach and also to realize what are the orders of magnitude of the corrections we must consider that

$$\begin{cases} \tilde{V}[\tilde{\underline{x}}] - V_s[\underline{x}_c] \cong \dfrac{\partial V_s[\underline{x}_c]}{\partial \underline{x}} \delta\tilde{\underline{x}} + \delta\tilde{V}[\underline{x}_c] \\ \\ \nabla\delta u[\tilde{\underline{x}}] - \nabla\delta u[\underline{x}_c] \cong \{\nabla\nabla\delta u[\underline{x}_c]\}\ \delta\tilde{\underline{x}} \quad : \end{cases} \tag{3.8}$$

the first is the correction of the matrix of coefficients and controls the convergence since it multiplies the vector $\underline{\xi}$, while the second is a fixed term and can be considered as a real error (or correction) in the known term of (3.6).

Remark 3.1. We are here in a situation very similar to what we experience in least squares theory where we learn that the design matrix can be much more roughly approximated than the known terms: we can expect the same to hold in this case.

Now if we analyze the first equation (3.8), we see that we can perform some elementary computations of the orders of magnitude for instance for the radial components. So we first notice that in $\delta\tilde{V}$ the biggest term is the one coming from the J_2 coefficient, so that

$$(\delta\tilde{V})_{radial} \sim \frac{\partial^2}{\partial r^2}\{J_2\ \mu\ \frac{a^2}{r^3} P_2(\cos\theta)\}\Big|_{r=R} \sim 12\ J_2\ \mu\ \frac{a^2}{R^5} P_2(\cos\theta) \quad , \tag{3.9}$$

while for the spherical potential one has

$$u_s = \mu/r \quad ; \quad |U_s| = |-\frac{\mu}{r^3}[I - 3P_r]| = |V_s|^{(5)} \quad : \tag{3.10}$$

in terms of orders of magnitude then

$$(\delta V)_{radial} \sim 6\ J_2(\tfrac{a}{R})^2\ |V_s| \quad .$$

Analogously

$$(\frac{\partial V_s}{\partial \underline{x}})_{radial}\ (\delta\underline{x})_{radial} \sim 3\ \frac{\delta\tilde{R}}{R}\ |V_s| \tag{3.11}$$

---

5 Pr is the projection in the radial direction, so as operation it is independent on the reference system: in components one has $[Pr]_{i,k} = x_i x_k/r^2$, and obviously the components x_i will change from one to another reference system, making the difference between U_s and V_s.

Since $J_2 \sim 10^{-3}$ and $\delta\tilde{R}/R \sim 10^{-3}$ at most (at least for short arcs as shown in the next paragraph), we see (for the radial components) that in the end

$$|\tilde{V}[\underline{\tilde{x}}] - V_s[\underline{x}_c]| = O(10^{-2}|V_s|) \tag{3.12}$$

so that changing $\tilde{V}[\underline{\tilde{x}}]$ into $V_s[\underline{x}_c]$ in the dynamical equations amounts to neglecting 1% of the correction vector $\underline{\xi}$.

Remark 3.2. This rough computation seems to suggest that an iterative solution of (3.6) could be achieved with an improvement of about two orders of magnitude at each iteration. Whence, if for instance we are sure from the beginning that $|\underline{\xi}|$ is only a few meters, even the simple first solution could be accepted as good.

As for the second factor, described by the second of (3.8), things are different. For instance considering only the effect of radial corrections one perceives that the error responds differently to the various frequencies: so assuming δu to be a harmonic of degree 1, one can write

$$\{\nabla\delta u[\underline{x}] - \nabla\delta u[\underline{x}_c]\}_{radial} \sim (\frac{\partial}{\partial r} \nabla\delta u)_{r=R} \cdot \delta\tilde{R} \sim (1+2)|\nabla\delta u| \frac{\delta\tilde{R}}{R} \quad . \tag{3.13}$$

For instance for degrees as high as $1=10^2$ one can arrive at an error of 10% in the corresponding known term: this should in principle not be neglected at least in the analytical solutions. However one should consider that already at the height corresponding to a/R=0.97 the effect of the degrees over 100 are significantly damped due to the exponential factor $(a/R)^{1+2}$ $(0.97^{100} = 0.05)$ so that their dynamic effect is in any way very small. Furtheron, since our main purpose is the computation of the covariance propagation through the analytical solution of (3.6), we could even accept a relative error of 10% since we never need such an accurate knowledge of the covariance function itself.
Finally one should consider the spectral character of relation (3.13): since the operator relating $\underline{\xi}$ to $\nabla\delta u$ is a smoother (we integrate twice along the orbit, so to say), we expect the curve of relative errors as function of 1 to be twisted from an increasing straight line as in (3.13) to a decreasing line, like a hyperbola.
As for the other two non-radial components of $\nabla\delta u$, the following remark holds.

Remark 3.3. As we know from its very definition the covariance function of the "anomalous" potential δu has the general rotationally invariant form

$$C(P,Q) = (\frac{\mu}{a})^2 \Sigma_1 \sigma_1^2 (21+1) (\frac{a^2}{r_P r_Q})^{1+1} P_1(\cos\psi_{PQ}) \quad , \tag{3.14}$$

where σ_l^2 are the degree variances of the potential defined as

$$\sigma_l^2 = \frac{1}{2l+1} \sum_{-l}^{l} {}_m \, |\delta u_{lm}|^2 \tag{3.15}$$

and ψ_{PQ} represents the angle at the origin between the two rays through P and Q. As we can see any similar radial error $\delta r_P = \delta r_Q$ will produce a cumulated effect on C(P,Q), while any "horizontal" shift of the two points, $\delta \underline{x}_P = \delta \underline{x}_Q$, will leave ψ invariant and therefore will not affect the covariance function.

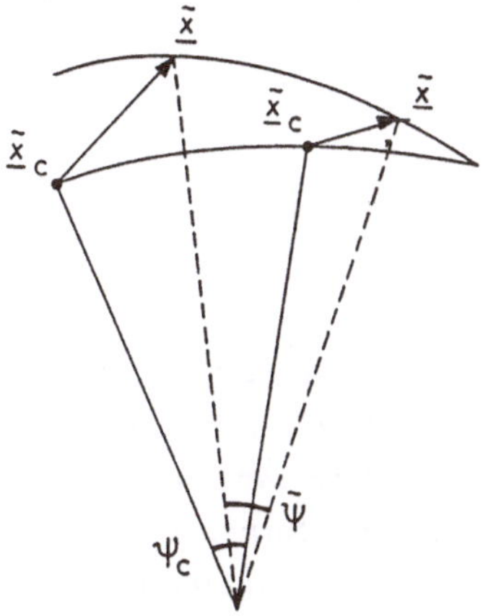

FIG. 3.2

Based on this remark and observing that the correction $\delta\tilde{\underline{x}} = \tilde{\underline{x}}-\underline{x}_c$ is very smooth (cfr. fig. 3.2), we may expect that $\delta\psi = \tilde{\psi} - \tilde{\psi}_c$ be small and correspondingly the horizontal correction of the covariance function be small too. Whence we have that the radial component of the error $\nabla\delta u[\tilde{\underline{x}}]-\nabla\delta u[\underline{x}_c]$ should be the more significant for covariance computations and for it the discussion already done is valid.

We may conclude this discussion by saying that, at least for the purpose of computing a covariance propagation, the very simple dynamic model (3.6), with spherical reference field and circular reference motion, could be applied. We believe that the covariance functions resulting from this model are meaningful, at least for short arcs, and that a further improvement could be obtained first of all by correcting $\underline{x}_c$, in the argument of $\nabla\delta u$, at least for the linear effects of J_2 in $\delta\underline{x}$, specially for the radial component.

We are thus left with the problem of solving (3.7): this solution is perfectly known and we will recall it in the next paragraph. As a byproduct we will also get the solution of (3.3) and will show that the orders of magnitude of $\delta\underline{x}$ used in this paragraph are correct.

Closing this paragraph we shall not fail to make once more an important remark on the nature of the equation (3.7).

Remark 3.4. It is important to stress that by solving (3.7) we do not try to compute the correction between the actual motion $\underline{x}(t)$ and circular reference motion $\underline{x}_c(t)$: in fact this is

$$\underline{x}(t) - \underline{x}_c(t) = \delta\tilde{\underline{x}}(t) + \underline{\xi}(t) \tag{3.16}$$

and it is a very large vector due to the presence of the effects of J_2 in $\delta\tilde{\underline{x}}(t)$. The vector $\underline{\xi}(t)$ is in fact only the correction to the already good reference motion $\tilde{\underline{x}}(t)$, so we expect it to be not larger than a few tens of meters. It is only in the variational equation (3.6) that we apply some simplifications, like taking the reference field as spherical and the reference motion as circular. This is prefectly analogous to what we do in the boundary value problem theory when, after computing the gravity anomaly with respect to the normal (ellipsoidal) potential, we modify the geometry of the boundary from the ellipsoid to the sphere and the coefficients in the boundary operator, computing them with a simple spherical potential.

4. THE SOLUTION OF HILL'S EQUATION IN THE CIRCULAR MOTION APPROXIMATION

What we want to do in this paragraph is to write the equation (3.7) in terms of components: this will provide us with the so called Hill's equations, but with the considerable simplification of being referred to a uniform circular motion.
Subsequently we shall solve these equations in various forms. In this respect we anticipate that being mainly interested to the functional relation between $\underline{\xi}(t)$,$\underline{\xi}(t)$ and the known term

$$\underline{f} = \nabla\delta u[\underline{x}_c] \quad , \tag{4.1}$$

we are free to choose whatever initial conditions we prefer, since these, as we know, determine only the specific integral of the associated homogeneous equation which by definition does not depend on $\underline{f}$. Whence we can reduce to the equation

$$\underline{\xi} = V_o\underline{\xi} + \underline{f} \tag{4.2}$$

$$\underline{\xi}(0) = 0 \quad , \quad \dot{\underline{\xi}}(0) = 0 \quad .$$

To resolve (4.2) into components we introduce first a mobile (Cartan) frame $\underline{E}(t)$ with the origin at the point P_c that represents the satellite at time t along the circular reference motion, $\underline{x}_c(t)$, and three axis with unit vectors

$$\underline{E}(t) = \begin{vmatrix} \underline{e}_o(t) \\ \underline{e}_a(t) \\ \underline{e}_r(t) \end{vmatrix} \quad \begin{matrix} \text{out-of-plane: orthogonal to the plane of the orbit} \\ \text{along track : tangent to the circular orbit} \\ \text{radial} \end{matrix} \tag{4.3}$$

We shall then resolve all the vectors in the frame $\underline{E}(t)$, i.e.

$$\begin{cases} \underline{\xi}(t) = \xi_o\, \underline{e}_o + \xi_a\, \underline{e}_a + \xi_r\, \underline{e}_r \\ \underline{f}(t) = f_o\, \underline{e}_o + f_a\, \underline{e}_a + f_r\, \underline{e}_r \quad . \end{cases} \tag{4.4}$$

Furthermore, since

$$\begin{cases} u_s = v_s = \mu/r \;^{(6)} \\ V_s = -\,\mu/r^3\,[I-3P_r] \qquad (P_r = \text{radial projection}) \quad , \end{cases} \tag{4.5}$$

we have also

$$V_s[\underline{x}_c]\underline{\xi} = -\,\mu/R^3\,\{\xi_o\,\underline{e}_o + \xi_a\,\underline{e}_a - 2\,\xi_r\,\underline{e}_r\} \quad . \tag{4.6}$$

Finally, taking into account that $\underline{x}_c(t)$ is a uniform circular motion with period, say, T and with mean motion $n=2\pi/T$, we easily see that the following differentiation table holds

$$\frac{d}{dt}\begin{vmatrix} \underline{e}_o \\ \underline{e}_a \\ \underline{e}_r \end{vmatrix} = \begin{vmatrix} 0 \\ -\underline{e}_r \\ \underline{e}_a \end{vmatrix} \cdot n \quad . \tag{4.7}$$

To illustrate this and to introduce the variables of the problem, we can inspect fig. 4.1.

The origin of the observation time, t=0, is chosen so that the corresponding point $\underline{x}_c(o)$ is in the middle of the observed arc: the relations between the various cynematic quantities are

$$\begin{array}{ll} F = F_o + \dot{F}t & \dot{F} = n = 2\pi/T \\ G = G_o + \dot{G}t & \dot{G} = \text{earth's spin} \\ L = \Omega - G & \\ L = L_o + \dot{L}t & L_o = \Omega - G_o \\ & \dot{L} = -\dot{G} \\ n^2R^3 = \mu & \text{(3rd Kepler's law)} \quad . \end{array} \tag{4.8}$$

---

6 u_s in the earth fixed system, but it coincides with the field v_s in the inertial system because it depends only on r.

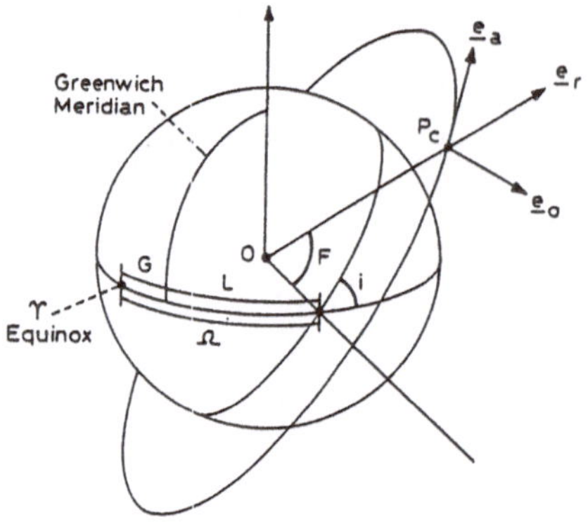

Oγ : fixed direction in inertial frame

G : Greenwich siderial time

L : terrestrial longitude of the node

Ω : celestial longitude of the node (constant of motion)

i : inclination (constant of motion)

F : anomaly along the orbit

FIG. 4.1

Now it is an elementary exercise to show that the dynamic system (4.2) can be written in terms of components as

$$\begin{cases} \ddot{\xi}_o + n^2\xi_o = f_o \\ \ddot{\xi}_a + 2n\,\dot{\xi}_r = f_a \\ \ddot{\xi}_r - 2n\,\dot{\xi}_a - 3n^2\xi_r = f_r \end{cases} \quad ; \tag{4.9}$$

this should be complemented for instance with zero initial conditions: this choice however is not strictly compulsory since what we really need is any particular solution of (4.9) for arbitrary force functions f_o, f_a, f_r.

Remark 4.1. From the dynamical point of view we see that the out-of-plane component is decoupled from others; this will however change in computing the covariance and cross-covariance functions since $f_o = \underline{e}_o \nabla \delta u$ will not be uncorrelated to the other components.

Remark 4.2. The second and third equation (4.9) can be manipulated, after one integration, to give

$$\begin{cases} \dot{\xi}_a + 2n\xi_r = \int_0^t f_a(\tau)\, d\tau \\ \ddot{\xi}_r + n^2\xi_r = f_r + 2n \int_0^t f_a(\tau)\, d\tau \end{cases} \quad . \tag{4.10}$$

A first form of the solution of (4.9) can be written by solving two simple harmonic oscillators and then integrating once more the first of (4.10). We get then

$$\xi_o = \frac{1}{n}\int_o^t \sin n(t-\tau)\; f_o(\tau)\; d\tau \tag{4.11}$$

$$\xi_a = -\frac{2}{n}\int_o^t [1-\cos n(t-\tau)]\; f_r(\tau) d\tau + \frac{4}{n}\int_o^t \sin n(t-\tau)\; f_a(\tau) d\tau - 3\int_o^t (t-\tau)\; f_a(\tau)\; d\tau$$

$$\xi_r = \frac{1}{n}\int_o^t \sin n(t-\tau)\; f_r(\tau)\; d\tau + \frac{2}{n}\int_o^t [1-\cos n(t-\tau)]\; f_a(\tau)\; d\tau \quad ;$$

it could be a useful exercise to verify that (4.11) satisfies (4.9) with homogeneous conditions.

As one can see the non-local character of the functional relation between f and ξ is apparent in (4.11) and it is, this, one important point that makes so difficult the application of the integrated approach to satellite geodesy.

Remark 4.3. We can see already from (4.11) that constants or very long wavelength accelerations will produce secular deviations only in ξ_a, ξ_r; in particular we would have for f_a, f_r constant, beyond the oscillating integrals, terms like

$$\begin{cases} \xi_a = -\frac{2}{n} f_r t - \frac{3}{2} f_a t^2 \\ \xi_r = \frac{2}{n} f_a t \quad . \end{cases} \tag{4.12}$$

Remark 4.4. From the solution in the form (4.11) we can infer a simple rule to evaluate the order of magnitude of a perturbation $\underline{f}$, on condition that its wavelength is comparable with the integration length[7]. In this case in fact we can simply say that

$$|\underline{\xi}| = O(|\underline{f}|\; \frac{t}{n}) \quad . \tag{4.13}$$

If we express t in terms of a fase

$$t = \phi \frac{T}{2\pi} = \frac{\phi}{n}$$

7 For much shorter wavelengths functions like $\sin n\tau$, $\cos n\tau$ appear as almost constant and the integrals (4.11) tend to average on t.

and recall that $n^{-2} = R^3/\mu = R/g_s$ (g_s being the modulus of the spherical gravity field felt by the satellite) we derive also

$$|\underline{\xi}| = O(|\underline{f}|/g_s \cdot \phi R) \quad . \tag{4.14}$$

If we take $|\underline{f}|/g_s = J_2/(a/R)^2 \sim 10^{-3}$ and $\phi_{max} \sim 1$, we see that the J_2 perturbation can produce as large displacements as several kilometers, as we claimed in §3.

Another way to find a solution of (4.9) is first to resolve the forcing field in terms of Fourier components and then to find the corresponding solution $\underline{\xi}$, frequency by frequency, the general solution being the sum of the single solutions by virtue of the linearity of our system.

For reasons to be found in the Appendix, the forcing field can be expanded in a series of terms of the form

$$\begin{aligned} f_o(t) &= f_{okm}\, e^{j(kF+mL)} \\ f_a(t) &= f_{akm}\, e^{j(kF+mL)} \\ f_r(t) &= f_{rkm}\, e^{j(kF+mL)} \end{aligned} \tag{4.15}$$

$$(F = F_o + nt \quad , \quad L = L_o - \dot{G}t) \quad ;$$

for the sake of simplicity we shall suppose that the ratio between n and G is not a rational number

$$n/G \neq \text{rational} \quad . \tag{4.16}$$

We can treat separately the out-of-plane and the other two components.

a) Out-of-plane component: setting

$$\bar{\xi}_o(t) = \xi_{okm}\, e^{j(kF+mL)}$$

one gets from (4.9)

$$\begin{aligned} \xi_{okm} &= \{n^2-[kn-m\dot{G}]^2\}^{-1} f_{okm} = s_{km}\, f_{okm} \\ m &\neq 0 \quad , \quad k \neq \pm 1 \end{aligned} \tag{4.17}$$

For m=0 and k = ±1 the forcing terms are in resonance with the proper frequency of the oscillator, so an unbounded solution corresponding to increasing oscillations is found, namely

$$\xi_o(t) = \pm f_{0,\ 1,0} \cdot \frac{t}{2jn} e^{\pm jF} \tag{4.18}$$

Remark 4.5. We can observe that due to the condition (4.16) we can never have

$$n^2 - [kn - m\mathring{G}]^2 = 0 \tag{4.19}$$

unless for the above studied case. Obviously the expression (4.19) could be close to zero for suitable values of k,m thus causing an increase in the corresponding term (4.17) giving rise to a quasi resonance.

Remark 4.6. As it has been pointed out at the beginning of the paragraph the "particular integral" we are looking for can satisfy an arbitrary initial condition: this however is true when the arbitrary condition is "fixed" in the sense that it doesn't depend of the force coefficients as otherwise we introduce also in the integral of the homogeneous equation a covariance propagation.

Choosing, as we did in the integral representation (4.11), the conditions $\xi_o=0$, $\dot{\xi}_o=0$, we find the correct particular integrals corresponding respectively to (4.17), (4.18) by adding a suitable integral of the homogeneous equation, i.e.

$$\begin{cases} \xi_o(t) = \xi_{okm} \{e^{j(kF+mL)} - e^{j(kF_o+mL_o)} \cdot \left[\frac{(k+1)n-m\mathring{G}}{2n} e^{jnt} - \frac{(k-1)n-m\mathring{G}}{2n} e^{-jnt}\right]\} \\ \xi_o(t) = \pm f_{oko}\{ \frac{t}{2jn} e^{\pm jF} - \frac{e^{\pm jF_o}}{4n^2} (e^{jnt} - e^{-jnt}) \} \qquad (k = \pm 1) \end{cases}$$

where ξ_{oklm} is given by (4.17).

We summarize the results in a synthetic form for future use, namely

$$\begin{cases} \xi_o(t) = \{e^{j(kF+mL)} - h_{km}\} s_{km} f_{okm} \qquad (k=0.2) \\ \qquad (k,m) \neq (\pm 1,0) \quad , \\ h_{km} = e^{j(kF_o+mL_o)} \left[\frac{(k+1)n-m\mathring{G}}{2n} e^{jnt} - \frac{(k-1)n-m\mathring{G}}{2n} e^{-jnt} \right] \\ s_{km} = [n^2 - \Delta_{k,m}^2]^{-1} \ , \quad \Delta_{k,m} = kn-m\mathring{G} \end{cases} \tag{4.20}$$

$$\begin{cases} \xi_o(t) = \{P_{ko} - h_{ko}\}\, f_{oko} & (k = \pm 1) \\ h_{ko} = \pm \dfrac{e^{\pm jF_o}}{4n^2}\,(e^{jnt} - e^{-jnt}) & \\ p_{ko} = \pm \dfrac{t\, e^{\pm jF}}{2jn} & \end{cases} \tag{4.21}$$

b) <u>Along-track and radial components</u>: setting

$$\begin{vmatrix} \bar{\xi}_a(t) \\ \bar{\xi}_r(t) \end{vmatrix} = \begin{vmatrix} \xi_{akm} \\ \xi_{rkm} \end{vmatrix} e^{j(kF+mL)}$$

one can derive the particular integral

$$\begin{vmatrix} \bar{\xi}_{akm} \\ \bar{\xi}_{rkm} \end{vmatrix} = S_{km} \begin{vmatrix} f_{akm} \\ f_{rkm} \end{vmatrix} = \frac{1}{\Delta^2_{k,m}(\Delta^2_{km} - n^2)} \begin{vmatrix} -\Delta^2_{km} - 3n^2 & -2nj\Delta_{km} \\ 2nj\,\Delta_{km} & -\Delta^2_{km} \end{vmatrix} \begin{vmatrix} f_{akm} \\ f_{rkm} \end{vmatrix} \tag{4.22}$$

$$(\Delta_{km} \neq 0 \quad , \quad \Delta_{km} \neq \pm n)$$

with

$$\Delta_{k,m} = kn - m\dot{G}$$

We have now two classes of irregular solutions: one corresponds to

$$\Delta_{k,m} = \pm n \quad \rightarrow \quad m = 0 \qquad k = \pm 1 \tag{4.23}$$

and again we have the phenomenon of resonance with particular solutions

$$\begin{vmatrix} \bar{\xi}_a(t) \\ \bar{\xi}_r(t) \end{vmatrix}_{\pm 1,0} = P_{\pm 1,0} \begin{vmatrix} f_{a,\pm 1,0} \\ f_{r,\pm 1,0} \end{vmatrix} = \tag{4.24}$$

$$\begin{vmatrix} \mp \dfrac{2jte^{\pm jF}}{n} + \dfrac{e^{\pm jF}}{n^2} - \dfrac{e^{\pm jF_o}}{n^2} \mp \dfrac{3jte^{\pm jF_o}}{n} & \dfrac{te^{\pm jF}}{n} \pm \dfrac{j}{n^2} e^{\pm jF} \mp \dfrac{j}{n^2} e^{\pm jF_o} \\ \pm \dfrac{2j\, e^{\pm jF_o}}{n^2} - \dfrac{te^{\pm jF}}{n} & \pm t \dfrac{e^{\pm jF}}{2jn} \end{vmatrix} \begin{vmatrix} f_{a,\pm 1,0} \\ f_{r,\pm 1,0} \end{vmatrix}$$

as one can verify directly.
The other set of unbounded solutions corresponds to

$$\Delta_{km} = 0 \quad \rightarrow \quad k = 0 \quad , \quad m = 0 \quad : \tag{4.25}$$

but in this case the form of the force function (4.20) is necessarily constant

$$\begin{vmatrix} f_a \\ f_r \end{vmatrix} = \begin{vmatrix} f_{aoo} \\ f_{roo} \end{vmatrix} \quad . \tag{4.26}$$

By going back to the original system we find the particular integral

$$\begin{vmatrix} \bar{\xi}_a \\ \bar{\xi}_r \end{vmatrix} = P_{oo} \begin{vmatrix} f_{aoo} \\ f_{roo} \end{vmatrix} = \begin{vmatrix} -\frac{3}{2}t^2 & -\frac{2}{n}t \\ \frac{2}{n}t & \frac{1}{n^2} \end{vmatrix} \begin{vmatrix} f_{aoo} \\ f_{roo} \end{vmatrix} \tag{4.27}$$

Now the three types of integrals (3.21), (3.24), (3.27) are to be reduced to homogeneous initial conditions by adding, in each case, a suitable integral of the homogeneous system, i.e.

$$A \begin{vmatrix} 2j \\ 1 \end{vmatrix} e^{jnt} + B \begin{vmatrix} -2j \\ 1 \end{vmatrix} e^{-jnt} + C \begin{vmatrix} 1 \\ 0 \end{vmatrix} + D \begin{vmatrix} t \\ -\frac{2}{3n} \end{vmatrix} \quad . \tag{4.28}$$

Results are collected in the next formulas

$$\left\{ \begin{array}{l} \begin{vmatrix} \xi_a(t) \\ \xi_r(t) \end{vmatrix} = \{I\, e^{j(kF+mL)} - H_{km}\} \cdot S_{km} \begin{vmatrix} f_{akm} \\ f_{rkm} \end{vmatrix} \\ \qquad (k,m) \neq (\pm 1,0) \quad , \quad (k,m) \neq (0,0) \\ H_{k,m} = e^{j(kF_o+mL_o)} \begin{bmatrix} +\frac{2\Delta}{n}(e^{jnt}-e^{-jnt})+1-3j\Delta t & -j(3-\frac{\Delta}{n})e^{jnt}+j(3+\frac{\Delta}{n})e^{-jnt}-2j\frac{\Delta}{n}-6nt \\ -j\frac{\Delta}{n}(e^{jnt}+e^{-jnt})+2j\frac{\Delta}{n} & -(\frac{3}{2}-\frac{\Delta}{2n})e^{jnt}-(\frac{3}{2}+\frac{\Delta}{2n})e^{-jnt}+4 \end{bmatrix} \end{array} \right.$$

$$S_{km} = \text{cfr. } (4.22) \qquad (\Delta = \Delta_{km} = kn - m\dot{G}) \tag{4.29}$$

$$\left\{ \begin{array}{l} \begin{vmatrix} \xi_a(t) \\ \xi_r(t) \end{vmatrix} = \{P_{\pm 1,0} - H_{\pm 1,0}\} \begin{vmatrix} f_{a\pm 1,0} \\ f_{r\pm 1,0} \end{vmatrix} \qquad (k,m) = (\pm 1,0) \\ \qquad P_{\pm 1,0} \ \text{cfr. } (4.24) \\ H_{\pm 1,0} = \begin{bmatrix} -\frac{3}{n^2}e^{\pm jnt} + \frac{e^{\mp jnt}}{n^2} + \frac{2}{n^2} & \mp\frac{j}{2n^2}e^{\pm jnt} \mp \frac{j}{2n^2}e^{\mp jnt} \pm \frac{j}{n^2} \\ \pm\frac{3j}{2n^2}e^{\pm jnt} \pm \frac{j}{2n^2}e^{\mp jnt} & -\frac{e^{\pm jnt}}{4n^2} + \frac{e^{\mp jnt}}{4n^2} \end{bmatrix} e^{\pm jF_o} \end{array} \right. \tag{4.30}$$

$$
\left\{
\begin{array}{l}
\begin{vmatrix} \xi_a(t) \\ \xi_r(t) \end{vmatrix} = \{P_{oo} - H_{oo}\} \begin{vmatrix} f_{aoo} \\ f_{roo} \end{vmatrix} \qquad (k,m) = (0,0) \\
P_{oo} \quad \text{cfr. } (4.27) \\
H_{oo} = \begin{bmatrix} (e^{jnt} + e^{-jnt} - 2)2 & j(e^{jnt} - e^{-jnt}) \\ -j(e^{jnt} - e^{-jnt}) & \frac{1}{2}(e^{jnt} + e^{-jnt}) \end{bmatrix} \frac{1}{n^2}
\end{array}
\right. \tag{4.31}
$$

With the (rather ugly) formulas (4.29), (4.30), (4.31) we achieve the goal of representing the same integral (4.11) of the variationals, with the integrations performed in spectral form.

We conclude the paragraph by giving a unified form to (4.20), (4.21), (4.29), (4.30), (4.31) since this will be needed in the next paragraph.

We agree to extend our previous definitions according to the rules:

$$
s_{km} = \left\{ \begin{array}{ll} s_{km} & \text{(cfr. (4.17)) in general} \\ 1 & (k,m) = (\pm 1,0) \end{array} \right. \tag{4.32}
$$

$$
S_{km} = \left\{ \begin{array}{ll} S_{km} & \text{(cfr. (4.22)) in general} \\ I & (k,m) = (\pm 1,0) \quad , \quad (k,m) = (0,0) \end{array} \right.
$$

$$
p_{km} = \left\{ \begin{array}{ll} e^{j(kF+mL)} & \text{in general} \\ p_{ko} & \text{(cfr. (4.21)} \quad (k,m) = (\pm 1,0) \end{array} \right.
$$

$$
h_{km} = \left\{ \begin{array}{ll} h_{km} & \text{(cfr. (4.20)} \quad \text{in general} \\ h_{ko} & \text{(cfr. (4.21))} \quad (k,m) = (\pm 1,0) \end{array} \right.
$$

$$
P_{km} = \left\{ \begin{array}{ll} I\, e^{j(kF+mL)} & \text{in general} \\ P_{ko} & \text{(cfr. (4.24))} \quad (k,m) = (\pm 1,0); \text{ (cfr. (4.27))} \quad (k,m) = (0,0) \end{array} \right.
$$

$$
H_{km} = \left\{ \begin{array}{ll} H_{km} & \text{(cfr. (4.29))} \quad \text{in general} \\ H_{ko} & \text{(cfr. (4.30))} \quad (k,m) = (\pm 1,0); \text{ (cfr. (4.31))} \quad (k,m) = (0,0) \end{array} \right.
$$

integrated with the definitions

$$
e_{km} = p_{km} - h_{km} \quad , \quad E_{km} = P_{km} - H_{km} \quad .^{(8)} \tag{4.33}
$$

---

8 The letters p,h,e stem respectively for particular integral, "homogeneous" integral, evolution: the same holds for the corresponding capital letters.

We observe that e,E depend on time directly or through F,L, whereas s,S are constants.
With these conventions all the integrals of Hill's equations can be written as

$$\underline{\xi}_{km}(t) = \begin{vmatrix} \xi_{okm}(t) \\ \xi_{akm}(t) \\ \xi_{rkm}(t) \end{vmatrix} \quad ; \quad \underline{f}_{km}(t) = \underline{f}_{km}\, e^{j(kF+mL)} \quad ; \quad \underline{f}_{km} = \begin{vmatrix} f_{okm} \\ f_{akm} \\ f_{rkm} \end{vmatrix} \tag{4.34}$$

$$\underline{\xi}_{km}(t) = \begin{bmatrix} e_{km}(t) & 0 & 0 \\ 0 & & \\ 0 & & E_{km}(t) \end{bmatrix} \begin{bmatrix} s_{km} & 0 & 0 \\ 0 & & \\ 0 & & S_{km} \end{bmatrix} f_{km} \quad . \tag{4.35}$$

5. THE COVARIANCE FUNCTION AND THE INTEGRATED SCHEME

Let us go back now to our general scheme of §2: to make it "simple" let us consider the linearized equation (2.10) for purely geometrical observations, i.e. when $F_u = 0$.
If we summarize our analysis of that paragraph by saying that in the evolution of the satellite motion $\underline{\xi}(t)$, $\dot{\underline{\xi}}(t)$ we can distinguish various parts (integrals) that depend respectively on general parameters δq, on the correction of the initial state $\underline{\xi}_o, \dot{\underline{\xi}}_o$, on the part of the anomalous potential parametrized by a finite summation of spherical harmonics and finally on the residual anomalous potential we arrive at a formulation of the type

$$\begin{aligned} \delta Q_o &= F_p \delta p + F_x \{ \frac{\partial \underline{\xi}}{\partial q} \delta q + H \underline{\xi}_o + \Sigma \frac{\partial \underline{\xi}}{\partial u_{lm}} \delta u_{lm} + \underline{\xi}_{res} \} + \\ &+ F_x \{ \frac{\partial \dot{\underline{\xi}}}{\partial q} \delta q + \dot{H} \dot{\underline{\xi}}_o + \Sigma \frac{\partial \dot{\underline{\xi}}}{\partial u_{lm}} \delta u_{lm} + \dot{\underline{\xi}}_{res} \} + \nu \quad , \end{aligned} \tag{5.1}$$

where $H, \dot{H}$ represent the evolution of the general integral of the homogeneous Hill's system.
Essentially, collecting all the parameters in a single vector η, we are back to the customary model of observation equations

$$\delta Q_o = A\eta + s + \nu \tag{5.2}$$

where s, the signal, has components

$$s = F_x\, \underline{\xi}_{res} + F_x\, \dot{\underline{\xi}}_{res} \quad . \tag{5.3}$$

Since the solution of (5.2) is (cfr. Moritz, 1980)

$$\eta = \{A^+[C_{ss} + C_{\nu\nu}]^{-1}A\}^{-1}A^+[C_{ss} + C_{\nu\nu}]^{-1}\delta Q_o \quad ,$$

one problem is to find the covariance matrix C_{ss}, which in turn is obtained from (5.3) by covariance propagation.

What propagates is the effect of the anomalous residual (or non modelled) potential δu_{res} on the solution $\underline{\xi}_{res}(t)$ of Hill's equations through their known term $\nabla\delta u_{res}$. The functional relations between $\underline{\xi}_{res}$ (and then $\dot{\underline{\xi}}_{res}$ too) and the known term $\underline{f}$ have been discussed in §4: the way in which $\nabla\delta u_{res}$ is seen in the inertial system as a function of the orbital variables (θ=0,F,r=R) is discussed in the Appendix.

We shall mainly refer to the equation (A1.21) that we rewrite in the form (degree by degree)

$$\delta\underline{g}_l = \sum_{-l}^{l}{}_m \, \delta\underline{g}_{l,m} = \sum_{-l}^{l}{}_m \sum_{-l}^{l}{}_k \, \underline{f}_{l,m,k} \, e^{j(kF+mL)} \tag{5.4}$$

$$\underline{f}_{l,m} = \gamma_l \, u_{lm} \, \bar{W}_{lmk} \, \underline{a}_{lk}$$

$$\gamma_l = \mu \frac{a^l}{R^{l+2}}$$

$$\underline{a}_{lk} = \begin{vmatrix} -\bar{P}_{l,|k|+1}(o)\, \sigma_{l,|k|} \\ \bar{P}_{l,|k|}(o)\, jk \\ -\bar{P}_{l,|k|}(o)(l+1) \end{vmatrix}$$

$$\sigma_{lk} = (1 - \frac{\delta_{ok}}{2})(l-k)(l+k+1)$$

$\bar{W}_{lmk}$ = Wigner coefficients for a rotation through an angle -i around x .

We shall notice immediately that the vector $\underline{a}_{lk}$ has the form

$$\begin{vmatrix} \alpha_{lp} \\ 0 \\ 0 \end{vmatrix} , \quad (k = l-2p-1) \quad ; \begin{vmatrix} 0 \\ \beta_{lp} \\ \beta_{lp} \end{vmatrix} , \quad (k = l-2p) \quad , \tag{5.5}$$

so that its non-null components are the out-of-plane when l-k is odd, and the other two when l-k is even.

According to (4.35) the solution $\underline{\xi}(t)$ corresponding to the force field (5.4) is

$$\underline{\xi}_l(t) = \sum_{-l}^{l}{}_{m,k} \, \gamma_l \, u_{lm} \, W_{lmk} \begin{vmatrix} e_{km}(t)\, s_{km} & 0 \\ 0 & E_{km}(t)\, S_{km} \end{vmatrix} \underline{a}_{lk} \quad . \tag{5.6}$$

Now we assume u_{lm} to be a set of uncorrelated random variables with variances depend ing only on l (degree variances).

$$E\,\{u_{lm}\,u^*_{l'm'}\} = \sigma_l^{\,2}\,\delta_{ll'}\,\delta_{mm'} \quad : \tag{5.7}$$

as it is shown in the Appendix, this is a direct consequence of the usual rules to compute degree variances from the real potential coefficients.
Basically (5.7) can be interpreted by saying that we want to derive results averaged over all possible configurations of the anomalous gravity field obtained from the real one by a purely random rotation (cfr. Sansò, 1986).
From the rule (5.7) it is easy to derive the formal expression of the autocovariance matrix of $\underline{\xi}(t)$, namely

$$C_{\underline{\xi}_l\underline{\xi}_{l'}}(t,t') = E\,\{\underline{\xi}_l(t)\underline{\xi}^*_{l'}(t')\} \overset{(10)}{=} \delta_{ll'} \sum_{k,n,m=-l}^{l} \gamma_l^2\,\sigma_l^2\,W_{lmk}\,{}^*W_{lmn}\,\cdot \tag{5.8}$$

$$\cdot \begin{vmatrix} e_{km}(t)\,s_{km} & 0 \\ 0 & E_{km}(t)\,S_{km} \end{vmatrix} \underline{a}_{lk}\,\underline{a}^+_{ln} \begin{vmatrix} e_{nm}(t')^*\,s^*_{nm} & 0 \\ 0 & S^*_{nm}\,E^*_{nm}(t') \end{vmatrix} \quad .$$

Similar expressions can be easily computed by differentiating (5.8) with respect to t and/or to t' to get $C_{\dot{\underline{\xi}}_l\underline{\xi}_{l'}}$, $C_{\dot{\underline{\xi}}_l\dot{\underline{\xi}}_{l'}}$. In this way formally the problem of computing the sought covariances is solved, though we obviously do not dare to claim that it is solved practically: we can only maintain that (5.8) is the natural formula to represent the covariance of $\underline{\xi}$ when the dynamics of this vector is analyzed in a spectral way.

Remark 5.1. It could very well turn out that for computational purposes it is convenient to compute directly the covariance of $\underline{\xi}(t)$ etc., by applying a (fast) Fourier technique directly to (4.11), so as to take advantage of the convolution form of the integrals.

Remark 5.2. As we expected from a remark at the end of the Appendix, the covariance of $\underline{\xi}(t)$ with $\underline{\xi}(t')$ is not stationary, i.e. it doesn't depend on t-t' only. This however is not only due to the presence of the rotation of the earth. As a matter of fact if we could put $\dot{G}=0$, we would get a simplification of (5.8), as in this case e_{km}, s_{km}, E_{km}, S_{km} depend on k, but not on m: subsequently we could perform the summation over m getting:

---

10 By $\underline{\xi}^*$ we mean the adjoint of the complex vector $\underline{\xi}$, i.e. the conjugate of the transpose of $\underline{\xi}$. The same definition applies to complex matrices.

$$\Sigma_m \bar{W}_{lmk} \bar{W}^*_{lmn} = \delta_{kn} \quad ,$$

so that the triple sum (5.8) would reduce to a single one.
Even in such a simplified situation however we would have expressions like $e_{km}(t)e^*_{km}(t)$ etc., which do not depend on t-t' not only because of the presence of some non-stationary integrals for particular values of k,m, but also because $e_{km}(t)$ includes the term $h_{km}(t)$ which oscillates at the frequency of the satellite $\pm n$, while $p_{km}(t)$ has as frequency kn-mG, i.e. kn if $\dot{G}$ is set to zero.

Remark 5.3. The sum (5.8) has no particular symmetry in general and even the separation between the out of plane component and along-track, radial components typical of Hill's equations, is not any more respected.
In fact the shape of the non-zero terms in $\underline{a}_{lk}\underline{a}^+_{ln}$ can be schematically represented as in fig. 5.1:

Shape of $\underline{a}_{lk}\underline{a}^+_{ln}$

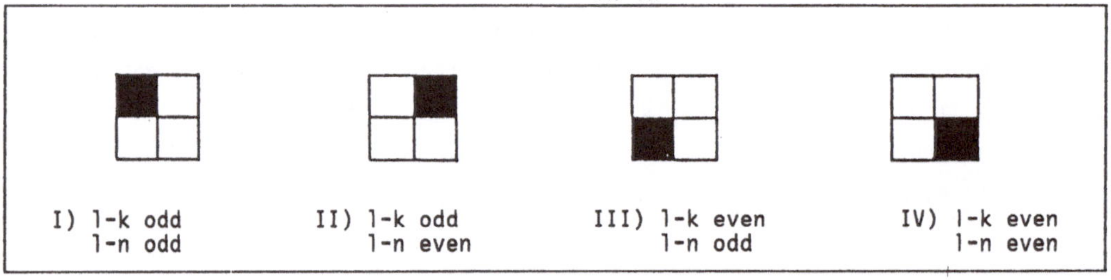

FIG. 5.1

It is easy to recognize that the matrix in (5.8) follows exactly the same patterns. Consequently in general $C_{\xi_o\xi_a}(t,t')$, $C_{\xi_o\xi_r}(t,t')$ are different from zero.

Example 5.1. (Equatorial orbit) The only case where we can elaborate analitically a little further the equation (5.8), corresponds to an equatorial reference orbit, i.e. when the inclination i is zero.
In this case in fact, we do not have the i rotation of the system, so that (cfr. Appendix)

$$\tilde{W}_{lmk} = \delta_{mk} \tag{5.9}$$

$$\bar{W}_{lmk} = e^{jm\Omega} \delta_{mk}$$

and accordingly (5.8) becomes

$$C_{\underline{\xi}_l\underline{\xi}_{l'}}(t,t') = \delta_{ll'} \sum_{-l}^{l}{}_m \gamma_l \sigma_l^2 \begin{vmatrix} e_{mm}(t)\, s_{mm} & 0 \\ 0 & E_{mm}(t)\, S_{mm} \end{vmatrix} \underline{a}_{lm}\underline{a}_{lm}^{+} \cdot$$

$$\cdot \begin{vmatrix} e_{mm}^{*}(t')\, s_{mm}^{*} & 0 \\ o & S_{mm}^{*}\, E_{mm}^{*}(t') \end{vmatrix} . \tag{5.10}$$

Since now the product $\underline{a}_{lk}\underline{a}_{ln}$ appears only in the case k=n=m, the corresponding matrix can have only the form I) or IV) of fig. 5.1, when l-m is respectively odd or even: in this case therefore, the out-of-plane component is uncorrelated with the other two.

Formula (5.10) is much simpler than (5.8) in that the summation is over one index only, but still the stationarity the covariance doesn't hold for the same reasons discussed in Remark 5.2.

It is instructive, for instance, to compute $\sigma^2[\xi_o(t)]$ and verify that it is a time dependent, but bounded function.

Remark 5.4. It is important to realize that, as already mentioned, our way of averaging on the anomalous potential coefficients implicitely defines a model of stochastic process obtained by random rotations of the actual potential. In this model the averaging is performed on a population of earths that at t=0 would appear as randomly tilted one respect to the other, but maintaining fixed in space the spin axis and the satellite's orbit, so that in the subsequent instants the models all rotate with parallel axes and equal angular velocity $\dot{G}$. The situation is illustrated in fig. 5.2.

It comes natural to the mind the question whether, when treating satellite observations, it wouldn't be more natural to average over the mission parameters: in our case these are essentially the inclination i, the constant L_o related to the longitude of the node, the constant F_o which in our circular case takes the place of the argument of the perigee.

Obviously we are free to average on what parameters we prefer, with the only proviso that we know that the results will be less and less representative of each specific situation.

On the other hand if we are treating other terrestrial data together with satellite data, and we want to be consistent, we cannot avoid taking the average on rotated configurations of the gravity field. A good idea could be that on satellite data we take a double average, the first rotating the gravity field, the second on the mission parameters (an average on a population of missions).

In this respect, however, we would prefer to distinguish between inclination and the two parameters L_0, F_0: in fact usually we are really interested to know what is the signal propagation to an orbit at a given inclination i, while things are different for L_0, F_0 since these constant are in any way undergoing slow variations along the satellite mission, due to the effects of J_2 which forces a secular precession of the longitude of the node and of the argument of perigee.

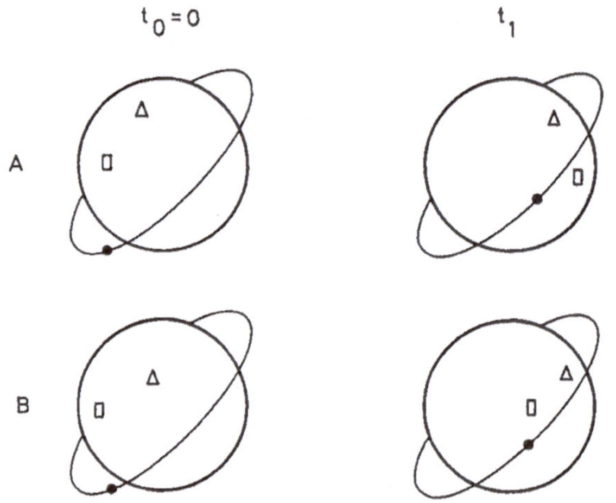

FIG. 5.2 - Model B is obtained by tilting model A at t=0 ; but then both models rotate around the same axis with equal angular velocity.

With this in mind, we first trace back the functions where L_0, F_0 enter, namely e,E, and ascertain that things are always so that

$$e_{km}(t) = e^{j(kF_0 + mL_0)}\, \bar{e}_{km}(t)$$

$$E_{km}(t) = e^{j(kF_0 + mL_0)}\, \bar{E}_{km}(t) \quad , \tag{5.11}$$

where $\bar{e}, \bar{E}$ do not depend any more on L_0, F_0.

Already that shows that in reality $C_{\xi_1 \xi_1'}$ does not depend on L_0, so that only the the dependence on F_0 is left for the presence of a factor $e^{j(k-n)F_0}$: averaging this function over the whole circle gives δ_{kn} so that (5.8) in any way is slightly simplified

$$M_{F_o}\{C_{\underline{\xi}_l \underline{\xi}_{l'}}(t,t')\} = f_{ll'} \sum_{k,m\,=-l}^{l} \gamma_l^2 \sigma_l^2 |\bar{W}_{lmk}|^2 \cdot \tag{5.12}$$

$$\cdot \begin{vmatrix} e_{km}(t)\, s_{km} & 0 \\ 0 & E_{km}(t)\, S_{km} \end{vmatrix} \underline{a}_{lk} \underline{a}^*_{lk} \begin{vmatrix} e^*_{km}(t)\, s^*_{km} & 0 \\ 0 & S^*_{km}\, E^*_{km}(t) \end{vmatrix} .$$

Since $\underline{a}_{lk}\underline{a}^*_{lk}$ is always of the type I) or type IV) of fig. 5.1, depending on the parity of l-k, we see that in the further averaged covariance (5.12) the out-of-plane component decouples again from the other two, thus getting a simplified structure which might be more suited to real computations.

6. SUB-OPTIMAL SOLUTIONS

The approach to data reduction presented in §1 and §5 is with no doubt the straightforward application of the general concept of integrated geodesy to satellite observations.
The critical point that forced us to make the lengthy computations of §4, §5 and of the Appendix is in that the satellite observations are generally functionals of $\underline{\xi}, \dot{\underline{\xi}}$ and these vectors in turn are non local functionals of the anomalous gravity vector because they are obtained from the latter by integrating a system of differential equations.
The authors then got the idea of trying to act on the side of the observations by differencing rather than the opposite, in order to arrive at equations where only local functionals of the anomalous field appeared: we called this the localization problem. After all the same idea is happily applied in positioning problems with the GPS technique, the only difference being that here we want to apply a differential operator rather than a finite difference scheme.
Two problems come immediately to our mind: a) when differentiating, some information is lost from the data; b) how can we differentiate noisy data and what should after that be the stochastic model of the noise part?
We can answer question a) that obviously we loose information on the long wave length side, and this is why such an approach can only be sub-optimal; as for question b), the answer is that all the streams of satellite data we have ever seen display a very distinct behaviour of the smooth part and of the irregular noise (e.g., cfr. Betti et al., 1987). Accordingly we can always make a pure empirical analysis to separate (estimate) the signal from the noise with some very simple technique like spline or polynomial interpolation.

In this way each sequence of data of a short arc is transformed via a least squares adjustment in a smooth curve controlled by the estimate of a finite number of purely empirical parameters, of which we can rigorously derive the covariance matrix as well[11]. It is this function that we can differentiate, also inheriting the correct covariance of the differentiated values. At that point we ambitiously tried to prove the guess that every type of observations could be reduced by a suitable differential operator to the form of a local functional of δu; but the goal was not achieved. Nevertheless we found some interesting examples where the idea could be applied, so for the rest of the paragraph we shall mainly concentrate on one of them to see how it works.

Example 6.1. Let us consider a short arc of altimetric data and suppose to model them as (cfr. Example 2.2)

$$\delta H_o = \xi_r - \frac{1}{\gamma} \delta u \, [\underline{x}_G] + p(t) + \nu \tag{6.1}$$

where ξ_r is the radial correction of the orbit, $\underline{x}_G$ is the foot print of the observation, p(t) is a low order polynomial (typically of the first order) taking into account several empirical effects like unmodelled tides, stationary sea surface topography etc.

Assume now to make a purely empirical analysis of δH_o and split it into signal and noise, where the signal is represented, e.g. by a suitable combination of Tchebychef polynomials:

$$\begin{aligned} \delta H_o &= s(t) + \nu \\ s(t) &= \Sigma\, c_i\, T_i(t) \\ C_{ss}(t,t') &= \Sigma_{i,j}\, T_i(t)\, T_j(t')\, \sigma(c_i c_j) \quad . \end{aligned} \tag{6.2}$$

We consider s(t) as an estimate of the signal part of (6.1) and then continue the reduction process by setting

$$s(t) = \xi_r - \frac{1}{\gamma} \delta u \, [\underline{x}_G] + p(t) + \varepsilon \quad , \tag{6.3}$$

where ε has a covariance matrix given by C_{ss} in (6.2).

Now the point is that ξ_r is a non local functional of δu, since it has to be derived (in our circular approximation) by integrating the equation

---

11 This is the covariance of the pure noise part propagated to the estimate of the empirical parameters.

$$\begin{cases} \ddot{\xi}_r + n^2\xi_r = \delta g_r + c^{(12)} + 2n \int_0^t \delta g_a \, d\tau \\ (\delta \underline{g} = \nabla \delta u \ , \quad \delta g_r = \underline{e}_r \cdot \delta \underline{g} \ , \quad \delta g_a = \underline{e}_a \cdot \delta \underline{g}) \quad , \end{cases} \tag{6.4}$$

which is already in a non local form due to the term $\int \delta g_a d\tau$.
But from (6.4) it is clear that after one differentiation, we get a new "localized" equation, namely

$$\dddot{\xi}_r + n^2\dot{\xi}_r = nR \ \underline{e}_a \cdot \delta U \underline{e}_r + 3n \ \delta \underline{g} \cdot \underline{e}_a \quad , \tag{6.5}$$

where R is the orbital radius, n the mean motion of the satellite, δU the matrix of second derivatives of δu, and we have taken into account that $\dot{\underline{e}}_r = n\underline{e}_a$.
Whence we can think of applying to (6.3) the same differential operator as in (6.5) to obtain a new equation in a localized form.
To this aim we first compute separately the time derivatives of the term $(1/\gamma)\delta u[\underline{x}_G]$, considering γ as constant.
Noting that $\underline{x}_G$ follows the satellite at the ground level, so that $\dot{\underline{x}}_G = na \ \underline{e}_a$, we derive

$$\begin{aligned} &\frac{d}{dt} \delta u[\underline{x}_G] = \delta \underline{g} \cdot \underline{e}_a \ na \\ &\frac{d^2}{dt^2} \delta u[\underline{x}_G] = -\delta \underline{g} \cdot \underline{e}_r \ n^2 a + \underline{e}_a \cdot \delta U \ \underline{e}_a \ n^2 a^2 \\ &\frac{d^3}{dt^3} \delta u[\underline{x}_G] = -3 \ \underline{e}_a \cdot \delta U \underline{e}_r \ n^3 a^2 - \delta \underline{g} \cdot \underline{e}_a \ n^3 a + \underline{e}_a \cdot (\delta U_{\underline{x}} \cdot \underline{e}_a)\underline{e}_a \ n^3 a^3 \end{aligned} \tag{6.6}$$

Using (6.5) and (6.6) in (6.3) we get

$$\begin{aligned} \dddot{s}(t) + n^2\dot{s}(t) = {} & \{nR \ \underline{e}_a \cdot \delta U \ \underline{e}_r + 3n \ \delta \underline{g} \cdot \underline{e}_a\}_S + \\ & - \frac{n^3 a^2}{\gamma} \{-3 \ \underline{e}_a \cdot \delta U \ \underline{e}_r + \underline{e}_a(\delta U_{\underline{x}} \cdot \underline{e}_a)\underline{e}_a \ a\}_G + \\ & + \dddot{p}(t) + n^2\dot{p}(t) + (\dddot{\varepsilon} + n^2\dot{\varepsilon}) \quad , \end{aligned} \tag{6.7}$$

where the quantities in the parentheses $\{ \ \}_S$, $\{ \ \}_G$ refer respectively to the satellite and ground level: the covariance structure of the "noise" $\dddot{\varepsilon}+n^2\dot{\varepsilon}$ has to be derived by covariance propagation from C_{ss} in (6.2). The goal achieved with (6.7) is

---

12 This constant accounts for possible non-null initial conditions in integrating $\ddot{\xi}_a + 2n\dot{\xi}_r = \delta g_a$.

that in this equation the functionals of δu are purely local, so that the application of an integrated approach becomes, if not practical, at least less cumbersome.

Example 6.2. Just to show that altimetry is not the unique case in which the idea of differencing to localize can be applied, we sketch another example.
Assume a satellite is tracked by SLR from three ground stations simultaneously, so that we can write the observation equations

$$\delta L_i = \underline{e}_i \cdot \xi \qquad i=1,2,3$$
$$\underline{e}_i = \frac{\underline{x}_s - \underline{x}_{Gi}}{|\underline{x}_s - \underline{x}_{Gi}|} \quad : \tag{6.8}$$

for the sake of simplicity we have held the station coordinates as fixed.
We can consider (6.8) as an algebraic transformation of the vector $\underline{\xi}$ into the vector $\delta\underline{L}^+ = |\delta L_1 \delta L_2 \delta L_3|$ through the matrix

$$E = \begin{vmatrix} \underline{e}_1^+ \\ \underline{e}_2^+ \\ \underline{e}_3^+ \end{vmatrix} \quad ,$$

i.e.

$$\delta\underline{L} = E\underline{\xi} \quad . \tag{6.9}$$

Now if we have observations $\delta\underline{L}_0$ we can first make an empirical analysis to estimate $\delta\underline{L}$ and then we can take a linear combination of $\delta\underline{L}$, $\delta\dot{\underline{L}}$, $\delta\underline{L}$ with matrices A, B, C:

$$A\delta\ddot{\underline{L}} + B\delta\dot{\underline{L}} + C\delta\underline{L} = (AE)\ddot{\underline{\xi}} + (A2\dot{E} + BE)\dot{\underline{\xi}} + (A\ddot{E} + B\dot{E} + CE)\dot{\underline{\xi}} \quad . \tag{6.10}$$

Since the system of Hill's equations can be written as

$$\ddot{\underline{\xi}} + \Lambda\,\dot{\underline{\xi}} + K\underline{\xi} = \delta\underline{g}$$

for suitable Λ and K, it is enough to choose A, B, C, so as to satisfy

$$AE = I$$
$$2A\dot{E} + BE = \Lambda$$
$$A\ddot{E} + B\dot{E} + CE = K$$

to get from (6.10) the system of observation equations

$$A\delta\ddot{\underline{L}} + B\delta\dot{\underline{L}} + C\delta\underline{L} = \delta\underline{g} \quad , \tag{6.11}$$

i.e., to obtain the sought localized equations.

We conclude the paragraph by saying that this approach, though sub-optimal, could be a brake through for the application of the integrated approach.
Whether this would be really necessary or only useful, we are not able to say, although there are cases, like the problem of altimetry, where we can see that specially to combine satellite data with terrestrial observations it could be nice and desirable to have a unique tool capable of treating the data all together.

APPENDIX

Purpose of this Appendix is to give a representation of the anomalous gravity vector suitable for the computation of its covariance function.
As always when developing an analytic theory, there is the critical point of choosing the coordinate system in which the calculations are to be performed.
In turns out that if we like to make very easy the description of the unperturbed motion it becomes convenient the use of dynamic elements (classical approach followed in Kaula's book (1966) with Laplace or Hamilton equations or with Hamilton-Jacobi theory, etc.), however the computation of the force function is cumbersome; the opposite happens if we choose to leave the force function in its simplest form, i.e. in spherical coordinates which are very suited for a spherical harmonic expansion.
Hill's equations (which have been recently chosen by several authors, e.g. Colombo, 1984) are somewhere midway, trying to take advantage of both procedures. These equations become really simpler and exactly handable when the reference motion is taken as uniform circular as shown in §4. If this is the case in fact, all what we have to do it to rotate the reference system from the usual one, with equator coinciding with earth's equator and origin of the longitudes at the Greenwich meridian, to the one where the equator coincides with the orbit and the origin of the longitudes is for instance at the ascending node: the geometry is shown in fig. A1.1. In spite of the claimed "simplicity" naturally to rotate the spherical harmonics is never straightforward.

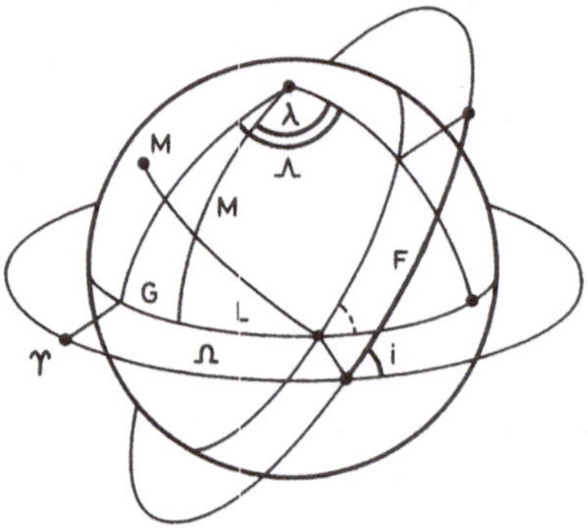

	Equinox
G	Greenwich siderial time
Ω	Long. of the ascending node (Ω=G+L)
λ	Longitude from Greenwich
Λ	Longitude from γ
i	inclination of the orbit
M	Greenwich meridian
M'	Origin meridian of the system (θ,F)

FIG. A1.1

Probably the most elegant way to do it is to use the so called Wigner coefficients, (cfr., for instance, the book by Miller W., Simmetry groups and their applications, Academic Press, 1972) which, given a rotation in terms of three Euler angles (α,β,γ) between the systems (ϕ',λ') and (ϕ,λ), give the connection between the two corresponding systems of spherical harmonics

$$Y_{lm}(\phi,\lambda) = \sum_{-l}^{l}{}_{m'} W_{l,mm'}(\alpha,\beta,\gamma)\, Y_{l,m'}(\phi',\lambda') \quad . \tag{A1.1}$$

We must underline that in (A1.1) the definition of the coefficients $W_{l,mm'}$ depends on the definition of the spherical harmonics: here the complex formalism will be assumed, with

$$Y_{lm}(\phi,\lambda) = \bar{P}_{l|m|}(\phi)\, e^{jm\lambda} \quad . \tag{A1.2}$$

With the help of (A1.1) we can perform the rotation from (ϕ,λ) to (θ,F) in two steps:

$$(\phi,\lambda) \rightarrow (\phi,\Lambda) \quad \text{and then} \quad (\phi,\Lambda) \rightarrow (\theta,F) \quad . \tag{A1.4}$$

In fact from

$$\delta u = \Sigma_l \mu \, \frac{a^l}{R^{l+1}}\, \delta u_l(\phi,\lambda) \tag{A1.3}$$

$$\delta u_l(\phi,\lambda) = \sum_{-l}^{l}{}_m \, u_{lm} \, Y_{l,m}(\phi,\lambda) \tag{A1.4}$$

$$u_{l,-m} = u^*_{l,m} \quad (13) \quad , \tag{A1.5}$$

we easily go to

$$\delta u_l(\phi,\Lambda) = \sum_{-l}^{l}{}_m \, u_{lm} \, Y_{l,m}(\phi,\Lambda) \, e^{-jmG} \tag{A1.6}$$

since $\Lambda = \lambda + G$.

Subsequently we can go from (ϕ,Λ) to (θ,F) by means of a rotation around z of the angle Ω and then a rotation around x of the angle i: whence we can write (A1.1) in the form

$$Y_{lm}(\phi,\Lambda) = \sum_{-l}^{l}{}_{m'} \, W_{l,mm'}(\Omega,i) \, Y_{l,m'}(\theta,F) \quad . \tag{A1.7}$$

Substituting (A1.7) in (A1.6) one gets finally

$$\delta u_l(\theta,F) = \sum_{-l}^{l}{}_{m'} \, Y_{l,m'}(\theta,F) \sum_{-l}^{l}{}_m \, u_{lm} \, e^{-jmG} \, W_{l,mm'} \, (\Omega,i) \quad . \tag{A1.8}$$

Now the three components of the force field

$$\begin{cases} \delta g_o = - \dfrac{1}{R} \dfrac{\partial}{\partial \theta} \; \Sigma_l \; \mu \dfrac{a^l}{r^{l+1}} \, \delta u_l(\theta,F) & \text{(out-of-plane)} \\[2ex] \delta g_a = \dfrac{1}{R \cos \theta} \; \dfrac{\partial}{\partial F} \; \Sigma_l \; \mu \dfrac{a^l}{r^{l+1}} \, \delta u_l(\theta,F) & \text{(along-track)} \\[2ex] \delta g_r = \dfrac{\partial}{\partial r} \; \Sigma_l \; \mu \dfrac{a^l}{r^{l+1}} \, \delta u_l(\theta,F) & \text{(radial)} \end{cases} \tag{A1.9}$$

can be computed along the reference motion

$$\begin{aligned} r &= R \\ F &= F_o + \dot{F}t \\ \theta &= 0 \quad . \end{aligned} \tag{A1.10}$$

---

13 (A1.5) is a condition that guarantees the reality of the sum $\delta u_l(\phi,\lambda)$: the coefficients u_{lm} are related to the real coefficients $\bar{C}_{l,m}$, $\bar{S}_{lm}$ according to $u_{lm} = 1/2\ (C_{lm} - jS_{lm})$ when $m>0$ and $u_{l,m} = u^*_{l,-m}$ when $m<0$.

Recalling also that $G = G_o + \dot{G}t$ we get in this way a natural Fourier representation of the forcing terms, what simplifies the search of a particular integral of the motion equations.

It is to be underlined that with our choice the Wigner coefficients are pure functions of the two constants of motion Ω,i and do not depend on time: or better by splitting the rotation into two steps we have shown the explicit time dependence of the total Wigner coefficients.

Before continuing with the computations we must be more specific on the definition of the Wigner coefficients, and in particular we must formulate some warnings before the expressions usually available in literature (e.g. in books on the quantum treatment of angular momentum), can be used:

a) the coefficients performing the transformation

$$Y(\phi,\Lambda) = \Sigma_{m'} W_{lmm'}(\Omega,i) \; Y_{lm'}(\theta,F)$$

are attached usually to the rotation from (θ,F) to (ϕ,Λ), i.e. to the sequence of the two rotations

$$(\theta,F) \quad \xrightarrow{} \quad \boxed{R_z(-\Omega)\; R_x(-i)}^{\text{apply}} \quad \xrightarrow{} \quad (\phi,\Lambda) \quad .$$

Correspondingly the Wigner coefficients can be split in a sequence

Rotation (-i) around x = $\Sigma_{m'} \bar{W}_{lmm'}(i) \; Y_{lm'}(\theta,F)$ -> Harmonic of order m

$Y_{lm}(\phi,\Lambda)$ = Rotation $(-\Omega)$ of the above harmonic = $e^{jm\Omega} \; \Sigma_{m'} \bar{W}_{lmm'}(i) \; Y_{lm'}(\theta,F)$.

In this way we have the decomposition

$$W_{l,mm'}(\Omega,i) = e^{jm\Omega} \; \bar{W}_{l,m,m'}(i) \quad , \tag{A1.11}$$

where the set $\bar{W}_{l,mm'}(i)$ represents a rotation of (-i) around x;

b) what is more commonly found in literature is the expression of the coefficients for a rotation around y, due to the customary definition of Euler angles. However, one can be reduced to the other by considering that (cfr. fig. A1.2)

$$R_x(-i) = R_z(\pi/2)\; R_y(-i)\; R_z(-\pi/2) \quad : \tag{A1.12}$$

attention should be paid to the fact that in (A1.12) the first rotation around z acts on a spherical harmonic of order m', while the last acts on a harmonic of order m. Accordingly we can write

$$\tilde{W}_{lmm'} = e^{-jm\pi/2}\, \tilde{W}_{lmm'}(i)\, e^{jm'\pi/2} = \qquad (A1.13)$$
$$= (j)^{m'-m}\, \tilde{W}_{lmm'}(i)$$

where finally $\tilde{W}_{lmm'}(i)$ represents the rotation through an angle -i around y;

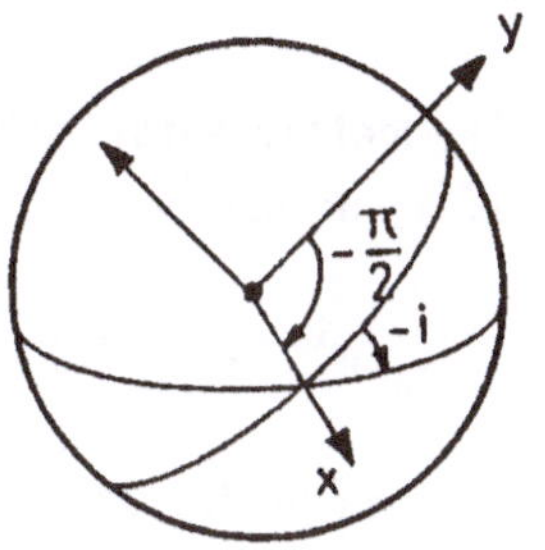

FIG. A1.2

c) before we use for instance the explicit Wigner formula to define (A1.13) we still have to take care of a phase problem related to our definition (A1.2) of the spherical harmonic functions. Going back to the definition of the associated Legendre functions we realize that in literature we meet $P_{l,m}(\phi)$ such that

$$P_{l,|m|}(\phi) = (-1)^{(m+|m|)/2}\, P_{l,m}(\phi) \quad ,$$

i.e. that coincide with P for m<0 and are phase-shifted by $(-1)^m$ when m>0. It follows that also in the definition of $\tilde{W}_{l,mm'}(i)$ we must multiply the usual expression by a phase factor $(-1)^{(1/2)(m+|m|+m'+|m'|)} = f_{mm'}$.

Keeping all that in mind, we find

$$\tilde{W}_{l,mm'} = f_{mm'}\, \Sigma_q (-1)^q \frac{\sqrt{(l+m)!(l-m)!(l+m')!(l-m')!}}{(l+m-q)!(l-m'-q)!q!(q-m+m')!} \left(\cos\frac{i}{2}\right)^{2l+m-m'-2q} \cdot$$
$$\cdot \left(-\sin\frac{i}{2}\right)^{2q-m+m'} \qquad (A1.14)$$

where q ranges from 0 to the smallest of the four numbers $l\pm m$, $l\pm m'$. In practice (A1.14) must be used for $m' > m$ and then the symmetry relation

$$\tilde{W}_{l,mm'}(i) = \tilde{W}_{l,m'm}(-i) \tag{A1.15}$$

can be exploited.
Another important symmetry of the coefficients $\tilde{W}_{l,m,m'}(i)$ is expressed by the equation

$$\tilde{W}_{l,-m,-m'} = \tilde{W}_{l,m,m'} \quad : \tag{A1.16}$$

this contrasts the analogous relation for the more usual coefficients where a factor $(-1)^{m-m'}$ appears in (A1.16), which however in our case is killed by the factor $f_{-m,-m'}/f_{m,m'}$.
After this digression on Wigner's coefficients, taking advantage of (A1.11) and recalling that Ω-G=L (cfr. fig. A1.1), we can rewrite (A1.8) in the form

$$\delta u_l(\theta,F) = \sum_{-l}^{l}{}_{m'} Y_{l,m'}(\theta,F) \sum_{-l}^{l}{}_{m} u_{lm}\, e^{jmL}\, \bar{W}_{lmm'}(i) \quad , \tag{A1.17}$$

with the inclination functions $\bar{W}$ defined by (A1 13) and (A1.14).
Now we are in a position to compute (A1.9). Taking (A1.10) into account, setting

$$\gamma_l = \mu \frac{a^l}{R^{l+2}} = \frac{1}{g_s} \left(\frac{a}{R}\right)^l \tag{A1.18}$$

and recalling that (e.g. cfr. Colombo, 1984, page 70),

$$\begin{aligned} &\frac{\partial}{\partial\theta} \bar{P}_{lm}(\theta) = -m \operatorname{tg} \theta \bar{P}_{lm}(\theta) + \bar{P}_{lm+1}(\theta)\, \sigma_{lm} \\ &\sigma_{lm} = (1 - \frac{\delta_{om}}{2})(l-m)(l+m+1) \qquad\qquad (\sigma_{l\cdot} = 0),\ (m>0) \end{aligned} \tag{A1.19}$$

we find for

$$\delta u_{lm}(r,\theta,F) = \left(\frac{\mu a^l}{r^{l+1}}\right) u_{lm} \sum_{-l}^{l}{}_{m'} \bar{P}_{l,|m'|}(\theta)\, e^{j(m'F+mL)}\, \bar{W}_{lmm'}(i) \tag{A1.20}$$

the corresponding force field

$$\begin{cases} \delta g_{olm}(R,0,F) = \gamma_l\, u_{lm} \sum_{-l}^{l}{}_{m'} - \bar{P}_{l,|m'|+1}(o)\, \sigma_{l|m'|}\, \bar{W}_{lmm'}\, e^{j(m'F+mL)} \\[2em] \delta g_{alm}(R,0,F) = \gamma_l\, u_{lm} \sum_{-l}^{l}{}_{m'} \bar{P}_{l|m'|}(o)\, jm'\, \bar{W}_{lmm'}\, e^{j(m'F+mL)} \\[2em] \delta g_{rlm}(R,0,F) = \gamma_l\, u_{lm}(l+1) \sum_{-l}^{l}{}_{m'} - P_{l,|m'|}(o)\, W_{lmm'}\, e^{j(m'F+mL)} \end{cases} \tag{A1.21}$$

these expressions can be somehow simplified by considering that

$$\begin{cases} \bar{P}_{l,|m'|+1}(o) = 0 \quad \text{only for} \quad m' = l-2p-1 \qquad 0 \leq p \leq l-1 \\ \bar{P}_{l,|m'|}(o) = 0 \quad \text{only for} \quad m' = l-2p \qquad 0 \leq p \leq l \quad ; \end{cases} \tag{A1.22}$$

then setting

$$-\bar{P}_{l,|l-2p-1|+1}(o) = \alpha_{l,p} \quad , \quad \bar{P}_{l,|l-2p|} = \beta_{l,p} \quad ,$$

we get

$$\begin{cases} \delta g_{olm} = \gamma_l \, u_{lm} \sum_{0}^{l-1}{}_p \, \alpha_{lp} \, \sigma_{l,|l-2p-1|} \, \bar{W}_{l,m,l-2p-1} \, e^{j[(l-2p-1)F+mL]} \\ \delta g_{alm} = \gamma_l \, u_{lm} \sum_{0}^{l}{}_p \, \beta_{lp} \, j(l-2p) \, \bar{W}_{l,m,l-2p} \, e^{j[(l-2p)F+mL]} \\ \delta g_{rlm} = \gamma_l \, u_{lm}(l+1) \sum_{0}^{l}{}_p - \beta_{lp} \, \bar{W}_{l,m,l-2p} \, e^{j[(l-2p)F+mL]} \quad . \end{cases} \tag{A1.23}$$

To (A1.20), (A1.21), (A1.23) we can also give back a real form: we perform explicitely the computation for (A1.20) leaving the others as an exercise.
We first af all recall the reality condition (A1.5) and the subsequent relation between real and complex coefficients

$$u_{l,m} = 1/2 \, (\bar{C}_{lm} - j\bar{S}_{lm}) \qquad (m > 0) \quad ; \tag{A1.24}$$

moreover we notice that, owing to (A1.13), (A1.16), the analogous relation holds

$$\bar{W}_{l,-m,-m'} = (\bar{W}_{l,mm'})^* \quad . \tag{A1.25}$$

Whence we can write for every m>0, along the reference orbit (θ=0, r=R),

$$V_{lm} = \delta u_{l,m}(R,0,F) + \delta u_{l,-m}(R,0,F) =$$

$$= R \, \gamma_l \, u_{l,m} \sum_{-l}^{l}{}_{m'} \, \bar{P}_{l,|m'|}(o) \, e^{j(m'F+mL)} \, \bar{W}_{l,mm'} +$$

$$+ R \, \gamma_l \, u_{l,-m} \sum_{-l}^{l}{}_{m'} \, \bar{P}_{l,|m'|}(o) \, e^{-j(m'F+mL)} \, \bar{W}_{l,-m,-m'} =$$

$$= 2\ \mathrm{Re}\ \{R\ \gamma_l\ u_{lm} \sum_{m'=-l}^{l} \bar{P}_{l,|m'|}(o)\ e^{j(m'F+mL)}\ \bar{W}_{lmm'}\} =$$

$$= R\ \gamma_l\ \Sigma_{m'}\ \bar{P}_{l,m'}(o)\ \{\bar{C}_{lm}[\cos(m'F+mL)\ \mathrm{Re}(\bar{W}_{lmm'}) - \sin(m'F+mL)\ \mathrm{Im}(\bar{W}_{lmm'})] +$$

$$+ \bar{S}_{lm}[\cos(m'F+mL)\ \mathrm{Im}(\bar{W}_{lmm'}) + \sin(m'F+mL)\ \mathrm{Re}(\bar{W}_{lmm'})]\} \quad .$$

Setting as before $m'=l-2p$, $P_{l,|l-2p|}(o)=\beta_{l,p}$ and on account of the relations (cfr. (A1.13) and remember that $\tilde{W}$ is real)

$$\mathrm{Re}(\bar{W}_{lmm'}) = \begin{cases} (-1)^{p+\frac{l-m}{2}}\ \tilde{W}_{l,m,l-2p} & l-m \text{ even} \\ 0 & l-m \text{ odd} \end{cases}$$

$$\mathrm{Im}(\bar{W}_{lmm'}) = \begin{cases} 0 & l-m \text{ even} \\ (-1)^{p+\frac{l-m-1}{2}}\ \tilde{W}_{l,m,l-2p} & l-m \text{ odd} \end{cases}$$

we eventually obtain (<a> denoting the integer part of a)

$$V_{lm} = R\ \gamma_l \sum_{0}^{l}{}_p (-1)^{p+\langle\frac{l-m}{2}\rangle}\ \tilde{W}_{l,m,l-2p}\ \beta_{lp} \cdot \{ \begin{bmatrix} \bar{C}_{lm} \\ \bar{S}_{lm} \end{bmatrix} \cos\ [(l-2p)F + mL] +$$

$$+ \begin{bmatrix} \bar{S}_{lm} \\ -\bar{C}_{lm} \end{bmatrix} \sin\ [(l-2p)F + mL]\ \} \quad , \qquad \text{(A1.26)}$$

with the first raw being valid when l-m is even and the second when l-m is odd. This is by no means a purely academical computation, but rather a check of the approach presented with the more traditional real theory developped for instance in Kaula's book (1966): in fact we can verify that (A1.26) is essentially identical with (3.61), provided that

$$(-1)^{(l-m)+p+[\frac{l-m}{2}]}\ \tilde{W}_{l,m,l-2p}(i)\ \beta_{lp} = \bar{F}_{lmp}(i) \quad . \qquad \text{(A1.27)}$$

This relation, which has been positively tested[14] for some values of l,m,p, creates

---

14 In reality, there is a difference in the multiplying factor with respect to Kaula's Table 1, page 34, which, however, is due to the fact that in Kaula's formula the simple coefficients C_{lm}, S_{lm} and not the normalized ones $\bar{C}_{lm}, \bar{S}_{lm}$ are used.

the necessary bridge between the two approaches and demonstrates the essential character of Wigner coefficients of the inclination functions $\bar{F}_{lmp}(i)$.
The same procedure as above can obviously be applied to $\delta g_{olm} + \delta g_{ol,-m} = 2\,\mathrm{Re}\,\delta g_{olm}$ and to the other components of the force field (A1.23).
It is interesting to observe that in this way we could have performed the computation of the two components

$$\frac{\partial}{\partial r} V_{lm} \quad , \quad \frac{1}{R}\frac{\partial}{\partial F} V_{lm}$$

directly from the real expression (A1.26), as it has been done for instance in Colombo (1984): however a direct computation of the out-of-plane component would have been difficult[15]. The proposal of Colombo (1984), to compute the out-of-plane component by the smart formula

$$\frac{1}{R \sin F} \frac{\partial}{\partial i} V_{lm}$$

(cfr. fig. A1.3), runs into the difficulty that apparently a singularity is created at the nodes where $F=0,\pi$. Indeed this has to be a simple matter of coordinates since there is no physical meaning for it. As a matter of fact we thought it worthwhile to make the exercise of repeating Kaula's reasoning (cfr. Kaula, 1966, §3.3), resulting in this case in the formula

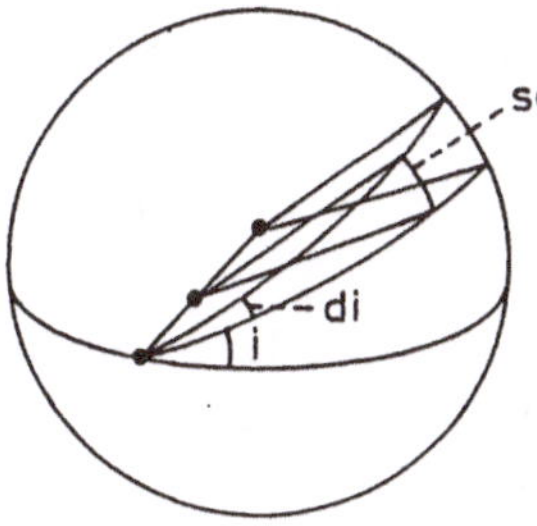

FIG. A1.3

$$\frac{1}{R \sin F} \frac{\partial}{\partial i} V_{lm} = \gamma_l \sum_{0}^{l-1}{}_{p} \overset{*}{F}_{lmp}(i) \left\{ \begin{bmatrix} C_{lm} \\ S_{lm} \end{bmatrix} \cos[(l-2p-1)F+mL] + \begin{bmatrix} S_{lm} \\ -C_{lm} \end{bmatrix} \sin[(l-2p-1)F+mL] \right\} \qquad \text{(A1.28)}$$

which is the real counterpart of the first of (A1.23).
The definition of a new system of inclination functions $\overset{*}{F}_{lmp}(i)$ springs out of the computation, namely:

---

15 As a matter of fact (A1.26) describes the field only along the circular orbit; however, we could observe that the field can be continued transversally (in the θ direction) only by multiplying the expression in { } by $\bar{P}_{l,|l-2p|}(\theta)$.

$$F^*_{lmp}(i) = \sum_{0}^{h_p}{}_t \sum_{0}^{m}{}_s \sum_{p\geq t-l+m}^{p-t}{}_c \; T_{lmt}(-1)^{k-t+c} \binom{m}{s} \binom{l-m-2t+s-1}{c} \binom{m-s}{p-t-c} \cdot \tag{A1.29}$$

$$\cdot \; (\sin i)^{l-m-2t-1} (\cos i)^{s-1} [(l-m-2t) \cos^2 i - s \sin^2 i]$$

where

$$h_p = \min(k, l-p-1) \quad , \quad k = \left\langle \frac{l-m}{2} \right\rangle$$

$$T_{lmt} = \frac{(-1)^t (2l-2t)!}{2^l t! \, (l-t)! \, (l-m-2t)!} \quad .$$

These functions are to be compared, apart from the normalization factor (in (A1.28) the normalized coefficients are considered), with $\alpha_{l,p} \, \sigma_{l,|l-2p-l|} \tilde{W}_{l,m,l-2p-l}(i)$, in analogy with (A1.27): this has been done for a few low values of l,m,p achieving a positive answer.

In any way, the relative simplicity of the expressions in (A1.23) talks in favour of the approach with the Wigner coefficients.

We conclude this Appendix by recalling the rules of covariance propagation for the potential coefficients in complex form, since these are needed to compute the global covariance.

If we have a system of real random variables $\{\bar{C}_{lm}, \bar{S}_{lm}\}$ satisfying

$$\begin{aligned} &E\,\{\bar{C}_{lm}\bar{C}_{lm'}\} = E\,\{\bar{S}_{lm}\bar{S}_{lm'}\} = \sigma_l^2\, \delta_{mm'} \\ &E\,\{\bar{C}_{lm}\bar{S}_{lm'}\} = 0 \end{aligned} \tag{A1.30}$$

and a system of complex random variables defined as

$$\begin{aligned} u_{lm} &= 1/2\,(\bar{C}_{lm} - j\bar{S}_{lm}) && m > 0 \\ u_{lo} &= \bar{C}_{lo} && \\ u_{lm} &= 1/2\,(\bar{C}_{lm} + j\bar{S}_{lm}) && m < 0 \quad , \end{aligned} \tag{A1.31}$$

then we have also, from the error propagation rule,

$$E\,\{u_{lm}\, u^*_{l'm'}\} = \sigma_l^2\, \delta_{ll'}\, \delta_{mm'} \quad . \tag{A1.32}$$

We wont fail to remark a crucial effect of the transformation of reference system. from terrestrial to inertial: <u>the potential seen from the satellite along its trajectory, is not a stationary process</u> (in time), due to the fact that the rotation between the two systems is time dependent, as a consequence of the rotation of the earth.

To see this explicitely, we first recall an important orthogonality property of Wigner's coefficients, i.e.

$$\sum_{-l}^{l}{}_m W_{lmm'} W^*_{lmm''} = \delta_{m'm''} \quad (16). \tag{A1.33}$$

We note also that (A1.33) holds when both $W_{lmm'}$ and $W^*_{lmm''}$ refer to the same rotation, as otherwise it is false.

Now let us go back to (A1.8) and let us try to compute the covariance of $\delta u_l(0,F)$ with $\delta u_l(0,F')$[17]: the important point here is to observe that the two points (along the orbit) F,F' refer also to different times t,t' and consequently to two different values G,G'. Whence if we apply (A1.32) we get

$$E\{\delta u_l(0,F)\ \delta u_l(0,F')\} = \Sigma_{m',m''} Y_{lm'}(0,F)\ Y_{l,m'}(0,F') \cdot$$
$$\cdot\ \sigma_l^2 \cdot \left[\Sigma_m e^{-jm(G-G')} W_{lmm'}(\Omega,i)\ W^*_{l,m,m''}(\Omega,i)\right] . \tag{A1.34}$$

Should G be constant (i.e. should the earth be non rotating) the expression in parenthesis would simply be $\delta_{m'm''}$ and the whole expression would assume the costumary simple form

$$E\{\delta u_l(0,F)\ \delta u^*_l(0,F')\} = \sigma_l^2\ \Sigma_{m'} Y_{lm'}(0,F)\ Y_{lm'}(0,F') =$$
$$= (2l+1)\ \sigma_l^2\ P_l(\cos(F-F')) \quad .$$

This however is not true, reflecting the physical fact that between the two positions (0,F), (0,F') of the satellite in the inertial reference frame, the gravity field under it has got a rotation G-G' so that the transformations between the terrestrial reference systems and the inertial system are different and the orthogonality rule doesn't apply.

---

16 This is most easily understood by going back to the very definition (A1.1) and observing that for every (ϕ,λ), (ϕ',λ'), the sommation rule of spherical harmonics yields

$$\Sigma_{m'',m'}(\Sigma_m W_{lmm'} W^*_{lmm''})\ Y_{lm'}(\phi',\lambda')\ Y^*_{lm''}(\phi',\lambda') \equiv (2l+1) \quad :$$

this, being an identity in (ϕ',λ'), compared with the same sommation rule for $Y_{lm}(\phi,\lambda)$, entails (A1.33).

17 From the rule (A1.32) we see that, each degree l being transformed by the rotation into the same degree, every process with uncorrelated degrees is transformed back into a process with uncorrelated degrees, whence we need only to care of the covariances degree by degree.

The same happens obviously to the gravity vector, which is attached to the rotating gravity potential, and it is precisely for this reason that covariance computations are so cumbersome in satellite geodesy.

REFERENCES

Betti B, Carpino M, Migliaccio F, Sansò F (1987) Signal and noise in SLR data. Bull Géod, vol 61, n 3

Brouwer D, Clemence G M (1971) Methods of Celestial Mechanics. Academic Press, New York

Colombo O (1984) The global mapping of gravity with two satellites. Netherlands Geodetic Commission, Publications on Geodesy, New Series, vol 7, n 3, Delft

Eeg J, Krarup T (1975) Integrated Geodesy. In: "Mathematical Geodesy", Methoden und Verfahren der matematischen Physik, Band 12, B I Wissenschafts-verlag, Zurich

Eissfeller B, Hein G W (1985) The Observation Equations of Satellite Geodesy in the Model of Integrated Geodesy. In: "A Contribution to 3-D Operational Geodesy, Part 4, Schriftenreihe des Wissenschaftlichen Studienganges der Bundeswehr, Munchen

Eissfeller B (1985) Orbit improvement using local gravity field information and least squares prediction. Manuscripta Geodaetica, vol 10, n 2

Giacaglia G.E.O. (1980) Transformations of Spherical Harmonics and Applications to Geodesy and Satellite Theory. Studia Geoph. et Geod. 24

Heiskanen W, Moritz H (1967) Physical Geodesy. W H Freeman

Kaula W M (1966) Theory of satellite geodesy. Blaisdell Publ Co, Waltham, Massachusetts

Morando B (1974) Mouvement d'un satellite artificiel de la terre. In: "Cours et documents de mathematique physique", Gordon and Breach, Paris

Moritz H (1980) Advanced Physical Geodesy. H Wichamm Verlag, Karlsruhe

Rummel R (1987) Satellite Gradiometry. In: "Mathematical and numerical techniques in Physical Geodesy", ed H Suenkel, Lecture Notes in Earth Sciences, Springer Verlag, Berlin-Heidelberg

Sansò F (1986) Statistical Methods in Physical Geodesy. In: "Mathematical and numerical techniques in Physical Geodesy", ed H Suenkel, Lecture Notes in Earth Sciences, vol 7, Springer Verlag, Berlin-Heidelberg

Tscherning C C (1977) A note on choice of norm when using collocation for the computation of the approximations to the anomalous potential. Bull Géod, vol 51, n. 2

DETERMINATION OF A LOCAL GEODETIC NETWORK BY MULTI-ARC PROCESSING OF SATELLITE LASER RANGES

A. Milani - E. Melchioni

Dipartimento di Matematica, Università di Pisa
Via F. Buonarroti 2, I-56100 Pisa, Italy

1. Introduction and summary

The difficulty currently encountered in the determination of geodetic networks by Satellite Laser ranging (SLR) can be summarised as follows. The laser ranging stations have achieved a remarkably high precision and accuracy. The consistency of the station calibrations performed before and after each pass, and even between different passes, is of $1 \div 2\ cm$ for the best stations. A well established atmosferic model (Marini and Murray, 1973) is believed to be capable to give the tropospheric light propagation correction to the same level of accuracy. The synchronisation of the station clocks at the microsecond level is not any more a problem. Since no other source of systematic observational errors is known, the best laser stations are believed to be capable of an accuracy of about 2 *cm* or even better; for the third generation stations the precision is so good and the repetition rate so high that normal points with random errors averaged out to the same level could in principle be generated every few seconds.

The main question could then be stated in a somewhat naïve way as follows: if the observational errors in the station–to–satellite distances are of $1 \div 2\ cm$, should the stations positions be determined with both a precision and a real accuracy of $1 \div 2\ cm$? Since the answer to this question, on the basis of the repeatibility tests and of the systematic error sensitivity tests, appears to be negative, at least for continental and intercontinental baselines, is the SLR technique a valid one?

A less naïve analysis of the error budget shows that the main limitation to the accuracy of satellite geodesy by SLR is not due to the observational errors, but rather to the errors in modelling the orbit. Although the list of the perturbations acting upon a satellite is quite long, the sources of these model errors are essentially only four –for a high cannon ball satellite of the *LAGEOS* class–; namely, the poor knowledge of the gravitational field, the non–gravitational perturbations, the tides, and the reference system uncertainty. These problems are not discussed here, but in Afonso et al., 1986; Barlier et al., 1986; Carpino et al., 1986; Afonso et al., 1988; for a review see Milani et al., 1987. We report only the main conclusions this analysis allows to draw upon the choice of the optimal length of the orbital arcs to be used for the determination of geodetic networks in Section 1. Our conclusion is that the orbit can be modelled to the 1 *cm* level of accuracy only over a very short arc, e.g. less than one orbital period of *LAGEOS*.

Two more difficulties arise in this kind of short arc technique. The first one is that the data from many orbital arcs have to be processed together. This multi–arc algorithm is somewhat less established than the conventional single arc differential correction procedure, but it is known and to be implemented in practice it needs only some caution in handling possible numerical problems. This is discussed in Section 4. The second problem is that the traditional theory of the degeneracy of the SLR geodetic network, and the classical method to remove it by fixing three parameters, does not work well for very short arcs. It is found that additional *approximate rank deficiencies* arise, and their elimination is essential to achieve our goal. This is discussed in Section 3.

Section 5 describes the simulations we have performed with our multi–arc simulation–correction software package *ORBIT6*. We would like to stress that the goal of this work was to show if, and under which conditions, it is possible to achieve an accuracy of the station positions as small as the systematic errors in the observations. Thus our simulations have been performed as follows. We have assumed observations with a given r.m.s. error, say 1 *cm*; these are not meant to be raw observations, but normal points produced in such a way that the systematic errors are of the same size of the random errors. Our code allows for these normal points to be correlated, if they are generated by a preprocessing which also gives a covariance matrix for the normal points; however this has not yet been used in the simulations presented in Section 5. We have then tried to fit different sets of solve for parameters, including always the initial conditions for each arc, under different hypothesis on the number of arcs, the number of observing stations for each arc and the systematic model errors. Since we perform a simulation with what we assume to be the real values of the parameters, then solve for some of them by giving either the same or different values to the others, the results include both a formal variance–covariance matrix for the solve for parameters and a table of *real* errors of the solution; if the latter are larger than the formal ellipsoid of error would suggest, they are essentially systematic errors and both the precision and the accuracy of the solution can be assessed. Of course the precision is proportional to the assumed precision of the normal points, while the systematic errors are proportional to the assumed model errors, and our results have to be considered as applicable to a real situation if and only if the hypotheses on both the observations and the model are applicable. The conclusions are presented in Section 5, but they are summarized here. All of them can be geometrically understood on the basis of the symmetry and rank deficiency arguments of Section 3.

The short arcs contain informations from which the geodetic network can be determined only if they are observed by at least three stations; otherwise the information contained in the data are not even enough to solve for the arc initial condition. Thus by number of arcs in the rest of this section we always mean the arcs observed by three stations.

The *approximate rank deficiency* of such a problem is four, that is four station coordinates have to be fixed (e.g. either not determined or stiffened in their known value by some a priori covariance) to allow for a good determination of the others. Thus for the minimal network of 3 stations there are 5 station coordinates to be solved for; a good determination of these 5 parameters requires 6 arcs. Moreover, the latter cannot be neither all ascending nor all descending, but must cross. When the number N_s of stations is increased, the number of arcs required grows approximately in proportion to the number of station coordinates to be solved for, that is to $3N_s - 4$. When all these conditions are satisfied, the station positions are determined with a precision of $\simeq 1$ *cm* if the observational precision was 1 *cm*, or anyway with about the same prescision.

The systematic error analysis, as already mentioned above, allows to conclude that the position of the rotation axis does not matter, i.e. it does not introduce errors above the 2σ level of the random error even if the rotation axis is *wrong* by much more than the present level of knowledge. Thus the pole positions during each arc need not to be in the list of solve for parameters, but they can be taken from existing models. On the other hand, the short arc technique cannot be used to determine the instantaneous position of the rotation axis.

The gravitational field, on the contrary, matters. When simulation and correction are performed with two gravitational fields which are *random clones*, i.e. such that the differences in the coefficients are random numbers distributed in a way consistent with the variance of some solution for the gravitational field (we have used Reigber et al., 1986), the residuals are somewhat larger than they would be by the effect of the observation errors only. This increase in the residuals does not result in significant systematic errors in the station coordinates, thus from the point of view of the purely geometrical solution the gravitational field could be used as it is known today. However, some information which could be used to improve the local gravitational filed is hidden in the observational data. That is, some parameters describing the gravitational anomalies on the observed spherical cap should be in the list of solve for parameters. The techniques to improve the local gravitational field on the observed spherical cap remain an important open problem.

2. Choice of the arc length

Since the errors in the dynamical model of the spacecraft orbit appear to be the main limitation to the accuracy of satellite geodesy, the question arises of whether it is possible to choose the length of the orbital arcs in such a way that the residuals are reduced to the observational noise level. For the purpose of this discussion, we shall call *long* an arc such that the largest orbital errors arise from the secular perturbations (the ones arising from effects which do not average out over an orbital period).

The largest errors in the orbit propagation of *LAGEOS* result from the non-gravitational perturbations, because their effects accumulate in a secular way, in particular in the form of a semimajor axis decay with average speed $\simeq 1\ mm/day$. The actual value of the semimajor axis decrease within a given span of time (of the order of a few weeks to a few months) can change by more than a factor 2 and is essentially unpredictable, because this phenomenon is the result of many effects related to the interaction of the satellite with the particles and photons encountered on its path. E.g. the radiation pressure resulting from light reflected by the Earth's surface(Anselmo et al.,1983; Barlier et al., 1986), the temperature differences among different parts of the satellite surface resulting in anisotropic emission (Rubincam, 1987; Afonso et al., 1988), the charging of the spacecraft resulting in electromagnetic interactions with the magnetospheric plasma (Afonso et al., 1986), all contribute in a significant way. Each of these phenomena is now understood in a qualitative sense, but to model all these effects to the accuracy needed to compute the orbit of *LAGEOS* over a long arc (e.g. one month) with an accuracy of few *cm* is hopeless. A realistic estimate of the non-modelled and non-modellable effects could be of the order of $10^{-10}\ cm/s^2$, with an along track effect of the order of $1\ cm/day^2$. The problems resulting from the difficulties in modelling the dynamical effect of tides, in particular the oceanic component, are similar.

The above quoted figures easily explain the difficulties in determining the station positions with *cm* accuracies, as well as in fitting the observations with residuals below the 10 *cm* r.m.s. level, when long arcs are used. However this still leaves open the possibility of using *medium* arcs, by which we mean arcs longer than one orbit but shorter than about a week.

The main objection to the use of medium arcs is that they are strongly affected by reference system problems. Very little is known about the motion of the angular velocity vector of the Earth over timescales of a few days. Some encouraging progresses have been done precisely by using SLR data, but the changes in the position of the Earth's rotation axis with periods close to one day are not really measured (however, see Caporali et al., these proceedings). This point is sometimes hidden in the technical astronomical jargon, in which the Earth's pole is not defined as the intersection of the Earth surface with the instantaneous rotation axis, as one would expect from elementary mechanics (see Mueller, these proceedings). The uncertainties in the position of the rotation axis are still well above the *cm* level, and over a medium arc the effect in the observational residuals in range is of the order of the displacement of the intersection of the axis with the Earth surface. This theoretical analysis is confirmed by the practical experience done with *LAGEOS* in the last few years; many authors (e.g. Tapley, these proceedings) agree that the only way to improve the determination of the pole and to reduce the size of the residuals is to use arcs of few days, by computing for each one a correction to the initial conditions and one or more pole positions.

The effect of the uncertainties of the geopotential coefficients is relevant and is of course discussed in many other places in these proceedings (see Wagner, Reigber, Rapp). We would only point out that whenever an arc longer than one revolution is used, the satellite position depends upon the value of the gravitational anomalies along all its path on the sphere with radius equal to the semimajor axis. Thus it is not possible to solve for the gravitational field over a given region, along with the station positions of a local geodetic network; only global solutions for the geopotential field are meaningful. While the method of global long and medium arc solutions, with all the positions of the observing stations and all the geopotential coefficients up to a given degree as solve for parameters, has given very important results especially in the geopotential solutions, it has not proven to be capable of achieving an accuracy of the order of 1 *cm*.

We think that these simple arguments lead in an forceful way to the need for a *short* arc technique, in which the orbit can be modelled without any systematic error larger than a few *cm* because the orbit propagation time is so short that the perturbing effects of the most complicated interactions are negligible. The effect of a perturbing acceleration F upon a short arc of a nearly circular orbit is given by the simple formula:

$$\Delta x \simeq -\frac{3}{2} F_T t^2 \tag{2.1}$$

where F_T is the tangential component of F. E.g. if an arc coincides with a single passage of the satellite over the local geodetic network –e.g. the European network of SLR stations– to be solved for, all the perturbations whose instantaneous values are smaller than $10^{-7}\ cm/s^2$ can be entirely neglected. Among the non–gravitational perturbations, this leaves only the main component of the radiation pressure, due to the direct sunlight, to be modelled, while many others, such as the Earth reflected radiation pressure and all the possible thermal emission effects, are negligible. Direct radiation pressure for a spherical satellite is easy to model; the equivalent reflectivity coefficient of *LAGEOS* is already known with an accuracy better than 10%, and is anyway a single solve for parameter (plus maybe a drift coefficient). Tidal models have to be used, but an accuracy of between 5 and 10% should be feasible.

The main problem to be solved before such a short arc technique can be used in a reliable way is to show that it is not subject to the same limitations of the medium arc technique; namely, that the present inaccuracy in the knowledge of the pole position and of the geopotential coefficients either does not matter –i.e. introduces systematic errors smaller than the observational noise– or can be improved by means of the observations contained in the short arc data themselves. In this paper we discuss the problem about the pole both by simulations and by means of a theoretical analysis of the rank deficiencies and we find that the position of the pole is known well enough.

The problem about the geopotential is more difficult. From our simulations, as well as from simple order of magnitude calculations, it is possible to conclude that the data of a large enough number of very short arcs observed by a local network contain some informations on the gravitational anomalies on the observed spherical cap (which is the portion of the sphere with radius equal to the semimajor axis of the *LAGEOS* satellite visible from the stations of the network). However the amount of information appears to be marginal with respect to the present knowledge of the low harmonics ($\ell \leq 5$); this is not surprising, since the SLR data about *LAGEOS* have a preponderant weight in the current solutions for the low harmonics. Moreover it is not possible to fit a set of parameters which describe the field over a complete sphere –such as the spherical harmonics coefficients up to a given degree– on the basis of local data. A different set of parameters better suited to a local representation of the gravitational anomalies should be used; we do not know yet how many of them. We have conjectured (Carpino et al., 1986) that the number of parameters needed to represent the gravitational anomalies over a cap is proportional to the ratio of the cap area to the area of the complete sphere; e.g., if to represent the gravitational anomalies to the 10^{-7} cm/s^2 level on the entire sphere we would use the harmonics up to $\ell = 10$, that is 121 parameters, to represent the anomalies to the same level of accuracy on an observed cap 12 times smaller in area we should need only about 10 parameters. Unfortunately not only this conjecture has not been proven in theory, but in practice we do not know which base functions we should use to represent the field on the cap with a reasonable number of parameters. This appears as a negative conclusion, but to some it might appear as a positive feature of the problem, since this shows that the SLR technique poses a challenge belonging to the class of *integrated geodesy*. In the latter the purely geodetic problem of station positioning cannot be separated and solved for independently from the problem of a better representation of the gravitational field.

3. Symmetries and rank deficiency

Let us suppose we want to adjust the parameters of a model to fit a set of observation r_i made at time t_i, characterized by a variance-covariance matrix Γ_ξ. In the following, with *observations* we mean already preprocessed data, so that an estimation of Γ_ξ is available from the preprocessing. This estimation is supposed to describe correctly the stochastic properties of the observations. Let μ be the set of the parameters of the dynamical model (i.e. those appearing in the right hand side of the equation of motion) and $\mathbf{x}_0$ the vector whose six components are the initial state of the satellite in some reference system. Let then $\mathbf{x} = \mathbf{x}(\mathbf{x}_0, \mu, t)$ be the state of the satellite at time t as computed with a given orbital model.

The observed quantity r is a known function of the position and velocity of the satellite, of the coordinates ν of the stations in the observing net, and of the time, namely $r = r(\mathbf{x}(\mathbf{x}_0, \mu, t), \nu, t)$; for each observation we can compute the residual $\xi_i = r_i - r(\mathbf{x}(\mathbf{x}_0, \mu, t_i), \nu, t_i)$. By the weighted least square method we must adjust the parameters of the model so that the quantity $Q = \xi^t W \xi$ is minimized; the positive–definite weight matrix W is chosen in a way consistent with the stochastic properties of the observations, i.e. $W = (\Gamma_\xi)^{-1}$.

The derivatives of Q with respect to the quantities we want to fit must be equal to zero; since W is symmetric we obtain the following equation:

$$B^t W \xi = \underline{0} \tag{3.1}$$

where $\underline{0}$ is the null vector. If $p = [\mathbf{x}_0, \mu, \nu]$ is the vector whose components are the quantities to be solved for, $B = -\partial \xi / \partial p = +\partial r / \partial p$.

To solve equation (3.1) we need to use the Newton iterative method. If Δ is the vector whose components are the *differential corrections* –to be added to $[\mathbf{x}_0, \mu, \nu]$ in order to obtain a better estimation of the solve for parameters– we have:

$$\left[\frac{\partial}{\partial p}\left(B^t W \xi\right)\right] \Delta + B^t W \xi = \underline{0}.$$

If the metod converges, ξ becomes small, but its derivative can be regarded as constant, so that we can neglect the term containing the second derivative and rewrite the previous equation as:

$$[B^t W B]\Delta - B^t W \xi = \underline{0} \tag{3.2}$$

which is called *normal system*. The solution of the (3.2) is obtained by inverting the normal matrix, the vector Δ is added to the parameters and the process is iterated by recomputing the residuals. In the following we will represent the normal system in a more compact way as $J\Delta - D = \underline{0}$ where we have put $J = B^t W B$ and $D = B^t W \xi$.

A property of the normal system is that $J^{-1} = \Gamma$, the variance-covariance matrix of the solve for parameters. This can be proven by performing the average on the stochastic space and neglecting the nonlinear terms, that is by assuming that the relationship (3.2) defines the correspondence between the normal distributions in the space of the solve–for parameters and in the observation space.

To optimize the determination of a set of parameters by means of a least square fit we should be sure that the quantities we are looking for are truly independent degrees of freedom. From an algebraic point of view, we should avoid that the design matrix B takes its null value on a (non trivial) subspace of the parameter space. Avoiding rank deficiency in B is of fundamental importance to obtain good results while analysing short arcs. What happens is that –given a set of solve for parameters– some can be determined with relative formal accuracy much worse than others. This fact can be seen from two points of view: we can say that some parameter cannot be well determined or, on the other hand, that we do not need to determine it, provided we know its value with an accuracy better or equal to the formal error of the fit to the observations (see Section 5).

To cope with the presence of various effects of this type and to distinguish them from effective lack of information in the observational data, we developed a mathematical formalism which helps to understand the symmetries that cause degeneracy and to take the proper action to avoid them.

To understand the problem we shall start by giving a geometrical interpretation of the least square fit. Let us suppose we are analizing S observations that we shall consider as the components of an S-dimensional vector r_0 belonging to the space $\mathcal{O}$. We can recognize two part in r_0: the true value $\bar{r}$ and the observational noise ε, so that $r_0 = \bar{r} + \varepsilon$. The satellite orbit is perturbed as a result of a large number of different physical phenomena (see Table 2.1 in Milani et al., 1987) which can be described –by a finite number of parameters– with an accuracy comparable to that of the observations. Thus the true observations $\bar{r}$ cannot be anywhere in $\mathcal{O}$, but they will be close to a submanifold $\mathcal{R}$, whose dimension is $M \leq S$. Because of the noise, and also because of the smaller effects neglected in the model, in general r_0 will not belong to this manifold. To obtain the fit of the observations we should bring r_0 in $\mathcal{R}$. Since we do not know $\bar{r}$, the best thing we can do is to project r_0 onto $\mathcal{R}$, and find an estimation $\hat{r}$ of the true value of the observable. The projection is performed by minimizing the residuals, i.e. by choosing $\hat{r}$ as the point of $\mathcal{R}$ closest to r_0. The fact that we use the weighted least square fit points out that the metric in the space of the observables is given by the weight matrix. The norm of a vector $r \in \mathcal{O}$ is consequently defined as

$$|r| = \sqrt{\frac{r^t W r}{S}}.$$

Rank deficiency problems arise when we parametrize $\mathcal{R}$ to find the best fit. Since we do not know exactly all the accelerations that act on the satellite, we can reconstruct $\mathcal{R}$ only up to the accuracies of the observational and orbital models; the choice of these is not easy. From here on, we shall suppose that $\mathcal{R}$ is the submanifold into which we can find $\bar{r}$, as a result of a given choice of the model and of the parameter space $\mathcal{P}$. We also suppose that the model is accurate to a level lower than the observational noise, thus the difference between $\bar{r}$ and $\hat{r}$ is mostly due to the observational noise.

With this definition of $\mathcal{R}$, if N_p is the dimension of the parameter space $\mathcal{P}$, we shall have *exact* rank deficiency if $N_p > \dim \mathcal{R}$. Besides this, however, we can have another form of degeneracy, the *approximate* rank deficiency, which is much more difficult to recognize and to deal with than its exact counterpart. In the simulation we made, we found that the approximated degeneracies are one of the most important sources of bad convergence of the differential corrections if short arcs are considered (Section 5).

We shall begin by recalling two definitions of rank deficiency given for the exact degeneracy. We shall then try to extend them to the approximate case.

The computation of predicted observations from an assumed value of the model parameters can be abstractly represented by a function $F : \mathcal{P} \to \mathcal{R}$ so that $r = F(p)$. In general $\mathcal{R}$ is not a linear manifold and F is not a linear function of the parameters. We linearize F by considering the tangent plane to $\mathcal{R}$ at the solution $\hat{r}$ and the tangent plane to $\mathcal{P}$ at the solution $\hat{p}$:

$$\begin{gathered} r = \hat{r} + B(\hat{p})(p - \hat{p}) + O((p-\hat{p})^2)\ , \quad B(p) = \left.\frac{\partial F(z)}{\partial z}\right|_{z=p} \\ \xi = r_0 - F(p). \end{gathered} \tag{3.3}$$

First we can give an algebraic definition (Betti and Sansò, 1983): the rank deficiency r_a is the dimension of the null space of the normal matrix $J = B^t W B$ evaluated at the solution $\hat{p}$, i.e.

$$r_a = N_p - \operatorname{rank} J\ ; \tag{3.4}$$

r_a is the multiplicity of the null eigenvalues of J: if $r_a \neq 0$ it is impossible to invert J to find the differential corrections. We can give a straightforward geometrical interpretation of the fact that J is not invertible. In the parameter space the *fibers* of F are the equivalence classes $\Phi(r) \equiv \{p \in \mathcal{P} | F(p) = r\}$. Since rank J = rank B, by the implicit function theorem their dimension at $\hat{p}$ is the rank deficiency of the problem.

By the algebraic definition, to evaluate the rank deficiency of a particular problem we should simply find the null eigenvalues of the J matrix. It however does not give any hint on how the degeneracy arises, nor how to avoid it. In order to find some help in this direction, we can give a definition of rank deficiency using symmetries. Let H be a Lie group of transformations $g : \mathcal{P} \to \mathcal{P}$ which has the property of being transitive, i.e. every point of $\mathcal{P}$ can be moved into any other point of $\mathcal{P}$ by the action of some element $g \in H$. We shall also suppose $\dim H < \infty$. Next consider two subgroups G and $G_{\hat{p}}$, such that $G_{\hat{p}} \subset G \subset H$. G is the biggest group of symmetries that can be found in H:

$$G \equiv \{g \in H | F(g(p)) = F(p)\}. \tag{3.5}$$

$G_{\hat{p}}$ is the isotropy subgroup of G in $\hat{p}$, the solution. The isotropy subgroup in a point x is defined as:

$$G_x \equiv \{g \in G | g(x) = x\}. \tag{3.6}$$

With this notation we can define the rank deficiency as:

$$r_s = \dim G - \dim G_{\hat{p}} \tag{3.7}$$

i.e. r_s is the dimension of the quotient space $G/G_{\hat{p}}$, which is also the dimension of the orbit of a point under the effect of the group of symmetries. To understand this definition we remark that all the transformations in $G_{\hat{p}}$ do not change the solution of the problem, so that they can be considered uneffective for our purpose. By definition, the orbit of r is contained in the fiber $\Phi(r)$ defined above. Thanks to the hypothesis of transitivity of the transformations of H, every point in the fiber can be reached by means of some symmetry in G and the orbit coincides with the fiber, thus $r_a = r_g$.

As an example, we can show that were the Earth perfectly spherically symmetric the rank deficiency would be 3 for a geodetic network determined by observing N arcs with SLR (of course we neglect lunisolar perturbations and radiation pressure). The solve for variables would then be

$$p = [x_0, v_0, \omega_\oplus, \nu] \tag{3.8}$$

with x_0, v_0 the vectors of length $3N$ containing the initial positions and velocities of each arc, $\omega_\oplus$ the vector of length $3N$ containing the angular velocity of the Earth during each arc, and ν the $3N_s$ station coordinates. The existence of a group of symmetries by rotation, that is $SO(3)$ acting upon p diagonally (i.e. each 3–vector is rotated at once), is obvious and its dimension 3 corresponds to the classic rank deficiency of ground–based geodetic networks (in this case, the isotropy subgroup is trivial). However, to show that the rank deficiency is not more than 3 we have to show that the diagonal action of $SO(3)$ is the largest such symmetry group inside some larger group H acting on $\mathcal{P}$ in a transitive way. The choice of H is largely arbitrary; we can e.g. chose H as the full group of affine transformations acting on $\mathcal{P}$.

To simplify the proof, we can make the further assumption that each arc is observed at the initial conditions, and often enough close to the initial conditions times to allow for a computation of the first and second derivative of the distance at the initial time; this corresponds to the usual procedure of fitting the observations to some polynomial function of the time. Then the argument could go as follows: to keep the observed distance at the initial times of each arc, the set of vectors $[x_0, \nu]$ can only move in a rigid way. However translations would change the gravitational acceleration acting upon the satellite at the initial conditions, hence would change the second derivative of the distance. In the same way, v_0 must be rotated exactly as $[x_0, \nu]$ not to change the first derivative of the distance. $\omega_\oplus$ must also be rotated in the same way not to change the centrifugal and Coriolis force at the initial time, since the latter would also change the second derivative of the distance. If the Earth is not spherical, but only axially symetric, and rotates around the symmetry axis, the rank deficiency is only 1; it is also 1 if the angular velocity vector is fixed, even for a spherical Earth. These two forms of symmetry breaking are further discussed below.

To deal with exact degeneracies is rather easy, because their effect are so sharp that the underlying symmetries can be detected with little efforts. A more subtle problem arises because of *approximated* degeneracies, for which even the definition is troublesome. In order to cast some light on the matter we tried to obtain useful definitions by modifying the ones given for the exact degeneracy.

A qualitative geometrical description of the situation can be easily given for the approximated degeneracy as well. If the random errors in the observations follow the normal distribution $N(0, \sigma_\varepsilon)$, the ellipsoid of error of the estimation $\hat{r}$ will be the M-dimensional sphere of radius σ_ε. Approximated rank deficiency arises when this sphere is mapped, by the parametrization, into a very flattened ellipsoid.

To define the exact rank deficiencv we never used the metric on $\mathcal{R}$ nor we needed any metric on $\mathcal{P}$. On the contrary, metrics will be a crucial point in the definition of approximate degeneracies. We shall therefore endow $\mathcal{P}$ with a suitable metric. If we know previous estimations of the parameters, along with their covariance matrix, a natural choice for the metric W_p of the space of the parameters is the inverse of some known covariance matrix. Thus we can define the norm in $\mathcal{P}$ as:

$$|p| = \sqrt{\frac{p^t W_p p}{N_p}} \tag{3.9}$$

However this is not the only interpretation of the metric W_p; it can express the goal of the determination. When the error in the determination of p is large according to the metric (3.9), if W_p represents the previous determination this means that the information contained in the observations used for the fit does not allow to improve the previous knowledge. If on the contrary we choose W_p in such a way that a norm 1 of the error corresponds to a significant improvement, a large norm of the errors means that such an ambitious goal has not been achieved. As an example, if we chose W_p in such a way that the station positions are measured in cm, the norm (3.9) measures the feasibility of the ambitious goal of 1 cm accuracy station positioning.

We shall suppose we made all the required base transformations to have W_p diagonal, and all the necessary changes of scale so that the metric is the identity matrix. We shall apply the same kind of transformations to the observation space as well, so that both in $\mathcal{R}$ and in $\mathcal{P}$ we can deal with isotropic euclidean metrics.

As previously we begin with the algebraic definition. We define the algebraic rank deficiency approximated by a factor k, as the number of eigenvalues of the normal matrix $J(\hat{p})$ that are *sufficiently small.* If $\lambda_1, \cdots, \lambda_{N_p}$ are the eigenvalues of J we have:

$$R_a(k) = \#\{\lambda_i | \lambda_i \leq \frac{1}{k^2}\} \tag{3.10}$$

where $\#$ is the counting operator. Since rank (J) is the number of $\lambda_i \neq 0$, it is easy to see that, for $k \to +\infty$, $R_a(k) \to r_a$ and that, for any value of k, $R_a(k) \geq r_a$. Remembering that J^{-1} is an estimation of the covariance matrix of $\hat{p}$, the smallest eigenvalues of J are the inverses of the squares of the longest axes in the ellipsoid of error of $\hat{p}$. Thus the above definition simply means that some parameters –or combinations of parameters– are determined with a formal accuracy which is worse by a factor k with respect to our expectations and/or our previous knowledge.

Of course we can always eliminate this kind of degeneracy by rescaling the value of the parameters, since the metric we have chosen is somewhat arbitrary. This, however, does not mean that we can eliminate the asymmetries in the achievable accuracies of the determinations of different parameters.

Let us give, as in the exact case, the geometrical interpretation of the above definition. Let us consider in the space $\mathcal{P}$ of the parameters the flattened ellipsoid of error corresponding to the sphere of radius $1/k$ in $\mathcal{R}$. As a limit for $k \to \infty$ we obtain the fiber $\Phi(\hat{r})$. The rank deficiency is the number of the longest axes of the ellipsoid. In the exact case, this number is the dimension of the fiber itself, since there are only infinitely long axes. Thus we can define the subspace containing the longest axes:

$$\Phi_k(\hat{r}) \equiv \{q \in \mathcal{P} | \ |\frac{d}{d\varepsilon}[F(\hat{p} + \varepsilon q) - \hat{r}]\big|_{\varepsilon=0}| \leq \frac{|q|}{k}\} \tag{3.11}$$

(with $\hat{r} = F(\hat{p})$) as the analogue of the fiber in the approximate case. In the exact degeneracy case the fiber was a manifold in the $\mathcal{P}$ space; here, on the contrary, we restrict ourselves to the tangent plane in $\hat{p}$ and we consider the subspace generated in it by the longest axes of the ellipsoid. We could have used the tangent space in the exact definition as well.

Let us give the definition of rank deficency –in the case of approximated degeneracy– in terms of approximate symmetries. As in the exact case we consider a Lie group H whose action is transitive on the space of the parameters and whose dimension is finite. Every element of H belongs to some one parameter subgroup $g_\varepsilon = \exp(\varepsilon\gamma)$, where γ is an infinitesimal transformation in the Lie algebra TH of H. We then consider $G_{\hat{p}} \subset G \subset H$, where G is the biggest group of *approximate* symmetries one can find in H and $G_{\hat{p}}$ is the isotropy subgroup *of approximation* k of G in the solution vector $\hat{p}$. In terms of Lie algebras TG and $TG_{\hat{p}}$ we have:

$$TG \equiv \{\gamma \in TH | \ |\frac{d}{d\varepsilon}[F(g_\varepsilon(\hat{p})) - F(\hat{p})]\big|_{\varepsilon=0}| \leq \frac{\|\gamma\|}{k}\} \tag{3.12}$$

$$TG_0 \equiv \{\gamma \in TG| \ |\frac{d}{d\varepsilon}[g_\varepsilon(\hat{p}) - \hat{p}]|_{\varepsilon=0}| \leq \frac{\|\gamma\|}{k}\}. \tag{3.13}$$

Since γ is an infinitesimal transformation, its action on $\mathcal{P}$ is a vector field; the norm $\|\gamma\|$ is the maximum size of this vector in the portion of $\mathcal{P}$ we are considering.

The meaning of those definitions is as follows: g is an approximated simmetry if we must change the parameters of a *big* amount to obtain a sensible variation in the position of r in $\mathcal{R}$. In the same way, a transformation is an approximate isotropy in $\hat{p}$ if the solution is displaced by a small amount with respect to the metric adopted on TH. As in the exact case, the rank deficiency defined by simmetry is:

$$R_s(k) = \dim G - \dim G_0$$

We can see that $R_s(k)$ also depends upon the metric adopted on TH.

The two definitions of rank deficiency given for the approximate case are related, in the sense that $R_s(k) = R_a(k')$ with k' some monotone function of k; however we do not know in an explicit way the functional relationship between k and k', that is we are not yet able to predict the eigenvalues of the variance–covariance matrix of the result on the basis of approximate symmetry computations. Nevertheless these two definitions can be used in a qualitative way: whenever an approximate rank deficiency is found in the simulations, an approximate symmetry can be found on the basis of simple geometrical arguments, and this indicates the procedure to be followed to remove the degeneracy, e.g. indicates which parameters have to be removed from the list of the solve for variables. To this purpose it is not necessary to find the exact relationship between k and k', because this requires only an order of magnitude estimate.

Let us illustrate this procedure by the main example of the approximate degeneracy of a multi–arc short arc analysis. Let us assume the solve for parameters are $[x_0, v_0, \omega_\oplus, \nu]$, i.e. only the initial conditions of each observed arc, the angular velocity of the Earth and the station coordinates. The Earth's gravity coefficients (in a given reference frame) are assigned.

If the Earth was axially symmetrical, there would be an exact rank deficiency of 1; since there are tesseral harmonics, the largest being J_{22}, a rotation by a small angle θ changes the acceleration acted by this harmonic upon the satellite by a fraction 2θ of its value, while moving the stations by $\theta R_\oplus$. Since the acceleration on *LAGEOS* due to J_{22} is 6×10^{-4} cm/s^2, the effect after $T/2 \simeq 1500$ s is $\simeq \theta \times 4000$ cm, that is smaller than $\theta R_\oplus$ by a factor $\simeq 1.5 \times 10^5$. Moreover the standard deviation of the residuals due to this effect is smaller by a factor $\sqrt{5}$ than the peak value. Thus we expect $R_a(3 \times 10^5) = 1$, which is confirmed by the computation of the eigenvalues of the Γ matrix in numerical simulations (see Section 5 and Table 1e).

The Earth oblateness has much larger effects, the acceleration on *LAGEOS* being of 0.1 cm/s^2; by the same computation done above, the effect of a rotation around an axis perpendicular to the polar axis on the J_2 perturbation is smaller than the displacement of the stations by a factor $\simeq 2,000$. However, if the rotation axis is held fixed the effect, of such a rotation on the orbit is larger, because the apparent forces due to the rotation of the Earth turn out to be more sensitive, giving a k factor of the order of 10 (see Section 5); thus we expect $R_a(10) \geq 3$. The k factors found in the numerical simulations (Table 1e) turn out to be somewhat larger. This can be due to the fact that finding a symmetry with a given k factor does not prove that there is not a better symmetry with a larger k which we have been unable to find; e.g. it is possible that by choosing a suitable rotation axis the effects of the J_2, J_{22} and the apparent forces cancel out somewhat.

So far the discussion of approximate symmetries has not provided any very important new result, since the existence of approximate degeneracies –corresponding to approximate symmetries which replace the exact ones when the shape of the Earth is taken into account– is intuitively obvious. However in our simulations we find, for 3 stations determined by 6 short arcs, $R_a(k) = 4$ with k as high as 28 (Table 1e). Where comes the additional approximate symmetry from? The question is far from being of purely mathematical interest. Unless we find which is the motion resulting in the approximate symmetry, we do not know how to remove the degeneracy. E.g. if the rank deficiency is 3 and results only from rotations around the center of the Earth, the classical method of fixing the latitude and longitude of one station and the longitude of another one effectively eliminates the degeneracy, allowing for a good determination. $R_a(28) = 4$ means that this classical method does not work.

When simulations with the classical method of fixing three station coordinates are performed, both the formal r.m.s. errors and the actual errors with respect to the *true* positions show that the whole network can move *up* or *down* in an approximately vertical direction by as much as 25 cm without any significant increase in the residuals above the 1 cm noise level. Of course when this happens the initial conditions of each arc are also moved in the same direction. We have to conclude from this that the group of approximate symmetries with approximation $k \simeq 28$ is the whole group of rigid motions of the stations and the initial positions. The reason for the existence of only 4 large axes of the ellipsoid of error, that is $R_a(28) = 4$, is that there is an approximate isotropy group of dimension 2 because the translations of the geodetic network along an horizontal direction are not very different from a rotation, the network being small with respect to the radius of the Earth.

However this approximate description of the symmetry is not enough to understand how it really works. It needs to be pointed out that the action of the 6–dimensional group of rigid motions upon the initial velocities of each arc is not simply by the action of the rotation component. If an orbit was simply displaced upward, without changing the modulus of the velocity, this would result in a change in the semimajor axis and of the mean motion which would result in residuals of the same order of the displacement after $\simeq 1,000$ s. Thus the initial velocities must also be adjusted to allow for a good fit to the observation from the displaced network. How this happens is far from being clear; nevertheless the discussion of this Section is enough to indicate what has to be done to remove the approximate degeneracy and to allow for a good determination of the local geodetic network. A fourth station coordinate has to be fixed, in such a way that the displacement of the network in the vertical direction is not possible; e.g. we can fix the three coordinates of one station, including the radius from the center of the Earth, and one coordinate –e.g. the longitude– of another station. This procedure works very well and actually allows to solve for the remaining station coordinates with accuracies at the *cm* level, as discussed in the next Section. The weak point of this procedure is that the absolute height of the network –e.g. with respect to the reference ellipsoid– cannot be found. On the other hand the fact that the determination of the height by satellite geodesy is often troublesome is not new, but occurs in a variety of different tracking methods and orbital geometries.

4. The multi–arc algorithm

In Section 3 the least square fit problem and its solution by means of the Newton iterative method has been outlined in the general case. In the multi–arc approach all the observations taken during one passage of the satellite over the local geodetic net are processed as if they were independent from those taken during the other passages. Many arcs are then merged so that the parameters of the model can be determined.

The set of parameters can be divided in a natural way into two subsets. The former, that we shall call *global*, is made up by all those parameters that do not change significantly while considering different arcs. On the contrary, the *local* set contains all those quantities that are typical of a single short arc, and that may change sharply from one arc to the other. For short arcs and a satellite like *LAGEOS* the two sets of parameters are as follows.

global parameters: the coordinate of the tracking stations, the parameters modelling the gravitational field, the dynamical effect of the tides and the parameters for the radiation pressure perturbation.

local parameters: the initial conditions of each passage, the parameters modelling the angular velocity vector of the Earth.

In the following we shall call $l = [l^{(1)}, \cdots, l^{(N)}]$ the vector whose components are the local parameters $l^{(i)}$, $i = 1, \cdots, N$ of the processed arcs, ordered by arc, and g the vector containing the global parameters.

As we consider only the orbital arc visible from the observing stations of the geodetic net and we determine new initial conditions for each passage, the solutions we find for the global parameters are not to be extended outside the observed cap and this is expecially true for the gravitational field. Since for each arc we determine initial conditions even if the passages are consecutive, each arc will be locally very well determined, but it is possible that two successive passages do not connect in a smooth way. This because we are overparametrizing the problem: two consecutive passages of the same satellite do not have two completely independent sets of initial conditions. On the other hand, the use of separate initial conditions allows us to neglect all those sources of systematic errors that prevents the methods that analyse medium and long arcs from reducing the residuals at the level of the observational noise. Moreover, the amount of information extracted from a passage is not in any way reduced when the passage belongs to another satellite; this could become a significant advantage in a near future, when more than one *LAGEOS* will be available.

As we suppose observations made during different passages to be independent the derivatives of the residuals of the observations of one passage with respect to the local parameters of another passage vanish. So the matrix of the partial derivatives $B = -\partial\xi/\partial p$ written for a batch of N passages has the block structure shown below (here and in the following matrices rappresentations we shall suppose the number of arcs to be $N = 3$):

$$B = \begin{bmatrix} B_{l_1 g} & B_{l_1 l_1} & \underline{\underline{0}} & \underline{\underline{0}} \\ B_{l_2 g} & \underline{\underline{0}} & B_{l_2 l_2} & \underline{\underline{0}} \\ B_{l_3 g} & \underline{\underline{0}} & \underline{\underline{0}} & B_{l_3 l_3} \end{bmatrix}$$

where the $B_{..}$ blocks are:

$$B_{l_i g} = -\frac{\partial \xi^{(i)}}{\partial g} \qquad B_{l_i l_i} = -\frac{\partial \xi^{(i)}}{\partial l^{(i)}}$$

with $\xi^{(i)}$ the vector of the residuals of the i-th passage.

As we supposed that each passage is processed independently to produce the normal points, the Γ_ξ and the W matrices are block diagonal. Thanks to that, when the matrix product WB is performed in order to evaluate J, the blocks do not mix together, and the shape of WB is the same as B, and by performing the triple matrix product, we find that $J = B^t W B$ is an *arrow matrix* of the form:

$$J = \begin{bmatrix} J_{gg} & J_{gl_1} & J_{gl_2} & J_{gl_3} \\ J_{l_1 g} & J_{l_1 l_1} & \underline{\underline{0}} & \underline{\underline{0}} \\ J_{l_2 g} & \underline{\underline{0}} & J_{l_2 l_2} & \underline{\underline{0}} \\ J_{l_3 g} & \underline{\underline{0}} & \underline{\underline{0}} & J_{l_3 l_3} \end{bmatrix}$$

where:

$$\begin{aligned} J_{gg} &= \sum_{i=1}^{N} (B_{l_i g})^t W_i B_{l_i i g} \\ J_{gl_i} &= (B_{l_i g})^t W_i B_{l_i l_i} \\ J_{l_i g} &= (J_{gl_i})^t \\ J_{l_i l_i} &= (B_{l_i l_i})^t W_i B_{l_i l_i}. \end{aligned} \tag{4.1}$$

Here W_i is the diagonal block of W which refers to the i-th passage.

The J_{gg} block is constructed by adding the contribution of *all* the arcs. This, however, is not true for the remaining blocks, for which only one passage is relevant. This allows important simplifications in the algorithm for the inversion of J. J is an arrow matrix, but its inverse Γ, the variance-covariance matrix of the determined parameters, does not have the same structure. So even if we supposed that the observations were independent, this is not the case of the parameters that model the single arcs, which appear to be correlated *a posteriori* both to the global parameters and to the local parameters of the other arcs. This can be understood by a simple example: if there is a symmetry such that the same observations could be obtained by moving all the stations and all the initial conditions along some coordinate axis, the components of the initial conditions along that axis have correlation 1 if the stations positions are determined. Indeed we performed several simulation which showed that we can by no means neglect the cross correlations between different arcs, if we want to keep the method rapidly convergent.

To solve the normal system by taking advantage of the arrow structure of J let us rewrite the equation (2.2) by splitting the blocks (Monti and Sansò 1983, Fukushima 1987):

$$\begin{cases} J_{gg}\Delta_g + \sum_{i=1}^{N} J_{gl_i}\Delta_{l_i} - D_g = \underline{0} \\ J_{l_ig}\Delta_g + J_{l_il_i}\Delta_{l_i} - D_{l_i} = \underline{0} \qquad i = 1, N \end{cases} \tag{4.2}$$

where:

$$D_g = \sum_{i=1}^{N} B^t_{l_ig} W \xi^{(i)} \qquad D_{l_i} = B^t_{l_il_i} W \xi^{(i)},$$

Δ_g is the vector of differential corrections for the global parameters, while Δ_{l_i} contains the differential corrections for the local parameters of the ith arc.

Equation (4.2) is the normal system (3.2) explicitly written for the multi-arc algorithm. To obtain its solution we first solve for Δ_{l_i}, $i = 1, \cdots N$ in the last N equations and we subsitute those vectors in the first of the (4.2), so that:

$$\begin{aligned} \Delta_g &= \left[J_{gg} - \sum_{i=1}^{N} J_{gl_i} J^{-1}_{l_il_i} J_{l_ig} \right]^{-1} \left[D_g - \sum_{i=1}^{N} J_{gl_i} J^{-1}_{l_il_i} D_{l_i} \right] \\ \Delta_{l_i} &= J^{-1}_{l_il_i} D_{l_i} - J^{-1}_{l_il_i} J_{l_ig}\Delta_g \qquad i = 1, N. \end{aligned} \tag{4.3}$$

To evaluate the corrections vector for the local parameters of the i-th passage we need the value of the correction Δ_g; the latter is known only after all the $J_{l_il_i}$ have been evaluated. Therefore, even if we split the set of solve for parameters in subsets, the solution can be obtained only for the whole set and having considered all the arcs of the batch. However, as the summation over the arcs extends to all the terms in the first of the (4.3), we can add the contribution to Δ_g one passage at a time.

To compute the solution (4.3) we have to invert only small symmetric matrices, whose dimensions are those of the subsets into which we divided the parameters. To optimize these matrix inversions we perform a *normalization* of the matrices to be inverted. This prevents the conditioning index from growing because of scale differences in the system of units adopted for the parameters. If A is the normalizing matrix for one block J_i, the nomalized matrix J_i^n is given by $J_i^n = AJ_iA$. A is a diagonal matrix whose elements are the the root mean square of the part of the column of B corresponding to the considered passage and parameter.

Since $\Delta = \Gamma D$, if we express the equations (4.3) as some matrix multiplying the vector D and if we put :

$$\tilde{J}_{gg} = J_{gg} - \sum_{i=1}^{N} J_{gl_i} J^{-1}_{l_il_i} J_{l_ig}$$

we obtain the following expressions for the blocks of the variance-covariance matrix of the solve for parameters:

$$\begin{aligned} \Gamma_{gg} &= \tilde{J}_{gg}^{-1} \\ \Gamma_{gl_i} &= -\tilde{J}_{gg}^{-1} J_{g_l i} J_{l_i l_i}^{-1} \\ \Gamma_{l_i l_j} &= J_{l_i l_i}^{-1} J_{l_i g} \tilde{J}_{gg}^{-1} J_{g l_j} J_{l_j l_j}^{-1} \qquad \text{for } i \neq j \\ \Gamma_{l_i l_i} &= J_{l_i l_i}^{-1} J_{l_i g} \tilde{J}_{gg}^{-1} J_{g l_i} J_{l_i l_i}^{-1} + J_{l_i l_i}^{-1}. \end{aligned} \tag{4.4}$$

Elsewhere in these proceedings (see, for example, Reigber) there are illustrations of the use of the differential correction method when the goal is a global solution, obtained from many arcs of different satellites and with observations of different nature. In this case the first step is to improve the orbit by itself (i.e. determining only the local parameters) by performing several iterations, while the last correction combines information from many arcs to determine the global parameters.

For a short arc multi–arc technique, the presence of high cross–correlations suggests that the best way to process information is to iterate the simultaneous determination of global and local parameters. In a global solution both the number of unknowns and the number of arcs are so large that it would be very difficult to perform more than one global iteration. On the contrary, in the case of a short arc analysis as outlined here, it turns out that it is possible to iterate the determination of both global and local corrections by using reasonable computer resources.

To improve the determination *a priori* observations of the parameters are used (the same procedure is often called *collocation*). If the value of some parameter is known, the fit must be constrained in such a way that the change is not bigger than the accuracy of the previous determination. To prevent this we can 'stiffen' the model by introducing fictitious observations in which one of the model parameters is directly observed. A priori observation of the global parameters are specially useful for the geopotential model: in this way we can easily assign the proper degree–variance to each coefficient to be solved for. A priori observations of some station positions, or functions thereof such as the longitudes, can be used to eliminate degeneracies. A priori observation of local parameters can be used to connect two consecutive arcs: in this case quantities such as the semimajor axis, that do not change very much from one passage to the next, could be constrained. The use of a priori knowledge stabilizes the solution and, in the case of the local parameters, it can partially avoid the negative effects of the overparametrization.

Let us suppose we want to make an a priori observation of the k-th global parameter p_{gk} whose known value is $\tilde{p}_{gk} \pm \sigma_k$. To preserve the peculiar shape of B and J we shall suppose this is the first observation. Therefore the line vector $V = [0, \cdots, 1, \cdots, 0]$ must be put as first line of B. The dimension N_p of V is the total number of determined parameters, local and global. The non vanishing component in V corresponds to the column of B that contains the derivatives with respect to the k-th global parameter. We must also add a null line and a null row to W and put $1/\sigma_k^2$ in the new diagonal component. With those changes, the first of the

(4.1) becomes:

$$(J_{gg})_{r,s} = \delta_{r,k}\delta_{s,k}\frac{1}{\sigma_k^2} + \left(\sum_{i=1}^{N} (B_{l_i g})^t W_i B_{l_i ig}\right)_{r,s} .$$

To observe the local parameter $p_{l_j k}$ we should simply put the additional line in B and W at the beginning of the l_j-th block, so that only the fourth of the (4.1) is modified, giving:

$$(J_{l_i l_i})_{r,s} = \delta_{r,k}\delta_{s,k}\frac{1}{\sigma_k^2} + \left((B_{l_i l_i})^t W_i B_{l_i l_i}\right)_{r,s} .$$

Finally, in both cases we must add to ξ a component in the proper position, containing the difference between the value of the parameter and the constraint.

A slightly different procedure must be adopted if we want to observe a priori a quantity that is not explicitly a parameter of the model, but a function of them (e.g. the latitude of one station or a keplerian element of the orbit). Let us suppose we are observing the quantity $q = q(p_{k_1}, \cdots, p_{k_s})$, where the p_{k_i} are some parameters appearing in the model. Thus, for the components of V we can write:

$$(V)_k = \frac{\partial q}{\partial p_k}.$$

$(V)_k$ can be non vanishing if $k \in \{k_i \; i = 1, s\}$. We must then add to J the matrix $(VV^t)/\sigma_q^2$ and modify W and ξ as above.

As the time required to evaluate the differential corrections can be long it is useful to have an efficient automated system to stop the iterations. The residuals ξ can be decomposed into a systematic part and a stochastic part, so that $\xi_i = \bar{\xi}_i + \varepsilon_i$, where ε_i follows the normal distribution $N(0, \sigma_{\xi_i})$. If we suppose our model is correct, i.e. the value of the parameters is the proper one and all the relevant perturbation are taken into account (a perturbation is considered as relevant if its effects over the lenght of the passage are comparable to the observational noise), then $\bar{\xi}_i = 0$. As ε_i contains no information on the solve for parameters, there is no increase in accuracy corresponding to a reduction of the residuals when $\xi \simeq \sigma_{\xi_i}$ already. Therefore, from the definition of Γ, the *correction norm:*

$$R = \sqrt{\frac{\Delta^t \Gamma^{-1} \Delta}{N_p}}$$

is $\simeq 1$ when $\xi \simeq \sigma_{\xi_i}$ and the iterations can be stopped whenever $R \leq 1$.

However, this is not the only quantity to be monitored in order to have a criterion to stop the iterations. Let us suppose there is some degeneracy, so that Q takes its minimum value on a subspace. If the correction Δ belongs to this subspace, R can be much bigger than 1, but Q does not change significantly, as the minimum has already been reached. Therefore also the variations of the target function Q must be kept under control to have a reliable way to detect the end of iterations. Moreover in some cases the Newton method is unstable and the total number of iterations already performed should be checked.

5. Experimental results.

Before analysing real data with a new technique, it is useful to test both the theory and the software by performing simulations. To exploit the method described in Section 2 we developed a software package, called *ORBIT6*, nearly 5000 *FORTRAN 77* statements long. Beside being able to apply differential corrections to a set of global and local parameters, the code also allows to simulate an observation campaign for a geodetic satellite like *LAGEOS*. The results provided in this work have all been obtained by means of this program. Common features of all the simulations are as follows.

The tracking method is the Satellite Laser Ranging (S.L.R.), the geodetic net is made up by the European S.L.R. stations and the tracked satellite is specifically *LAGEOS,* although the results can be applied to any probe of approximately the same area to mass ratio and semimajor axis as the one used. Finally, for any arc used, the initial time (i.e. the time of the initial conditions) is placed in the baricenter of the times of observation.

For the European net and a satellite like *LAGEOS* each passage lasts roughly $T \simeq 3000$ *sec*, i.e. less than one fourth of a complete orbit of the satellite around the Earth.

The first part of the simulation campaign was dedicated to single short arc analysis (Melchioni, 1987). We tryed to find the minimum number of stations that must track one passage of the satellite in order to determine the initial conditions of the orbit with the same accuracy of the observations (normal points). It was found that at least three stations are required, positioned with a favorable geometry with respect to the arc.

To be able to compare the eigenvalues of the covariance matrices and compute the proper k factors we chosed as common metric unit the centimeter. Therefore the positions (i.e. the initial position of the satellite or the stations coordinates) need no transformations, while for the initial velocities we considered the effect at the end of the observed arc, in the hypotesis that the initial conditions were in the center of the arc. To rescale the covariance matrix we multiply both the rows and the columns corresponding to the components of the initial velocity by a factor $T/2$. For the components of the angular velocity of the Earth the rescaling factor is the ratio between the radius of the planet and its nominal angular velocity $\omega_{\oplus}$. That is we are considering the pole position at the surface of the Earth. In the formalism of Section 3, this defines the metric W_p.

By diagonalizing the covariance matrices obtained while determining the initial conditions of an arc tracked by two stations we find that there is approximated rank deficiency (see Table 1, columns a and b). If the ground track is roughly perpendicular to the baseline (Table 1a), the approximated algebraic rank deficiency is 1 and the approximation factor $k \simeq 16$. If the passage is parallel to the baseline(Table 1b), the situation is even worse: there are two eigenvalues much bigger than the others, with k factors of about 20 and 12.

TABLE 1

Eigenvalues of the covariance matrix of the determined parameters. The matrices have been rescaled and the values can be directly compared. The solve for parameters are as follows:

a) Initial conditions of an arc observed by 2 stations. Ground track almost perpendicular to the baseline.
b) Initial conditions of an arc observed by 2 stations. Ground track almost parallel to the baseline.
c) Initial conditions of an arc observed by 3 stations.
d) Initial conditions and angular velocity vector of the Earth for an arc observed by 3 stations.
e) Coordinates of 3 stations tracking 6 arcs. Degeneracy not removed.
f) Coordinates of 3 stations tracking 6 arcs. Degeneracy removed.

a)	b)	c)	d)	e)	f)
0.04	0.04	0.03	0.03	0.06	0.11
0.07	0.07	0.05	0.05	0.09	0.23
0.36	0.37	0.26	0.26	0.38	0.83
2.62	9.08	2.56	2.55	0.99	1.50
8.90	137.4	9.63	9.63	1.80	3.28
263.8	385.5	12.9	12.89	817.1	
			1261.8	1341.3	
			1754.6	2064.6	
			15827.9	1.5×10^{13}	

TABLE 2

Total value of corrections and standard deviations obtained by determining the initial conditions of an arc tracked by 2 stations. Positions in *cm*, velocities in $cm\,sec^{-1}$. Baseline in the $\hat{x}\hat{z}$ plane.

a) arc almost perpendicular to the baseline.

	correction	*r.m.s*
x	−0.711	1.585
y	−13.596	9.124
z	−0.013	0.407
$\dot{x}$	−0.0051	0.0026
$\dot{y}$	−0.0006	0.0006
$\dot{z}$	−0.0142	0.0095

b) arc almost parallel to the baseline.

	correction	*r.m.s*
x	0.482	1.699
y	−4.340	17.091
z	0.287	0.778
$\dot{x}$	0.0009	0.0025
$\dot{y}$	0.0005	0.0079
$\dot{z}$	0.0011	0.0049

Since the initial values of the parameters used in the correction procedure were the *true* ones, the corrections introduced by the fit are the absolute errors; we find (Table 2) that the biggest displacements in the initial conditions are perpendicular to the baseline for both passages. This can be explained as a rotation around the baseline itself.

Adding one station to the observing net to form a roughly equilateral triangle we eliminate the approximate degeneracy described above (Table 1c). The last eigenvalue is still big compared to the smallest one. This means that the ellipsoid of error is rather flattened. However we are not interested in the shape of the ellipsoid, but in the lenght of its axes compared to the accuracy we want to achieve. Since the biggest eigenvalues corresponds roughly to 3.5 cm, (i.e. $k = 3.5$), and such an error in the orbit translates in a somewhat smaller error in the station positions in the multi–arc procedure, we do not worry if the Γ matrix is not well conditioned.

Established that for short arcs and satellites like *LAGEOS* each passage should be tracked by at least 3 stations to determine the initial conditions with about the same accuracy of the observations, we turned to the determination of the short periodic motion of the pole.

Using one passage observed by 3 stations, we solved for the initial conditions and the instantaneous angular velocity vector of the Earth at the initial time (the observations were simulated with a fixed pole and a constant angular velocity $\omega_\oplus$). We found (Table 1d) that in this case there is approximated algebraic deficiency $R_a(k) = 3$, with approximation factors k of 36, 42 and 125 respectively.

To explain this, let us suppose the motion is described in a non inertial reference frame, fixed with the Earth. If $\Delta\omega$ is the error in a component of the angular velocity vector of the Earth, the error in some component of the initial velocity in the inertial frame is $\Delta v_0 \simeq a\Delta\omega$, where a is the semimajor axis of the orbit; we can always change the angular velocity and the initial speed to compensate each other at the initial time; however the effect of this exchange will grow with the time elapsed from the initial conditions. If T is the total duration of the passage, the effect at the ends of the observed portion of the orbit will be $\Delta x \simeq a\Delta\omega T/2$; the standard deviation of this displacement over the observation interval will be smaller by a factor $\sqrt{3}$. Considering the displacement of the pole $\Delta P = R_\oplus \Delta\omega/\omega_\oplus$ we find:

$$\frac{\delta P}{\delta x_{RMS}} \simeq \frac{R_\oplus}{a} \frac{2\sqrt{3}}{T\omega_\oplus}. \qquad (5.1)$$

For short arcs, such as the ones considered in our simulations, this ratio is $\simeq 7$. Since the r.m.s. displacement in the orbit translates into a somewhat smaller change in the r.m.s. of the residuals (the displacements are mostly in the horizontal directions), the k factors found in the simulations are explained at least in a qualitative sense.

This means that, with a single short arc, the instantaneous position of the pole is determined with an error larger than that of the observations by one order of magnitude. On the other hand if we use some model for the motion of the pole, the effect of its systematic error on the accuracy of the determination of the initial conditions would be correspondingly smaller.

The basic question to answer when testing a multi-arc technique applied to short arcs is: how many arcs do we need to be able to determine the coordinates of the stations tracking the passages with the same accuracy of the observations? The simplest simulations assume that there are no parameters to be determined apart from the station coordinates and the initial conditions for each arc; the minimum network includes only 3 stations.

To be able to answer, we must first eliminate the 4 degeneracies whose existence is suggested by the considerations on the approximate symmetries discussed in Section 3. Table 1e shows the eigenvalues of the covariance matrix of the station coordinates obtained by analysing the data of six arcs. The biggest eigenvalue refers to the rotation around the polar axis of the Earth, that is an almost exact simmetry. The other three large ones can be referred to the other two independent rotations and to the translations.

We experimented on many ways to avoid this rank deficiencies, such as assigning an a priori covariance matrix or determining fewer coordinates. The best results were obtained in the following way. The reference system was rotated so that the $\hat{z}$ axis passed through one station location and a second station was in the $\hat{x}\hat{z}$ plane. Only 5 coordinates were then determined, i.e. the x an z coordinates of the second station and the entire position vector of the last station. The results of this operation are shown in Table 1f. It is easy to see that the degeneracy is fully removed. We adopted this solution as the standard for all the following simulations.

To find the number of passages required to determine as accurately as it is possible a given geodetic net of observing stations we made several simulations in which we varied the number of arcs analysed and we computed the root mean square of the formal standard deviations of the station coordinates determined. Figure 1 shows the plot of these values versus the number of arcs analysed. The values in abscissa are divided by the standard deviation of the observational noise. The r.m.s strongly decreases by adding arcs until six passages are considered, then the value remains roughly constant at a value about equal to the r.m.s. of the observational noise.

We therefore conclude that the number of arcs to be used with three stations is six and we derived the following empirical formula to determine the number N of arcs to be used if a net of N_s stations is to be determined:

$$N = c \times (3 \times N_s - 4), \quad c \simeq 1.25. \tag{5.2}$$

This particular value of c means that each passage, after its initial conditions have been determined, gives a contribution of somewhat less than one station coordinate which can be determined.

Formula (5.2) has been checked against an observing network of 4 and 5 stations, in the pessimistic hypothesis that each passage was tracked by only three stations simultaneously, and a reásonable agreement has been found.

Particular attention should be paid to the reciprocal geometry of the passages. In all the simulations described above we analysed an equal –or almost equal– number of ascending and

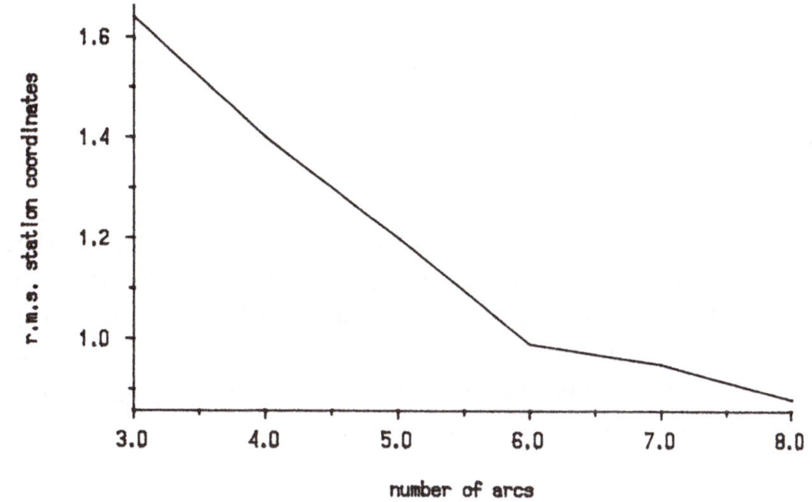

FIGURE 1: Root mean square values of the standard deviations of the coordinates of the stations versus the number of arcs analysed. The observing net contains 3 stations. The r.m.s. are divided by the standard deviation of the observational noise contained in the normal points.

TABLE 3

a) Number of arcs observed by fixed and mobile stations within the Wegener/ Medlas campaign. In the last three columns the count refers to the passes tracked by at least 2 other stations. Italic names are mobile stations sites. In the 1986[1] column only the fixed stations are considered.

station	*total*	*number*		≥ 3		*station*	*tot.no.*	≥ 3
name	1985	1986	1985	1986	1986[1]	*name*	1986	1986
RGOL	382	501	73	93	60	*DYON*	49	8
MATE	437	246	119	96	76	*ROUM*	76	16
POTZ	55	50	37	26	24	*KRIS*	25	10
GRAZ	88	126	46	77	69	*KATT*	58	22
GRAS	219	108	93	47	41	*ASKI*	60	11
WETZ	177	122	93	73	63	*KARI*	46	18
ZIMM	0	43	0	24	23	*BASO*	29	10

b) Number of passages tracked by 3, 4, 5 and 6 european stations simultaneously in 1985 and 1986. In 1986[1] the contribution of the mobile stations is not considered.

number of stations	1985	1986	1986[1]
3	96	100	75
4	32	36	25
5	9	15	9
6	0	2	0

descending passages. If the passages considered all have similar ground tracks the system shows approximate degeneracy. For example, by using 6 descending arcs with closely spaced ground tracks, it was found that the r.m.s. of the standard deviations for the determined coordinates is roughly 7 times the observation error and that there is algebraic rank deficiency whose approximation is $k = 12$. However, as soon as an ascending arc is introduced in the batch (i.e. 1 ascending and 5 descending passages are considered) the degeneracy disappears and the formal errors are again comparable to the observing noise.

This fact could be a problem for the stations which cannot track the satellite during the day. The period of the node of *LAGEOS* is roughly 1050 *days* so that the solar hour of the node (i.e. the angle between the line of nodes and the Sun) has a cycle of 560 *days*. Since our aim is to detect the annual continental drift, data shold be grouped in batchs of less than a year. Special attention should therefore be made in order to have passages with ground tracks in general positions in the set of arcs to be analysed.

Statistical analysis based on the passages tracked by the european network during 1985 and 1986 shows that for each year there are more than 100 passages observed by at least 3 station simultaneously, and that quite often happens that the same passage is tracked by 4, 5, and (in 1986) even 6 stations (Table 3). By listing the number of arcs observed by each station together with at least two others it is found that for the fixed stations it is possible to perform a multi–arc determination with few months of tracking data. For some mobile laser stations, the number of arcs which can be used with our method can be a problem. As suggested by Sinclair (1987) it is possible to increase the useful tracking data for the mobile stations by considering also the arcs observed by two stations; however, this requires to use a reference medium arc solution and to constrain the short arc initial conditions to it. The approximate degeneracy arising in the arcs observed by two stations only can be removed with either one of the techniques already described: either less than six components of the initial conditions are solved for (as proposed by Sinclair), or by constraining some components (or functions thererof, such as the semimajor axis) by a priori observations as described in Section 4.

Since each arc is treated as independent from the others, nothing prevents from considering arcs of *different* satellites, provided their dynamical models are roughly similar (e.g. *LAGEOS* and *STARLETTE* do not match well). We therefore welcome the launch of further *LAGEOS*–like satellites: this could easily solve the problem of having a good reciprocal geometry among the passages of the batch and would increase the number of arcs which can be processed with this method allowing more solutions in a year.

From the error analysis of the previous simulations was also clear that the position and velocity of the satellite are determined, on the average, a few times worse than the coordinates of the stations. Thus the orbital errors do not necessarily produce errors of the same size in the station positions. To investigate the sensibility of the station coordinates and baselines to the orbital errors we performed tests with systematic errors in the orbital model due to biases in the polar motion and in the gravitational field .

TABLE 4

Effects on the coordinates of the stations of various systematic errors in the model of the angular velocity. Values of dimensional quantities in cm and $cm\,sec^{-1}$. Standard deviation of random observational noise: 1 cm. Subscripts in the coordinates refer to different stations.

a) Simulation without bias. Residual r.m.s. before fit 0.99 cm, after fit 0.89 cm.

	error	r.m.s.	e./r.m.s.
X_m	−0.013	1.086	−0.02
Y_m	0.350	1.394	0.25
Z_m	0.379	1.247	0.30
X_{sf}	0.299	0.651	0.46
Z_{sf}	−0.644	0.924	−0.70

b) $\Delta\omega = 5 \times 10^{-12}\; rad\,sec^{-1}$ in the 3 components of the angular velocity vector of the Earth ($R_\oplus \Delta\omega/\omega_\oplus \simeq 40\; cm$). Residual r.m.s. before fit 2.09 cm, after fit 0.99 cm.

	error	r.m.s.	e./r.m.s.
X_m	−1.381	1.025	−1.35
Y_m	1.161	1.277	0.91
Z_m	−2.463	1.188	−2.07
X_{sf}	−0.342	0.560	−0.61
Z_{sf}	0.637	0.807	0.78

c) $\Delta\omega = 1 \times 10^{-11}\; rad\,sec^{-1}$ in the 3 components of the angular velocity vector of the Earth ($R_\oplus \Delta\omega/\omega_\oplus \simeq 80\; cm$). Residual r.m.s. before fit 3.67 cm, after fit 1.05 cm.

	error	r.m.s.	e./r.m.s.
X_m	−2.471	1.025	−2.41
Y_m	1.223	1.277	0.96
Z_m	−3.521	1.188	−2.96
X_{sf}	−0.179	0.560	−0.32
Z_{sf}	−0.053	0.807	−0.07

d) Bias in the geopotential harmonics of degree $\ell = 2, 3, 4, 5$. Random clone with normal distribution and r.m.s. taken from Reigber et al., 1986, Table 3. Residual r.m.s. before fit 1.24 cm, after fit 0.90 cm.

	error	r.m.s.	e./r.m.s.
X_m	0.247	1.086	0.23
Y_m	0.297	1.394	0.21
Z_m	1.659	1.247	1.33
X_{sf}	0.523	0.651	0.80
Z_{sf}	−0.028	0.924	−0.03

e) Bias in the geopotential as before, with r.m.s. larger by a factor 10. Residual r.m.s. before fit 7.82 cm, after fit 1.46 cm.

	error	r.m.s.	e./r.m.s.
X_m	2.589	1.086	2.38
Y_m	−0.176	1.394	−0.13
Z_m	13.178	1.247	10.57
X_{sf}	2.546	0.651	3.91
Z_{sf}	5.508	0.924	5.97

First we considered the effect of a systematic error in the model of the polar motion on the determination of the minimum network of 3 stations, when 6 arcs are considered. To simulate the observations we assumed an uniformely rotating Earth. In the differential correction process we modified the components of the angular velocity and we did not solve for them, thus producing a systematic effect. Since we know the "true" values of the station coordinates, it is possible to compare the systematic error with the formal precision. Table 3 b, c summarizes the results obtained: we can see that, for a systematic error of 5×10^{-12} $rad\,sec^{-1}$ (i.e. ~ 40 cm in the position of the pole at the surface of the Earth) in each component of the angular velocity and for each passage and with random errors in the observations whose standard deviation was 1 cm, the induced error in the coordinates of the stations is no more than 2 times the formal accuracy of the determination. Similar results have been obtained if an angular acceleration bias was introduced. Substantial errors (i.e. ~ 3 standard deviations) are obtained only for systematic errors of 10^{-11} $rad\,sec^{-1}$ in all the three components of the angular velocity.

At present the position of the pole and the orientation of the Earth reference system can be measured with few 10^{-3} *arcsec* accuracy (i.e. $\sim 10 \div 20$ cm at the surface of the Earth) every few days -see for example Mueller and Tapley in these proceedings. The daily motion of the angular velocity vector, although poorly known, can be easily modelled to the level of accuracy of $10 \div 20$ cm at the surface of the Earth. Thus when the multi-arc technique is applied to short arcs the instantaneous angular velocity vector of the Earth can be assigned for each arc without significantly affecting the determination of the station coordinates.

We also tested how much this method of data analysis is sensitive to the gravitational signal contained in the observations. For this purpose we introduced systematic errors in the gravitational model adopted to simulate the observations and we tested how the multi-arc technique could absorb the corresponding signal. We found that the initial conditions of the arcs are rather sensitive to the gravitational signal, while the coordinates of the stations are roughly one order of magnitude less sensitive. The sensitivity of the multi-arc technique applied to short arcs is comparable to the accuracy of the present gravity models. This can be tested by generating a *random clone* of the geopotential model, by changing each coefficient by a random number generated by a simulator of a normal distribution with r.m.s. equal to the one of the geopotential determination. We have taken the r.m.s. values for the errors in the coefficients of degree $\ell = 2, 3, 4, 5$ from Reigber et al., 1986, Table 3. The residuals generated by this change of the geopotential field -which is within the range of values compatible with the accuracy of the geopotential solution used, mostly based upon *LAGEOS* long arc analysis- are just above the observational error level, if the latter is assumed to be at the 1 cm level (Table 3d). When this technique is used with random changes of the geopotential 10 times larger, the signal is well above the observational error and the adjustment of the initial conditions and the station positions resulted in a reduction of the residuals at about the noise level, but with values of the station coordinates "wrong" by more than 10 cm (Table 3e).

As far as baselines are concerned, it turns out that the systematic errors induced by the orbital biases are much smaller than the errors in the station coordinates; the effect of the systematic errors in the geopotential is rather a tilt of the geodetic net.

Although more simulations should be done, our tentative conclusion is that –if the variances of the geopotential coefficients given by Reigber et al., 1986 can be relied upon– the station positions can be determined with *LAGEOS* short arc analysis without improving the geopotential. On the other hand, the signal produced by the anomalies of the geopotential with respect to the current models is significant and should be recovered, when a suitable technique becomes available.

Finally, a simulation was made to test the combined effect of biases in the pole position and in the gravitational field. The bias in the pole position was set to about 20 *cm* for each arc, while the gravitational field error was randomly chosen according to the present knowledge (i. e. comparable to the accuracy of *GEMT1*, Marsh et al. 1987). The result was that, although the errors in the station coordinates reached in one case the amplitude of 6 standard deviations, the bias in the baselines was well below 1 sigma.

We would like to make a further comment on the problem of the most suitable length of the arcs to be used in the analysis of the *LAGEOS* data, for the purpose of achieving a 1 *cm* accuracy in the station positions. If the short arc analysis of the *LAGEOS* data cannot provide accurate instantaneous positions of the Earth angular velocity vector, it might be necessary to wait for a significant improvement in the observing techniques before any reliable information can be gathered on the motions of the rotation axis with periods close to one day or smaller. Until such data are available, we have to assume that the systematic errors induced by short periodic polar motions cannot be eliminated from either medium or long arc analysis; over arcs longer than one orbit, the effect of a displacement of the pole by 10 *cm* is of the order of 10 *cm*, not one order of magnitude less as in the case of a short arc; thus an accuracy of $\simeq 1$ *cm* is unlikely to be achieved by medium arc analysis. A similar argument applies to the effect of the unknown anomalies of the gravitational field. If their effect is of the order of 1 *cm* in the station positions over a short arc, it is likely to be larger over any arc of longer duration. We think that these arguments, as well as the positive results of our simulations, strongly support the need for short arc analysis. The constraints on the number of arc observed by at least 3 stations, as well as the requirements on the observational errors, which result from our simulations, might pose some problems, and the data processing problems still require to be studied in more detail ; but these difficulties have to be overcame, if the *LAser GEOdynamic Satellite* and its followers have really to earn their name.

Acknowledgements

The drive to find techniques suitable for solving local geodetic networks by SLR, of which the present work is one step, resulted from a joint research program between the Space Mechanics Group of the University of Pisa, the Institute of Topography and Geodesy of the Polytechnic of Milan and the Astronomical Observatory of Milan. This research program has been funded by the National Space Program of the italian National Research Council; in particular the authors have used funds from contracts PSN-86-051 and PSN-87-075. The computer resources (IBM 6150 workstations) have been provided by IBM Italy under a joint study contract. The authors would like to thank many people for useful discussions and suggestions, and especially F. Barlier, B. Bertotti, M.Carpino, P. Farinella, A.M. Nobili, F. Sacerdote, F. Sansò, A. Sinclair, C. Wagner and S.Zerbini.

References

Afonso, G., Barlier, F., Berger, C., Mignard, F. and Walch, J.J. : 1986, Reassessment of the Charge and Neutral Drag of LAGEOS and Its Geophysical Implications, *JGR* **90B**, 9381–9398

Afonso, G., Barlier, F., Carpino, Farinella, P., Mignard, F., Milani, A. and Nobili, A.M., 1988: Orbital effects of the LAGEOS' seasons and eclipses, submitted

Barlier, F., Carpino, Farinella, P., Mignard, F., Milani, A. and Nobili, A.M., 1986: Non-gravitational perturbations on the semimajor axis of LAGEOS, *Annales Geophysicae* **4 A**, 193–210

Betti, B. and Sansò, F. : 1983, A detailed analysis of rank deficiency in Hipparcos project, in *First FAST Thinkshop*, Bernacca ed.

Carpino, M., Farinella, P., Milani, A., Nobili, A.M. and Sacerdote, F.: 1986, Determination of a local network from satellite laser ranging: problems and perspectives, in *Proceedings of the International Symposium on Space Techniques for Geodynamics, Sopron, Hungary, 1985*, Vol. 2, 33–42

Carpino, M., Farinella, P., Milani, A. and Nobili, A.M. : 1986, Sensitivity of LAGEOS to changes in Earth's (2,2) gravity coefficients, *Celest. Mech.* **39**, 1–13

Fukushima, T. : 1987, Treatment of Multi-Level Parameters in the Method of Least Squares, in *Proceedings of the twentieth symposium on* " Celestial Mechanics" *held in Tokyo*, Kinoshita, Nakai and Yoshikawa eds., 116–123

Marini, J.W. and Murray, C.W. : 1973, Corrections of laser range tracking data for atmospheric refraction at elevations above 10 degrees, *Document* **X–591–73–351**, Goddard Space Flight Center, Greenbelt, Maryland

Marsh, J.G., et al. : 1987, An Improved Model of the Earth's Gravitational Field: GEM T1, *NASA Tech. Memorandum* **4019**, Goddard Space Flight Center, Greenbelt, Maryland

Melchioni, E. : 1987, Thesis, University of Pavia.

Milani, A., Nobili, A.M. and Farinella, P. : 1987, *Non-Gravitational Perturbations and Satellite Geodesy*, A. Hilger, Bristol

Monti, C. and Sansò, F. : 1983, Tecniche e metodi avanzati per il posizionamento dei punti: il principio di funzionamento, in *XXVIII Convegno Nazionale SIFET*

Reigber, C., Balmino, G., Müller, H., Bosch, W. and Moynot, B. : 1986, GRIM Gravity Model Improvement Using LAGEOS (GRIM3-L1), *JGR* **90B**, 9285–9299

Rubincam, D.P.: 1987, LAGEOS orbit decay due to infrared radiation from Earth, *JGR* **92**, 1287–1294

Sinclair, A.T. : 1987, The determination of coordinates and motions of SLR stations by short–arc methods, preprint, Royal Greenwich Observatory

BOUNDARY VALUE PROBLEMS AND INVARIANTS OF THE GRAVITATIONAL TENSOR IN SATELLITE GRADIOMETRY(*)

P. Holota

Research Institute of Geodesy, Topography and Cartography

250 66 Zdiby 98, Praha-Vychod, Czechoslovakia

Summary

The purpose of the paper is to discuss the use of the theory of boundary value problems for partial differential equations in satellite gradiometry. An approximation of an energetic level of the satellite orbit by a geo-centric sphere is treated in connection with the problem of a boundary and boundary data definition. The choice of basic observables and their reduction to the sphere of approximation as well as possibilities to substitute the knowledge of the instrument frame orientation in space by means of invariants of the gravitational tensor are discussed. Within the framework of a linear theory the separation of field and orbit perturbation is investigated.

1. INTRODUCTION

Starting with classical physical geodesy, we know well that the solution of its fundamental goal to determine the external gravitational field and figure of the earth from astrogeodetic, gravimetric and levelling measurements is essentially connected with a particular type of boundary value problem, the famous problem of Molodensky. In principle data are given on the earth's surface and the problem is to determine the shape of the single connected solution domain outside the earth and the gravitational potential in the earth's exterior.

Obviously the relatively simple problem like this cannot cover all possible cases of geodetic relevance, particularly at present. Modern techniques, especially those based on a variety of principles of spaceborne instruments stimulate progress and new directions in the development of the relevant mathematical formalism in geodesy. The field of boundary value problems of mathematical physics proved to be in a closet connection with the new developments in geodesy.

---

*) Presented at the Int. Symp. on Instrumentation, Theory and Analyses for Integrated Geodesy. Sopron, Hungary, May 16-20, 1988 and also at the Int. Summer School of Theoretical Geodesy on "Theory of Satellite Geodesy and Gravity Field Determination". Assisi, Italy, May 23-June 3, 1988.

Take for example satellite altimetry and mixed boundary value problems. This interesting topic has been intensively studied during the last eight years and very valuable results have been achieved. The goal is again the determination of the external gravitational field and figure of the earth. However, the situation is little more complicated in comparison with the classical case described above. The earth's surface is no more covered by the same type of data but it is split into disjoint parts covered by boundary data of different types and origin. The data used derive non only from astrogeodetic, gravimetric and levelling measurements on the continents (eventually from a gravimetric survey on the seas if it is sufficiently dense) but also from a direct measurement of the altitude of a satellite flying over the sea. The free part of the boundary is exactly the continental surface of the earth. The geometry of the remaining oceanic part is known through satellite altimetry.

At present very much is expected from satellite gradiometry. It is interesting that the theory of boundary value problems for partial differential equations plays an important part also here. However, the problem of a rigorous boundary and boundary data definition has not reached its definitive solution yet and calls for further investigations. Its our aim to discuss it here before we treat an integration of data coming from satellite gradiometry with other geodetic measurements.

2. DIFFERENTIAL ACCELEROMETRY

A gradiometer can be based on a variety of principles and operate in various environments. It can measure forces or torques. (Historically seen, the torsion balance was the starting point of gradiometry.) Differential micro-accelerometry appears to be the most promising principle for satellite gradiometry. A spaceborne gradiometer consists of several ultra-sensitive three-axis accelerometers on different planes. In the GRADIO project eight accelerometers are located at the corners of a cubic structure, half a meter side (see Balmino, 1986; Satellite GRADIO, 1987). The differential measurement between two accelerometers is independent of the external (linear) non-gravitational force acting on the spacecraft (resultant of the effects of the atmospheric drag and of the radiation pressures).

Acceleration differences are measured in a moving (non-inertial) frame where the physical theory as, e.g., in Landau and Lifshits (1973, §39), has to be applied. In principle, by taking the measurement differences between individual accelerometers in various combinations, the following nine components of the tensor L_{ij} can be derived

$$L_{ij} = - V_{ij} + \Omega^2_{ij} + \dot{\Omega}_{ij} \tag{2.1}$$

see Rummel (1985, 1986), Bernard and Touboul (1987). The anti-symmetric part $\dot{\Omega}_{ij}$ of the tensor L_{ij} corresponds to the angular acceleration of the spacecraft while the symmetric part $-V_{ij} + \Omega^2_{ij}$ corresponds to the centrifugal acceleration and to the gravitational tensor

$$V_{ij} = \partial^2 V/\partial x_i \partial x_j \tag{2.2}$$

where V is the gravitational potential. [Because of, e.g., restraining springs, there is no Coriolis term in (2.1).] Obviously, the gravitational tensor is not directly observable. Nevertheless by simple manipulations we can separate the symmetric and anti-symmetric part of L_{ij}

$$(L_{ij} + L_{ji})/2 = - V_{ij} + \Omega^2_{ij} \quad , \tag{2.3}$$

$$(L_{ij} - L_{ji})/2 = \dot{\Omega}_{ij} \quad . \tag{2.4}$$

Moreover, following Rummel (1985, 1986), with some moderate additional support, e.g. coming from a star tracker, it is even possible to separate the gravitational tensor from the rotational part.

By now we will assume that a full tensor gradiometer is available that furnishes us with the six second-order derivatives V_{ij} of the gravitational potential V. The V_{ij} are expressed in the instrument frame of mutually orthogonal Cartesian coordinates x_1, x_2, x_3,. The observation point P is located at the satellite. All observation points P make up the actual orbit trajectory (of altitude 160-240 Km for a typical expected gradiometer mission). Our aim is to show what do V_{ij} tell us in terms of the gravitational field and orbit determination.

Following literature, we usually assume that it is possible to transform the gravitational tensor from the instrument triad to the local orbital triad (with axes corresponding to the along track, cross track and vertical directions). Using the redundancy of observations, the methods discussed so far in literature usually eliminate the orbit improvements from perturbation equations and the measured perturbations of the gravitational tensor components are averaged within equi-angular volume cells which divide the space between two geo-centric spheres containing all orbits of a gradiometric satellite. For a nearly circular orbit of a gradiometer mission the thickness of the space between the inner and outer sphere (and also of the equi-angular cells) is typically less than 10 Km (see Rummel and Colombo, 1985; Rummel, 1986). In the next step the observations averaged inside each cell (each component separately) are usually taken for a kind of boundary data for the determination of the disturbing potential. Unfortunately the boundary surface is defined in a rather vague way in such a case.

To avoid a drawback like this we try to treat the observed components V_{ij} of the gravitational tensor in some other way which would make it possible to use the theory of boundary value problems in a more rigorous sense. Moreover, we will not discuss the question of whether and how it is possible to transform the components V_{ij} measured in the instrument frame to some global frame of reference. We will rather assume that the orientation of the instrument frame is not known.

3. INVARIANTS OF THE GRAVITATIONAL TENSOR

Problems related to the determination of the instrument frame (sensor triad) orientation show the importance of the invariants of the gravitational tensor in a convincing way. The invariants are independent of the orientation of the instrument frame. Therefore the idea to use them in the way of basic observables appears quite naturally (see also Spaceborne Gravity Gradiometers, 1983).

The gravitational tensor V_{ij} is a symmetric, second-order tensor and for its trace holds

$$V_{11} + V_{22} + V_{33} = 0 \tag{3.1}$$

outside the attracting masses. Thus V_{ij} is a traceless tensor or a deviator. In order to find invariants of the gravitational tensor we will first search for its principal directions (eigenvectors) $\underline{\mu}$ which leads to the relations

$$V_{ij}\,\mu_j = \lambda\,\mu_i \quad (i,j = 1,\ 2,\ 3) \tag{3.2}$$

for the components. The necessary condition for the existence of a non zero solution $\underline{\mu}$ is

$$\det(V_{ij} - \lambda\,\delta_{ij}) = 0 \tag{3.3}$$

where δ_{ij} is the Kronecker delta. The characteristic polynomial

$$\lambda^3 + I_1\,\lambda^2 + 2I_2\,\lambda + I_3 = 0 \tag{3.4}$$

follows from (3.3) after some simplification. (3.4) has three real roots λ_1, λ_2, λ_3 in general (Note: the eigenvalues of a symmetric tensor with real elements are real.) The numbers λ_i are independent of the choice of the coordinate system, see e.g. Faddeev and Faddeeva (1963). Hence, the coefficients I_1, I_2 and I_3 are invariants and we observe that

$$I_1 = V_{11} + V_{22} + V_{33} \quad , \tag{3.5}$$

$$I_2 = \sum_{i=1}^{3}\sum_{j=1}^{3} (V_{ij}^2 - V_{ii}V_{jj}) \quad , \tag{3.6}$$

$$I_3 = \det(V_{ij}) \tag{3.7}$$

where

$$I_1 = 0 \tag{3.8}$$

in view of (3.1).

Remark. In order to show a deeper relation of the invariants to the geometry of the gravitational field, we consider the tensor V_{ij} expressed in a local system of mutually orthogonal Cartesian coordinates y_1, y_2, y_3 with the origin at the point P, y_1 and y_2 aligned with the principal directions of the equipotential surface of the gravitational field passing through the point P and y_3 oriented along the line of force. Following Marussi (1985, Ch. VII, Eq. 23), in the local system of coordinates as above

$$V_{11} = - gk_1 = - g/R_1 \quad , \tag{3.9}$$

$$V_{12} = 0 \quad , \tag{3.10}$$

$$V_{13} = - gf_1 \quad , \tag{3.11}$$

$$V_{22} = - gk_2 = - g/R_2 \quad , \tag{3.12}$$

$$V_{23} = - gf_2 \quad , \tag{3.13}$$

$$V_{33} = gH \tag{3.14}$$

where g is the gravitational acceleration, k_1 and k_2 are the principal curvatures of the equipotential surface (R_1 and R_2 are its principal radii of curvature), $H=k_1+k_2$ is the mean curvature of the same surface whereas f_1 and f_2 denote the y_1 and y_2 components of the curvature vector of the line of force. All the above quantities are referred to the point P and for the invariants II_2 and I_3 we easily find that

$$I_2 = g^2(H^2 - K + f_1^2 + f_2^2) \quad , \tag{3.15}$$

$$I_3 = g^3(HK + f_1^2k_2 + f_2^2k_1) =$$

$$= g^3[\ HK + (f_1^2 + f_2^2)H/2 - (f_1^2 - f_2^2)(k_1 - k_2)\ ] \tag{3.16}$$

where $K = k_1k_2$ is the Gaussian curvature.

4. REDUCTION AND LINEARIZATION

In order to develop our ideas as outlined in the preceding sections we start with the observed data. The components of the gradiometric tensor V_{ij} are observed at a point P on the orbit, i.e. on an energetic level

$$V + T = \text{const.} \tag{4.1}$$

where T is the kinetic energy. However, the energetic level does not represent a closed two-dimensional manifold which could be taken (in a usual sense) for a boundary surface in the solution of a boundary value problem for some domain in the three-dimensional Euclidean space. Nevertheless we know that all orbits are contained in a space situated between two geo-centric spheres of relatively close radii (which do not differ more than 10 Km for a typical mission, see above). Therefore it is natural to try to approximate the energetic level related to a definite gradiometer mission by means of a (geo-centric) sphere.
Using the analyticity of the gravitational potential V in the earth's outer space, we can reduce the components V_{ij} measured at a point P of the energetic level to a point P' of the geo-centric sphere of a suitably chosen radius R. For the point P' we take the radial projection of P on the geo-centric sphere. Its position vector is

$$\underline{r}' = \underline{r} + h\underline{r}/|\underline{r}| \tag{4.2}$$

where, obviously,

$$|\underline{r}'| = R \tag{4.3}$$

and

$$h = R - |\underline{r}| \tag{4.4}$$

is the distance between P and P'. Now we can write

$$\partial^2 V(\underline{r}')/\partial x_i \partial x_j - \tag{4.5}$$

$$- h \sum_{k=1}^{3} [\, \partial^3 V(\underline{r}')/\partial x_i \partial x_j \partial x_k \,] \, r_k'/R + O(2) = V_{ij}(P)$$

where O(2) indicates the influence of terms which are non-linear in h.
Unfortunately a practical use of (4.5) is rather difficult since we have not made any assumption which could mean that we know the orientation of the instrument triad of coordinates x_1, x_2, x_3. As we have already indicated our aim is to show whether the orientation problem may be solved by means of invariants of the gravitational tensor.

Following our idea to take the invariants in the nature of basic observables, we immediately see that only I_2 and I_3 can be effectively used in practice since $I_1 = 0$ as an identity. Nevertheless similarly as in (4.5) we can write

$$I_i(\underline{r}') - h\,\partial I_i(\underline{r}')/\partial r + O(2) = I_i(P) \tag{4.6}$$

where i = 2, 3.

In order to show what do the invariants I_2 and I_3 tell us in terms of the gravitational field and orbit determination, we will first examine the linear equations. We start with a nominal case (i.e. with a model), supposing that I_i, $\underline{r}$, $\underline{r}'$, h are as above which satisfy (4.6) and depend smoothly on a parameter t. Denoting the derivatives with respect to t by a dot, we obtain from (4.6) linear equations for perturbations resulting from an infinitesimal deformation of the initial model (to bring it closer to reality). We have

$$\begin{aligned}\dot I_i(\underline{r}') &+ \sum_{j=1}^{3} \dot r_j'\, \partial I_i(\underline{r}')/\partial y_j - \dot h\, \partial I_i(\underline{r}')/\partial r - h\, \partial \dot I_i(\underline{r}')/\partial r - \\ &- h \sum_{j=1}^{3} \dot r_j'\, \partial^2 I_i(\underline{r}')/\partial r \partial y_j - h \sum_{j=1}^{3} (r_j/|\underline{r}|)^{\cdot}\, \partial I_i(\underline{r}')/\partial y_j + \\ &+ O(2) = (I_i)^{\cdot} \quad, i = 2, 3\end{aligned} \tag{4.7}$$

where the derivatives $(I_i)^{\cdot}$ represent the observed data and y_1, y_2, y_3 are mutually orthogonal Cartesian coordinates in some global frame of reference (where the model quantities may simply be computed). Obviously

$$\partial I_i(\underline{r})/\partial y_j = \partial I_i(\underline{r}')/\partial y_j - h\, \partial^2 I_i(\underline{r}')/\partial r \partial y_j + O(2) \tag{4.8}$$

and

$$\dot r_j' = \dot r_j + h (r_j/|\underline{r}|)^{\cdot} + \dot h\, r_j/|\underline{r}| \quad . \tag{4.9}$$

Thus, keeping the accuracy of the original Taylor's expansion,

$$\begin{aligned}\dot I_i(P') &- h\, \partial \dot I_i(P')/\partial r + \\ &+ \sum_{j=1}^{3} \dot r_j\, \partial I_i(P)/\partial y_j + O(2) = (I_i)^{\cdot}\end{aligned} \tag{4.10}$$

for i = 2, 3.

The internal structure of I_i and $\dot I_i$ may simply be obtained from (3.6), (3.7). In view of the fact that

$$4V_{11}V_{22} + (V_{11} - V_{22})^2 = V_{33}^2 \tag{4.11}$$

[which is an immediate consequence of (3.1)] we easily deduce that

$$I_2 = (3/4)V_{33}^2 + (1/4)(V_{11} - V_{22})^2 + V_{12}^2 + V_{13}^2 + V_{23}^2 \quad , \tag{4.12}$$

$$\begin{aligned} I_3 = {} & (1/4)V_{33}^3 - (1/4)(V_{11} - V_{22})^2 V_{33} - \\ & - V_{11}V_{23}^2 - V_{22}V_{13}^2 - V_{33}V_{12}^2 + 2V_{12}V_{13}V_{23} \quad . \end{aligned} \tag{4.13}$$

In consequence

$$\begin{aligned} \dot{I}_2 = {} & (3/2)V_{33}\dot{V}_{33} + (1/2)(V_{11} - V_{22})(\dot{V}_{11} - \dot{V}_{22}) + \\ & + 2(V_{12}\dot{V}_{12} + V_{13}\dot{V}_{13} + V_{23}\dot{V}_{23}) \quad , \end{aligned} \tag{4.14}$$

$$\begin{aligned} \dot{I}_3 = {} & (1/4)[\, 3V_{33}^2 - (V_{11} - V_{22})^2 - V_{12}^2 \,]\, \dot{V}_{33} - \\ & - (1/2)[\, (V_{11} - V_{22})V_{33} + V_{23}^2 \,]\, \dot{V}_{11} + \\ & + (1/2)[\, (V_{11} - V_{22})V_{33} - V_{13}^2 \,]\, \dot{V}_{22} + \\ & + 2(V_{13}V_{23} - V_{33}V_{12})\dot{V}_{12} + 2(V_{12}V_{23} - V_{22}V_{13})\dot{V}_{13} + \\ & + 2(V_{12}V_{13} - V_{11}V_{23})\dot{V}_{23} \quad . \end{aligned} \tag{4.15}$$

Over a mission duration of several months a large set of observations is accumulated that covers the earth at a nearly constant altitude. However, equations (4.10) cannot be directly used in the nature of boundary conditions (on the geo-centric sphere of radius R) for the determination of $\dot{V}$. It satisfies the Laplace equation

$$\Delta\dot{V} = 0 \tag{4.16}$$

but, as we observe, (4.10) depend on the orbit uncertainty $\dot{\underline{r}}$. Analogously to the geodetiic free boundary value problem the components $\dot{r}_j$ (j = 1,2,3) have to be eliminated from (4.10). However, (4.10) represents two equations only.

In order to give equations (4.10) some specific form and to throw more light upon the problem of the orbit uncertainty elimination, we choose a particular model of the earth's gravitational field.

5. SEPARATION OF THE FIELD AND ORBIT PERTURBATIONS

The most simple model is the dominant part of the earth's gravitational field given by

$$V = MG/r \tag{5.1}$$

where GM is the geocentric gravitational constant and r is the geo-centric radius. We easily deduce that

$$V_{33} = V_{rr} = 2\ GM/r^3 \tag{5.2}$$

in a local system of mutually orthogonal Cartesian coordinates y_1, y_2, y_3 with the origin at the point P' and the y_3-axis oriented in the direction of the position vector of the point P'. Thus

$$I_2 = 3\ (GM)^2/r^6 \quad , \tag{5.3}$$

$$\partial I_2/\partial y_j = 0 \quad \text{for} \quad j = 1,\ 2, \tag{5.4}$$

$$\partial I_2/\partial y_3 = -\ 18\ (GM)^2/r^7 \quad , \tag{5.5}$$

$$\dot{I}_2 = 3\ (GM/r^3)\ \dot{V}_{33}(P') \quad , \tag{5.6}$$

$$\partial \dot{I}_2/\partial r = -\ 9\ (GM/r^4)\ \dot{V}_{33}(P') + 3\ (GM/r^3)\ \partial \dot{V}_{33}/(P')/\partial r \tag{5.7}$$

and similarly

$$I_3 = 2\ (GM)^3/r^9 \quad , \tag{5.8}$$

$$\partial I_3/\partial y_j = 0 \quad \text{for} \quad j = 1,\ 2, \tag{5.9}$$

$$\partial I_3/\partial y_3 = -\ 18\ (GM)^3/r^{10} \quad , \tag{5.10}$$

$$\dot{I}_3 = 3\ (GM/r^3)^2\ \dot{V}_{33}(P') \quad , \tag{5.11}$$

$$\partial \dot{I}_3/\partial r = -\ 18\ (GM)^2/r^7\ \dot{V}_{33}(P') + 3\ (GM/r^3)^2\ \partial \dot{V}_{33}(P')/\partial r \tag{5.12}$$

In consequence, equations (4.10) have the following form

$$\begin{aligned} &3\ (GM/r'^3)[\ (1 + 3h/r')\dot{V}_{33}(P') - h\ \partial \dot{V}_{33}(P')/\partial r - \\ &- 6\ (GM/r^4)(r'/r)^3 \dot{r}_3\ ] + O(2) = (I_2)^{\cdot} \quad , \end{aligned} \tag{5.13}$$

$$\begin{aligned} &3\ (GM/r'^3)^2[\ (1 + 6h/r')\dot{V}_{33}(P') - h\ \partial \dot{V}_{33}(P')/\partial r - \\ &- 6\ (GM/r^4)(r'/r)^6 \dot{r}_3\ ] + O(2) = (I_3)^{\cdot} \quad . \end{aligned} \tag{5.14}$$

Multiplying (5.13) by GM/r^3 and subtracting (5.14), we obtain

$$(GM/r^3)(I_2)^{\cdot} - (I_3)^{\cdot} = 0 + O(2) \tag{5.15}$$

which is but a condition for the observed quantities $(I_2)^{\cdot}$ and $(I_3)^{\cdot}$.
To clarify the meaning of (5.13) and (5.14) more precisely we first take into consideration the fact that

$$\dot{V}_{33}(P') - h\,\partial\dot{V}_{33}(P')/\partial r - 6\,(GM/r^4)\,\dot{r}_3 + O(2) = (V_{33})^{\cdot} \tag{5.16}$$

may easily be obtained from (4.5) in view of the tensor character of V_{ij}, provided that the position of the y_3-axis of our local system of coordinates is known with respect to the instrument frame. Comparing (5.16) with (5.13) and (5.14), we immediately see that

$$3\,(GM/r^3)(V_{33})^{\cdot} - (I_2)^{\cdot} = 0 + O(2) \quad , \tag{5.17}$$

$$3\,(GM/r^3)^2(V_{33})^{\cdot} - (I_3)^{\cdot} = 0 + O(2) \tag{5.18}$$

which are again but conditions for the observed quantities. Consequently in the case V=GM/r we can state that the perturbations of invariants used in the nature of basic observables are able to substitute the knowledge of the position of the y_3-axis within the instrument frame. Inversely the use of perturbations as above implies an information on two orientation parameters of the instrument frame in space. However, no pair from the three equations (5.13), (5.14) and (5.16) may effectively be used to separate the field and orbit perturbations.
The elimination of $\dot{r}_3$ may be achieved by one of the following combinations

$$6\,(GM/r^3)(V_{ii})^{\cdot} + (I_2)^{\cdot} \quad , \tag{5.19}$$

$$6\,(GM/r^3)^2(V_{ii})^{\cdot} + (I_3)^{\cdot} \tag{5.20}$$

for i=1 or 2. This, however, needs to know the position of the y_1-axis (or the y_2-axis) within the instrument frame (which means an information about an additional parameter of orientation of the instrument frame in space). Analogously to (5.16) we obtain from (4.5)

$$\dot{V}_{11}(P') - h\,\partial\dot{V}_{11}(P')/\partial r + 3\,(GM/r^4)\dot{r}_3 + O(2) = (V_{11})^{\cdot} \tag{5.21}$$

or

$$\dot{V}_{22}(P') - h\,\partial\dot{V}_{22}(P')/\partial r + 3\,(GM/r^4)\dot{r}_3 + O(2) = (V_{22})^{\cdot} \quad . \tag{5.22}$$

Thus, e.g.,

$$3\,(GM/r^3)[\,2\dot{V}_{11}(P') + \dot{V}_{33}(P')\,] -$$

$$- 3\,(GM/r^3)\,h\,[\,2\,\partial\dot{V}_{11}(P')/\partial r + \partial\dot{V}_{33}(P')/\partial r\,] + \qquad (5.23)$$

$$+ O(2) = 6\,(GM/r^3)(V_{11})^{\dot{}} + (I_2)^{\dot{}}$$

which does not contain $\dot{r}_3$ as desired. There is an analogy between Rummel (1986) or Rummel and Colombo (1985) and the elimination procedure described here. On the other hand the essential difference is that the procedure (5.19)-(5.23) was developed so as to be applicable for perturbations of invariants [instead of $(V_{33})^{\cdot}$] in combination with $(V_{11})^{\cdot}$ or $(V_{22})^{\cdot}$.

In case we would like to achieve the elimination of the orbit perturbations by means of equations (4.10) only, we immediately see that instead of V=GM/r some more precise model of the earth's gravitational field has to be used in the nature of a reference point of the linearization in order to be able to make full use of the difference of information contained in $(I_2)^{\cdot}$ and $(I_3)^{\cdot}$. It is a subject of a further research to show whether the difference will be large enough for an effective separation of the field and the orbit perturbation in practice.

REFERENCES

Balmino, G (1986): Present Status and Future Improvements in Measuring the Gravity Field of the Earth and Planets. In: Anderson, AJ and Cazenave, A (1986) - eds.: Space Geodesy and Geodynamics. Academic Press Inc., London, pp 19-54

Bernard, A and Touboul, P (1987): GRADIO: An Electrostatic Spaceborne Gravity Gradiometer. Proc. of the IAG Symp. at the XIX Gen. Assembly of the IUGG, Vancouver, 1987, Tome I, pp 225-238. Publ. by Bureau Central de l'A.I.G., Paris

Faddeev, DK and Faddeeva, VN (1963): Computational Methods of Linear Algebra. Fizmatgiz Publishers, Moscow-Leningrad (in Russian)

Landau, LD and Lifshits, EM (1973): Theoretical Physics, Part I: Mechanics. Nauka Publishers, Moscow (in Russian)

Marussi, A (1985): Intrinsic Geodesy. Springer-Verlag, Berlin-Heidelberg-New York-Tokio

Rummel, R (1985): Satellitengradiometrie. ZfV 6, pp 242-257

Rummel, R (1986): Satellite Gradiometry. In: Suenkel, H (1986) ed: Mathematical and Numerical Techniques in Physical Geodesy. Lecture Notes in Earth Sciences, Vol 7, pp 318-363, Springer-Verlag, Berlin-Heidelberg

Rummel, R and Colombo, OL (1985): Gravity Field Determination from Satellite Gradiometry. Bull. Géod., 59, pp 233-246

Spaceborne Gravity Gradiometers (1983): Proc. Workshop held at NASA Goddard Space Flight Center, NASA Conf. Publ. 2305, Greenbelt

Satellite GRADIO pour measure le champ de gravité terrestre. PL Air et Cosmos, 1987, 25, No 1156, 37

A POSSIBLE APPLICATION OF THE SPACE VLBI OBSERVATIONS FOR ESTABLISHMENT OF A NEW CONNECTION OF REFERENCE FRAMES

József Ádám
Institute of Geodesy, Cartography and Remote Sensing;
Satellite Geodetic Observatory
H-1373 Budapest, Pf. 546.
Hungary

SUMMARY: Three dedicated space VLBI projects are currently in preparation to launch one or more VLBI radio telescopes in orbits between 1992-1996. One in the Soviet Union called RADIOASTRON which is already an approved mission. The second one is a Western European mission called QUASAT with potential international participation. A consortium in Japan is studying an orbiting VLBI mission called VSOP. Therefore, it is to be expected that space VLBI will be a reality in the next decade. However, the main goals of all three current space VLBI projects are astrophysical purposes.

In this paper it will be pointed out that a space VLBI system offers a good opportunity to connect two types of Conventional Inertial System (CIS): a direct tie between reference frames of the so-called Radio Source-CIS and Dynamic (Satellite Orbit)-CIS inherent in the space VLBI system can be established. On the basis of the ground-based VLBI network coordinates adopted in the Conventional Terrestrial System (CTS), and related to the mass center of the Earth by the space VLBI system itself, the relationship between frames of the CTS and the two above-mentioned CIS can also be established. Therefore, a space VLBI system may make a considerable contribution to geodynamics as well.

A fundamental element of the space VLBI observation geometry is illustrated. Some characteristics of the space VLBI network with respect to the different types of satellite geodetic networks will be emphasized from the viewpoint of theoretical geodesy. Some problems involved in space VLBI network design are discussed. Results from a preliminary estimability (rank defect) analysis of simultaneous space VLBI observations are also reviewed.

1. Introduction

Three dedicated space VLBI projects are currently in preparation to launch one or more VLBI radio telescopes in orbits for the years 1992-1996. One in the Soviet Union called RADIOASTRON which is already an approved and funded mission. The second one is a Western European mission called QUASAT with potential participation of Canada, Australia and NASA. The latter project is now in Phase A Study at ESA. Finally, a consortium in Japan is studying a Japanese orbiting VLBI mission called VSOP.

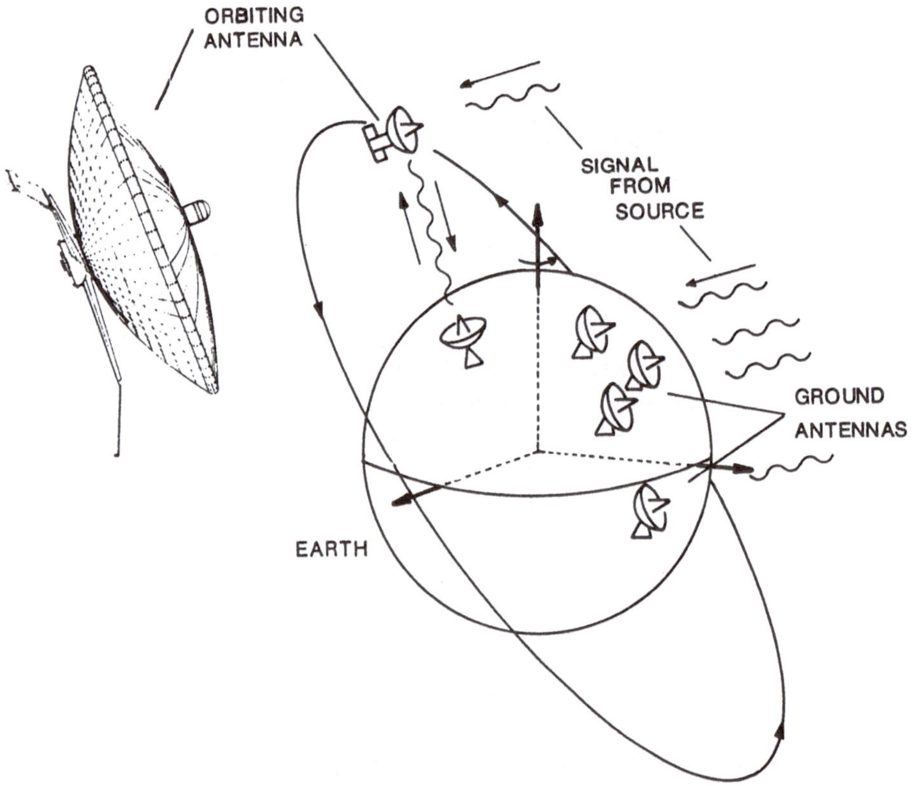

Fig.1 Schematic view of the space VLBI concept
(QUASAT; Schilizzi et al. (1984))

The Soviet Union has announced the Radioastron space VLBI program and expects to launch two or three Radioastron satellites in orbits of up to 76 000 km. The first Radioastron satellite with a 10 m diameter VLBI antenna will fly on a highly elliptical 24 h orbit with apogee 75 530 km, perigee 7 470 km and inclination 65°. The anticipated launch date is 1993. The international participation is invited.

It has been proposed that ESA and NASA jointly launch a satellite called QUASAT, which would carry about a 15 meter antenna into earth orbit by the middle of the next decade. This Earth-orbiting radio telescope will observe the same celestial objects simultaneously with the ground based VLBI stations (Figure 1). The spacecraft will have its frequency reference transferred from ground-based hydrogen masers via a microwave uplink. The astronomical data will be relayed back to the ground on a wideband downlink, to be recorded on a VLBI terminal and then shipped to a central processing facility. There the spacecraft data will be correlated with data from the ground VLBI arrays. The Doppler of the phase transfer signal will also be used to provide the high-accuracy orbit determination needed. The QUASAT operational orbits are optimized to have apogee/perigee altitudes of 36 000 km and 22 000 km/5 000 km, and a low inclination (30°), cf. Frisk et al. (1988).

Note that the feasibility of and potential for using a dedicated VLBI observatory in space has already been demonstrated, cf. Levy et al. (1987). The Tracking and Data Relay Satellite System (TDRSS) was succesfully used as the orbiting element with a nearly 5 m antenna of the VLBI demonstration. The longest observed baseline was 17 800 km, about 1.4 Earth diameters.

It is to be expected that space VLBI will be a reality in the next decade. The main goals of all three space VLBI projects are to carry out astrophysical investigations. The current missions are devoted to improve imaging quality and angular resolution of compact galactic and extragalactic radio sources.

However, even astrophysical space VLBI data can be utilized for geodetic and geodynamical purposes as well. Namely, an Earth-orbiting radio telescope will observe the same extragalactic radio sources simultaneously with the ground-based VLBI networks as their integral elements. This extended network of VLBI stations will have a novel characteristics which was neglected until now: because of the orbiting element it will be naturally connected to the mass centre of the Earth. This connection is missing at the purely ground-based VLBI networks. Therefore, the unique opportunity of the space VLBI in a new geocentric tie of reference frames inherent in the space VLBI system should be exploited for geodynamics and related fields. This new technique offers a possibility of a new connection of reference frames in addition to that reference frame ties which exist among various reference frames of the present space geodetic techniques.

Potential application of space VLBI in geodynamics has been pointed out by Fejes et al. (1985, 1986, 1987).

2. The role of space VLBI in reference frames tie

The capability of the new space techniques to provide valuable information on the systematic differences between the frames of the various Conventional Inertial Systems (CIS) and Conventional Terrestrial Systems (CTS) is now widely investigated, see Dickey et al. (1986), Kovalevsky and Mueller (1981), Moritz and Mueller (1987), Mueller (1985, 1988), Wilkins and Mueller (1986). It is important from the viewpoint of geodesy and geodynamics that all reference frames be interconnected and unified.

In the following, it will be pointed out that a space VLBI system offers a good opportunity to connect the reference frames of the CTS with two types of CIS; Radio Source - CIS and Dynamic (that is Satellite Orbit) - CIS inherent in a space VLBI system. The establishment of the relationship between them would be of great scientific interest.

2.1. It is well-known that the best CIS is the one attached to extra-galactic radio sources playing a major role among conventional quasi-inertial systems. The Radio Source - CIS which is currently used in the frame of International Earth Rotation Service (IERS) is based on observations performed by the network of VLBI stations on the Earth surface. This ground-based geodetic VLBI technique is a purely geometric method and these measurements are independent from the mass center of the Earth. The space VLBI station in orbit at the other hand is an integral element of this ground-based network and at the same time is tied dynamically (via its orbit) to the mass center of the Earth. Therefore, a space VLBI system can relate the existing ground-based VLBI networks to the mass center of the Earth.

The orbital plane of the space VLBI satellite telescope defines the so-called Satellite-CIS, a special type of the Dynamic-CIS through the dynamics expressed in the equations of the satellite motion. Since this orbiting radio telescope will observe extragalactic radio sources with a ground-based VLBI networks as its integral element, the

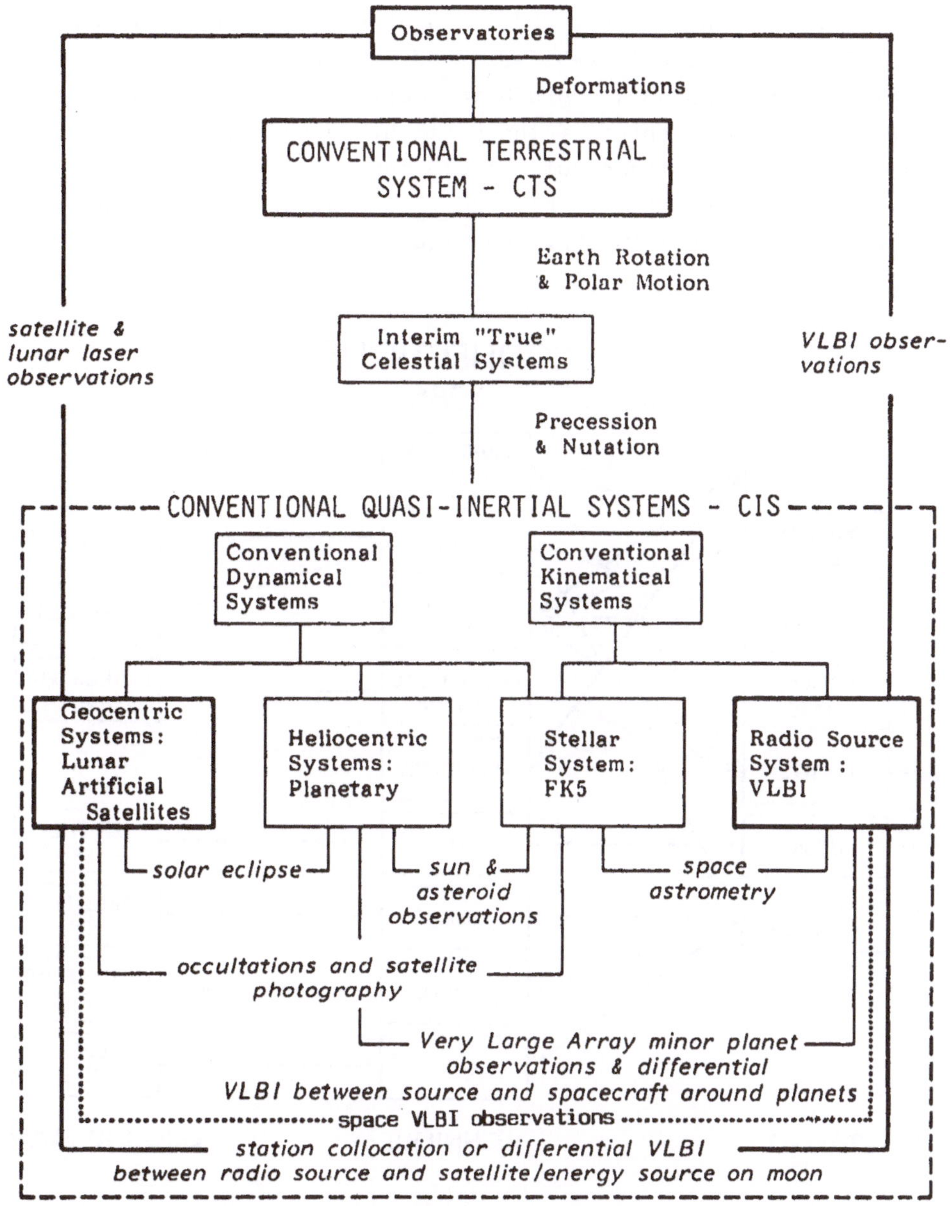

Fig. 2. Conventional terrestrial and quasi-inertial reference systems with examples of possible connections (Kovalevsky and Mueller, 1981; Moritz and Mueller, 1987; Mueller, 1988). A new connection offered by space VLBI is indicated by dotted line.

geodetic evaluation of space VLBI data provides a direct connection between frames of Satellite-CIS and Radio Source-CIS. Through the ground-based VLBI network coordinates defining a CTS (and which can be related to the mass center of the Earth by space VLBI itself), the relationship between frames of the CTS and the two CIS inherent in the space VLBI system can be established. The new reference frame tie offering by space VLBI is shown on Figure 2 and 3.

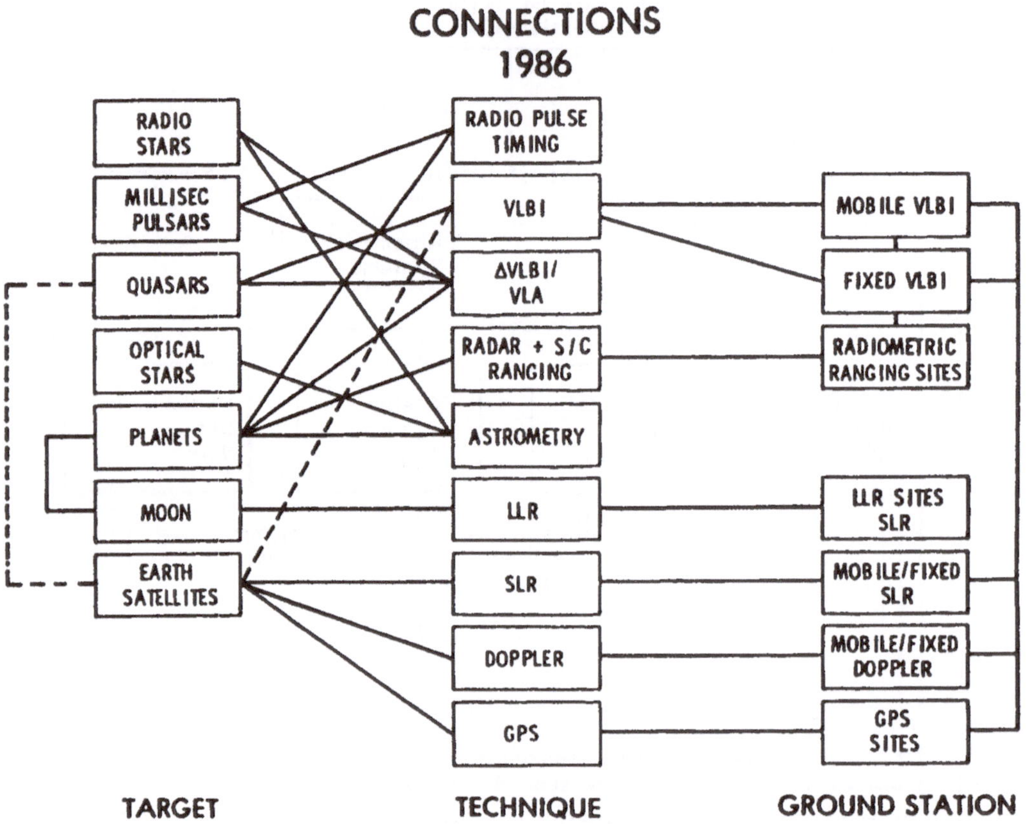

Fig. 3. Reference frame connections 1986 (Dickey, 1986; Mueller, 1988). A new connection offered by space VLBI is indicated by dotted line.

2.2. Since 1984, on the basis of the experiments of MERIT/COTES projects (Wilkins and Mueller, 1986), independent sets of terrestrial frames (CTS), quasi-inertial frames (CIS) and series of Earth Rotation Parameters (ERP) determined by various space geodetic networks (including different sets of VLBI stations) are used simultaneously in the new realization of the BIH Terrestrial System (BTS), cf. Boucher and Feissel (1984), Boucher (1987), Feissel (1986). The ground-based VLBI networks in this implementation are related to the center of mass of the Earth by colocation of a number of VLBI, Satellite Laser Ranging (SLR), Lunar Laser Ranging (LLR) and Transit Doppler stations. Four annual realizations of BTS have been achieved for 1984, 1985, 1986 and 1987, whose results are published in the corresponding BIH Annual Reports. The combination algorithm is described by Boucher and Feissel (1984).

The procedure of the use of space VLBI measurements in reference frames tie is under preparation.

3. Rank deficiencies within a space VLBI network

In order to utilize successfully and efficiently the space VLBI observations, a detailed description of the very complex problems involved in space VLBI network design is desirable. A detailed estimability (rank defect) analysis of geodetic, astrometric and geodynamic parameters within a space VLBI network has to be carried out. The results of such an investigation can be utilized not only for space VLBI network design work in order to plan observation sessions for geodynamical purposes, but also for final processing of this new type VLBI observations in a space geodetic network adjustment.

Note that there has been done a lot of rank defect analysis in different type of satellite geodetic networks, see for instance Delikaraoglou (1985), Grafarend and Livieratos (1978), Grafarend and Heinz (1978), Grafarend and Müller (1985), Grafarend et al. (1979, 1982), Pachelski (1983), Wells et al. (1987).

For a detailed estimability (rank defect) analysis of ground-based VLBI networks we refer to Aardoom (1972), Dermanis (1980), Dermanis and Grafarend (1981), Dermanis and Mueller (1978).

3.1. A fundamental element of the space VLBI observation geometry is illustrated on Figures 4. In the ground-based VLBI network, two ground radio telescopes (i,j) form a ground-to-ground baseline (Figure 4a). The signal from the radio source arrives to the ground-based VLBI stations with time delay (τ_{ij}), and the changing geometrical configuration gives rise to the delay rates $\dot{\tau}_{ij}$; cf.e.g. Dermanis (1980):

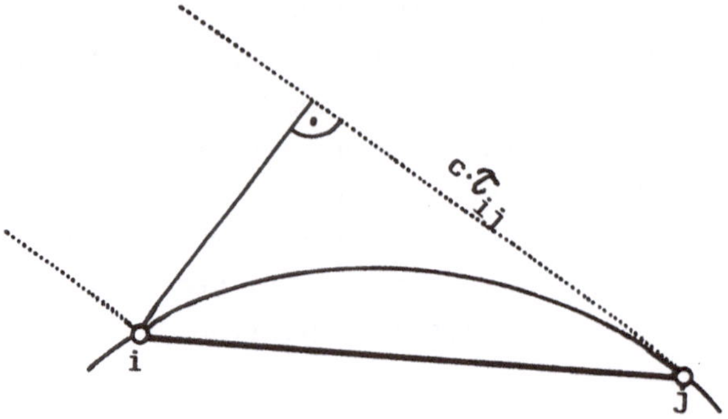

Fig. 4a. Ground-based VLBI geometry

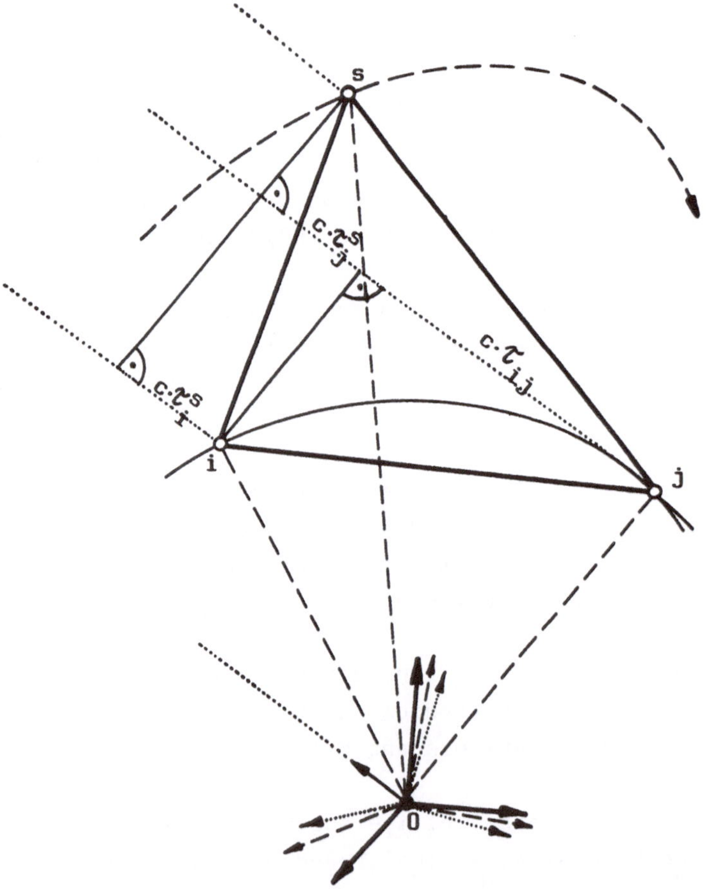

Fig. 4b. Basic geometry for space VLBI

$$\tau_{ij} = c^{-1}(X_i - X_j)^T M(t)\, e + C_{ij} + C'_{ij}\, t \qquad (1a)$$

$$\dot{\tau}_{ij} = c^{-1}(X_i - X_j)^T M(t)\, e + C'_{ij} \qquad (1b)$$

$$i,j=1,2,3..,g.$$

where M(t) is an orthogonal rotation matrix, X_i and X_j are the station coordinates in an Earth-fixed frame (CTS), C_{ij} is the relative offset and C'_{ij} is the relative frequency drift between the clocks at stations i and j.
An Earth-orbiting VLBI telescope with two ground VLBI stations at a moment t_s forms in addition two ground-space baselines. The corresponding time delays (τ_i^s) and delay rates ($\dot{\tau}_i^s$) can be expressed by similar observation equations:

$$\tau_i^s = c^{-1}(X^s - X_i)^T M(t)\, e + C_i^s + C_i'^s\, t \qquad (2a)$$

$$\dot{\tau}_i^s = c^{-1}(X^s - X_i)^T M(t)\, e + C_i'^s \qquad (2b)$$

$$s=1,2,3,...,h.$$
$$i=1,2,3,...,g.$$

Here the delay rates ($\dot{\tau}_i^s$) play a more important role than at the ground-based network as has been pointed out by Fejes (1988).

Suppose that ground-based VLBI stations of number g and an Earth-orbiting VLBI telescope co-observe simultaneously a radio source at each epoch. Figure 5 is a scheme of such a space VLBI network, which is a very special type of ground-space geodetic network. It can be considered as a combination of a ground-based VLBI network and a satellite geodetic network of ground-space baselines like in SLR. Therefore, this type of network comprising ground-to-ground and ground-space baselines is very interesting from the viewpoint of theoretical geodesy.
The processing of the measurements within a ground-based VLBI network is well-known. A possible method for determination of the ground-space baselines is based on the postprocessing of fringe phase and fringe rate as given e.g. by Guoqiang (1984), Fejes et al. (1986). The method assumes that the interferometric group delay and delay rate are function of the ground radio telescope's position, the spacecraft's position, the source's position, the Earth's rotation parameters, etc. On the basis of the comparison of observed and calculated fringe phase

and fringe rate values, the assumed unknown parameters can be determined by the least squares adjustment. The so-called calculated fringe phase and fringe rate values can be derived with a theoretical model of the spacecraft motion and Doppler tracking data. (It may also be considered other type of tracking data, e.g. PRARE, SLR, etc.)

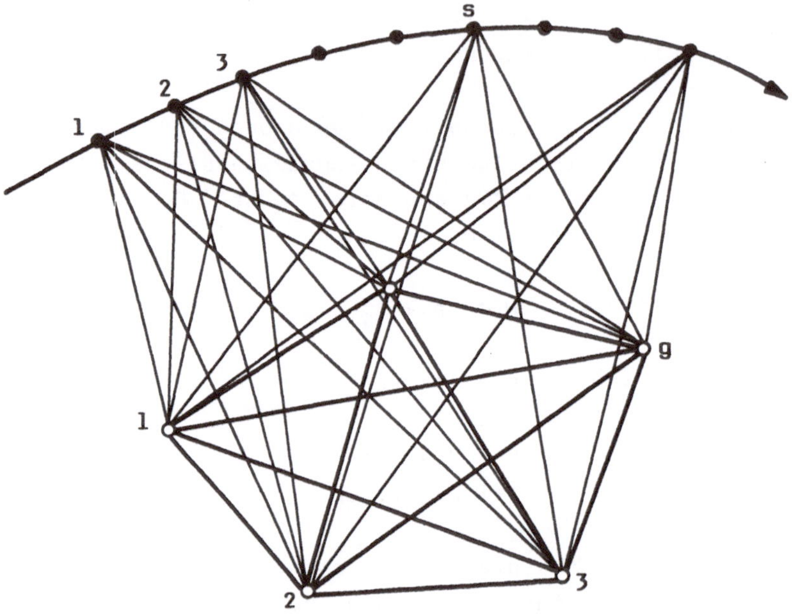

Fig. 5. Space VLBI geodetic network

3.2. Following Dermanis and Grafarend (1981), the situations of the available unknowns and the observations, and the conditions occuring within such a space VLBI network are shortly summarized.

At each epoch of observation, the direction of the observed radio source with respect to the network is described by two parameters, e.g. its angles with two network sides. If the stations co-observe for h epochs, 2h parameters are totally needed for the instantaneous radio source directions. The observations of delay derivatives depend also on the instantaneous rotation vector of the network at each observation epoch. This vector can be referred to the rigid network by three parameters, two for its relative orientation and one for its magnitude and therefore a total of 3h additional parameters are introduced.

Only differences of clock offsets appear in the observations. The number of the complete and functionally independent relative clock offsets is g. These parameters may become estimable if a sufficient number of observations is available.

The number of the independent complete set of observations for an epoch is g. The total number of independent delay observations for h observational epochs is gh and the same number holds for observations of delay derivatives.

The determination of all considered estimable parameters, e.g. the ground-based VLBI station coordinates, satellite positions, instantaneous radio source directions and rotation vector values for all observation epochs, is secured once the number of observations n is not less than the number of unknowns m, i.e. $n \geq m$.

Note that two other side inequalitis have to be satisfied, cf. Dermanis and Grafarend (1981):

(i) Since only delay observations are sensitive to clock offsets their number must be at least equal to the number of clock offsets parameters. This leads to the trivially satisfied inequality $hg \geq g$.

(ii) Since only observations of delay derivatives are sensitive to the rotation vector, their number must not be less than the number of rotation vector parameters. This leads to the inequality $h\,(g-1) \geq 3h$ yielding $g \geq 4$.

In order to avoid configuration defects in a space VLBI network, the simultaneous observations have to fulfill certain inequalities, the so-called fundamental Diophantine inequalities which are to be reviewed very thoroughly. In the geometric mode of simultaneous observations there are three types of situations to be examined with respect to configuration defects. That is considering

(a) ground stations unknown, satellite points known;
(b) ground stations known, satellite points unknown;
(c) both ground stations and satellite points unknown (free network).

In each case, the number of ground stations is denoted by g. Each ground-based VLBI satellite and the Earth-orbiting VLBI station is assumed to observe the same quasar.

In the first situation we assume to have known satellite positions of the VLBI satellite from its orbit determination or its "known" orbit which defines completely the reference frame as geocentric. This leads to m=3g as the number of estimable unknown ground station coordinates for observation equations.

For the three situations above, we end up with the following inequalities. The results of these inequalities for the various values of g are summarized in Tables 1-3. For example, in the model a(1) the adjustment model leads to the fundamental inequality:

$$gh \geq 3g+2h$$

which characterizes the set of configuration constraints. For $g \geq 3$ a set of solution is shown in Table 1, wich gives the number of ground stations and satellite points required in order to avoid configuration defects.

Situation a:

a(1). Observations of delays only, no clock offsets

Unknowns: $m=3g+2h$

Observations: $n=gh$

$n \geq m \longrightarrow gh \geq 3g+2h \longrightarrow$

$h \geq 3g\ (g-2)^{-1}$, always $g \geq 3$ and $h \geq 4$.

a(2). Observations of delays only with clock offsets

Unknowns: $m=3g+2h+g=4g+2h$

Observations: $n=gh$

$n \geq m \longrightarrow gh \geq 4g+2h \longrightarrow$

$h \geq 4g\ (g-2)^{-1}$, always $g \geq 3$ and $h \geq 5$.

a(3). Observations of delays and delay rates, no clock offsets

Unknowns: $m=3g+2h+3h=3g+5h$

Observations: $n=2gh$

$n \geq m \longrightarrow 2gh \geq 3g+5h \longrightarrow$

$h \geq 3g(2g-5)^{-1}$, always $g \geq 3$ and $h \geq 2$.

a(4). Observations of delays and delay rates with clock offsets

Unknowns: $m=3g+2h+g+3h=4g+5h$

Observations: $n=2gh$

$n \geq m \longrightarrow 2gh \geq 4g+5h \longrightarrow$

$h \geq 4g(2g-5)^{-1}$, always $g \geq 3$ and $h \geq 3$.

Table 1.: Necessary number of observation epochs h for the determination of estimable parameters in a network of g stations co-observing one radio source at each epoch (situation a).

	No clock offsets g	h≥	With clock offsets g	h≥
Delay observations only	3	9	3	12
	4	6	4	8
	5-11	5	5	7
	≥12	4	6-9	6
			≥10	5
Delay and delay rate observations	3	9	3	12
	4	4	4	6
	5-9	3	5-7	4
	≥10	2	≥8	3

Situation b:

b(1). Observations of delays only, no clock offsets

Unknowns: $m=3h+2h=5h$

Observations: $n=gh$

$n\geq m \longrightarrow gh\geq 5h \longrightarrow$

$g>5$ always holds irrespective of the number of the observed satellite points.

b(2). Observations of delays only with clock offsets

Unknowns: $m=3h+2h+g=5h+g$

Observations: $n=gh$

$n\geq m \longrightarrow gh\geq 5h+g \longrightarrow$

$h\geq g(g-5)^{-1}$, always $g\geq 6$ and $h\geq 2$.

b(3). Observations of delays and delay rates, no clock offsets

Unknowns: $m=3h+2h+3h=8h$

Observations: $n=2gh$

$n\geq m \longrightarrow 2gh \geq 8h \longrightarrow$

$g>4$ always holds irrespective of the number of the observed satellite points.

b(4). Observations of delays and delay rates with clock offsets

Unknowns: $m=3h+2h+3h+g=8h+g$

Observations: $n=2gh$

$n\geq m \longrightarrow 2gh\geq 8h+g \longrightarrow$

$h\geq g(2g-8)^{-1}$, always $g\geq 5$ and $h\geq 1$.

Table 2.: Necessary number of observation epochs h for the determination of estimable parameters in a network of g stations co-observing one radio source at each epoch (situation b).

	No clock offsets g	h≥	With clock offsets g	h≥
Delay observations only	≥5		6 7 8-9 ≥10	6 4 3 2
Delay and delay rate observations	≥4		5 6-7 ≥ 8	3 2 1

Situation c:

c(1). Observations of delays only, no clock offsets

Unknowns: $m=3(g+h)-6+2h$

Observations: $n=gh$

$n \geq m \longrightarrow gh \geq 3(g+h)-6+2h \longrightarrow$

$h \geq (3g-6)(g-5)^{-1}$, always $g \geq 6$ and $h \geq 4$.

c(2). Observations of delays only with clock offsets

Unknowns: $m=3(g+h)-6+2h+g$

Observations: $n=gh$

$n \geq m \longrightarrow gh \geq 3(g+h)-6+2h+g \longrightarrow$

$h \geq (4g-6)(g-5)^{-1}$, always $g \geq 6$ and $h \geq 5$.

c(3). Observations of delays and delay rates, no clock offsets

Unknowns: $m=3(g+h)-6+2h+3h=3(g+h)-6+5h$

Observations: $n=2gh$

$n \geq m \longrightarrow 2gh \geq 3(g+h)-6+5h \longrightarrow$

$h \geq (3g-6)(2g-8)^{-1}$, always $g \geq 5$ and $h \geq 2$.

c(4). Observations of delays and delay rates with clock offsets

Unknowns: $m=3(g+h)-6+2h+3h+g=3(g+h)-6+5h+g$

Observations: $n=2gh$

$n \geq m \longrightarrow 2gh \geq 3(g+h)-6+5h+g \longrightarrow$

$h \geq (4g-6)(2g-8)^{-1}$, always $g \geq 5$ and $h \geq 3$.

Table 3.: Necessary number of observation epochs h for the determination of estimable parameters in a network of g stations co-observing one radio source at each epoch (situation c).

	No clock offsets g	h≥	With clock offsets g	h≥
Delay observations only	6	12	6	18
	7	8	7	11
	8-9	6	8	9
	10-13	5	9	8
	≥14	4	10-11	7
			12-18	6
			≥19	5
Delay and delay rate observations	5	5	5	7
	6-9	3	6	5
	≥10	2	7-8	4
			≥ 9	3

4. CONCLUSIONS

In sum, a space VLBI system offers a good opportunity to connect terrestrial and celestial reference frames. A direct relation of the existing VLBI networks to the center of mass of the Earth can be established by space VLBI system itself. Therefore, a space VLBI system would make a considerable contribution to geodesy and geodynamics in the future. However, the key condition is that the orbit of a VLBI satellite be determined with sufficient accuracy.

Only the first steps in the directions of the study of space VLBI application in the field of geodynamics have been done. Future works and investigations are necessary to work out in detail the algorithm of reference frames tie inherent in the space VLBI. Estimability (rank defect) analysis has to be extended for more realistic space VLBI network situations (including different number of radio sources observed, simultaneously coobserving subgroups of ground-based VLBI stations, etc.). However, this analysis is necessary (and very important), but still not sufficient. A complete error investigation (simulation model computations, tests, etc.) has to be carried out.

An extension of the research work mentioned above should be to investigate the possibility of the combined use of satellite laser ranging and space VLBI data (of an Earth-orbiting VLBI satellite with laser reflectors) for the reference frame unification.

ACKNOWLEDGEMENTS

Dr. I. Fejes provided the initial suggestion for this paper. His encouragement and useful comments on this work is gratefully acknowledged. A part of this work was started during the author's stay at the University of Stuttgart, Department of Geodetic Science as a research fellow of the Alexander von Humboldt-Foundation.

This paper was presented as a seminar at the International Summer School of Theoretical Geodesy on "Theory of Satellite Geodesy and Gravity Field Determination", Assisi (Italy), May 23-June 3, 1988 and as a contributed paper at the 6th International Symposium on "Geodesy and Physics of the Earth", Potsdam (GDR), August 22-26, 1988. I also wish to thank to Mrs. Brigitta Klein for her expert typing.

REFERENCES

Aardoom,L. (1972): On a Geodetic Application of Multiple-Station Very Long Baseline Interferometry. Netherlands Geodetic Commission, Publications on Geodesy, New Series, Vol.5, No.2, Delft.

Boucher,C. (1987): Definition and Realization of Terrestrial Reference Systems for Monitoring Earth Rotation. Paper presented at the IUGG General Assembly, Symposium No. 4, Vancouver, Canada.

Boucher,C.-Feissel,M. (1984): Realization of the BIH Terrestrial System. Proceedings of the Int. Symp. on Space Techniques for Geodynamics, Vol. 1, pp. 235-254, Sopron, Hungary.

BIH Annual Report for 1984, Sevres and Paris, July 1985.

BIH Annual Report for 1985, Sevres and Paris, June 1986.

BIH Annual Report for 1986, Sevres and Paris, June 1987.

BIH Annual Report for 1987, Sevres and Paris, June 1988.

Delikaraoglou,D (1985): Estimability Analyses of the Free Networks of Differential Range Observations to GPS Satellites. In: Optimization and Design of Geodetic Networks, ed. Grafarend E.W. and Sanso F., pp. 196-220, Springer, Berlin-Heidelberg.

Dermanis,A. (1980): VLBI: Principles and Geodynamic prospectives. Quat. Geod., 1(3), pp. 213-230.

Dermanis,A.-Mueller,I.I. (1978): Earth Rotation and Network Geometry Optimization for Very Long Baseline Interferometers. Bull. Géod., 52(1978), pp. 131-158.

Dermanis,A.-Grafarend,.E.W. (1981): Estimability analysis of geodetic, astrometric and geodynamical quantities in very long baseline interferometry. Geophys. J. R. astr. Soc., 64(1981), pp. 31-56.

Dickey,J.O.-Esposito,P.B.-Lestrade,J.F.-Linfield,R.P.-Melbourne,W.G.--Newhall,X.X.-Niell,A.E.-Preston,R.A.-Standish,E.M.-Williams,J.G.--Muhleman,D.O.-Berge,G.L.-Rudy,D.J.: Reference Frame Studies at JPL/CALTECH. Paper presented at the XIX. IAU General Assembly, New-Delhi, India, November 20, 1985. (JPL Geodesy and Geophysics Preprint No. 147.)

Feissel,M. (1986): Analysis of the Ties between Earth Orientation Determinations and the Related Reference Frames. XXVIth COSPAR General Meeting, Symposium 2 on Applications of Space Techniques for Geodesy and Geodynamics, Toulouse, France.

Fejes I.-Mihály Sz.-Ádám J.-Almár I.-Györgyei J. (1985): QUASAT Potential Geodynamic Applications. Paper prepared for the ESA QUASAT Assessment Study, p. 9. Budapest.

Fejes I.-Almár I.-Ádám J.-Mihály Sz. (1986): Space-VLBI: Potential Applications in Geodynamics. Adv. Space Res., Vol. 6, No. 9, pp. 205-209.

Fejes I.-Almár I.-Ádám J.-Mihály Sz. (1987): On astrometric and geodynamic aspects of space VLBI. Nablyudeniya ISZ, No. 25 (1987), pp. 502-516, Budapest.

Fejes I. (1988). Why is space-VLBI different? Paper presented at the Workshop on the Interdisciplinary Role of Space Geodesy, Erice, Italy.

Frisk,U.-Hawkyard,A.-Cornelisse,J.W. (1988): Quasat-A 50 000 km-diameter Quasar Probe. ESA Bulletin, No. 55, pp. 18-23.

Grafarend,E.-Livieratos,E. (1978): Rank Defect Analysis of Satellite Geodetic Networks I - Geometric and Semi-dynamic Mode. Manuscripta Geodaetica, Vol. 3(1978), pp. 107-134.

Grafarend,E.-Heinz,K. (1978): Rank Defect Analysis of Satellite Geodetic Networks II - Dynamic Mode. Manuscripta Geodaetica, Vol. 3(1978), pp. 135-156.

Grafarend,E.W.-Kleusberg,A.-Richter,B. (1979): Free Doppler Network Adjustment. Proceedings of the Second Int. Geod. Symp. on Satellite Doppler Positioning, Vol. II., pp. 1053-1069, Austin, USA.

Grafarend,E.W.-Kleusberg,A.-Kremers,H.Massmann,F. (1982): The processing of satellite Doppler observations in the free network mode. AVN, 89(1982), 7(286-296).

Grafarend,E.W.-Müller,V. (1985): The critical configuration of satellite networks, especially of Laser and Doppler type, for planar configurations of terrestrial points. Manuscripta Geodaetica (1985), 10:131-152.

Guoqiang,T. (1984): A short note on the high-precision navigation of QUASAT. Proc. of Workshop on Quasat, Gr. Enzersdorf, Austria, 18-22 June, 1984. (ESA SP-213, September 1984), pp. 185-186.

Kovalevsky,J.-Mueller,I.I. (1981): Comments on Conventional Terrestrial and Quasi - Inertial Reference Systems. Proceedings of the IAU Colloquium 56 on "Reference Coordinate Systems for Earth Dynamics", Sept. 8-12, 1980, Warsaw, Poland, D.Reidel, E.M.Gaposchkin and B. Kolaczek (eds.), pp. 375-384.

Levy,G.S. et al (1987): Results and Communications Considerations of the Very Long Baseline Interferometry Demonstration using the Tracking and Data Relay Satellite System. Acta Astronautica, Vol. 15, No. 6/7, pp. 481-487.

Moritz,H.-Mueller,I.I. (1987): Earth Rotation: Theory and Observation. The Ungar Publishing Company, New York.

Mueller,I.I. (1985): Reference Coordinate Systems and Frames: Concepts and Realization. Bull. Géod., 59(1985), pp. 181-188.

Mueller,I.I. (1988): Reference Coordinate Systems: an Update. Lecture Notes for the International Summer School of Theoretical Geodesy on "Theory of Satellite Geodesy and Gravity Field Determination", Assisi, Italy, May 23-June 3, 1988 (this volume).

Pachelski,W. (1983): Critical Configurations of Satellite Networks. Artificial Satellites-Planetary Geodesy, Vol. 18, No.1, pp. 73-88.

Schilizzi,R.T.-Burke,B.F.-Booth,R.S.-Preston,R.A.-Wilkinson,P.N.-Jordan,J.F.-Preuss,E.-Roberts,D. (1984): The QUASAT project. Proceedings of the IAU Symp. No. 110 on "VLBI and Compact Radio Sources", pp. 407-414, (Eds. R. Fanti et al.).

Wells,D.E.-Lindlohr,W.-Schaffrin,B.-Grafarend,E.W. (1987): GPS design: undifferenced carrier beat phase observations and the fundamental differencing theorem. Report of the Department of Surveying Engineering, University of New Brunswick.

Wilkins,G.A.-Mueller,I.I. (1986): On the Rotation of the Earth and the Terrestrial Reference System: Joint Summary Report of the IAU/IUGG Working Groups MERIT and COTES. Bull. Géod., 60(1986), pp. 85-100.

OPTIMIZATION OF THE REORDERING ALGORITHM FOR LEAST SQUARES PROBLEMS RELEVANT TO SPACE GEODESY

Mattia Crespi, Gianfranco Forlani, Luigi Mussio

Istituto di Topografia, Fotogrammetria e Geofisica
Politecnico di Milano, P.za Leonardo da Vinci 32, I-20133 Milano

SUMMARY

The satellite geodesy and the space photogrammetry are entering into the theater of the surveying and photogrammetry measurements. Their high importance and quick diffusion are constraning to assume a new point a view also for the simulation, the adjustment and the evaluation. Some statistic, numerical and informatic problems must be changed, e.g. the reordering algorithm. This paper shows that a least squares normal matrix with minimum reduced profile and right margin can be better then one with minimum profile only. The modification of the reordering algorithm is illustrated; a simulated example shows the performance of the new version. Real examples of geodesy and photogrammetry prove that good results may be achivied. The lists of the main program, of the master subroutines and of the other subroutines are enclosed in the appendix.

1. THE PROBLEM

A graph represents the topology of a geodetic network or of a photogrammetric block. A set of unknown parameters interesting the same geographic position corresponds to one vertex of the graph; a set of observation equations connecting two vertices corresponds one side of the same graph. A one to one correspondence exists between a connected graph and a square symmetric matrix.

A path is a set of adjacent sides; a tree is a sample of paths linked without loops. A level structure is a tree, in which the vertices of a level are connected only to the vertices of the preceding level and/or the following one; each level is a set of vertices of the graph.

The depth is the number of levels; the width is the size of the largest level i.e. the number of vertices contained in it. If the first level contains one vertex only, the level structure is called rooted.

A numeration allows for recognizing the vertices; a good numeration of the vertices follows the increasing or decreasing number of the levels.

The aim of a reordering algorithm is the search of a good numeration of the vertices. This numeration must minimize the profile and/or the bandwidth of the square symmetric matrix associated to the graph (named symbolical normal matrix) and then the profile of the normal matrix of a least squares problem. The minimazation of the profile is the true target, but the one of the bandwidth is easier to be achieved. Indeed the following inequality:

$$\beta \leq 2\alpha - 1$$

links the bandwidth of the symbolical normal matrix (β) and the width (α) of the level structure.

The minimization of the width of a level structure is obtained by searching the level structure of maximum depth.

Therefore a reordering algorithm gives generally a minimum profile normal matrix; nevertheless minimum bandwidth normal matrix doesn't correspond to a minimum profile normal matrix for some least squares problems.

This situation occours always:

- when the shape of a network presents some long connections coming from measurements of long distances or directions;
- when the pattern of a block contains also few images to a small scale, i.e. the elevation of their projection centres is very high;
- when several point coordinates are given in different reference frame, so that fictitious origines must be used;
- when a combination of the preceding cases occours.

These cases become frequent when measurements of satellite geodesy and space photogrammetry are processed together with the traditional surveying and photogrammetric measurements. Therefore a modification of the reordering algorithm must be sought.

2. THE METHOD

Two anomalous cases don't permit the correspondence between the minimum bandwidth normal matrix and the minimum profile normal matrix:

- inhomogeneous connections with respect to the average length of the sides;
- vertices connected by a very high number of sides (i.e. high degree vertices).

These cases contain all the examples of the proceding paragraph.

The new idea is to perform the original reordering algorithm with and without the

problematic sides and vertices and then to compare the two results.
The new procedure consists of the following steps:

1) to apply the reordering algorithm to the n vertices of a graph with all the m sides and compute the original profile;
2) to compute the degree of each vertex and build up the diagram of frequences of these degrees;
3) if the diagram of frequences is bimodal, to remove the h high degree vertices and the k correspondent sides (if some vertex or small subnetwork are connected only to the removed vertices, they are removed too);
4) to compute the lenght of each side and build up the diagram of frequences of these lengths;
5) if the diagram of frequences is bimodal to remove the l longest length sides from the graph;
6) to apply the same reordering algorithm to the $(n-h)$ remaining vertices of a graph with $(m-l-k)$ remaining sides and to compute the reduced profile;
7) to duplicate the $2l$ vertices correspondent to the l removed long connections and put them at the bottom of the list of vertices together with the h high and degree vertices;
8) to insert the $2l$ sides connecting each duplicated vertex with its original one; this operation corresponds to add $2l$ constraints to fix the equality between each duplicated vertex and its original one in the normal matrix of the least squares problem;
9) to compute the right margin: its dimensions are approssimatively equal to $(h+2l)$ (i.e. the number of the removed vertices plus the number of the duplicate ones) by n (i.e. the number of the original vertices) and add it to the reduced profile: their sum is called the extended profile;
10) to compare the original profile and the extended profile and choose the smaller profile as the operative one.

If the second minimum profile normal matrix is chosen, the least squares problem becomes a problem with $(m-2l)$ groups of observation equations and $(n+2l)$ groups of observation equations and $(n+2l)$ groups of unknown parameters.
Note that the number $l,h,k,2l$ are only indicative, because a case can accour together with another; this situation decreases suitably these numbers.
The procedure of removal of the vertices and the sides (respectively the steps 3) and 5) of the new procedure) consists in the following steps:

3.1) order the array containing the degrees of all vertices in decreasing way;
3.2) define the dimension of the set of the degrees (costitued by the highest degrees) where is sought the maximum difference between two consecutive degrees in the ordered array containing the set;

3.3) count of the number of the vertices to be removed;

3.4) verify if the distribution of the degrees is remarkably bimodal: this happends if the maximum difference is larger than critical value furnished by a logistic curve according to the number of vertices of the network;

5.1) order the array containing the lengths of the sides indecreasing way;

5.2) define the dimension of the set of the distances (costituted by the largest distances) where is sought the maximum difference between two consecutive distances in the ordered array containing the set;

5.3) count of the number of sides to be removed;

5.4) verify if the distribution of the distances is remarkably bimodal: this happends if the maximum difference is larger than critical value proportional to the standard deviation of a trimmed set of the distances;

The choise for the removal of sides and vertices could be refined. However if the number of the removed sides and vertices is low, the variation appears slight; therefore a better search can be omitted.

3. THE PROGRAM

As the new procedure is remarkably complex, the architecture of the master routine performing is complicated too. Indeed a lot of array variably (indicators, pointers, counters, etc.) are necessary to link the nomenclature of the vertices, and their original, reduced and extended numerations.

For this reason an indipendent service program has been implemented and the external compatibility with other programs for network and for blocks adjustment and digital modelling has been verified. In this case the data structure compatibility is easier to be verified than the internal software compatibility.

The program regarding the new version of the reordering algorithm is called NEWORD. Its package contains:

- the main program: NEWORD$, just for input/output data management;
- 2 principal subroutines:
  MASTER performs the original version (step 1),
  MARGIN performs the new version (steps 2, ..., 10);
- 16 secondary subroutines, common to both principal subroutines, that build up the graph, run the reordering algorithm and prepare the correspondences among the nomenclature and the numerations;
- 2 secondary subroutine, referred only by the second principal subroutine:
  ELIMI performs the removal (steps 2,3,4,5),
  DUPLI performs the duplication (steps 7,8);
- 2 service subroutines to process real data, indirectly referred by the second principal subroutine only.

Another subroutine is supplied too; it isn't collocted together with the outhers, but it is about the same arguments.
The program NEWORD is written in the language FORTRAN 77, it runs on the computer UNIVAC 1100/90 of the CILEA; all the array variables are declared as virtual memory.

4. THE TEST

A simulated example has been built up to test the new procedure. Its graph contains:

- a network with 22 vertices connected by 34 sides;
- 2 connections, i.e. long length sides;
- 1 high degree vertex, to which are linked many vertices in different parts of the graph.
- 1 additional vertex connected only to the high degree vertex (therefore this vertex won't belong to the graph, after the removal of the high degree vertex).

The fig. 4.1 shows the shape of the graph; the shape of the network is just the same used to test the original reordering algorithm six years ago.
The figs. 4.2, 4.3, 4.4, show respectively:

- the initial symbolical normal matrix;
- the original minimum profile symbolical normal matrix;
- the extended minimum profile symbolical normal matrix;
- the minimum profile symbolical normal matrix corresponding to the configuration of six years ago.

The fourth matrix and the third without the right margin and the additional vertices i.e. the reduced minimum profile symbolical normal matrix, are just the same. This means that the new procedure is able to recognize the anomalous cases, to separate them from the other parts of the graph, to run the reordering algorithm, and to build up the right margin and the linkage.

5. THE EXAMPLE

The investigation of real example of geodesy and photogrammetry proved that good results can be achieved and the performance of new procedure corresponds to the desiderata.

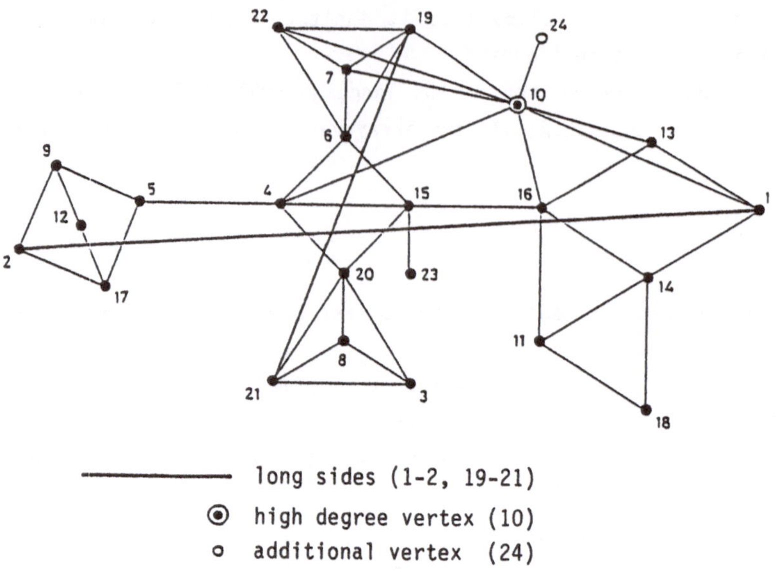

Fig. 4.1 - Shape of the graph.

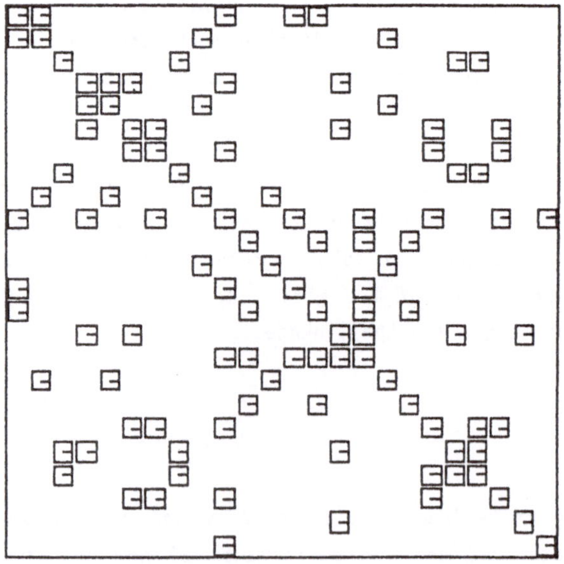

Fig. 4.2 - Initial symbolical normal matrix.

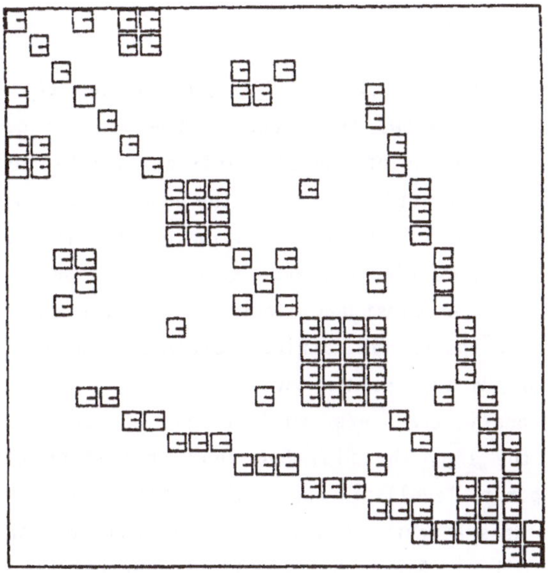

Fig. 4.3 - Original minimum profile symbolical normal matrix.

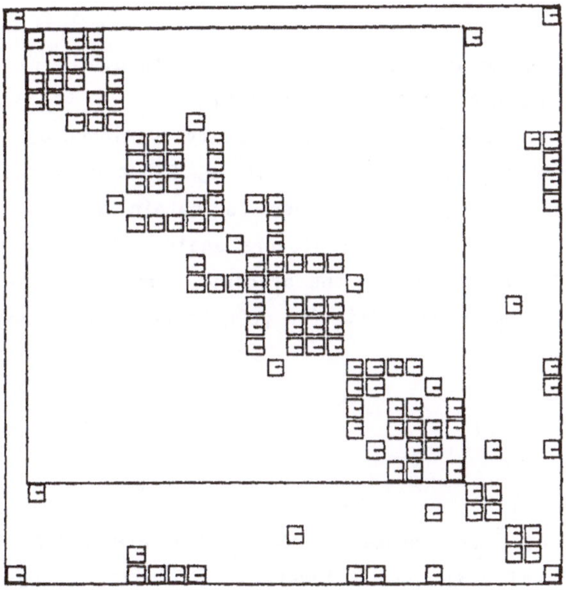

Fig. 4.4 - Extended minimum profile symbolical normal matrix; the submatrix corresponds to the minimum profile symbolical normal matrix.

a) Geodesy

Two years ago the Doppler campaign PDN85 was carried out in Italy by the IGMI (Florence). The fig. 5.1 shows the shape of the Italian geodetic network of the first order and the positions of the 18 points established by Doppler measurements. A new adjustment was performed to estimate the new coordinates of the points of the network and the parameters of the trasformation between the Doppler reference system and the network one. Since the reference system links several points in different parts of the network, the first anomalous case of the 2nd paragraph occours. The figs. 5.2 and 5.3 show the normal matrices obtained by using the original version of the reordering algorithm and the new one.
Next years a GPS geodetic transverse will be carried out in Italy, according to the recommendation of the IAG. The fig. 5.4 shows the shape of the Italian geodetic network of the spirit levelling and the positions of the 39 points will be established by GPS measurements. A new adjustment will be performed to estimate the new heights of the points. Since the geodetic transverse has long length sides, the first anomalons case of the 2nd paragraph occours. The figs. 5.5 and 5.6 show the normal matrices obtained by using the original version of the reordering algorithm and the new one.

b) Photogrammetry

In the last years images of space photogrammetry (e.g. the Metric Camera by ESA and the Large Format Camera by NASA) were available.
The fig. 5.7 shows a zone in the north of Italy overlayed by a model; infortunately neither a block nor some strips are available. Since many points are contained in this model and are necessary in order to obtain a good accuracy, the second anomalous case of the 2nd pragraph occours again. The figs. 5.8 and 5.9 show the normal matrices obtained by using the original version of the reordering algorithm and the new one. A block would limit the gain in profile; however the anomalous case would occour again.

6. REMARKS

The satellite geodesy and space photogrammetry are extending the range of the measurements from a few to many kilometers. A similar change occoured only three centuries ago in the transition from the land-surveying to the geodesy. The more recent improvements introduced new measurement techniques, but didn't change the range: consequently simulation, adjustment and evaluation must be modified, e.g. the reordering algorithm as shown in the 2nd paragraph.
The new version of the reordering algorithm procedures normal matrices with right

Fig. 5.1 - Shape of the Italian geodetic network of the first order and positions of the 18 points established by Doppler measurements.

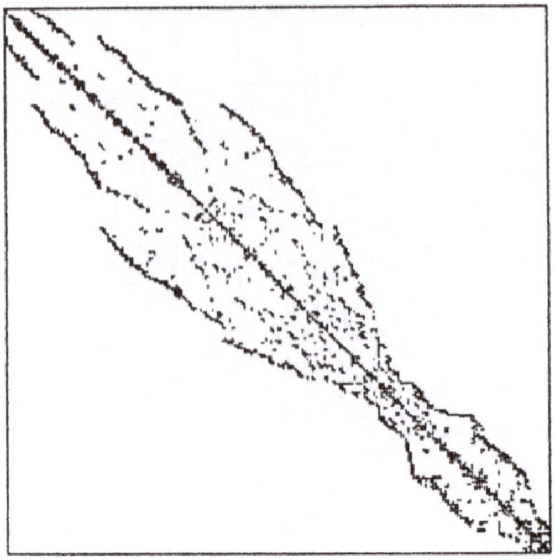

Fig. 5.2 - Original minimum profile normal matrix.

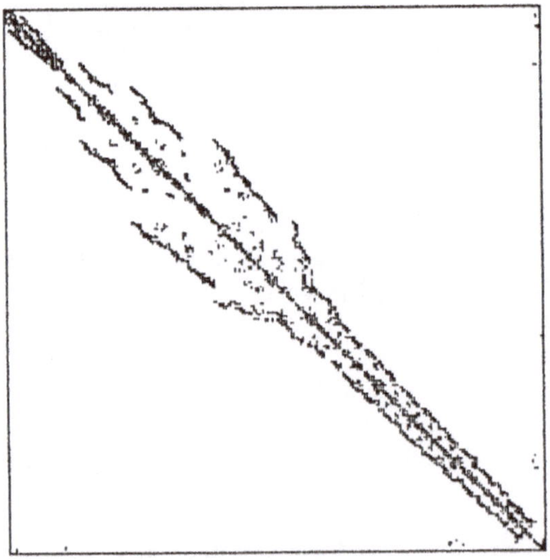

Fig. 5.3 - Extended minimum profile normal matrix.

Fig. 5.4 - Shape of the Italian geodetic network of the spirit levelling and positions of the 39 points will be established by GPS measurements.

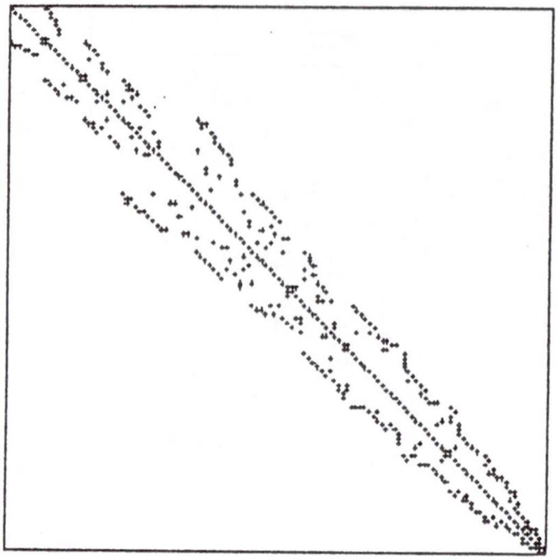

Fig. 5.5 - Original minimum profile normal matrix.

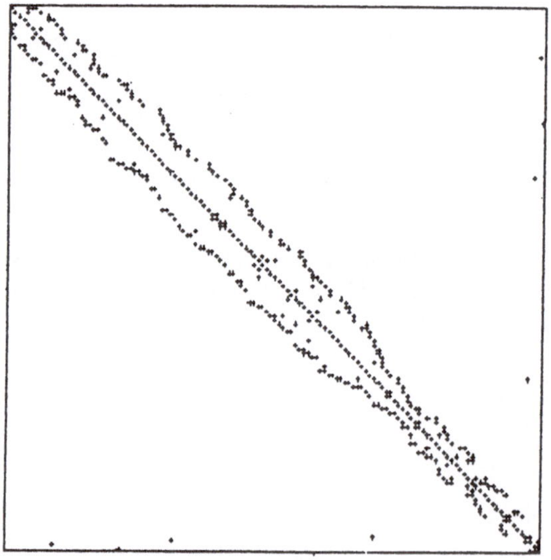

Fig. 5.6 - Extended minimum profile normal matrix.

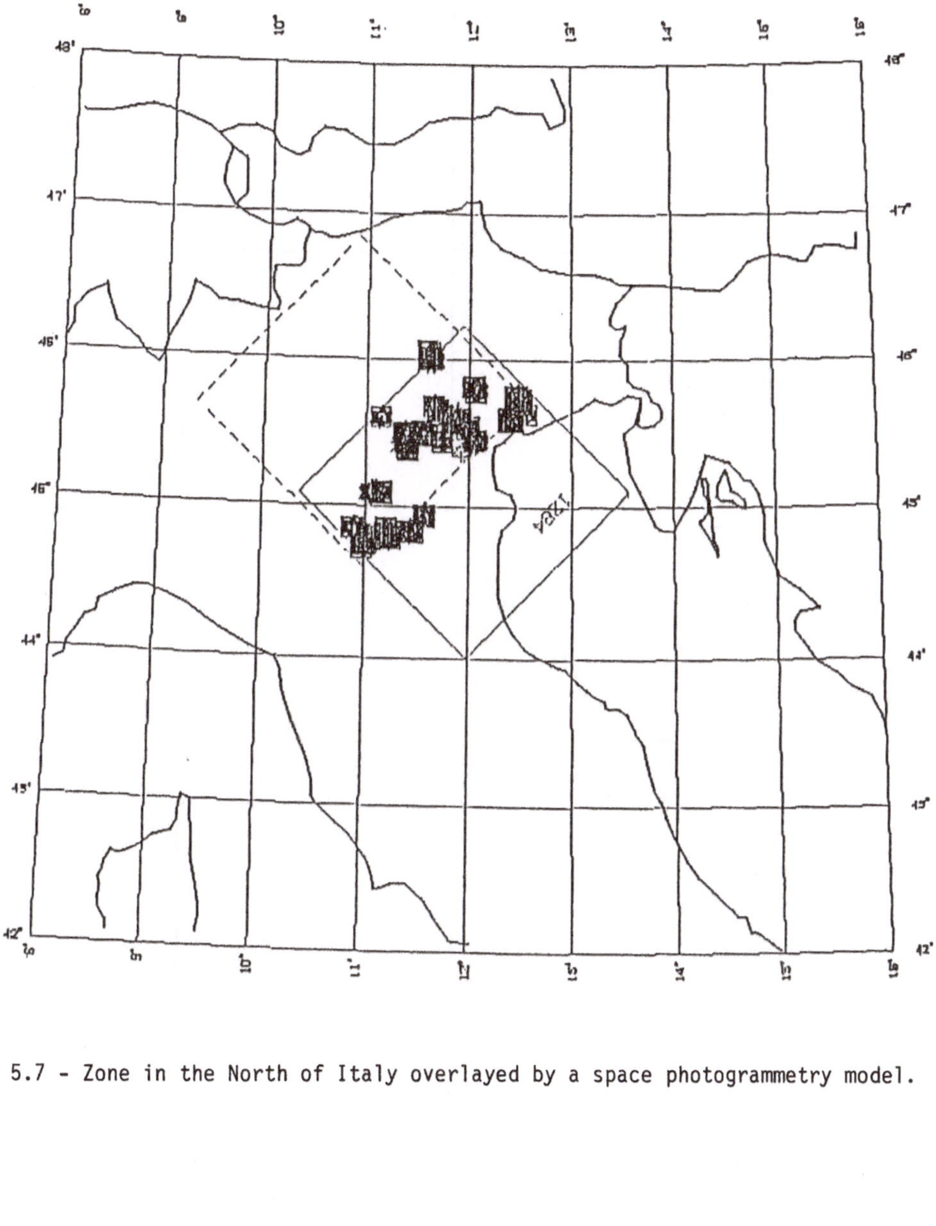

Fig. 5.7 - Zone in the North of Italy overlayed by a space photogrammetry model.

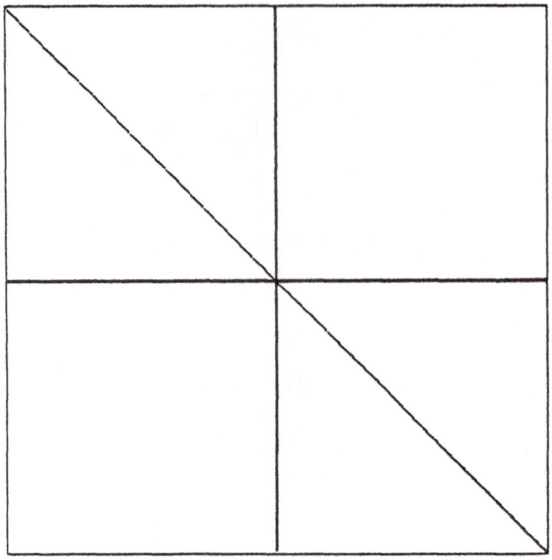

Fig. 5.8 - Original minimum profile normal matrix.

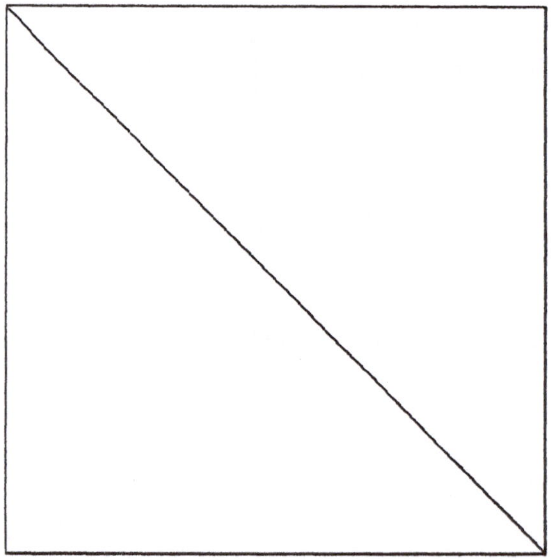

Fig. 5.9 - Extended minimum profile normal matrix.

margin.

The normal matrices with right margin has been used for a long time, e.g. in the systematic error compensation. However in this case the right margin is fixed, but the new procedure allows for searching the best dimension of the right margin in each example.

The fondamental hypothesis of a new procedure is the presence of a small number of the anomalous cases in the graph. Besides the reordering algorithm concerns the long length sides, as a network of the first order, and the short side follow, as a network of a lower order.

Note that the modified reordering algorithm doesn't perform always a better reordering than the classical reordering algorithm; e.g., the real example of GPS transverse, even if it seems to be the first anomalous case of the second paragraph, shows it is necessary to perform both the reordering algorithm and to compare their results.

The service program NEWORD causes some troubles: a principal program must be stopped, the service program must run and the principal program must be restarted. Nevertheless the reordering algorithm can be performed one true only; later the corrispondence among nomenclature and numerations can be supplied as additional input data.

The normal matrices with right margin ask for accurate linear algebra routines; particular care is necessary in the inversion routine. In such a way one spares the core storage request and avoids the waste of computing time.

REFERENCES

Benciolini B., Betti B., Mussio L.: "A remark on the application of the GPS procedure to very irregular graph"; Proc. Meeting of FIG Study Group 5B: Survey control Network, Scriftenreihe HSBw Heft 7, pagg. 97-102, Monaco di Baviera, 1982.

Benciolini B., MussioL.: "Test on a reordering algorithm for geodetic network and photogrammetric block adjustment"; Proc. International Symposium: Management of Geodetic data, GEodaetisk Institut, Meddelelse n. 55, pagg. 341-354, Copenhagen, 1981.

Cuthill E., McKee J.: "Reducing the bandwidth of sparse symmetric matrices"; Proc. ACM National Conference, pagg. 157-172, Association for comuting machinery, New York, 1969.

Gibbs N.E., Poole W.G., Stockmeyer P.K.: "An algorithm for reducing the bandwidth and profile of a sparse matrix"; SIAM Numerical analysis, vol. 13, n. 2, 1976.

Lecture Notes in Earth Sciences

Vol. 1: Sedimentary and Evolutionary Cycles. Edited by U. Bayer and A. Seilacher. VI, 465 pages. 1985.

Vol. 2: U. Bayer, Pattern Recognition Problems in Geology and Paleontology. VII, 229 pages. 1985.

Vol. 3: Th. Aigner, Storm Depositional Systems. VIII, 174 pages. 1985.

Vol. 4: Aspects of Fluvial Sedimentation in the Lower Triassic Buntsandstein of Europe. Edited by D. Mader. VIII, 626 pages. 1985.

Vol. 5: Paleogeothermics. Edited by G. Buntebarth and L. Stegena. II, 234 pages. 1986.

Vol. 6: W. Ricken, Diagenetic Bedding. X, 210 pages. 1986.

Vol. 7: Mathematical and Numerical Techniques in Physical Geodesy, Edited by H. Sünkel. IX, 548 pages. 1986.

Vol. 8: Global Bio-Events. Edited by O.H. Walliser. IX, 442 pages. 1986.

Vol. 9: G. Gerdes, W.E. Krumbein, Biolaminated Deposits. IX, 183 pages. 1987.

Vol. 10: T.M. Peryt (Ed.), The Zechstein Facies in Europe. V, 272 pages. 1987.

Vol. 11: L. Landner (Ed.), Contamination of the Environment. Proceedings, 1986. VII, 190 pages. 1987.

Vol. 12: S. Turner (Ed.), Applied Geodesy. VIII, 393 pages. 1987.

Vol. 13: T.M. Peryt (Ed.), Evaporite Basins. V, 188 pages. 1987.

Vol. 14: N. Cristescu, H.I. Ene (Eds.), Rock and Soil Rheology. VIII, 289 pages. 1988.

Vol. 15: V.H. Jacobshagen (Ed.), The Atlas System of Morocco. VI, 499 pages. 1988.

Vol. 16: H. Wanner, U. Siegenthaler (Eds.), Long and Short Term Variability of Climate. VII, 175 pages. 1988.

Vol. 17: H. Bahlburg, Ch. Breitkreuz, P. Giese (Eds.), The Southern Central Andes. VIII, 261 pages. 1988.

Vol. 18: N.M.S. Rock, Numerical Geology. XI, 427 pages. 1988.

Vol. 19: E. Groten, R. Strauß (Eds.), GPS-Techniques Applied to Geodesy and Surveying. XVII, 532 pages. 1988.

Vol. 20: P. Baccini (Ed.), The Landfill. IX, 439 pages. 1989.

Vol. 21: U. Förstner, Contaminated Sediments. V, 157 pages. 1989.

Vol. 22: I.I. Mueller, S. Zerbini (Eds.), The Interdisciplinary Role of Space Geodesy. XV, 300 pages. 1989.

Vol. 23: K.B. Föllmi, Evolution of the Mid-Cretaceous Triad. VII, 153 pages. 1989.

Vol. 24: B. Knipping, Basalt Intrusions in Evaporites. VI, 132 pages. 1989.

Vol. 25: F. Sansò, R. Rummel (Eds.), Theory of Satellite Geodesy and Gravity Field Theory. XII, 491 pages. 1989.